J. Lückel (Ed.)

Proceedings of the
Third Conference
on Mechatronics and Robotics

Proceedings of the

Third Conference on Mechatronics and Robotics

"From design methods to industrial applications"

October 4–6, 1995 Paderborn

Edited by Prof. Dr.-Ing. Joachim Lückel
Universität-Gesamthochschule Paderborn

 B. G. Teubner Stuttgart 1995

Die Deutsche Bibliothek – CIP-Einheitsaufnahme

Conference on Mechatronics and Robotics <3, 1995, Paderborn>:
Proceedings of the Third Conference on Mechatronics and
Robotics, "From Design Methods to Industrial applications" :
October 4–6, 1995, Paderborn / ed. by Joachim Lückel. –
Stuttgart : Teubner, 1995
    ISBN 3-519-02625-2
NE: Lückel, Joachim [Hrsg.]

© B. G. Teubner Stuttgart 1995
Printed in Germany
Druck und Binden: Hubert & Co. GmbH & Co. KG, Göttingen
Einband: Peter Pfitz, Stuttgart

Table of Contents

IV. Robot Control

V. Robot Applications

VI. Mobile Robots

VII. Modelling and Simulation

VIII. Vehicles

IX. General Aspects in Mechatronics

X. Real-Time Processing

XI. Robots - General Aspects

I. Introduction

J. Gausemeier, D. Brexel, T. Frank, A. Humpert
Integrated Product Development - A New Approach to the
Computer Aided Development in the Early Design Stages

B. Gombert, G. Hirzinger, G. Plank, M. Schedl, J. Shi
Modular Concepts for the New Generation of DLR's Light
Weight Robots

E. Kallenbach
Integrated Design of Shape and Function in Mechatronic
Systems

Integrated Product Development
A New Approach for Computer Aided Development
in the Early Design Stages

Jürgen Gausemeier, Dirk Brexel, Thorsten Frank, Axel Humpert
Heinz Nixdorf Institut, Paderborn, Germany

Abstract: The article presents the methodology of Integrated Product Development. Integrated Product Development offers a frame for structuring and integrating development acitivites and methods which are necessary for the development of complex industrial products. Conceptual Product Design is introduced and described as an essential part of Integrated Product Development in order to methodically support the development process. Conceptual Product Design aims to ensure the fulfilment of requirements, the basic workability, and the expected economic success as early as possible. Finally, information technology tools necessary to support Conceptual Product Design are explained.

1   Introduction

In many areas, the efficiency of the product development process[1] determines the ability of an industrial enterprise to compete. There are increasing requirements of the product development process. It is necessary to consider customers' demands fast and with low costs by clever product structuring. There is a change in technology. Products are getting more complex. Competition forces the periods of product development to become shorter. These challenges to product development can be mastered by two approaches:

Intensifying the "Early Stages"[2]: The early stages of product development determine the success or failure of a product. The product conception lays down the functions and the usability of a product as well as the manufacturing costs and the development time [1]. A principle solution which is inadequate leads to redundant change loops and will endanger the successful introduction on the market. Therefore, "doing it right the first time" is a crucial means to enhance the ability to compete. This means to plan thoroughly and to transfer the conceptual design fast and vigorously. The product conception must be thought over until

- the fulfillment of requirements,
- the basic workability of the product and

---

1   The product development process is a line of tasks ranging from the product idea to the successful introduction on the market. Thus, it includes product planning/product marketing, development/design, work scheduling an building the means of production.

2   The early stages of product development consist of product planning and conceptual product design. They result in a priciple solution.

- the expected economic success

have become clear.

System Composition: Since industrial products are getting more complex and innovation cycles are getting shorter, there will be a need for "system composition". This term implies the ability to quickly transfer requirements to products which are economically attractive, by consequently employing modules[1]. This transfer will only succeed if enterprises concentrate on their original competences by cleverly using existing modules instead of developing parts on their own.

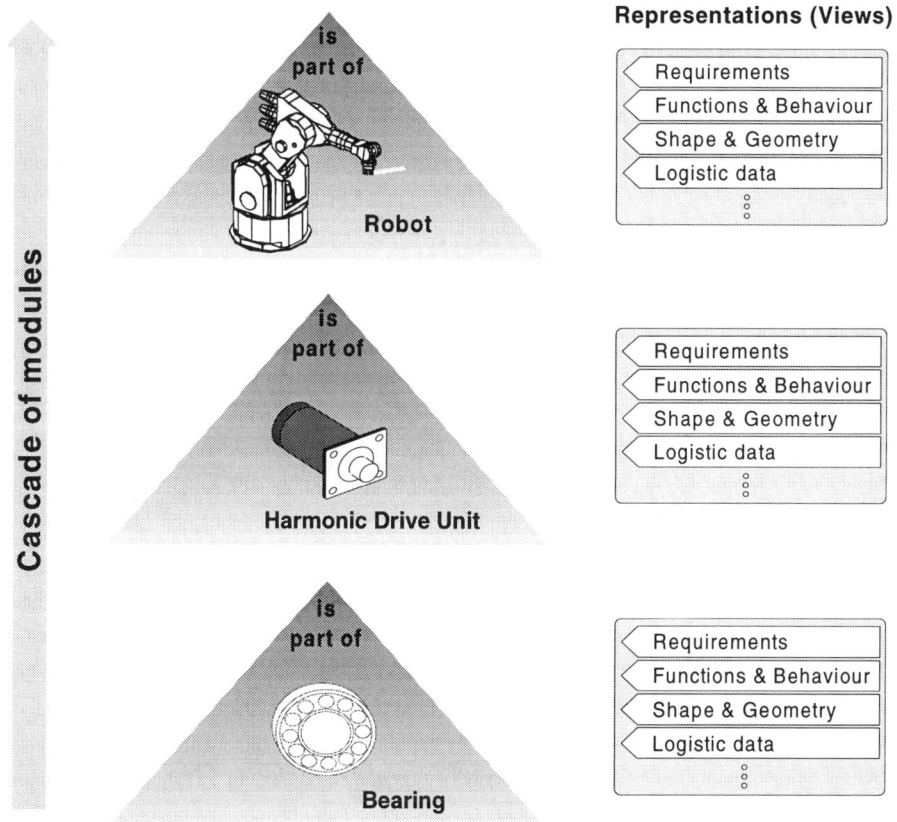

Figure 1: Hierarchy of Modules of an Industrial Product

Basically, every industrial product consists of a cascade of modules (Figure 1).

---

1 A module is an implemented solution for a specific function of an industrial product which is well-tried. This solution can be as elementary as a screw or a shaft/axle connection, but also as complex as a complete drive system consisting of mechanical subassemblies, electronic components and control technology/software. Depending on the degree of completion during the product engineering process, modules are represented by different views. Examples of views are requirements, behavior, function and shape. An industrial product in general consists of a cascade of modules.

Therefore, product development must lead to choosing and applying the approviate modules on each level of the hierarchy. To achieve this aim, the modules must be provided by the information and communication technologies as shown by Figure 1.

In order to realize these approaches to product development, an "all-embracing" development methodology[1] must be applied. This kind of development methodology systematizes the procedure of optimizing a product with respect to its functionality, usability, costs, and time of introduction on the market.

This article deals with the problems mentioned above. Starting with the level of technology and the requirements of an all-embracing development technology (Chapter 2), the "Integrated Product Development (IPD)" is introduced in Chapter 3, using a flexible and integrated structure of activities as it is main feature, and concentrating on supporting the early stages, the re-use of approved solutions, and the use of opportunities offered by modern data processing. Chapter 4 explains the basic structure of CAE tools[2] used for activities in the early stages of product development.

2   Level of Technology and Need for Action

Development methodologies which systematize the product development process have been pushed ahead in different fields of application during the past decades. These methodologies include schedules connecting work steps or development stages, rules and strategies for performance-oriented realization, as well as supporting development methods and tools. Common models of activities are divided into each stage. In every stage of a development methodology, problems specific to this stage must be solved. The solutions are found according to equal principles, but being adapted to each specific problem.

The practical application of the theoretical development methodologies showed a close relation between the model of activities, the product to be created, and information technology (IT). This relation, however, depends on how well the products can be described physically and be represented by data systems. This fact is pointed out by comparing development methodologies of the mechanics, electronics, and software development areas. Each of those methodologies emphasizes specific aspects.

Mechanical Development: The focal point of the existing development methodologies in the mechanics area is the creation of functional structures, principle solutions and scale delineations [3].

Electronical Development: The existing development methodologies in the electronics area concentrate on the functional and logical design of electronic circuits. The following stages are, to a large extent, systematically synthesized. This is because the

---

1   A development methodology consists of concrete instructions for the development of technical systems [2], [3]. With the help of development methodologies, development tasks can be solved. They can be used in every stage of the development process in order to generate (synthesis), judge (analysis), represent (modeling), show (presentation) and develop results.

2   CAE: Computer Aided Engineering, i.e. computer support of the product development process. Tools are used to support methods. CAE tools are information technology tools used for product development.

physical basic conditions for creating electronic products are generally known and can be automated.

Software Development: In this area, the functions and the behavior of products are described in a more abstract and formal manner [4], [5]. That is why there is a better connection between product development and product generation.

The following statements characterize the state of the development methodology and the need for action:

- Generally, all proceeding methods of the mechanics, electronics and software areas describe systematic procedures, from the registration of requirements up to the generation of manufacturing documents on data for chip lay-out, or to source coding of software. The sequential procedure neglects interdependences and reciprocal actions between the activities of product development. That is why the further development of the methodology has to be based on an integrated handling of all activities, as suggested by "Concurrent/Simultaneous Engineering".
- No proceeding method offers a possibility to join the mechanics, electronics, and software areas because of their different product documentations[1] which are physically influenced. But, looking at the increasing integration of mechanical, electronic and data processing components, just this joint point of view gains in significance.
- Existing development methodologies concentrate on the creation of technical solutions. In most cases, economic aspects are missing although they are important to the economic success of a product. These aspects include early and consequent cost management, cost-benefit optimization of the product variance [6], and development of a graded product offer, i.e. optimizing the structure of supply with respect to the course of time.
- The available information technology has a decisive influence on the configuration of the development methodology [7]. An example of this influence is given by the change that the introduction of the 3D CAD technology has contributed to the procedure of shaping. Furthermore, existing IT applications[2] do not make use of a joint product model[3], so that the product documentation will not be consistent, if one considers all stages of product development. That is why the consequent use of the potential offered by modern information technology is the central idea of a further development of the development methodology. Special emphasis is laid on the universal support of the whole development process by software, being based on an integrated product model.

---

1 Product documentation consists of documents or data which have their origins in different stages of product development. For example, engineering drawings or bill of material are parts of product documetation.

2 IT application: an information technology like CAD or FEM systems which helps solving a problem in the product development process by completer support. An IT application is also called IT tool or CAE tool.

3 A product model represents all the data of a product, covering the whole life cycle.

3 Integrated Product Development

3.1 Basic Principles

The Integrated Product Development (IPD) which is suggested in this article, is an advancement of known development mothodologies. This advancement is based on the following basic principles.

Flexible Structure of Acitivities

Integrated Product Development is based on the premise that there is no universally valid method to develop advanced engineering products. Instead, there is a basic generic model of activities containing all the activities that product development consists of. These activities may, again, be composed of a number of sub-activities. All activities concentrate on certain stage of the development process. Figure 4 gives an example of some main activities and of their grouping.

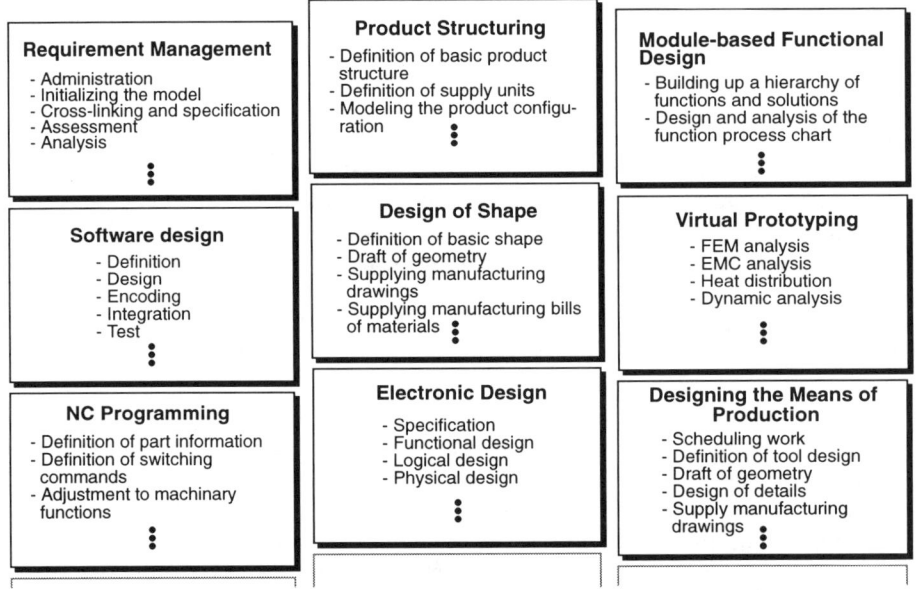

Figure 2: Generic Model of Activities

Considering this generic model of activities, a certain number of activities can be related to each development task which are required to solve this task. The number of activities depends on the industrial sector, the kind of product and its complexity, and on the concrete task. The exact order and frequency of these activities, as a rule, depends on the situation and on logical contraints as shown in Figure 3.

For example, it can be useful to start with certain activities of software design or design of electronics if basic conditions for the conceptual product design are expected to depend on the results of those activities. Furthermore, partial results of the mechanics development can be sufficient to start analyses. The result of these, again, will have influence on the further steps of synthesis.

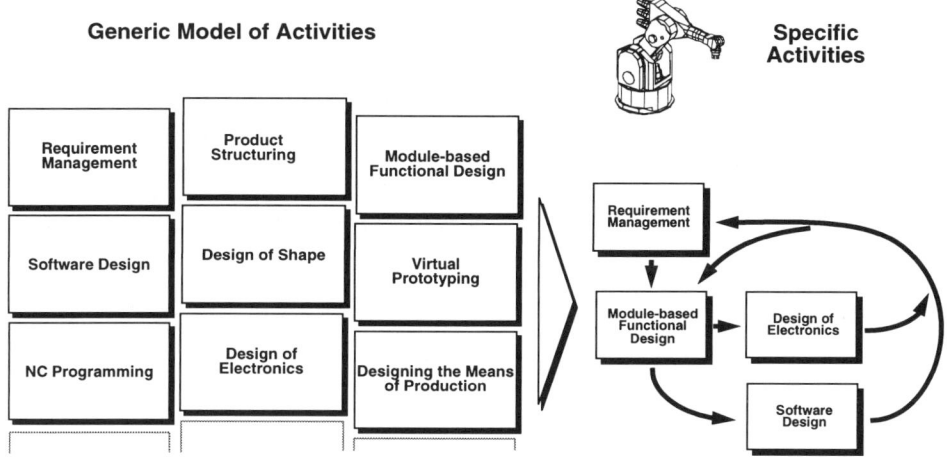

Figure 3: Network or Schedule of Activities

Horizontal Integration of Activities

The principle of horizontal integration contains two aspects:

- Process Integration of activities by dynamically modeling information flow and communication flow relations between the activities.
- Model integration of development results.

Process Integration: The process integration deals with the information flow and communication flow relations which are nessesary in order to co-ordinate activities which were started at any point of time. This includes relations to co-ordinate parallel and structurized, sequential activities. The process integration of Integrated Product Development is based on a dynamic determination of the input and output information which is basically available for a certain activity at a certain point of time, and of their state. This requires a close connection to the product model.

Model Integration: Model integration is the integration of development results which have been achieved be carrying out the activities. It makes use of the conception of a joint and integrated product model on a logical level. The product model, as shown in Figure 4, consists of a number of coherent partial models[1] which represent the number of IT tools, and of „inter-model relations" between elements of the partial models. These inter-model relations help keeping the product documents consistent throughout all stages of product development. Also, these relations can be used as a foundation for the

---

1 A partial model is a part of a model. Coherent partial models of a complete model do not have overlapping contents.

process integration.

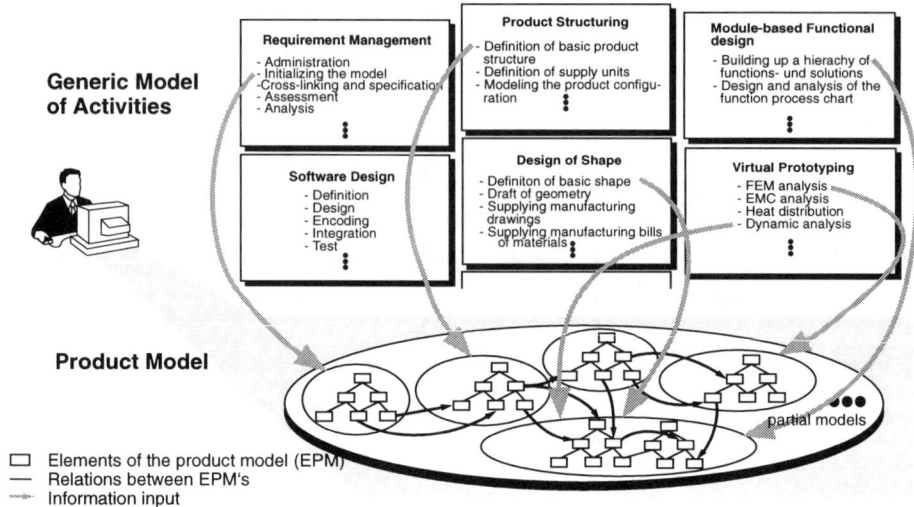

**Generic Model
of Activities**

Requirement Management
- Administration
- Initializing the model
- Cross-linking and specification
- Assessment
- Analysis

Product Structuring
- Definition of basic product structure
- Definition of supply units
- Modeling the product configuration

Module-based Functional design
- Building up a hierachy of functions- und solutions
- Design and analysis of the function process chart

Software Design
- Definition
- Design
- Encoding
- Integration
- Test

Design of Shape
- Definiton of basic shape
- Draft of geometry
- Supplying manufacturing drawings
- Supplying manufacturing bills of materials

Virtual Prototyping
- FEM analysis
- EMC analysis
- Heat distribution
- Dynamic analysis

Product Model

partial models

☐ Elements of the product model (EPM)
— Relations between EPM's
⌁ Information input

Figure 4: Integrated Product Model

Vertical Integration of Development Areas

Integrated Product Development offers an integration model that leads to co-ordinated procedures in the areas of developing mechanical, electronical and software components (Figure 5). The conception of Integrated Product Development is marked by the following aspects:

• A joint method for the early stages of development which is called Conceptual Product Design.
• The parallel area oriented development of product components using existing methods.
• The combination of analysis methods which cover all development areas. These methods describe certain results from different development areas with the help of a common model, and they prove the workability of a product by judging the technical co-operation of the components.

Conceptual Product Design: Compared to existing methodologies of development and design, Integrated Product Development deals with the stages product planning and conceptual design [2]. The Conceptual Product Design contains the tasks of requirement management, product structuring, and module-based functional design. Integrated Product Development results in a product documentation which has been adapted to customers' demands, and tested with repard to potential economic success and basic workability. This approach focuses on the choice of solution approaches[1] and modules in

---

1 Solution approaches are basic possibilities how to fulfil an function [7]. Solutions approaches can be e.g. technologies or method.

order to realize the required product functions and to model the co-operation of modules by discrete state descriptions.

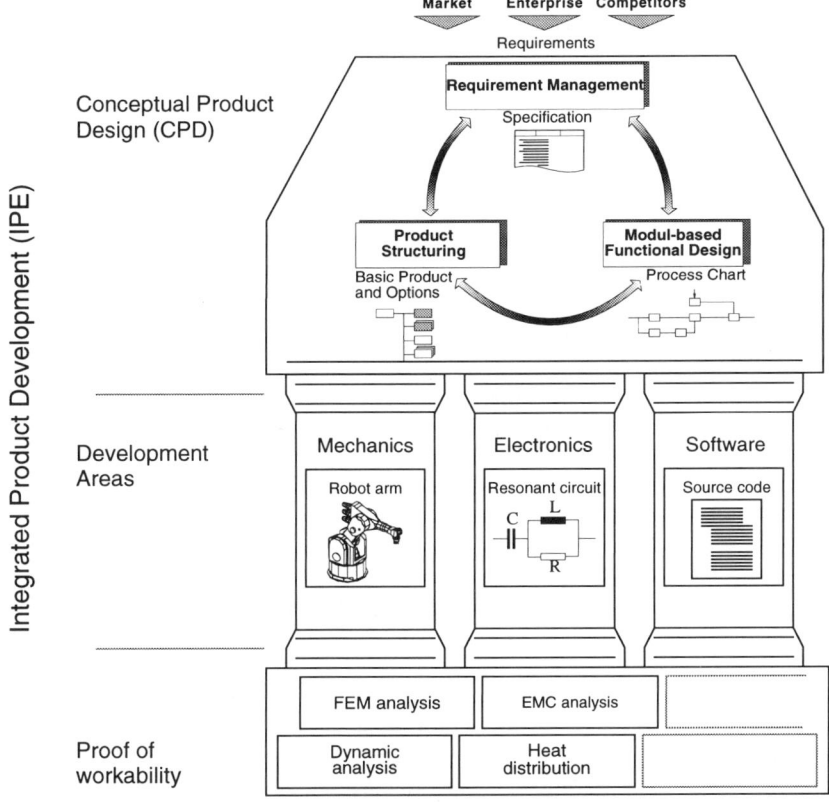

Figure 5: Basic Structure of the Integrated Product Development

Area-oriented Development of Components: Starting from the results of Conceptual Product Design, especially the chosen solution approaches and modules, the development of product components is separately continued by the development areas, using area oriented methods. This further development can be one of a new product, of an adaption, or of variations, dependings on the results of the Conceptual Product Design.

Analysis Methods Covering All Development Areas: The results of development are continued in order to be verified as a whole by virtual prototyping[1]. According to the chosen method, the analysis makes use of results coming from different areas of development. For example, modeling and analyzing the dynamic behavior of complex systems requires joining and co-ordinating the mechanical, electronical and information technology sub-systems [8], [9].

---

1   A prototype is a preproduction version of a product. „Virtual prototyping" means that development results are verified completely with the help of computer models.

Ensuring Economic Success

The success of a product is decided by parallel optimization of functions, usability, time of introduction on the market, and costs. This is why Integrated Product Development not only supports generating the „technical" solutions, but also a consequent cost and variation management which is integrated into the development activities.

Variation management concentrates on the realization of a product variance which maximises profit. Product structuring with respect to marketing as well as to technical aspects is an essential item of variation management. Product structuring with respect to marketing must set up a product offer which is economically sensible, i.e. it brings forth an optimal cost-benefit ratio. The offer can range from a standard solution to a modular structure with a maximum variation of components, depending on the market and the potential of the enterprise. Product structuring with respect to technical aspects means to realize the supplier's offer with a minimum of costs by standardizing, modularity, and products versions.

The cost management section of Integrated Product Development means consequent and integrating management of target costs using the technical and marketing product structures. Furthermore, Integrated Product Development supports the cost-oriented choice of solution approaches and modules when it comes to system composition within the module-based functional design.

4   Conceptual Product Design as a Part of Integrated Product Development

The Conceptual Product Development is a significant part of Integrated Product Development. The methodical support of the steps of requirement management, product structuring, and module-based functional design is a task that has to be solved first of all, regarding the importance of intensifying the early stages and of system composition. This applies to mechtronic systems all the more. Therefore the following chapters explain these steps with respect to mechatronic systems.

4.1 Requirement Management

The aim of Requirement management is to explain requirements formaly which are often expressed in every-day language, and to prepare them so that they can be used as guide-lines for the development of new product components, or for the selection models. Also, a permanent controll if the requirements are fulfilled has to be achieved by requirement management.

Requirement management is based on:

- A number of model constructs to formalize requirements and to present them by a computer.
- A model of activities to set up the model of requirements.

Model Constructs of the Requirement Model

Model constructs (see Figure 6) to set up the requirement model are used to describe

- information about requirements
- specifying relations between requirements during the product development process
- inter-model relations between requirements and development results

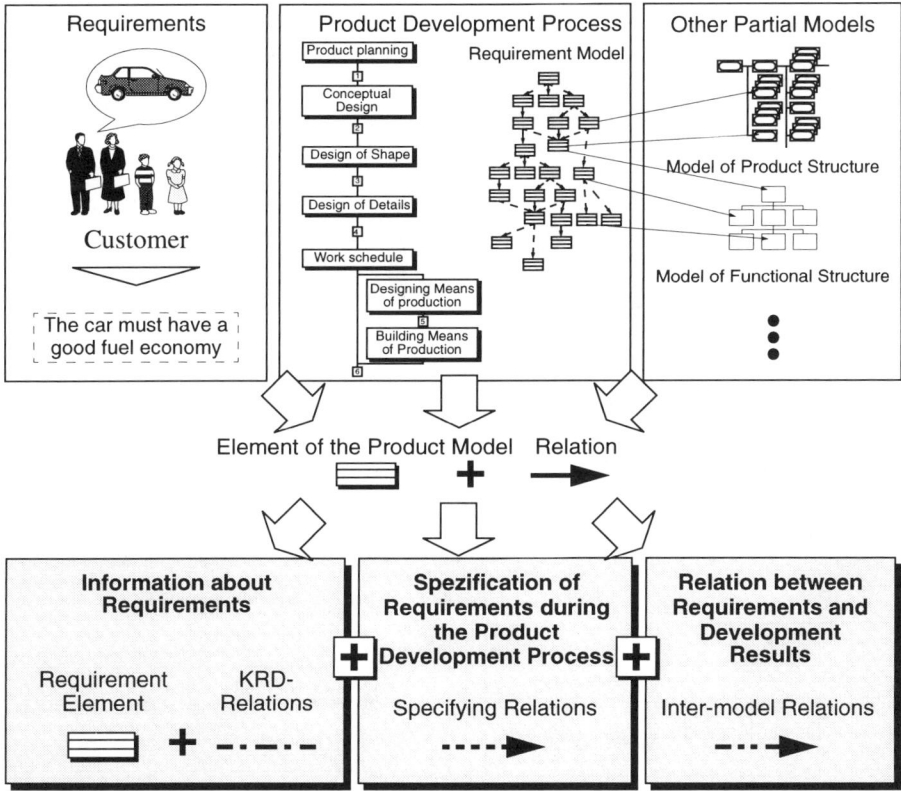

Figure 6: Model Constructs

Information about Requirements: In order to describe the information about requirements, the requirement element and relation are used as model constructs. Requirement elements are used to describe the semantics and the organizational features of a requirement. The relations between the requirement elements can be devided into three categories, depending on the information they bear. These categories serve to represent diffuse knowledge, rules, and dependences (KRD). Any requirement, given in natural language, can be formally represented by these contructs.

Specification of Requirements: The model construct „specifying relations" is made for the specification of requirements. This model construct supports the process of describing requirements more exactly in the course of product development.

Inter-model Relations: The third group of model constructs consists of inter-model relations. The requirement model can be combined with other partial models by inter-model relations. This way, it is possible to use requirements as guidelines for development and to judge the degree to which development results meet the requirements.

Acitivities of Requirement Management

Requirement management consists of the following activities, as shown in Figure 7:

- administration
- initializing the requirement model
- cross-linking and specifying the requirement model
- assessment of the requirement model
- analysis of the requirement model

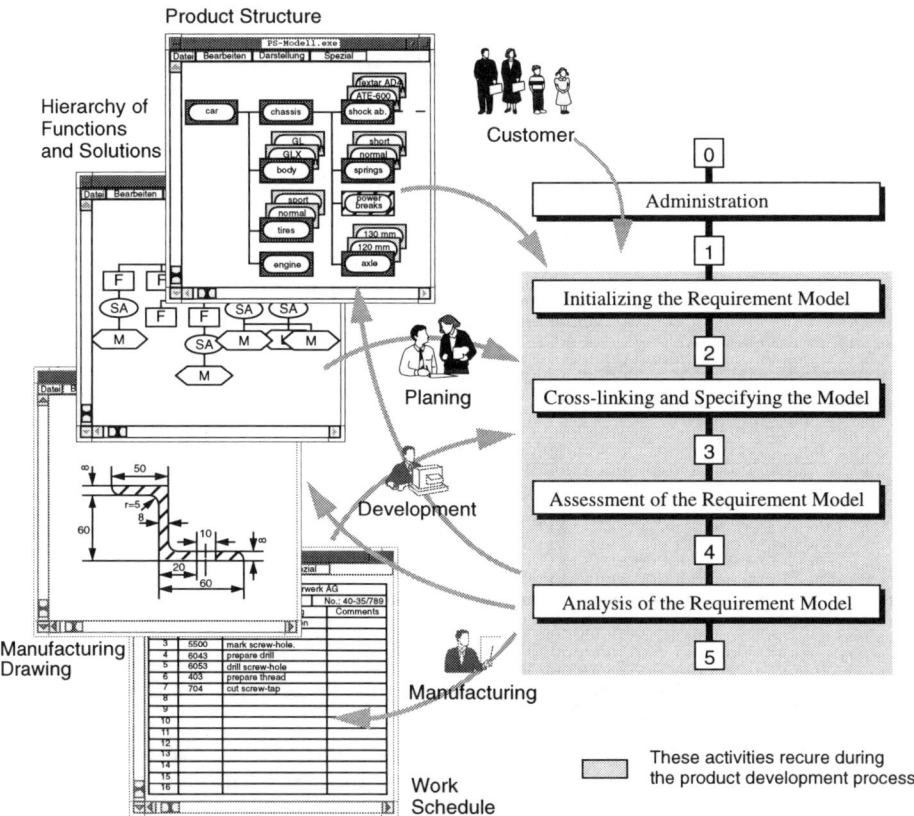

Figure 7: Activities of Requirement Management

Administration: The administration prepares all the modeling activities that are part of the requirement management. For example, modeling roles are defined or tools are pre-adjusted which find relations between elements of the product model. This makes it easier to model the requirements in the following steps.

Initializing the Requirement Model: Initializing the requirement model aims at creating the structure of the requirement model, planning the determination of requirements, and defining the basic structure of the requirement model by a product data model.

Cross-linking and Specification: This step is the focal point of requirement management. It combines all the activities that occur during product development and which have to be adapted to other activities like specification of requirements and building the inter-model relations to other partial models.

Assessment of the Requirement Model: On the basis of the cross-linking, the connections are assessed. This has to be done continuously so that the network can be analyzed at any time.

Analysis of the Requirement Model: The analyses that can be carried out at any time convey results which can be used to control the product development process. Analysis results can be for example the degree to which the requirements have been fulfilled, or the chronological development of fulfilled requirements. The design engineer will get the results in the shape of diagrams or text.

It is obvious that requirement management is applied throughout all stages of the product development process (Figure 7). So, the requirement model is a tool to control the whole process of product development.

4.2 Module-based Functional Design

It is the aim of module-based functional design to develop a principle solution for a technical system by performing the following activities:

- building up a hierarchy of functions and solutions
- proof of workability

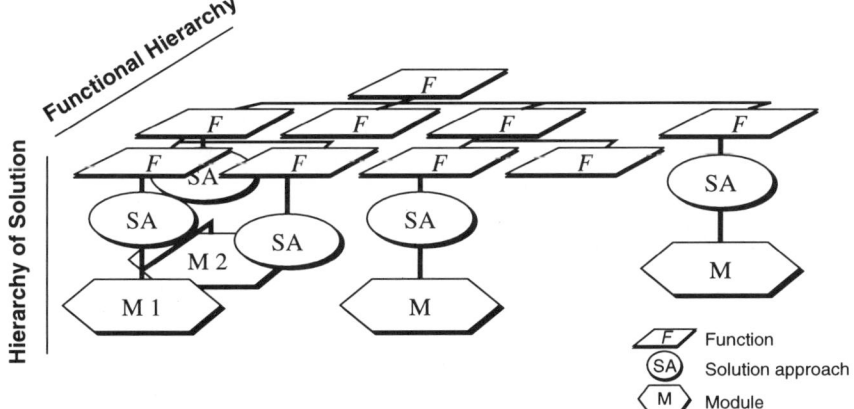

Figure 8: Hierarchy of Functions and Solutions

Hierarchy of Functions and Solutions

Proceeding from the functional requirements, the topmost functional level of the product is divided into its technical functions (F). This division forms the functional hierarchy (see Figure 8). As far as possible, solution approaches (SA) and the resulting modules (M) are related to the functions of the functional hierarchy. A function will be subdivided into its subfunctions if no appropriate module can be found. When relating solution approaches and modules are found, one has to check if these fulfill the conditions given by the requirement model.

Proof of Workability

In order to test if the system concept will work, a functional process chart (Figure 9) is set up. The functional process chart represents a network. The nodes of this network are the modules taken from the hierarchy of functions and solutions. There are relations between the nodes as well as between nodes and the environment. The relations can be energy, material and information. If no modules have been found yet, the solution approaches or functions belonging to the modules can be used temporarily instead. The input and output of those are to be regarded as boundary conditions for the choice of modules.

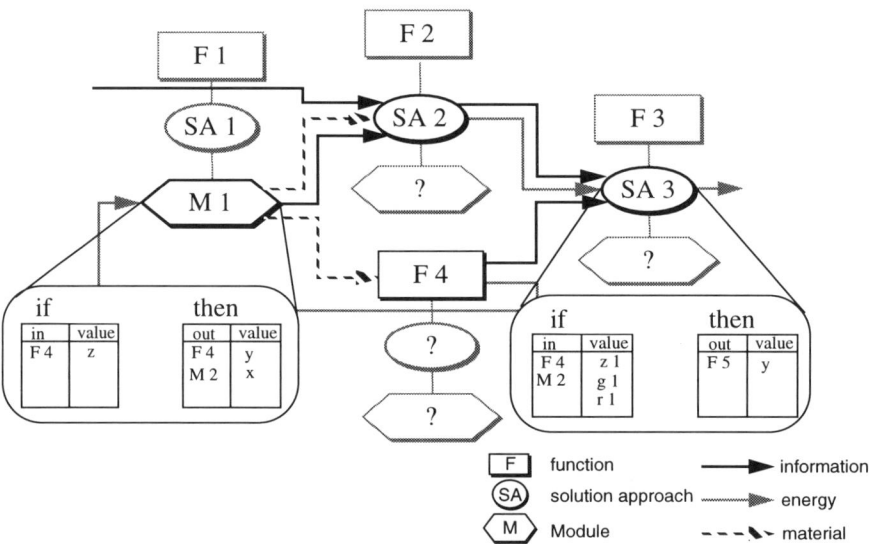

Figure 9: Functional Process Chart of One Level of the Hierarchy

In addition, to test the workability of the system, the behavior of each node has to be described no matter if the nodes are functions, solution approaches, or modules. According to the Automation Theorie of information Science, the behavior can be described using states and simple rules. All possible states of the input/output relations of a node are defined. This way, to each node, semantics based on rules are related which resemble the digital simulation of circuits. All states of the input/output relations are linked by if-then relations. After describing the behavior of all nodes, the basic behavior and the workability of the whole system can be judged.

4.3 Product Structuring

Product structuring economically optimizes product offer and product performance. It consists of the following activities:

- product structuring with respect to marketing
- technical product structuring.

Product structuring with respect marketing

The marketing product structure (Figure 10) describes product offer and product performance from the customer's point of view. Modeling this structure requires the following activities:

- definition of marketing features
- definition of supply units
- modeling the marketing configurations.

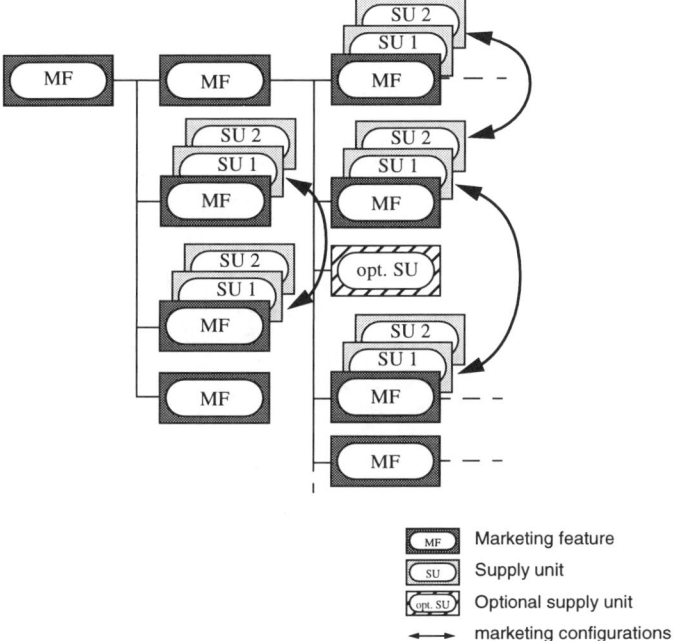

Figure 10: Marketing Product Structure

Definition of Marketing Features: Marketing features serve the customer to choose the configuration of his product. The features, therefore, have to be adjusted to the customer's comprehension and requirements. For this reason, a marketing-oriented definition of features offers a semantic level of product configuration which prevents the customer from unnecessary choosing from the level of modules or technical details. At the same time, this procedure is an elegant way of limiting the number and kinds of configuration and thus, limiting complexity.

Definition of Supply Units: There are one or several versions of each marketing feature. These versions are called supply units. The number of supply units depends on the range of requirements.

Modeling the Marketing Configurations: If there are many variations of a product, some of the configurations which are theoretically possible, are not practicable or desirable because there are supply units which depend on other supply units. These configurations are represented by the marketing product structure. The marketing product structure and the configuration logic define which configurations will be designed throughout the life cycle.

Technical Product Structuring

In the course of the development process, the marketing product structure and configuration logic are translated into a technical product structure. The necessary activities are:

- relating modules to supply units
- modeling the technical configuration.

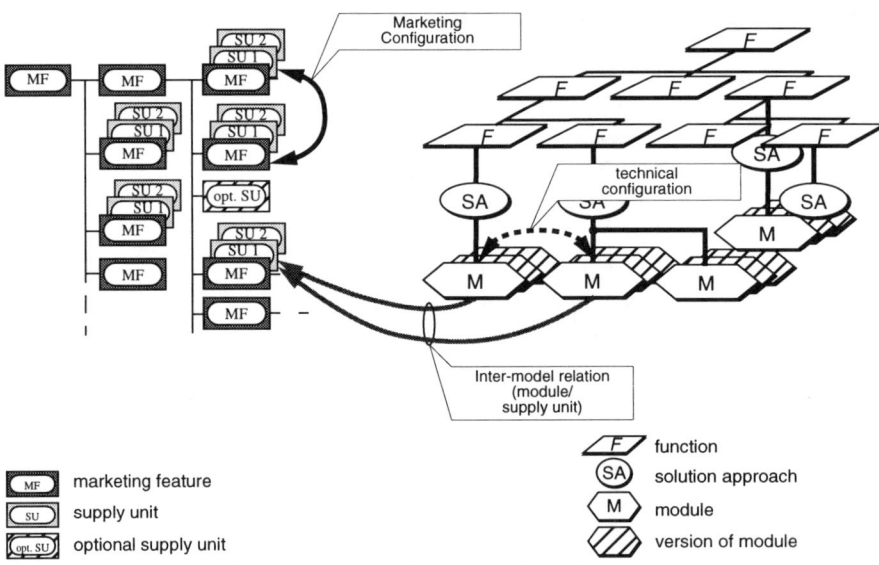

Figure 11: Technical Product Structure

Relating Modules to Supply Units: Supply units are regarded as "fill-ins" which are gradually replaced by modules. With Integrated Product Development, this is done by inter-model relations (Figure 11). According to the flexible activity model of Integrated Product Development, the marketing product structure can be modified if it cannot be realized by (combinations of) modules with regard to technical conditions or costs.

Modeling the Technical Configurations: The technical dependences between the chosen modules are modeled by another type of relation. The aim is to model the configuration history which can be used to represent (in)compatibility of versions. The technical product structure is a view[1] on the product model, including all partial models. This view includes features, supply units, solution elements, internal configurations and inter-model relations. The possibility of choosing and structuring pieces of information with regard to application can be used to meet the different views which different corporate sectors have on the product structure.

5 CAE Tools to Support the Early Stages of Product Development

The consequent computer representation of a product requires the supply of IT tools. For modeling and processing of results of other development acitivities, methods have to be provided which can be compared to the functions of existing CAD systems. This especially applies to Conceptual Product Design because there are no comparable CAE tools for these early stages of product development. Looked at more closely, the CAE tools which are required for Conceptual Product Design, mainly have to administer dependences on the level of product components. This includes relations between the whole and its parts (aggregational relations) to set up product structures and functional structures, as well as inter-model relations between the elements of different partial models.

Functions to Manipulate the Product Model: Manipulation functions let the product designer generate the product model by different elements (for example structure of requirements, product structure, functions) and different types of relations (for example specifying relations, inter-model relations).

Functions to Visualize and Form Views: In the early stages of product development, abstract and non-geometric information is generated. This information is visualized by tabular text and tree or network graphs. This visualization must bear the possiblity to emphasize views of certain correlations within the model. This is called "browsing" the contents of the model. Not only views within the partial models, but also views covering several partial models have to be visualized.

Functions to analyze the model: The analysis of model correlations is based on a continuous evaluation of the model by the product designer. Therefore, methods to analyze networks and methods which are related to the respective task, have to be supplied. They serve to judge the workability and technical consistence of the product model already in the early stages. The assessment tools can for example be based on the conjoint analysis [10].

The conceptional foundation of the horizontal model integration is a joint, integrated product model on a logical level. This results in a joint basic modeling tool which is offered to the CAE tools that support the early development stages. The basic modeling tool

---

1   Views combine correlations between information of different partial models. They help the product designer with his decisions.

is a universal class hierarchy which can be extended by specific methods for each CAE tool.

Figure 12 shows the class hierarchy of the basic modeling tool. It is divided into four levels. The classes[1] of **level 0** define attributes which all objects[2] of the basic modeling tool have in common, e.g. generating, deleting, editing, or attributes describing a definite object identitiy (OID).

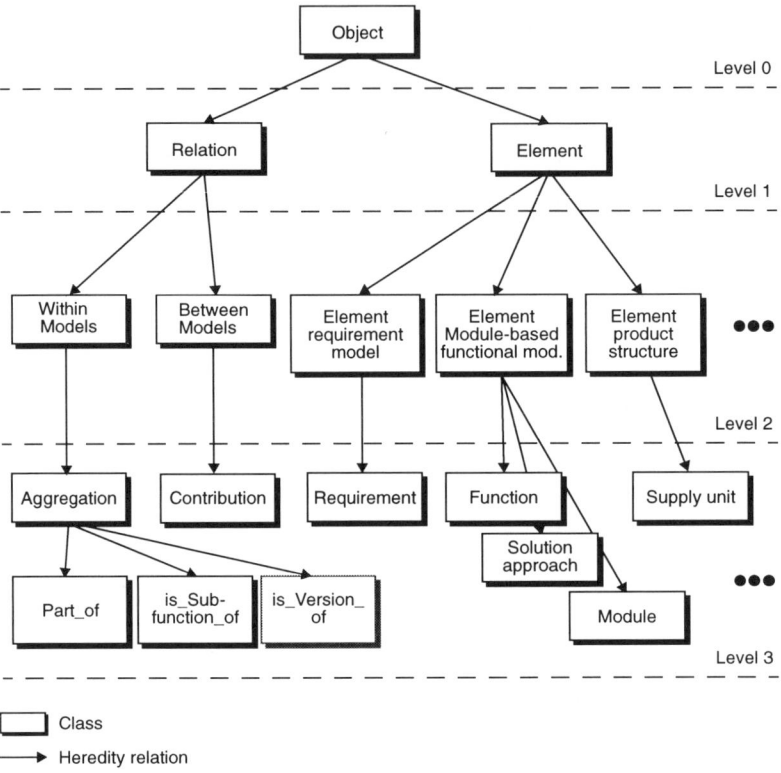

Figure 12: Class Hierarchy of a Basic Modeling Tool

The class of **level 1** defines attributes which are necessary in order to build up structures. These attributes can be operations to visualize tree or network graphs, or operations to realize standard algorithims for graphs, like "shortest way" or "maximum flow" [11].

---

1 A class is a general description of a set of objects well to distinguish with equal attributes and operations and therefore equal behavior. A class additionally contains the description of the operations that create or delete objects of this class. Every class has its definite name within an object-oriented system.

2 An object is a real or imaginary thing which has a definite identity. It contains attributes and values (data) belonging to these attributes, and operations. Operations can be public or private. There is no access to attributes or operations from the outside. One object can cause the public execution of another object. Every object is in a state which can be characterized by the current values of some or of all attributes.

The class "relation" additionally contains operations to control cardinalities [10].

On **level 2**, general relations within and between partial models are defined. Also, for every partial model an element class is defined. This way, special attributes and operations required by all objects of a partial model can be defined. The class hierarchy can be extended by new classes on this level as well as on the following levels. This measure will be necessary if for example a new partial model or a new general relation is required.

The classes of **level 3** describe the attributes of the elements belonging to a partial model and the relations between these elements. For example, in the class of aggregation the attribute cardinality is limited. There is no minimum or maximum number of elements but only one number which is minimum and maximum at the same time. This definition, ensures that the aggregate (left element) exists only if all of its parts (right elements) exist.

6 Summary and Outlook

This article presents the methodology of Integrated Product Development. Integrated Product Development offers a frame for structuring and integrating development acitivites and methods which are necessary for the development of complex industrial products.

Emphasis is laid on the fact that the development of such products requires universally valid methods to describe mechanical, electronical and information technology product components. This particularly applies to the early stages of product development and computer aided analysis.

Considering this background, Conceptual Product Design is introduced and described as an essential part of Integrated Product Development in order to methodically support the procedure. Conceptual Product Design aims to ensure the fulfilment of requirements, the basic workability, and the expected economic success as early as possible.

The conceptual foundation of Integrated Product Development and Conceptual Product Design is the generic model of acitivies. This model allows activities which follow certain tasks to be adjusted to applications. Thus, the model integration and process integration of development activities gain in significance. For this purpose, an integrating conception is introduced which is based on partial models being related to activities, and which, on a logical level, connects the partial models to an integrated product model by inter-model relations. Finally, the CAE tools being necessary to set up such a cross-linked product model are explained. Here, a basic modeling tool to generate, manipulate, and represent abstract, non-geometric product structures is the center of interest.

The aim of further research is, on the one hand, to verify the methodology of Integrated Product Development by industrial projects, and on the other hand, to specify the generic model of activities. There are also plans to provide the CAE tools which are nessesary to support Conceptual Product Design, on the basis of the existing basic modeling tool. They will be tested by chosen examples of use. In this area, future research will concentrate on the development of an open system architecture which will be suitable for CSCW[1]. For this purpose, the basic modeling tool must be extended by functions to coordinate distributed, parallel development activities, and to integrate commercial CAE tools [12].

7   References

[1]   Gausemeier, J./ Grafe, M./ Frank, T./ Humpert, A.: Integratives Konzipieren neuer Produkte mit CAD. In: Datenverarbeitung in der Konstruktion '94, Proceedings of the VDI Conference, 27-28 October, VDI-Verlag, Düsseldorf, 1994.

[2]   Beitz, W.: Konstruktionsmethodik aus der Praxis, published by Springer-Verlag, Konstruktion 41,1989, Berlin, Germany, p. 403-405.

[3]   Pahl, G./ Beitz, W.: Konstruktionslehre - Methoden und Anwendung, published by Springer-Verlag, 3rd edition, Berlin, 1993.

[4]   Heym, M.: Methoden-Engineering - Spezifikation und Integration von Entwicklungsmethoden für Informationssysteme. Doctoral thesis hochschule St. Gallen für Wirtschafts-, Rechts- und Sozialwissenschaften, Rosch-Buch Buchbinderei, Hallstadt, 1993.

[5]   Löhr-Richter, P.: Methodologie - Methodik - Methode. Was steckt dahinter?. In: EMISA FORUM Mitteilungen der GI-Fachgruppe - Entwicklungsmethoden für Informationssysteme und deren Anwendung -, Heft 1 (1993), p. 39-41.

[6]   Rathnow, P. J.: Integriertes Variantenmanagement. Göttingen, Vandenhoeck und Rupprecht, 1993.

[7]   Gausemeier, J./ Frank, T./ Genderka, M.: Erfolgspotentiale integrierter Ingenieursysteme. In: CAD'94 - Produktdatenmodellierung und Prozeßmodellierung als Grundlage neuer CAD-Systeme, Proceedings of the Conference of the Gesellschaft für Informatik, Paderborn 17-18 March, 1st edition, published by Carl Hanser Verlag, Munich, 1994.

[8]   Lückel, J./ Moritz, W./ Schütte, H./ Wittler, G.: Development of a Modular Mechatronic Robot System. In: Proceedings of the 2nd Conference on Mechatronics and Robotics, Duisburg/Moers, 27-29 September 1993.

[9]   Lückel, J./ Wittler, G./ Moritz, W./ Schütte, H./ Neumann, R.: Funktionsorientierter Entwurf und Softwarewerkzeuge für mechatronische Systeme am Beispiel eines modularen Robotersystems. In: CAD'94 - Produktdatenmodellierung und Prozeßmodellierung als Grundlage neuer CAD-Systeme, Proceedings of the Conference of the Gesellschaft für Informatik, Paderborn 17-18 march, 1st edition, Carl Hanser Verlag, Munich, 1994.

[10]   Schlageter, G./ Stucky, W.: Datenbanksysteme: Konzepte und Modelle. 2nd edition, B.G. Teubner, Stuttgart, 1983.

---

1   CSCW: Computer Supported Cooperative Work

[11] Cormen, T. H./ Leiserson, C. E., Rivest, R. L.: Introduction to Algorithms. The MIT Press Cambridge MA, London GB, 1991.

[12] Gausemeier, J./ Hahn, A./ Schneider, W.: Architekturprinzipien verteilter objektorientierter Ingenieursysteme. In: Proceedings of the Conference: Die Herausforderung eines globalen Informationsverbundes für die Informatik, Gesellschaft für Informatik (GI) und Schweizerische Informatiker Gesellschaft (SI), Zurich, 18-20 September, 1995.

Authors: Prof. Dr.-Ing. Jürgen Gausemeier
                  Dipl.-Ing. Dirk Brexel
                  Dipl.-Ing. Thorsten Frank
                  Dipl.-Ing. Axel Humpert

Address: Heinz Nixdorf Institut
                  Universität-GH Paderborn
                  Pohlweg 47-49
                  D-33098 Paderborn
                  Tel.: 05251/603262
                  Fax.: 05251/603241

Modular concepts for the new generation of DLR´s light weight robots

B. Gombert, G. Hirzinger, G. Plank, M. Schedl, J. Shi

D L R
(German Aerospace Research Establishment)
Institute for Robotics and System Dynamics
Oberpfaffenhofen, 82234 Wessling

Abstract: The paper outlines DLR's mechatronic developments in the robotics area over the last five years. They aim at designing a new multisensory, modularly configurable light weight robot generation in a unified and integral way. A first step in this direction has been the development of a complex multisensory gripper. It turned out to be a key element in ROTEX, the first real robot in space that flew with shuttle COLUMBIA in early 1993.

Sensors and actuators in DLR's new ultra-light-weight robot show up miniaturised integration of mechanics, electronics and microprocessor control. Joint torque control based on inductive sensing may be combined with different type of gearings: translational ones (the "artificial muscle" using a new spindle drive concept), originally developed for grippers, as well as compact, highly reducing rotational ones. With optimised carbon fibre grid structures and optical information transfer between the joints a new, extremely light-weight type of robot arises. All power and signal electronics is integrated into the arm which shows up a 1 : 1,5 ratio between load and own weight. Joint control concepts are outlined and a 7 dof version of such a light weight arm is described.

1 Introduction

Robots today in nearly all applications are purely position controlled devices with maybe some static sensing, but still far away from the human arm's performance with its amazingly low own weight against load ratio and its on-line sensory feedback capabilities involving mainly vision and tactile information, actuated by force-torque-controlled muscles.

Many groups have specialised on certain areas in robotics science but it is difficult to integrate the results, make them available to others and even more difficult to transfer them to industry. These observations have motivated us to design a new robot generation from bottom up in a modular, unified and integrated approach. Its main features are

- Ultra light carbon fibre grid structures for the links, with structurally optimised integration of new torque-controlled actuators, and with all electronics integrated into the arm, aiming at a 1:1 ratio of own weight/maximum load.

- Multisensory grippers that are not just based on an assembly of available sensory components, but are designed from scratch following a unified multisensory hard- and software design philosophy. Design and integration of articulated 3-finger-hands as an alternative has just started.

- Learning and self-adaptation concepts for internal and external (sensorbased) control loops, as well as for environment modelling.

- Tele-sensor-programming, a sensor-based off-line programming technique which is strongly related to teaching by showing elemental moves.

This paper focuses on the sensory and mechatronic concepts involved.

2 The mechatronics system concept

Our mechatronics approach in designing a new generation of multisensory light-weight robots is an integral one, based on miniaturised mechanical components with as much as possible electronics, software and control integrated. To be more specific: the new sensor and actuator generation we developed in the last years does not only show up a high degree of electronic and processor integration based on surface-mount-device (SMD)-technology, but also a fully modular hard- and software structure that allows to duplicate common concepts from one sensor or actuator system to another one. Analogue conditioning, power supply and digital pre-processing are typical subsystem modules of this kind. The 20 kHz power supply line connecting all sensor and actuator systems in a galvanically decoupled way (i.e. they may derive their individually requested voltages via tiny transformers), and the high-speed serial data bus, via which all systems communicate with a busmaster, are typical features of our multisensory (and Multi-actuator) concept for the new robot generation envisioned (see figure 1). For the endeffector sensory and actuator system a 625 kBaud serial bus system based on the so called μLAN-Concept of the 8051 microcontroller family has been implemented. But for the joint actuators we are using the SERCOS bus with optical fibres at a data rate up to 10 MBaud.

The modularity concept is pertained in the software of all sensors and actuators.

All sensory or joint information is written into dualport-RAM's or shared memories via the so-called busmaster, so that the robot controller has direct access to them without the need to handle time consuming protocols as is usual in present-day robot systems. A 1ms Cartesian control cycle (including external sensory feedback laws) is baseline for the design of the robot controller.

In order to start demonstrating these concepts a multisensory gripper with 16 sensory components has been developed in our lab and has been a key item in the space robot technology experiment ROTEX (see [1] and figure 1). This gripper (probably the most complex one that has been built so far) integrates a stiff (straingauge-based) and a compliant (optically measuring) ring-shaped 6 dof force-torque sensor, 9 tiny triangulation based laser range finders, one for longer ranges (3-30 cm) and 8 for small ranges (0-3 cm), two tactile arrays and a miniaturised stereo camera system. Thus this gripper for us serves as an excellent testbed for sensory fusion. For modular use in all of our actuators (in gripper and joints as well) we have developed highly dynamic brushless, flat disk DC motors with speed-dependent phase-lead compensation for speeds up to 20.000 rpm. For these motors we found that the electrical time constant in the range of a few milliseconds is dominant compared to the mechanical one!

We combine these motors either with our low-friction, rotational-translational gearings (the patented DLR-spindle concept) or with our extremely compact rotational-rotational high-reduction gearings. While for the gripper as well as for a new articulated 3 fingered hand (not described in more detail here) the spindle concept is used uniformly, our robot joints have been designed alternatively using the above mentioned gearings, both approaches however resulting in rotational joints. Comparative tests will be made till the end of this year.

An important feature of these new joint drives however is their joint-torque sensing and control capability, as all modern robot control approaches are based on commanding joint torques; due to friction this is not feasible with standard robots today. An inductive torque sensing system has been integrated into all joints of our new light weight robot.

Figure 6 shows, a first PUMA size six axis version off (with the electronics embedded inside the arm), with carbon fibre grid link structures (developed by our colleagues from DLR's institute of structural research and development in Stuttgart), optimally integrated torque-controlled joints, and a resulting load to weight ratio of nearly 1 : 1,5.

3 The gearing type alternatives

3.1 Rotational-translational gearings; the "artificial muscle"
In an electrical robot gripper as in figure 2 the problem is to transform a motor's high-speed rotational motion into a fairly slow axial motion to move the fingers. For this type of transmission a new mechanical spindle concept has been designed. The spindle fixed to the motor axis wears a very fine thread with extremely small pitch (e.g. 0,2 mm/revolution). Six planetary rollers arranged concentrically between spindle and the finger-driving nut show up fine grooves (no pitch) that fit precisely into the grooves of the spindle thread. However they also show up much coarser grooves that mate with corresponding coarse grooves inside the nut, so the rollers do not move axially relative to the nut nor do they change their mutual distance. Why not? The real trick has been to provide the rollers' fine grooves with a "phase shift" from one roller to the next, so that without using a multiple thread (not realisable with this pitch anyway) they simulate a nut with the desired low pitch. And, as the other crucial factor, by using rollers of this type sliding friction as prevailing in a pure spindle-nut concept has been replaced by much smaller rolling friction. These systems show up an efficiency of 96 meanwhile after some improvements in manufacturing.

What we have gained with this motor-gearing combination is a powerful small prismatic drive. Used as a two-finger-gripper drive as realised in ROTEX it allows to exert grasp forces of more than 300 N with a gripper weight of 5 N and a grasp speed of about 15 cm/sec. Without measuring and feeding back grasp forces we arrive at a feedforward grasp force control resolution of $\approx$ 1 N (0,5 % of max. force) with high repeatability. The reduction rate referring to the finger rotation is $\approx$ 1 : 1000.

However there are many other applications for miniaturised, integrated motor-spindle combinations (we call them "artificial muscles") outside the robotics area, e.g. in replacing hydraulic actuators.

3.2 Rotational-rotational gearings

The "phase-shift" ideas as outlined in the last chapter have been transferred to pure rotational gearings, too. The basic concept of our design is schematically depicted in figure 3, consisting of two simple planetary gearings that are connected via the planetary wheels e.g. via friction so that a friction clutch (very useful in case of joint overload) is inherent in the system. The phase-shift idea now is realised by letting the number of teeth in the second (moving) hollow tooth wheel z_5 differ by one compared to the first (fixed) one z_3. To assure that despite of this fact both wheels of a planetary wheel block simultaneously mate with their corresponding hollow wheels, they must show up a certain "phase shift"

$$\Delta\varphi = \frac{360°}{z_4 \cdot n_p} \cdot i, i = 1, \ldots n_p \qquad (1)$$

of their teeth (see figure 4), where z_4 denotes the number of teeth in the planetary wheels and n_p is the number of revolving blocks. Due to the friction clutch the phase shift is adopted automatically. Gear transmission ratio is

$$r = \frac{z_2 z_5 (z_3 + z_1)}{z_1 (z_2 z_5 - z_3 z_4)} \qquad (2)$$

A reduction rate of $\approx$ 1 : 600 turned out to be easily realisable (a factor of 1 : 100 being contributed by hollow wheels with 100 and 101 teeth), but higher reductions are no problem, too. Note that the gearing is very compact, has more teeth in contact (i.e. better force transmission) and is considerably stiffer than a harmonic drive.

4 Modular architecture in the arm

Design goals

The new DLR-robot (figure 6 and 7) was designed in a modular way, each module consisting of a joint, the (carbon fibre grid) link-structure-element and the embedded electronics. Each electronic module is of the same type and consists of a power inverter, the joint controller module and the joint-torque-sensor. Only the motors are adapted in size depending on their position and the expected load within the kinematic chain. The first and the last joint are of a different type for connecting them to either a fixup in a workcell or our quicklock-mechanism of the multisensory gripper. Both flanges are easily fixed by six bolts and can be adapted to any existing interface.

The design-philosophy of this light-weight-robot was to achieve a type of manipulator similar to the human arm, with a load to weight ratio of about 1:1, a total system- weight of about 100 N, seven degrees of freedom, no bulky wiring on the robot (and no electronics cabinet as it comes with every industrial robot), and a high dynamic performance to improve force control at constraint movements. On the other hand it was not necessary to achieve extremely high accuracy for feedforward positioning (as we prefer to rely on sensory feedback), not to build a very stiff structure nor to perform tasks with high loads at top-speed. From figure 7 it comes clear that a seven axis version may be assembled in different ways including alternatives that allow to fold the robot in a very small volume, e.g. for stowing it in a space mission.

The brushless motors originally are disk-type permanent magnet stepping motors supplied by Portescap. We converted them into synchronous motors by adding analogue hall sensors to derive the commutation signals and a position reference. These motors offer a high number of pairs of poles, so the motor-position can be evaluated with high resolution and no additional encoder is necessary. Additionally the inertia of the rotor is extremely low resulting in a highly dynamic behaviour of the motor. There are different types available ranging from 1.2Nm - 0.18 Nm torque corresponding to 0.8kg to 0.2kg of weight. It is possible to achieve noload speeds up to 20.000 rev/min by commanding a positive phase shift of the driving voltages relative to the commutation signals. Although the rotor's inertia is very low, it is the dominant part of the mass-matrix of the robot-system due to the high reduction-ratio of the gears. So the position-dependant part of the robot's mass-distribution is only 20% of the joint-inertia.

5 Electronic Hardware

The approach for the joint-level-control of the robot is a system of distributed joint-modules with only a few necessary connections between them. Since each joint has its own "intelligence" integrated, only electrical power and information have to be distributed between the joints and the robot-controller. Thus the rather big amount of usual wiring for signals and motor-connections can be avoided. In our robot the connecting bus between the joint- modules consists of a fiber-optical-ring of 0.5mm plastic-fiber for data-transmission, a shielded twisted pair connection for supplying the electronic subsystems, a highly flexible unshielded twisted pair for the 80V DC-supply for the power electronics and a single connection for case-ground. There are no analogue or digital electrical signals distributed on the robot. The same difficulties arise when in a mechatronic subsystem, e.g. a joint module, electronic modules of different types are integrated close to each other. The best way is to keep signal paths as short as possible, especially those which conduct switched power, and to transfer sensitive signals at the highest possible level, preferably in form of digital information.

5.1 Joint-torque-Sensor
The Joint-torque-sensor (figure 5) is placed between the motor and the gear to measure the actually exerted torque between the gear-connection and the following structure-element. The mechanical part is a six- spokes-wheel forming a rotational spring and a differential position measurement sensor built with two inductive sensors. The evaluation-electronics is mounted right beneath the sensors on a ring shaped board and contains the preamplifiers and evaluation circuitry. The torque-information is fed directly to the joint- processor via a three-wire synchronous serial connection at 2MBaud data-rate. The information contains a 13bit torque-value with over- and underrange and zero-flags. The electrical signal-bandwidth is set to 2kHz and the signal-to-noise-ratio is better

than the 13bit resolution. Apart from the serial connection the sensor is completely galvanically decoupled.

5.2 Power Inverter

The power inverter is located directly at the motor, so the motor-connections carrying the switched high voltage are extremely short. It is capable of switching 100V at 10A. The power-switches (MOSFETS) are galvanically decoupled via optocouplers to avoid disturbances of the analogue current-control-circuitry.

The current-controller is a simple fixed-frequency cycle-by-cycle switchmode controller. It is located on the power inverter-module and operated at 40 kHz switching-frequency. The module also contains a set of amplifiers for the Hall-sensor-signals of the motor and position-references and several thermosensors. The size of this inverter is equal to the motors in diameter and the cover, which closes the joint-assembly on the side opposite the gear is used as a cooling-device for the MOSFET-Transistors.

The inverter is wired to the joint-controller-module, which is located inside the carbon-fibre-structure on the motor-side. It receives the analogue current-references, the current-direction signal and a start-stop signal from the joint-controller and supplies hall-sensor-signals, position-references, current-values, high-voltage-dc-monitoring and thermal monitoring to the controller.

5.3 Joint Controller Modules

The Joint-controller consists of four circuit-boards of the same size (9cm x 9cm) piled up in a stack. One board carries the supply-circuitry where the necessary subsupplies are derived, galvanically decoupled via a tiny transformer. All analogue, digital and power circuits are supplied independently to avoid interference and yield a good analogue resolution and signal to noise-ratio.

The inner two boards each carry a 80c166 microcontroller operated at 40MHz with peripherals. The processor on the master-board is connected to the SERCOS-controller, where cyclic data (reference-values and commands) is received and cyclic measurement data (position, speed, torque) is sent. Its main task is to perform the desired joint control algorithm within the specified cycle-time of 1ms. Besides that it performs housekeeping-procedures like thermal-control and on-line monitoring of voltages and position-references or complex commands like testing or basic initialisation and referencing of the joint- module.

The master-processor communicates with the motor-processor via a dualport-memory. It commands the type of motions and the according references to the motor-controller. The motor-control- processor runs a motor-position-control at 8kHz cycle-rate. Its measurement-signals come from the power-inverter and an interface- circuitry which is placed on the last board in the stack.

On the interface-board a digital high-resolution motorposition-signal is generated and the motor current reference-values, which are generated by the processor via a two-channel digital-analogue converter, are rotated into the motor-system according to the hall-sensor-signals. Here it is possible to change the commanded current-vector phase-angle relative to the analogue hall-sensor information. We can achieve very high no-load-speeds by commanding a positive phase-shift, which means operating the synchronous motor in a capacitive way, or we achieve instantaneous braking by applying negative phase-angles. Additionally the motorprocessor can take full control over the motorcurrents, neglecting the hallsensor information and run the motor in microstepping mode without any ripple and at any arbitrarily low speed. This results in excellent micro-gravitational behaviour which is only limited by the small uncontrolled movements occurring due to the gear's backlash during load changes.

To avoid the reference-movement after power-on each processor is equipped with non-volatile memory to keep its position and configuration data after shutdown. The program-memory is integrated in flashmemory-technology to ensure flexibility in software- or parameter-changes. After booting the systems communication-cycle it is possible to access any memory location of both processors via the SERCOS interface. For diagnostic-purposes without the need of a SERCOS-Controller a simple status-display and a terminal interface is provided.

6 Dynamic Control

From the control point of view, the DLR light weight robot manipulator belongs to the category of flexible joint robots due to the structure of the gear box and the integrated torque sensor. The dynamic model can be established by applying Lagrange's Equation,

$$M(q_1)\ddot{q}_1 + V(q_1,\dot{q}_1) + G(q_1) + k(q_1 - q_2) = J^T(q_1)F, \tag{3}$$

$$J_2\ddot{q}_2 - k(q_1 - q_2) = \tau_m, \tag{4}$$

where

$M(q_1)$ - mass matrix,

$G(q_1)$ - gravitation vector ,

$V(q_1,\dot{q}_1)$ - vector corresponding to centrifugal and coriolis forces,

F - external force applied to the endeffector,

J - Jacobian matrix,

k - elasticity matrix,

τ_m - motor torque vector,

J_2 - motor and gear inertia matrix and

q_1, q_2 - link and motor angular position vector.

For the decoupling of the manipulator dynamics this model is transformed into a new coordinate in which the joint torque is treated as a state variable instead of motor position i.e. $q_1 = q_1$, $\tau = k(q_2 - q_1)$. This leads to the so-called singular perturbation formulation of the robot dynamics and enables one to separate the dynamics dependent on the motion rate of the related state variable. As the result, the fast motion corresponds to the joint-torque loop equation.(4) and the slow motion corresponds to the dynamic path concerned with the link position equation.(3):

$$M(q_1)\ddot{q}_1 + N(q_1,\dot{q}_1) = \tau + J^T(q_1)F, \tag{5}$$

$$A\ddot{\tau} + B(\tau,q_1,\dot{q}_1) = \tau_m, \tag{6}$$

where $A = J_2 k^{-1}$, $N = V + G$, $B(\tau,q_1,\dot{q}_1) = (I + J_2 M^{-1})\tau + J_2 M^{-1}(J^T F - N)$.

Because the arm dynamics and the joint dynamics (referred to the dynamics in the robot joints) can be clearly distinguished through such a motion-separation concept, the dynamic control problem is simplified significantly. Moreover, some measurements (or estimations) like link acceleration and jerk become no more necessary. This is the major reason why we decided to incorporate the joint-torque sensor into the robot design even if additional elasticity has been introduced.

For the design of the joint torque controller a perturbation estimator based on the theory of variable-structure systems (VSS) is involved to enhance the robustness of the pole-placement design using feedback-linearization technique. Because the switching action caused by the VSS does not occur in the real control path but rather in the estimation algorithms, the chatter and the adverse effects of conservative bounds on system perturbation, often encountered in conventional sliding-mode control system (SMCS) , are alleviated by this control strategy. Due to the unmodelled dynamics, parameter variation and non-linear dependence on the robot pose in the joint-torque loop, such a perturbation estimator is proven to be necessary through simulations and experiments.

As for the control problems concerned with the slow dynamic path (e.g. manipulator position) control problem, same techniques for the rigid-body robot such as computed torque method may be applied to the manipulator due to the fast convergence of the inner torque-control loop.

7 Robot-Controller Hardware Architecture

Most of all robot controllers use specialised and expensive hardware. In contrast the light weight robot is equipped with standard hardware components based on the VMEbus. i.e. CPU Boards and IO-Controllers. Another problem with conventional robot controllers is the long signal path between the motors including measurement systems for position, velocity, and torque at one end and the power amplifier with the motor controller at the other. The resulting measurement signals are disturbed by the environment. To overcome this problem the light weight robot - as mentioned above - has its amplifier and drive controllers located in its arms close to the motors. The distribution of the robot controller and the drive controllers, which are usually located in the same VME rack, requires a deterministic communication system between these units. Due to our goal using standard components a couple of "field bus" systems were evaluated. The best suited system for our purpose has turned out to be the "SERCOS".

8 SERCOS optical field bus system

SERCOS is a SErial Real-time COmmunication System for digital data exchange between controllers and drives of numerically controlled machines (NC). The SERCOS interface specification [3] was developed by a joint working group of the VDW (German Machine Tool Builders Association) and ZVEI (German Electrical and Electronic Manufacturer's Association). The main features of this system are high noise immunity and isolation, because the system transfers its data via an optical fiber. Data exchange is performed cyclically at short fixed intervals, programmable in the range from $62\mu s$ to 65ms).The SERCOS interface communication system consists of a master and several slaves. These units are connected via an optical fiber ring structure . The ring starts and terminates at the master. The slaves recover and repeat the data they receive or transmit their own data telegrams. All telegrams sent by the master are received by all slaves while the master receives data telegrams from the slaves. The optical fiber assures reliable high speed (2MBaud to 10MBaud) data transmission with excellent noise immunity.

The SERCOS interface includes three basic communication tasks:

- slave synchronisation with respect to the master,

- control and measurement data exchange between master and slaves (cyclic data transmission),

- sending parametric data and commands to the slaves and receiving detailed status information from the slaves on request by the master (acyclic data transmission).

A very useful feature of the slaves is the possibility to reload the complete controller software in the joint-modules via the acyclic data transmission protocol, without intervention to the joints hardware.

9 Software Architecture

Our robot control system is based on the industrial standard Real-time Multitasking Operating System VxWorks with its powerful development framework. The basic software structure idea is a data flow driven intertask communication approach. That means the necessary software modules (tasks) e.g. path interpolator, coordinate transformation, and drive controller are synchronised via the message passing mechanism supported by VxWorks. This modular concept allows e.g. a very easy and fast porting of software from a single to a multi-processor environment. The support of a distributed computing environment via a network is envisioned (see also [6]).

10 Conclusion

A design concept and the first realisation steps for a new generation of torque-controlled light-weight-robots based on a fully modular mechatronics concept have been presented. With all electronics integrated into the carbon fibre arm, a remarkable 1 : 1 weight / load ratio has been achieved; links and joints were not designed separately and combined afterwards, but in an integral, iterative and mutually interfering process. Different advanced joint torque control approaches (e.g. sliding mode control and learning control schemes) which include explicit treatment of joint elasticities are under investigation presently. With the multisensory gripper presented, the design of which was guided by the same mechatronics concepts, these kind of light, sensor-based robots as pushed forward by the needs of space technology, are kind of a starting shot for what we might call "next generation" robots. Controller design on all levels including sliding mode control and neural nets [4] are under investigation presently. The sensorbased programming concepts for these kind of robot systems are explained in different papers, e.g. [2]. Articulated three fingered hands based on the "artificial muscle" concept are under development in our lab and will bring the light weight robot's performance even closer to the human arm.

Technical Data

Gear		
reduction ratio		606
direction		equal
weight	g	700
dimensions	mm	ø83 x 38
max. input speed	min^{-1}	20.000
rated output torque	Nm	100
peak torque	Nm	160
backlash	min	< 6
transmission accuracy	min	5
efficiency	%	40

Joint		1	2	2a	3	4	5	6
type		Roll 1	Pitch 1	Roll 2	Pitch 2	Roll 2	Pitch 3	Roll 3
motor escap		P852	P852	P632	P632	P632	P532	P520
motor weight	g	860	860	420	420	420	175	145
total weight	g	2129	2070	1606	1657	1606	1423	1165
size	mm	Ø130	(Ø96)	Ø115	(Ø96)	Ø115	(Ø96)	Ø115/Ø96

References

[1] Hirzinger, G., Grunwald, G., Brunner, B. and Heindl, J. :
A Sensor-based Telerobotic System for the Space Robot Experiment
ROTEX, Proc. ISER'91 Second Int. Symposium on Experimental
Robotics/ISER, Toulouse, France, June 25-27, 1991.

[2] Hirzinger, G., Brunner, B. and Arbter, K. : Proc. IROS'94 Intelligent Robots and
Systems, Munich

[3] Foerdergemeinschaft SERCOS Interface e.V.: SERCOS Interface Specification,
Foerdergemeinschaft SERCOS Interface e.V.,Im Muehlenfeld 28 D-53123
Bonn

[4] Hirzinger, G., Baader, A., Koeppe, R., Schedl, M. : Towards a new generation of
multisensory light-weight robots with learning capabilities, IFAC'93
International Federation of Automatic Controll CONGRESS 1993, Sydney,
Australia, July 18-23, 1993.

[5] Freund, E: "Fast nonlinear control with arbitrary pole-placement for industrial robots
and manipulators", The International Journal of Robotics Research, p. 65-79,
1983.

[6] A. Beguelin, J. Dongarra, G. Geist, R. Manchek, K. Moore, V.S. Sunderam, : PVM and HENCE, MIT Press, December 1992

Figures

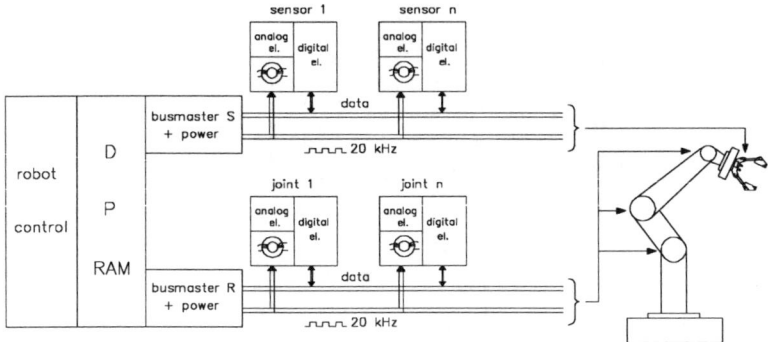

figure 1: Information and power transfer in the multisensory gripper and in the joints of the new DLR-light-weight-robot

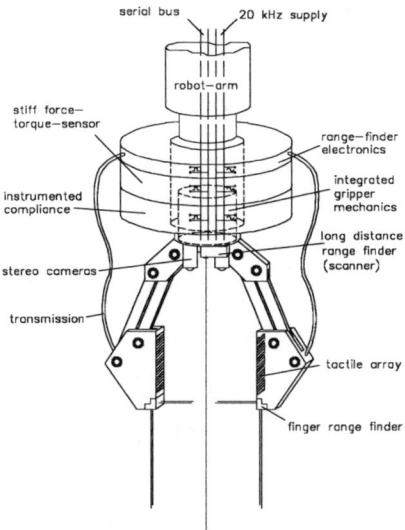

figure 2: Structure of DLR´s multisensory gripper including 16 sensory components

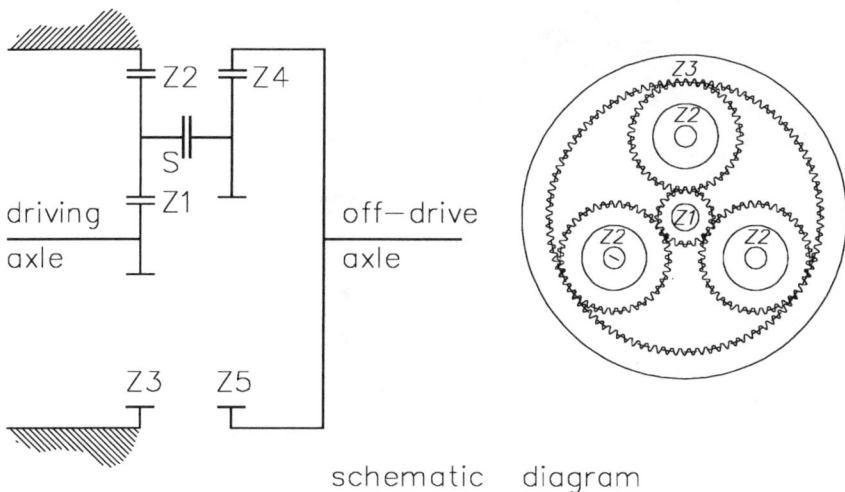

schematic   diagram

figure 3: The compact, high reduction gearing

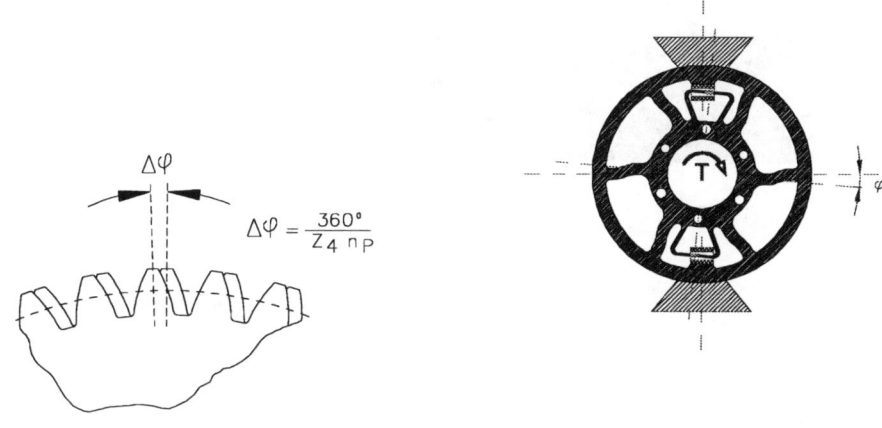

$$\Delta\varphi = \frac{360°}{Z_4 \, n_P}$$

"Phase shift between planetary
wheels Z2 and Z4

figure 4

**figure 5: Integrated torque measurement
with differential inductive sensors**

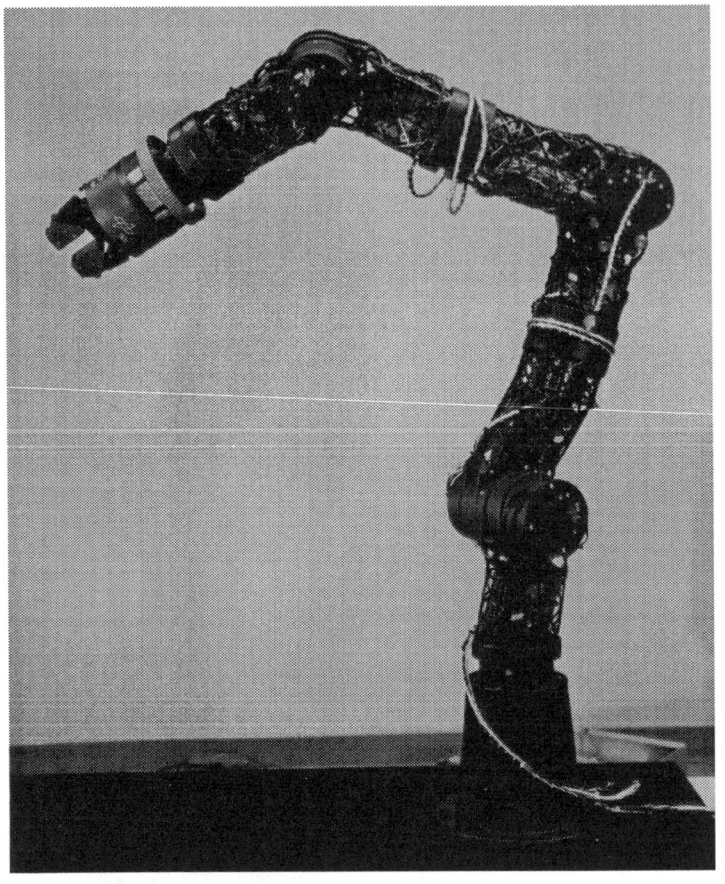

figure 6: The 7 axis version of the DLR light-weight-robot with integrated electronics.

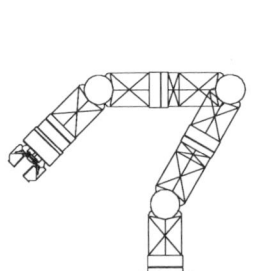

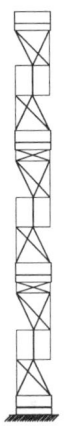

figure 7: kinematic structure and two different assemblies. The right type is optimally foldable for stowing in space applications.

B. Gombert, G. Hirzinger, G. Plank, M. Schedl, J. Shi
D L R (German Aerospace Research Establishment)
Institute for Robotics and System Dynamics
Oberpfaffenhofen, 82234 Wessling
Tel: +49 8153/28-2400
Fax: +49 8153/28-1134
e-mail df28@master.df.op.dlr.de

Integrated Design of Shape and Function in Mechatronic Systems

E. Kallenbach, K. König, E. Saffert, Chr. Schäffel, M. Eccarius

Technical University of Ilmenau, Ilmenau, Germany

Abstract: Increasingly more complex technical products have to be developed in shortening design cycles in industry. This task can only be solved if both the methodical basis and suitable design tools are created. Based on the design of an integrated multi-coordinate drive the opportunities for the design and configuration of heterogeneous systems are described whose functional features arise from the consistent application of the functional integration design principle in the mechanical domain and the functional separation design principle the automatic control domain. The integration of the components (actuator, sensor, control system) into an optimal configured system is crucial for its achievable technical parameters. During this process the prospects of decoupling by means of automatic control should be kept in mind.

1 Introduction

The technological development during the past decades is characterised by a rapid increase in functional complexity of modern technical products. With the market globalization and the increasing pressure of competition on one hand and the permanent build up of special knowledge and new highly efficient technologies on the other, new innovative products can be produced that increasingly better satisfy the requirements of the customer. Because of that the urge for new products rises faster than ever, while innovation is often achieved by designing products whose operation is based on the close interaction of physically heterogeneous subsystems. Seemingly irrelevant subfunctions are often crucial for the success on the market.

This development also has an effect on the modern natural and engineering sciences. The growing specialisation requires in terms of the knowledge transfer an ever more effective interdisciplinary cooperation of the individual disciplines of science and technology. The discipline oriented advancement of knowledge and the interdisciplinary integration e.g. in the analysis of complex systems or the design of new technical products mutually require each other. They are sources of stimulation for the advancement of sciences and the development of innovative potential within technology.

Since the mid-seventies with the development of electronics and microelectronics the tendency towards an integration of mechanical, electrical, and electronic components has increasingly found its way into mechanical engineering, car manufacturing, and space technology. To name this development the artificial word *Mechatronics* was introduced.

2 Characteristics of mechatronic systems

In comparison to classical products of mechanical engineering mechatronic products can be characterised through the features illustrated in *Figure 1* in particular:

- Mechatronic systems are distinguished from classical systems by a **greater number of subfunctions** (complexity), that can be realised on different physical levels which are supported by a many different technologies during the manufacturing (heterogeneity).

- The function of mechatronic systems is based on the **interaction of heterogeneous subsystems** and crucially effected by the automatic control.

- The utilisation of the opportunities given by the automatic control system for the implementation of required system features allows to **extend the area of operation significantly** and to adapt it flexibly to special requirements (tuning, adaptation, learning). In the interest of novel system features the main functions are complemented by structurally embedded support functions.

- The complexity requires the application of **methods of function and configuration oriented system analysis** like partitioning, hierarchical and modular structuring, integration.

- To achieve the desired functional features (e.g. high precision, fast dynamics) a **special functional structuring** of the mechatronic system is required.

- The **configurational system integration** allows the integration of system components (functional integration) in spatially concentrated systems (component/subsystem level) inducing reductions in the number of components and partially significant reductions in volume that usually also contribute to an **improvement in the operational limits.**

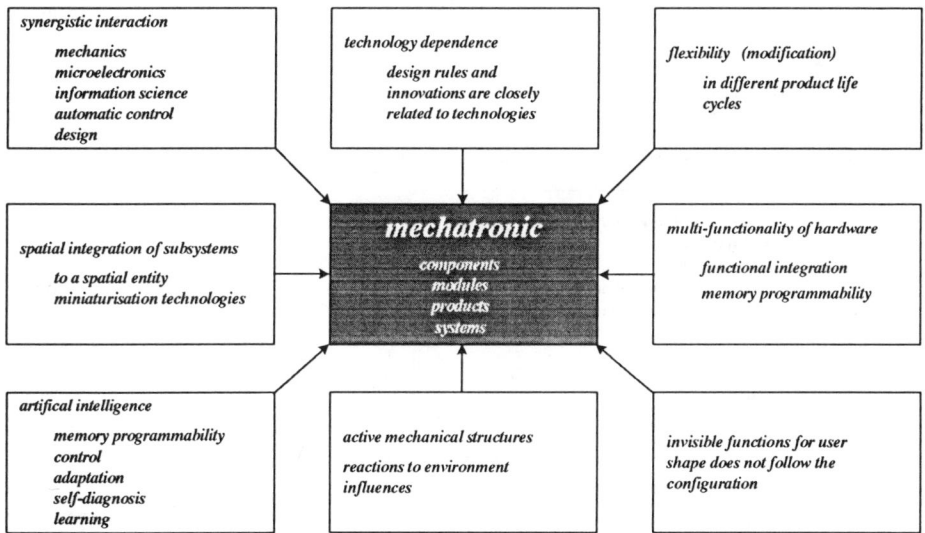

Figure 1: Characteristics of mechatronic products [1]

3 Design of mechatronic systems

3.1 Design phases of mechatronic systems

The design of mechatronic systems can be described by an analogue phase model on which the design of classical mechanical engineering products is based. However, a number of specific features of this specific phase model must be observed *(Figure 2)*.

Because of the high complexity the early restriction of the solution space with the help of an *early design phase* is of particular relevance. The refinement of the specification taking the requirements of the mechatronic product to be developed into well-balanced consideration can serve very early for a *reduction of the solution space* for the synthesis process. It corresponds with the experience of any designer that far too often information which is important to the design and can be derived from the problem situation is not or only insufficiently taken into consideration. This deficiency can be removed only if the *refinement of the specification* is given particular attention during all phases of the development process. For this the separation of the specification into areas of different weight (hard, minimal, maximal requirements, wishes) is essential to the proceeding and evaluation of individual development steps.

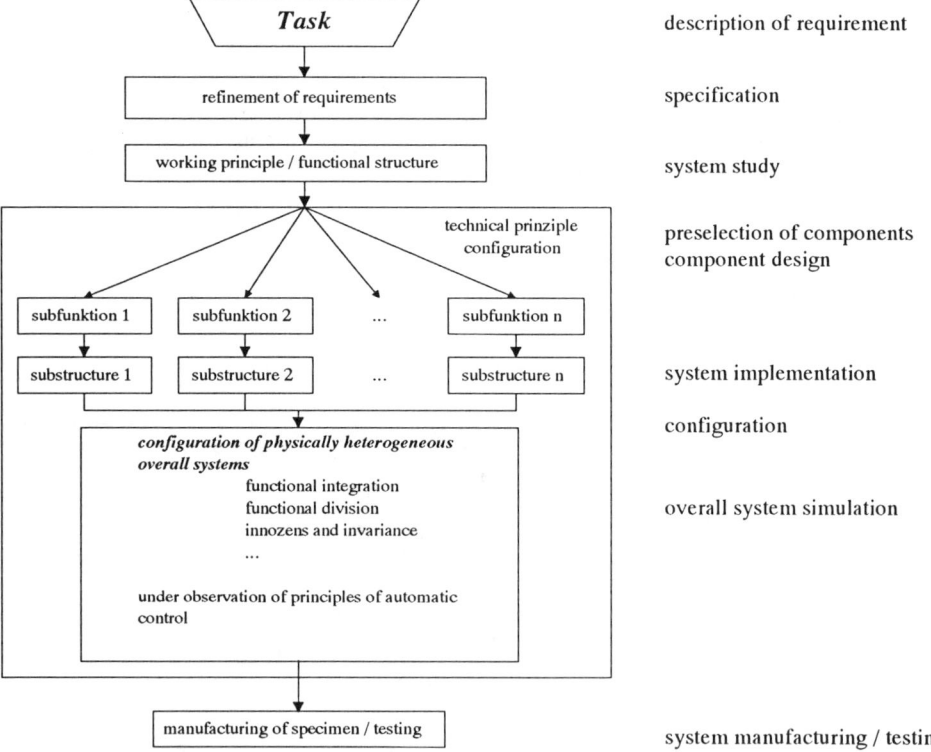

Figure 2: Specific features of the phase model for mechatronic systems

Function-structure-databases and knowledge-based methods can make it easier to trace a functional structure. Because function-structure-relationships are ambiguous, a suitable functional structure can often be found only by multiple simulations. Prior to that substructures and components meeting the functional requirements have to be assigned to the subfunctions. Doing that the areas of technology which can be expected to yield an optimal overall function and which are compatible in the manufacturing process have to be considered first. For this it is particularly important that partitioning, selection of the components and their integration are controlled in a way that a global optimum is reached and the number of system designs and the likelihood of error remains manageable. In face of the huge number of necessary tools a strategy for a systematic search for a nearly optimal functional and configurational structure is then effective if extensive analysis results on the function-structure-relationships are already available and tools are at hand. It is useful, if such analysis results are processed into design guidelines and design principles which are stored in databases ready for access. At the moment the existing knowledge is processed suitably only partially and in specific disciplines of science. Therefore theoretical and methodical work should be much more focused on synthesis appropriate knowledge pre-processing. This means e.g. to store the limits of physical principles in generalised but well manageable form.

3.2 Configuration of mechatronic products

A crucial step in the phase model of the design process is the definition of the technical principles, i.e. the transition from the functional description to the description of the mechatronic system as a material object. This process step, the configuration of a product includes the definition of geometric, material and state properties in a way that the technical product serves its purpose as optimal as possible and can be manufactured effectively.

Although the configuration is crucial to the competitiveness of a technical product and the work effort of the designer for configuring and refining is significantly greater than for drafting, configuring has often received too little attention within the design methodology in the past. This also applies to the design methods for mechatronic products. Based on the complexity of the configuration it is useful to divide the configuration phase into several steps, where with every configuration step that contributes to an ever better implementation of the overall function the amount of configurational detail increases. Even because of the high complexity of the functional structure of mechatronic products it is necessary to start from a functional structure in which a configurational element meeting the functional requirements is assigned to each subfunction. After that it has to be checked, to what extent new multi-functional configurational elements can be developed from single-functional configurational elements through integration (functional integration *Figure 3*). As a rule such an integration process is not only connected with a size reduction but can also lead to a significant improvement in the technical parameters and to an extension of the limits of a technical principle e.g. through the elimination of coupling elements, reduction of moving masses.

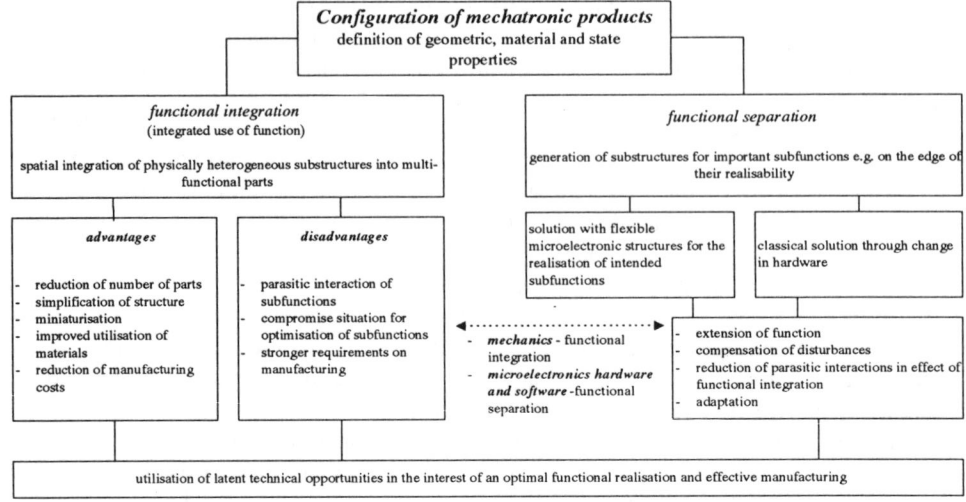

Figure 3: Influence of the information processing on the configuration

4 Design example

This chapter presents the design of a planar electrodynamic multi-coordinate drive (Figure 4) to illustrate some of the topics discussed in the previous chapters.

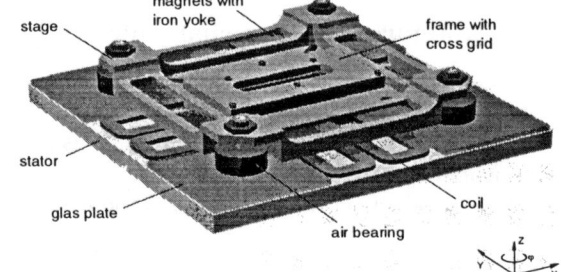

Figure 4: CAD picture of drive

Traditional designs start with a separation of the three axes which are configured individually and then put together into a serial arrangement (Figure 5). Because of clearance, mechanical hysteresis and elasticity this concept leads to a limited positioning accuracy. The need for motion transducers and the great masses that have to be moved often also limit the dynamic behaviour.

In the integrated design the hierarchy is removed and all subfunctions are put on the same level. They are then regrouped according to function (Figure 5). This leads to the design of multi-coordinate (multi-functional) components for the required functions.

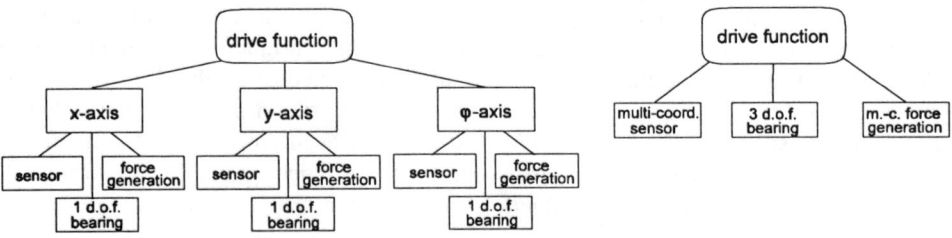

Figure 5: Functional structures of traditional (left) and integrated (right) design

4.1 Measuring system

Figure 6 illustrates how the multi-coordinate position sensor is implemented. It consists of three single-coordinate optic-incremental sensors in a special arrangement that allows for the determination of the rotational angle φ.

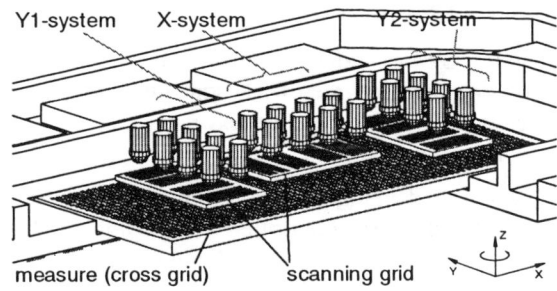

Figure 6: Measuring system

As all sensors scan the same cross grid, position accuracy depends only on the accuracy of that grid and the scanning systems but not on any mechanical guidance systems. This also reduces the number of parts as only one high precision measure is required.

The drawback is that the scanning grids must be specifically designed to be insensitive to rotation up to a certain maximum allowed angle. Also, the control system has to eliminate the influence of the rotation on the sensor signals and must ensure that the rotation stays within its limits. This requires that sensor signals are decode by software rather than by simple incremental encoders. So simplicity on the mechanical side is traded against a higher complexity on the electronic / software side.

4.2 Force and torque generation system

Figure 7 shows the multi-coordinate force and torque generation system. It consists of four force generation elements (f.g.e.) two for each linear axis. Torque around the z-axis is generated by generating two unequal forces along the same axis.

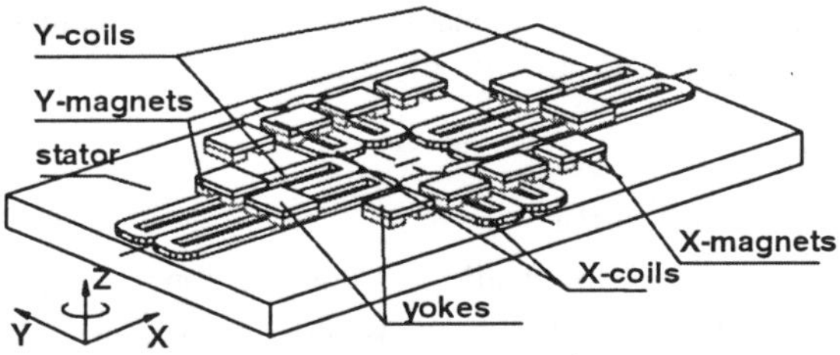

Figure 7: Force and torque generation system

Although a layout like that of the measuring system (only one force generation system for the x-axis) would have been sufficient the above layout was chosen as it provides room for the measuring system in the centre (see section 4.4).

The chosen moving magnet concept does both relieve the stage of trailing power supply cables and also generate a attracting reluctance force between stage and stator which permits overhead operation of the drive and preloads the air-bearings. Unfortunately, this force has also horizontal components which disturb the position control of the stage.

The resulting multi-coordinate force and torque generation system is structurally much simpler than a serial arrangement of three individual systems. However, there is now a lot of interaction (coupling) between the four f.g.e. which has to be compensated by coordinated control. Note, that each of the f.g.e. consists itself of two commutated coils making control of the eight individual coils even more complicated.

4.3 Guidance system

The integration allows to replace the three one-degree-of-freedom guidance systems by only one three-degree-of-freedom guidance system consisting of four air-bearings (see Figure 4). It contains a lot less parts and is far easier to manufacture as only one high precision flat surface is needed.

The major advantage of this is, however, that the guidance system does not affect the accuracy of the three controlled axes anymore. The stage is guided electronically in those axes by the control system with the full accuracy of the measuring system. The mechanical guidance system takes only care of the so far uncontrolled z-axis and prevents rotation around the x- and y-axes.

Because of their preload (see section 4.2) the air-bearings possess a very high stiffness. This further reduces any problems and ensures precise guidance.

4.4 Component integration

After multi-functional components according to the required multi-coordinate functions have been designed these components are even further integrated into subsystems. Stage and stator of the drive share parts of the components described before. For this integration it is vital that the layout of the components has been chosen such that they can now all be integrated easily. In this case the measuring system fits perfectly into the centre of the force and torque generation system while the air-bearings fit perfectly into its corners. The configuration is such that the stage carries the cross grid measure while the scanning grids and sensors are positioned in the stator. This relieves the stage also of the sensor signal cable. It only drags a flexible pipe for the air supply of the air-bearings.

5 Conclusions

Through the application of the functional integration design principle to the mechanical substructures and of the functional separation design principle to the microelectronic substructures and the software new mechatronic drive systems can be developed which exhibit both improved dynamic behaviour and higher precision. An additional advantage of the drive is its extremely slim design.

The construction of a demonstration model of the planar electrodynamic drive was founded by the German Research Council DFG in 1993/94.

6 References

[1] Hewit, J. R. (Editor): Mechatronics. Springer, Wien, New York 1993

[2] Salminen, V. / A. Verho: Systematic and Innovative Design of a Mechatronic Product. Mechatronics Vol. 2, No. 3, 1992, pp. 257 - 275

[3] Kallenbach, E. / R. Eick / P. Quendt: Elektromagnete - Grundlagen, Berechnung, konstruktive Anwendung. B. G. Teubner, Stuttgart 1994

[4] Luu van Thin: Untersuchungen zum Entwurf von Zustandsregelungen für dynamisch hochwertige mehrachsige Präzisionsantriebe. PhD-thesis, TU Ilmenau 1993

[5] Schäffel, Chr. / Schober, U. / Kallenbach, E. / König, K. / Eccarius, M.: Integrated Electrodynamic Multi-coordinate Drive - A Modern Component for Intelligent Motion. Joint Hungarian-British International Mechatronics Conference, Budapest 1994, Proceedings, pp. 419 - 429

Authors

Prof. Dr.-Ing. habil. Eberhard Kallenbach
Dipl.-Ing. Klaus König
Dipl.-Ing. Eugen Saffert
Dipl.-Ing. Christoph Schäffel
Dipl.-Ing. Mathias Eccarius

Technische Universität Ilmenau
Fakultät Maschinenbau
Institut für Mikrosystemtechnik, Mechatronik und Mechanik
PF 0565
D-98684 Ilmenau
Germany

Tel.:   +49 (0)3677 / 69-2485
Fax: +49 (0)3677 / 69-1802
e-mail: E.Saffert@Maschinenbau.TU-Ilmenau.de

II. Software Engineering in Mechatronics

M. Anantharaman, B. Fink, M. Hiller, S. Vogel
Integrated Development Environment for Mechatronic Systems

R. Kasper, W. Koch
Object-Oriented Behavioural Modelling of Mechatronic Systems

J. Bielefeld, G. Pelz, G. Zimmer
Analog Hardware Description Languages for Modeling and Simulation of Microsystems and Mechatronics

Integrated Development Environment for Mechatronic Systems

Martin Anantharaman, Bodo Fink, Manfred Hiller, Stefan Vogel
Institut für Mechatronik, IMECH GmbH, Moers, Germany

Abstract

An approach to the rapid development of complex mechatronic systems based on the comprehensive use of computer-simulation techniques is presented. Mechatronic systems with mechanical and non-mechanical components including discontinuous behaviour and varying structure are modelled as collections of dynamic transmission-elements using object-oriented programming techniques. The control system and user-interface, including the input-devices, are modelled in the same environment, allowing a realistic simulation of the complete controlled system at an early stage. Subsequently, the control system is ported to the final real-time processor without changes. This approach is demonstrated in an ongoing project to develop the control system of a balance-crane.

1 Introduction

The development of complex control systems for manipulators requires the use of powerful development tools. This is especially true for large-scale systems such as cranes, which are assembled and tested at the customer-site itself. Since no prototype exists for testing purposes and a short commissioning period is desired, the complete control system software must be developed and tested by simulation methods. This requires a comprehensive software package for modelling and simulation of mechatronic systems consisting of different components and subsystems, such as mechanics, hydraulics, control systems, user-interfaces, etc. A further requirement is that control system software developed with the help of simulation can be ported to the final real-time processor without major modifications. An ongoing project to develop the control system of a hydraulically-driven balance-crane is introduced in Section 2 as an example where these requirements are particularly obvious.

In this project several innovative software tools are being applied, and these tools together constitute an *integrated development environment for mechatronic systems*. The development environment presented here consists of two parts. The first part is for modelling and simulation, whereas the second allows the development of the control system. The simulation package may be used to model general dynamic systems, consisting of components like mechanical assemblies, hydraulics and feedback-control, including discontinuous behaviour and varying structure. The method is based on the concept of coupled dynamic transmission elements and makes use of object-oriented programming, as described in Section 3.

The second package, concerned with the development of control systems, also makes use of object-oriented programming to implement functional modules of the control system such as feedback-controllers, inverse-kinematics, path-planners and user-interfaces, which can be completely tested and demonstrated during simulation. This is described in Section 4.

2 Example: A hydraulically driven balance-crane

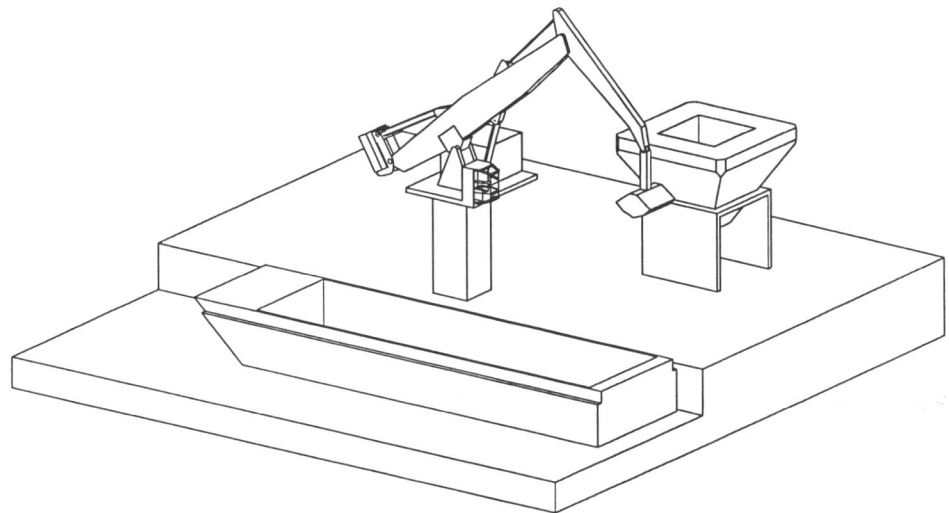

Figure 1: Hydraulically driven balance-crane

The methods introduced above are being applied to develop the control system for a heavy, hydraulically driven balance-crane of the company *Kransysteme Rheinberg*, which is mainly used for loading and unloading material from ships (see Figure 1). Distinguishing features of this crane are the hydraulic actuation and the use of a counter-weight to compensate the weight of the dipper-arm. The VME-based control system shown in Figure 2 provides Cartesian hand-control as well as automatic path-motion similar to that found in industrial robots.

The crane will be assembled at the final installation site itself and will then be equipped on site with the control system. Therefore, the control system has to be put into operation with no opportunity for testing, so that simulation plays a crucial role in its development. For that reason a detailed and realistic dynamic model is required, where the heavy mechanical assembly as well as the hydraulic system are adequately represented. Apart from the hybrid nature of the system, discontinuous behaviour arises from components such as pressure relief valves. Modelling such a system is a challenging task requiring a sound theoretical framework, robust numerical methods and powerful programming techniques.

A further aspect is that a control concept based on Cartesian control is being used for the first time in these cranes. Therefore, apart from developing and testing the control system, it is important to demonstrate the control concept and operation of

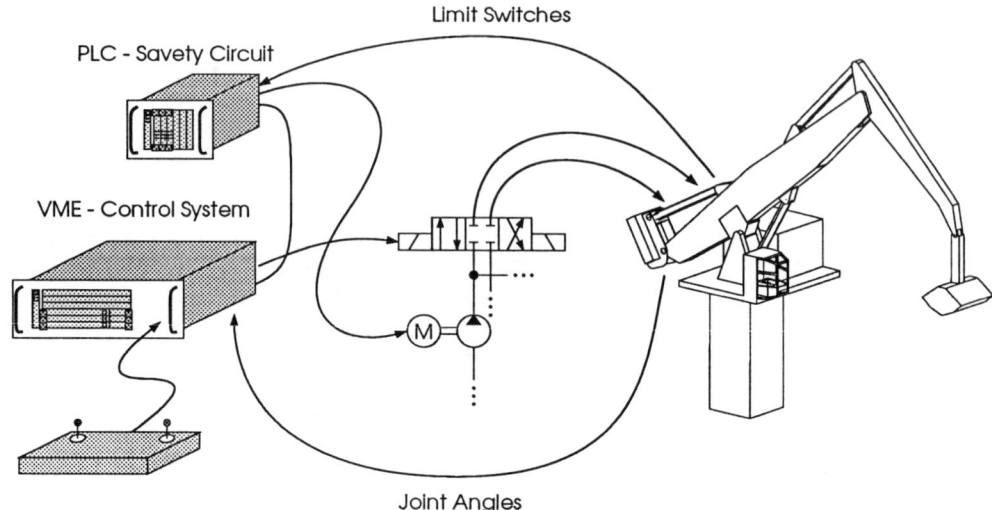

Figure 2: Overview of the control system

the crane in a realistic manner at an early stage. To achieve this, the control system including the user-interface is also modelled and integrated into the simulation, with realistic computer animation of the working environment allowing an evaluation of the user-interface and its operation.

3   Modelling and simulation using dynamic transmission elements

Modelling and simulation of hybrid systems like the balance-crane requires a synthesis of methods from various disciplines[1]. In view of the mechanical structure as a chain of links coupled by joints it is natural to take methods for multi-body systems as a base. Meanwhile several powerful general-purpose program packages such as ADAMS, DADS, MECHANICA [12] have gained industrial acceptance. But, these programs and the underlying multi-body methods do not allow to simulate general hybrid mechatronic systems. The use of general simulation packages such as ACSL, MATRIX$_X$ or MATLAB including interactive user-interfaces such as SystemBuild and SIMULINK is also not useful because the models have to be defined in terms of equations or block-diagrams, which is unnatural for complex mechanical and mechatronic systems. A promising approach based on the object-oriented modelling language DYMOLA is described in [5, 3, 4], where general physical systems can be modelled in a declarative manner. The potential of this approach for

---

[1]See [11] for an excellent overview of alternative simulation methodologies for mechatronic systems

mechatronic systems, in particular robots, has been demonstrated in [11], taking into account discontinuous and event-driven behaviour.

In the following an alternative approach is presented, which extends the concept of *transmission-elements* and *object-oriented programming* in C++ [7, 8] to mechatronic systems, and offers considerable advantages in connection with control system development. Since the scope of any method for modelling dynamic systems is intimately connected with the structure of the equations of motion and the numerical simulation procedures these equations support, Section 3.1 formulates a general framework sufficient for most mechatronic systems. The task of formulating such equations of motion for complex systems is facilitated by assembling the equations in modular fashion, considering the system to consist of coupled *transmission-elements* or *blocks* (see Section 3.2). A vital role in the development is played by object-oriented programming methods, in which generic properties and interrelationships of entities such as blocks of dynamic systems are expressed with the help of *abstract classes*, which may be specialised by *inheritance* to obtain representations of mechanical and hydraulic components such as joints, cylinders, etc. The model of a given system can then be assembled from instances of such components, as demonstrated in Section 3.3 for the mechanical and hydraulic subsystems of the balance-crane.

3.1 Structure of the equations of motion

Considering first the equations of motion of a mechanical system, the equations of motion of a constrained multi-body system are often written in the form

$$
\begin{aligned}
M\dot{v} + G^T\lambda_q &= Q(v, q, u), & (1) \\
\dot{q} &= V(q)v & (2) \\
\ddot{g}(\dot{v}, v, q) &= 0, & (3) \\
\dot{g}(v, q) &= 0, & (4) \\
g(q) &= 0, & (5)
\end{aligned}
$$

where q and v are position and velocity state-variables, λ_q are Lagrange-multipliers corresponding to the constraint equations and u are inputs. We note that Equation (1) is rarely evaluated as shown, using the mass matrix M and constraint-Jacobian G. In the following we only consider an implicit function $f_q(\dot{v}, v, q, \lambda, u)$ which is the residual of this equation.

Turning now to non-mechanical systems like the hydraulic system, the general approach is to formulate first-order differential equations

$$
\dot{x} = f_x(x, u), \tag{6}
$$

with state variables x and inputs u. This form is overly restrictive, as it cannot represent implicit differential equations and does not permit constraint-equations relating state-variables. Therefore, we use a form like those given for mechanical systems with an implicit formulation of the residual.

To take into account discontinuities and events we add a set of *event functions* f_e, whose sign-change indicates the occurrence of an event, and a corresponding set of discrete *event-flags* e, which are non-zero only just after the occurrence of the associated event. The equations of motion are now considered to depend on these event-flags to model state-transitions and varying structure. This formulation is obviously at a lower level of abstraction than the approach described in [6, 11], and essentially corresponds to the computational model of integration procedures with "root-finding" capabilities.

Putting all this together, we describe the dynamic behaviour of hybrid dynamic systems using the following state variables

Mechanical position variables	q,
Mechanical velocity variables	v,
Mechanical Lagrange-multipliers	λ_q,
State variables	x,
State Lagrange-multipliers	λ_x,
Event flags	e,

to formulate the following equations or functions

Equations of motion	$f_q(\dot{v}, v, q, \lambda_q, \dot{x}, x, \lambda_x, u, e) = 0$,	(7)
Velocity equations	$\dot{q} = V(q)v$,	(8)
Position constraints	$g_q(q, e) = 0$,	(9)
Velocity constraints	$\dot{g}_q(v, q, e) = 0$,	(10)
Acceleration constraints	$\ddot{g}_q(\dot{v}, v, q, e) = 0$,	(11)
State equations	$f_x(\dot{x}, x, \lambda_x, \dot{v}, v, q, \lambda_q, u, e) = 0$,	(12)
State constraints	$g_x(x, e) = 0$,	(13)
State derivative constraints	$\dot{g}_x(\dot{x}, x, e) = 0$,	(14)
Event functions	$f_e(\dot{v}, v, q, \lambda_q, \dot{x}, x, \lambda_x, u)$	(15)

with given input-functions u. Note that we allow the equations of motion and the state-equations to depend on variables from both mechanical and non-mechanical subsystems and the event-functions are allowed to depend on all state variables, including the Lagrange-multipliers. Further, all equations conceivably affected by discontinuities and structure-changes are formally dependent on the event-flags. This is a formidable system of equations, but except for the mechanical constraints *all* parts are required for the balance-crane, in particular the state-constraints!

All numerical solution procedures are developed for these equations, using algorithms similar to the inverse-dynamics-based method of [15]:

Constraint analysis ("kinematic analysis" in the purely mechanical case) solves the constraint equations (9), (10), (11), (13) and (14) for q, v, x, $\dot{x}$ and $\dot{v}$, respectively, in the given order, *and* the dynamic equations (7) and (12) for λ_x and λ_q;

Forward Dynamic Analysis simultaneously solves the dynamic equations (7) and (12) for $\dot{v}$, λ_q, $\dot{x}$ and λ_x;

Equilibrium Analysis ("static analysis", "set-point analysis") simultaneously solves the dynamic equations (7) and (12) for q, λ_q, x and λ_x;

Dynamic Simulation integrates the set of differential-algebraic Equation (7)–(14) to determine all state-variables, interrupting the integration when the event-functions f_e change sign and reconfiguring the system with corresponding event-flags e. An extended form of the method described in [10] is used.

3.2 Modular formulation of the equations of motion using transmission elements

The equations of motion described in the last section primarily define the interface between numerical solution procedures and a system model, and they can be formulated directly in closed form only for very simple systems. For practical cases like the balance-crane it is necessary to have a methodology for systematically assembling the equations of motion from the contributions of various subsystems or components of the system. Such components are individually modelled as *transmission elements* or *blocks*[2] and then assembled to the model of the complete system as described in Section 3.2.1. The general transmission element is specialised to mechanical and hydraulic components in Section 3.2.2 and Section 3.2.3, respectively.

3.2.1 General properties of transmission elements

The concept of modelling components of dynamic systems as transmission elements transmitting motion, force, energy, etc. can be derived and used in different ways. In [7, 8] transmission elements are derived from differential-geometric considerations. These elements are used in the C++ Toolkit M◻BILE to model and simulate complex mechanical systems by mapping forward the state of motion and backward the forces. On the other hand the approach of [5, 4] is based on the concept of energy flows in physical systems and the theory of Bond-graph [3]. A distinguishing feature of these approaches is that the sequence in which computational operations are performed is determined automatically ("causality assignment problem") from the coupling between individual elements in [5, 4], whereas in [7, 8] an appropriate ordering of transmission-elements has to be generated manually.

The approach used here is closely related to that of [7, 8] and extends it to general hybrid systems. Such an extension has already been presented in [13, 14] for a large hydraulically-driven manipulator, but in that case the hydraulic subsystem is not completely integrated into the transmission-based methodology. Uniform transmission-elements are used for mechanical and non-mechanical subsystems, but the differential-geometric interpretation is not pursued. Instead, the concept is motivated and justified essentially by analysing well-established algorithms from the areas of multi-body dynamics, hydraulics, etc. and restructuring them in a way suitable for object-oriented implementation.

Thus, consider the model of a complete dynamic system and the solution procedure used to analyse it. Their relationship or "interface" is easily specified in terms of the equations of motion of Section 3.1: The solution procedure has to be able to extract the state variables of the system from the model, change the values of the

---

[2] "Transmission element" and "block" are used interchangeably in the following.

variables and use the model to compute one or more of the Equation (7) (see Figure 3). Now we consider that the system actually consists of a collection of blocks and try to map system properties to block properties (see Figure 4). It is natural to consider the state variables of the complete system to consist of the state variables of the blocks, so that the operations of extracting and changing state variables are easily defined: Each block must have the same operations as the system and the system invokes these block operations in sequence.

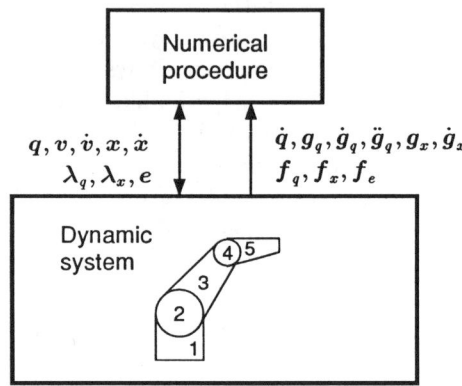

Figure 3: Interface between solution procedure and dynamic system

The evaluation of equations can be similarly defined in terms of contributions from individual blocks, but the crucial aspect is the coupling of the blocks. Following the block-diagram concept prevalent in many simulation methods and control theory we consider each block to define zero or more (output) *signals* s, such as position, velocity, acceleration, pressure, etc., which are used by other blocks as input-signals. By defining such couplings in the form of a directed acyclic graph (DAG) one can compute signals in the right order by traversing the block-diagram in topological-sort order (see e.g. [1]), but obviously such a traversal cannot take into account the reactions of blocks to their input signals. Exactly this kind of reaction characterises physical systems and is a precondition for evaluating the dynamic equations (7) and (12).

We motivate an extended form of signal-transmission by first considering algorithms from robot dynamics [9], which have greatly influenced multi-body dynamics. These algorithms[3] may be interpreted as evaluating one set of quantities (position, velocity and acceleration) on a forward traversal and a dual set (forces) on a backward traversal. Whereas the computed kinematic quantities are directly used by subsequent elements, forces are added at junctions before transmitting them. In the broader context of hybrid physical systems [3, 4] this corresponds to the duality of "effort" or "across" variables on the one hand and "flow" or "through" variables on

---

[3]In [7, 8] these algorithms are formalised in a differential-geometric context.

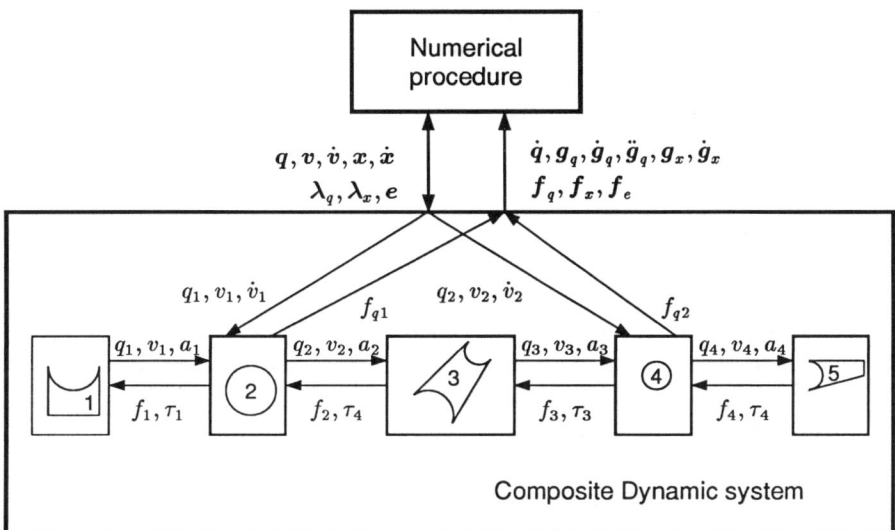

Figure 4: Interface between solution procedure and a composite dynamic system

the other hand: At junctions between components of the system the across variables are equated whereas the through variables must sum to zero.

Applying this to block-diagrams, we consider signals to be composite quantities with two parts: an *action-part* with components corresponding to state variables q, v, x and their time-derivatives $\dot{v}$, $\dot{x}$ and a *reaction-part* with components corresponding to the dynamic residuals f_q and f_x. The action-part of signals is evaluated on a forward traversal of the block-diagram (generally by mapping the action-part of input signals to output signals), whereas the reaction-part of the signals is evaluated on the backward-traversal (generally by accumulating and mapping the reaction-part of output signals to input-signals). The corresponding operations to be performed by a block are termed *action* and *reaction*.

Returning to the task of evaluating the equations (7)–(14) for the complete dynamic systems as part of the traversals of the block-diagram, we evaluate contributions to the velocity equations (8) and constraint residuals g_q, $\dot{g}_q$, $\ddot{g}_q$, g_x and $\dot{g}_x$ on the forward traversal as part of the *action*-operations, whereas the dynamic residuals f_q and f_x as well as the event-functions f_e have to be computed on the backward traversal in the *reaction*-operations (see Figure 4). Comparing this to the method of [4], we see that this assignment of operations to different traversals of the same graph solves the "causality assignment problem", once the coupling-structure of the block-diagram has been defined in the form of a DAG, and the resulting sequence of operations will be equivalent to that generated by equation-sorting in [4]. Note that the basic framework of block-diagrams with dual signals and bidirectional data-flow obviously includes the classical form of block-diagram as a special case, and this reduced form is the basis for the control system considered in Section 4.

In the object-oriented implementation we use the same interface, involving "action" and "reaction" operations for blocks of a system as well as for a complete system. In particular, we introduce an abstract base class DynamicBlock, with virtual methods act and react[4], and specialise it by inheritance to SimpleDynamicBlock for primitive (or atomic) blocks and DynamicSystem for composite dynamic systems, consisting of an ordered list of primitive blocks. Defining primitive blocks by inheritance from SimpleDynamicBlock requires definition of the signals they transmit and implementation of the act and react operations. The exact structure of signals and the kind of mapping performed in act and react depends on the physics of the system (mechanics, hydraulics, electromagnetism, etc.) and needs to be derived from appropriate dynamic principles in each case. These details will be discussed for mechanical systems in Section 3.2.2 and for hydraulic systems in Section 3.2.3.

3.2.2 Transmission elements for the mechanical subsystem

Deriving transmission elements for mechanical components by using the method described above is very similar to [7, 8] and will only be briefly recapitulated using our block-diagram oriented terminology. The first step is the definition of appropriate signal-types for communication between mechanical blocks. These signals have action-components corresponding to $q, v, \dot{v}$ and reaction components to f_q only. The simplest signal is a scalar, represented by MechanicalSignal which may be used for lengths, angles, etc. Apart from several vector signals the most important mechanical signal-type is the reference frame, represented by the class Frame, which describes a rigid-body motion relative to an unspecified inertial reference. It has action-components R (transformation matrix), r (translation), ω (angular velocity), v (velocity), α (angular acceleration), a (acceleration) and reaction components f (force) and τ (torque). All mechanical signal types can either be independent (playing the role of parameters or state variables) or dependent, i.e. they are computed by blocks as outputs. Some typical blocks using such signals in various ways are listed below:

1. FrameMap is the base class of all joints and links, it links an input Frame to an output Frame. It defines operators for mapping $R, r, \omega, v, \alpha, a$ forward in act and f, τ backward in react. A typical joint such as a RevoluteJoint reimplements a subset of FrameMap operations using an internal scalar signal as joint variable.

2. FrameMass applies inertial forces of a rigid body in react using the velocity and acceleration of the frame.

3. SpherialConstraint computes a vectorial output signal for the displacement between two input frames and exports this signal as a constraint. Further it applies forces computed from a Lagrange-multiplier vector.

In this manner classes for most joints, links, standard force elements and constraints have been implemented. Closed-form solutions of loops as in [7, 8] are also available, but go beyond the scope of this article.

---

[4]With arguments action and reaction specifying the subset of transmission operations required by the numerical solution procedure.

3.2.3 Transmission elements for the hydraulic subsystem

The previous approach to object-oriented simulation of hydraulic systems [13, 14], describes the modelling of a hydraulic circuit as a single force element in a mechanical system. In the following, hydraulic elements are modelled as transmission elements.

Similar to the implementation of the mechanical transmission elements we introduce a signal for communication from one hydraulic block to another. During the forward calculation the pressure p and its time-derivative $\dot{p}$ are mapped from one to the next hydraulic block, whereas the backward calculation adds a flow $Q = f(p)$. These three components are collected in the `HydraulicSignal`. As a elementary block the `DifferencePressure` is introduced, which has two `HydraulicSignals` as input variables. The function `act` calculates the difference pressure and stores it to the output `HydraulicSignal`, the `react` function applies the flow at the output to the input signals.

All hydraulic blocks use the signals as state, output or input variables. Some typical hydraulic blocks are listed below:

1. `Volume` implements the differential equation of the hydraulic capacity [2]

$$\dot{p} = \frac{E_{oil}(p)}{V} \cdot Q \quad , \tag{16}$$

   its output and state is a `HydraulicSignal`.

2. `ThrottleValve` is the simplest hydraulic valve, which applies a flow

$$\Delta Q = sign(\Delta p) \cdot \frac{\sqrt{|\Delta p|}}{R} \tag{17}$$

   to `DifferencePressure`.

3. `ServoValve34` calculates the oil flow between four `HydraulicSignals` depending on the position of the valve piston [13, 14].

In the same way several other hydraulic blocks are implemented, e.g. `PressureReliefValve`, `CheckValve`, `PressureCompensator`, so that it is possible to simulate most hydraulic circuits.

In order to couple mechanics and hydraulics the chambers of a hydraulic cylinder are used. The C++-class `Chamber` has a `MechanicalSignal` as input variable and a `Volume` as a reference. It applies a force $f_q = f_q(p,\text{piston area})$ to the `MechanicalSignal`, increases the flow $Q = Q(v,\text{piston area})$ and modifies the capacity $V = V(q,\text{piston area})$ of the `Volume`.

3.3 Object-oriented modelling of the balance-crane

The real balance-crane consists of several major subsystems: The mechanical system, the hydraulic system, the control system, the man-machine-interface and the power-supply. These major subsystems can be subdivided into several functional units. For example the mechanical system of the balance-crane (Figure 5) is built

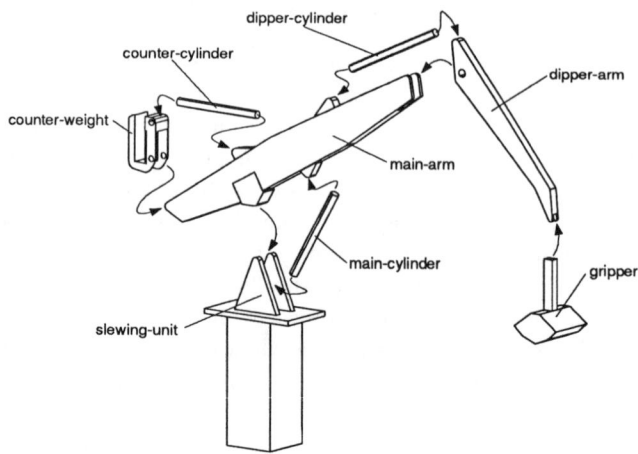

Figure 5: The mechanical system of the balance-crane

up from the slewing-unit, the main-arm, the dipper-arm, the gripper and the counterweight.

As an example the modelling of a mechanical subsystem MainArm (Figure 6) of the balance-crane is considered. Each subsystem is modelled in terms of a C++ class, so the following source code implements the class MainArm as a functional unit.

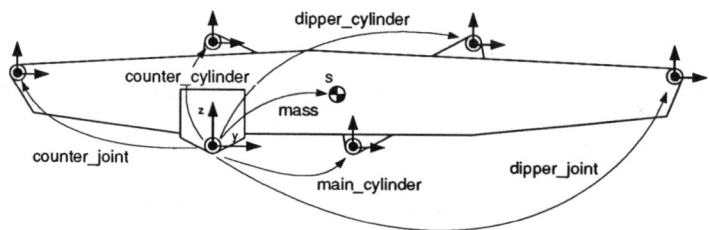

Figure 6: The mechanical model of MainArm

```
class MainArm : public DynamicSystem
{
public:
  RevoluteJoint main_joint;
  RigidTranslationLink dipper_joint, dipper_cylinder;
  RigidTranslationLink counter_joint, counter_cylinder;
  RigidTranslationLink main_cylinder;
  FrameMass mass;
```

```
/* The Constructor MainArm(Frame& from) intitializes the class MainArm.
   The main_joint is connected to the frame form, the other elements
   are connected to a frame which is rotated by the the main_joint. */

MainArm (Frame& from)
   : main_joint        (from),
     dipper_joint      (main_joint, dipper_joint_displacement),
     dipper_cylinder   (main_joint, dipper_cylinder_displacement),
     counter_joint     (main_joint, counter_joint_displacement),
     counter_cylinder  (main_joint, counter_cylinder_displacement),
     main_cylinder     (main_joint, main_cylinder_displacement),
     mass              (main_joint, m, mass_displacement, inertial_tensor)
 {
   append(main_joint, mass);
 }
};
```

The complete mechanical subsystem is finally built up from these functional units. The complete code for the whole mechanical system is given by the following lines:

```
class BalanceCrane : public DynamicSystem
{
public:
  SlewingUnit       slewing;
  MainArm           main;
  DipperArm         dipper;
  Gripper           gripper;
  CounterWeight     counter;
  HydraulicCylinder main_cylinder, dipper_cylinder, counter_cylinder;

  BalanceCrane (Frame& base)
    : slewing           (base),
      main              (slewing.main_joint),
      dipper            (main.dipper_joint),
      gripper           (dipper.gripper_joint),
      counter           (main.counter_joint),
      main_cylinder     (slewing.main_cylinder, main.main_cylinder),
      dipper_cylinder   (main.dipper_cylinder, dipper.dipper_cylinder),
      counter_cylinder  (main.counter_cylinder, counter.counter_cylinder)
  {
    append(slewing,counter_cylinder);
  }
};
```

The hydraulic major-subsystem is subdivided into small functional units in the same way as the mechanical system. So the following lines show the source code of the servo hydraulic circuit (Figure 7) for the main-arm.

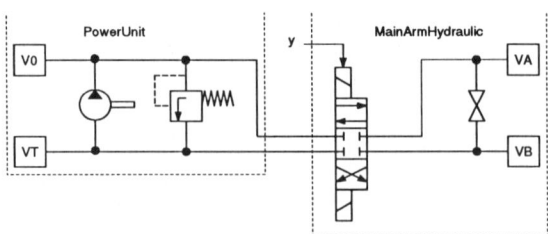

Figure 7: Main Arm Hydraulic System

```
class ServoHydraulic : public DynamicSystem
{
public:
  Volume va, vb;
  ThrottleValve throttle;
  ServoValve34 servo;
  Chamber ca, cb;

  ServoHydraulic(MechanicalSignal& cylinder, Signal& y,
                 Volume& v0, Tank& vt)
    : va(capacity_va, pressure_va_0), vb(capacity_vb, pressure_vb_0),
      throttle(va, vb, resistor_va_vb),
      servo(v0, vt, va, vb, y, geometry_servo, resistor_servo),
      ca(va, cylinder, piston_area, capacity_va),
      cb(va, cylinder, annulus_area, capacity_vb)
  {
    append(va,cb);
  }
};
```

4   Object-oriented design of the control system

The main idea of the integrated development environment is to develop, program and test the control system on a workstation, and then cross-compile it for the real-time-system without any changes. Therefore, libraries for real-time-programming and device-handling with exactly the same programming interfaces have been developed for HP-Workstations and real-time-systems. The main difference between these libraries is that the real-time-device drives an IO-board while the simulation-device is connected to a dynamic model of the real system.

The structure of a control system (Figure 8) shows that it consists of several blocks, e.g. Controller, PathGenerator, PathPlanner and FrontEnd. The blocks are communicating with signals from one block to another, like the models of mechanical and hydraulic systems. In contrast to the sequential order of computations

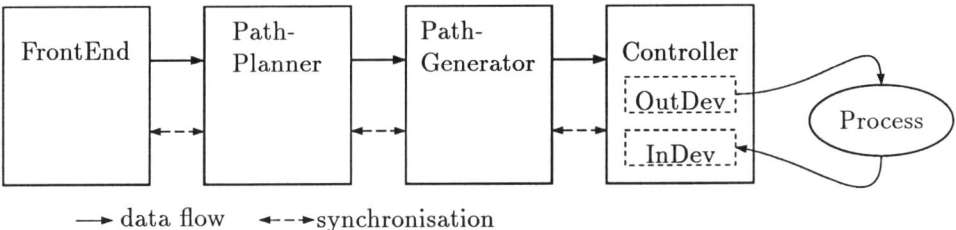

data flow   synchronisation

Figure 8: Control system

used for evaluation the equations of motion, blocks in the control system generally perform concurrent activities. Therefore, they are implemented as separate tasks on one or more processors. These tasks communicate with the help of signals and require synchronisation at the beginning or the end of important subtasks.

For the time discrete `Controller` it is necessary to have a constant cycle frequency. So this task has to run with the highest priority and it synchronises the `PathGenerator`. The `PathGenerator` calculates the desired joint angles, velocities and accelerations for the `Controller`. To get a high closed loop accuracy the `PathGenerator` should reach the same cycle frequency as the `Controller` so it works with a medium priority. The `FrontEnd` allows the interaction with the operator, and generates instructions for the `PathPlanner`. Both systems work off-line, so they have the lowest priority in the control system. For the manual control of the system the `PathGenerator` is extended by an interface to joysticks.

A special front-end for this balance-crane has been designed to decrease the cycle time. It allows Cartesian control of the crane for gripping new material. By pushing a buttom, it starts an automatic motion to unload the material at a desired position.

5   Conclusion

In this paper an approach to the rapid development of complex mechatronic systems based on the comprehensive use of computer-simulation techniques has been presented. Object-oriented programming techniques have been used to model mechatronic systems consisting of mechanical and non-mechanical components with discontinuous behaviour or varying structures as transmission-elements. The control system and user-interface are modelled in the same environment, allowing a realistic simulation of the complete controlled system at an early stage. Subsequently, the control system can be ported to the final real-time processor without changes. This approach was demonstrated for the control-system of a hydraulic-driven balance-crane, which is currently being developed.

References

[1] Alfred V. Aho, John E. Hopcroft, and Jefrrey D. Ullman. *Data structures and algorithms.* Addison-Wesley, 1987.

[2] D. Backé. Umdruck zur Vorlesung Servohydraulik. Technical report, Institut für hydraulische und pneumatische Antriebe und Steuerungen, RWTH Aachen, 1992.

[3] François E. Cellier. *Continuous System Modeling.* Springer-Verlag, 1991.

[4] François E. Cellier and Hilding Elmquist. Automated formula manipulation supports object-oriented continuous-system modeling. *IEEE Control Systems*, pages 28–38, April 1993.

[5] Hilding Elmquist. *A structured model language for large continuous systems.* PhD thesis, Department of Automatic control, Lund Institute of Technology, 1978.

[6] Hilding Elmquist, François E. Cellier, and Martin Otter. Object-oriented modeling of hybrid systems. In *Proceedings 1993 European Simulation Symposium*, Delft, NL, October 25–28 1993.

[7] A. Kecskeméthy. *Objektorientierte Modellierung der Dynamik von Mehrkörpersystemen mit Hilfe von Übertragungselementen.* Number 88 in VDI-Fortschritt-Berichte, Reihe 20,. VDI-Verlag, 1993.

[8] A. Kecskeméthy and M. Hiller. An object-oriented approach for an effective formulation of multibody dynamics. *Comp. Meth. Appl. Mech. Eng.*, 115:287–314, 1994.

[9] J.Y.S. Luh, M.W. Walker, and R.P.C. Paul. On-line computational scheme for mechanical manipulators. *J. Dyn. Syst. Meas. Contr.*, 102:69–76, June 1980.

[10] Martin Anantharaman and Manfred Hiller. Numerical simulation of mechanical systems using methods for differential-algebraic equations. *International Jornal for Numerical Methods in Engineering*, 32(8):1531–1542, December 1991.

[11] Martin Otter. *Objektorientierte Modellierung mechatronischer Systeme am Beispiel geregelter Roboter.* Number 147 in VDI-Fortschritt-Berichte, Reihe 20,. VDI Verlag, 1995.

[12] W. Schiehlen, editor. *Multibody Systems Handbook.* Springer, 1990.

[13] M. Schneider and M. Hiller. Modellbildung und Regelung eines elastischen, hydraulisch angetriebenen Großmanipulators. *Zeitschrift für angewante Mathematik und Mechanik*, 75:127–128, 1995.

[14] M. Schneider and M. Hiller. Modelling, simulation and control of a large hydraulically driven redundant manipulator with flexible links. In *Proceedings of the 9th World Congress on Theory of Machines and Mechanisms*, Milano, Italy, 30 August – 2 September 1995. IFToMM. (To appear).

[15] M.W. Walker and D.E. Orin. Efficient dynamic computer simulation of robotic mechanisms. *J. Dyn. Syst. Meas. Contr.*, 104:205–211, September 1982.

Dipl.-Ing. Martin Anantharaman
Institut für Mechatronik
IMECH GmbH
Bergwerkstraße
Moers 47445
Tel.: 02841/101-276
Fax: 02841/101-251
martin@mail.imech.uni-duisburg.de

Dr.-Ing. Bodo Fink
Institut für Mechatronik
IMECH GmbH
Bergwerkstraße
Moers 47445
Tel.: 02841/101-250
Fax: 02841/101-251
email: fink@mail.imech.uni-duisburg.de

Prof. Dr.-Ing. habil. Manfred Hiller
Institut für Mechatronik
IMECH GmbH
Bergwerkstraße
Moers 47445
Tel.: 02841/101-250
Tel.: 0203/379-3337
Fax: 02841/101-251
email: hiller@mail.mechatronik.uni-duisburg.de

Dipl.-Ing. Stefan Vogel
Institut für Mechatronik
IMECH GmbH
Bergwerkstraße
Moers 47445
Tel.: 02841/101-276
Fax: 02841/101-251
email: vogel@mail.imech.uni-duisburg.de

Object-oriented behavioural modelling of mechatronic systems

Roland Kasper, Walter Koch
Otto-von-Guericke-Universität
Institute of Machine Systems and Drive Technology
Magdeburg, Germany

Abstract: Mechatronic systems overcome the traditional borders of classical engineering disciplines. Behavioural modelling allows the developer to describe the behaviour of components and systems from arbitrary disciplines. This paper introduces a way to transform the functional models of those systems into an object-oriented description. The objects implementing this integration of behaviour and animation are called virtual objects. Ideally these objects can be recognised and handled by a user like their real world counterparts. Complex tasks like design, manufacturing etc. no longer have to be done with expensive and difficult to handle real components.

1 Introduction, state of the art of mechatronic system design

The modelling of mechatronic systems and components is a very important key strategy to optimise the development and manufacturing processes of this highly sophisticated class of products. The great advantage of mechatronic products is the combination of mechanics, electronics and information processing into one device to achieve an optimised function and construction. Mechatronic systems overcome the traditional borders of classical engineering disciplines and extract the best solution wherever found. From the modelling point of view the interdisciplinary approach however leads to a number of unsolved problems.

Traditionally the mechanical design engineer is faced with a design task that is concerned with questions about outline and shape of mechanical components. Consequently the modelling of mechanical systems in most cases is based on rigid or elastic bodies connected together via joints and linkages. There exists a number of modelling and simulation tools that assist this mechanical view during the specification of models and to calculate simulation results.

The electric engineer does his job in an equivalent way. His point of interest is to select the optimal electronic circuits and components and to connect them to complete electronic systems. Consequently the modelling of electric and electronic systems traditionally is based on standardised component models and net lists describing the connections between them. There exists a large number of powerful simulators that are very comfortably integrated into the CAE framework used by the design engineer for his work. The information processing part of mechatronic systems in many cases is realised by software.

The software engineer traditionally uses a programming language for specification and compilers, linkers and so on to produce final object code. Simulation is not supported by typical software development tools. Self made test programs are the common approach to validate the correct function of a design idea.

As a result of this separation it is not possible to develop and validate the function of a complete mechatronic system using one of the traditional modelling and simulation tools. One possibility to bring together all different parts of a mechatronic system in future will be offered by so called simulation backplanes that support the synchronisation and data exchange between different modelling and simulation tools. Simulation backplanes allow the reuse of existing simulation packages and support loosely coupled distributed simulation in cases where the simulation backplane is extended across multiple workstations of a computer network. The problems associated with this solution are the huge overhead of simulator coupling and the missing integrated specification of a mechatronic system. Another way is actually prepared in the area of electronic simulation, where the possibilities of behavioural modelling are strongly extended with the introduction of the simulation language standard VHDL-A [Berg 94]. Behavioural modelling allows to describe the behaviour of components and systems from arbitrary disciplines using the mathematical level of algebraic and differential equations combined with a flexible formulation of discrete events. So far this emerging standard can be seen as a powerful and flexible platform to formulate models of complete mechatronic systems and to use them for simulation. The problem is that the simulator has to deal with a mix of differential and algebraic equations during integration, which means a large overhead compared to the standard problem of solving only differential equations.

On the other hand electronic design today moves rapidly toward a straight forward top down approach: in a first design step only the demanded functions of the component under design are specified without restrictions concerning the method of implementation of one of these functions. During a second, ideally automated, synthesis or implementation step each function is built up by existing components and circuits. Only this second implementation level needs simulation support on the component level. The first design level only deals with functional models that are represented by their input output behaviour, comparable to well known control devices but with more complex interfaces. This means for simulation a mechanical component for example will no longer be represented by bodies and joints but by a dynamic transfer function and interfaces representing the behaviour of the joints. The transformation of today used component oriented descriptions into functional models with well defined input-output behaviour can be carried out by means of symbolic computation [Dyna 94, Wolf 91].

To simulate these functional models one well known method is to transform them into a block diagram and use a standard simulator to carry out the numerical integration. The components can be represented as blocks, the interfaces and connections can be formulated as signal lines between block ports. This method works quite well which can be proven by a large number of really complex models built up and used especially in the automotive industry [Eppi 90]. Building up larger models and libraries however shows that the simple transfer of data offered by signal lines in a block diagram doesn't guarantee a proper representation of the physical structure of the underlying system. The negative effects for the user are bad documented models and unnecessary computational overhead during simulation. To avoid these problems it is better to transform the functional models into an object-oriented description. Components, interfaces and connections are represented as objects. As usual the component objects carry the functional behaviour of the physical component.

It is very important that the object structure of a model is not restricted to the specification phase but will be conserved during simulation. Using the object structure even for simulation first will simplify the complicated task of debugging complex models. Second, the objects can be distributed across a number of processors for parallel simulation. This feature is very important in time critical applications such as real-time and Hardware-in-the-Loop experiments.

2 Features of objects and object-oriented systems

Objects and the art of object-oriented thinking have proven to be one of the most powerful concepts of software design. Following this paradigm a new generation of software development tools [Digi 93, Micr 94], database tools [Ahme 91] and graphical user interface tools [Obje 94] has been created during the last decade.

Before discussing the advantages offered by this technology for modelling and simulation of complex mechatronic components and systems, a short recapitulation of the base concepts of objects seems appropriate:

Managing complexity
Objects are always (with few exceptions at the lower end of hierarchy) build up from other objects. They are well suited to be used for the construction of more complex objects. The resulting object hierarchy, conserving the structure of the objects across all levels, is the key to manage (even validate and debug) very complex systems built up from ten thousands of individual objects.

Data encapsulation
Objects strongly encapsulate their inner structure using instance variables (see Fig. 1). They represent the state of an object that will be conserved during object lifetime. The structure of an object in terms of instance variables is fixed. Access to the contents of instance variables for change or readout is only possible inside the object.

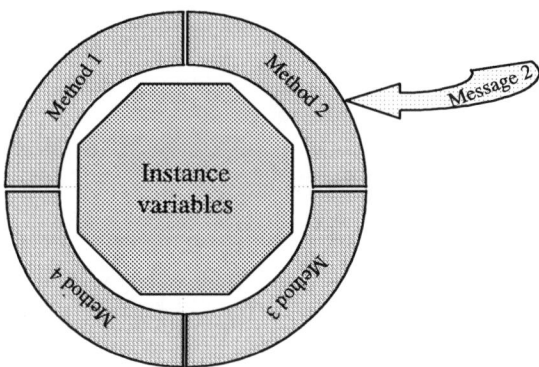

Fig. 1: Representation of an object with interface
and message

Strict interfaces

Methods are available to access the object's state from outside, or to implement a specific behaviour of the object. The methods support a flexible interface for the objects with a fixed number of parameters passed to the method but with no constraint to the type or class of a passed parameter. Methods are activated by sending messages to an object.

Interaction and message passing

Once created, an object has nothing to do than receiving messages from other objects, activating the corresponding method and sending messages to underlying objects from inside the activated method. From this point of view objects are very interactive individuals and therefore well suited to built up highly interactive systems.

Class hierarchy: abstraction and inheritance

The structure (instance variables) and behaviour (methods) of objects is defined using classes. Classes are abstract schemes that can be used to create an arbitrary number of similar objects called instances. Classes are organised in a hierarchy allowing parent classes to be used to define child classes. This dependency is called inheritance. Both structure and behaviour can be inherited.

Integration of graphics

One obvious plus of most objectoriented systems is the impressive integration of graphics. Although being first of all a consequence of the object-oriented paradigm and the flexibility of objects, this feature is very important for every user and therefore often regarded as a base concept.

The next step will be to use this base concepts of objects to improve modelling and simulation of dynamic components and systems.

3 Representing system behaviour

Continuous time systems
A very convenient form to represent the behaviour of non-linear, time-variant dynamic systems is offered by the state space description

state equation $\qquad\qquad \dot{x}_{(t)} = f_{(x,u,t,p)}$ $\qquad\qquad$ (1)

output equation $\qquad\qquad y_{(t)} = g_{(x,u,t,p)}$ $\qquad\qquad$ (2)

with u the vector of system inputs and p the vector of system parameters. The vector of state variables x is explicitly defined by its derivatives $\dot{x} = \dfrac{dx}{dt}$ in the state equation with the time t as independent variable. The vector of system outputs y is calculated in the output equation. In its explicit form the state space description is equivalent to a corresponding block of a block diagram.

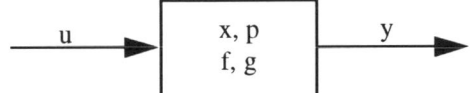

Fig. 2:   Block representation of a
state space model

The internal structure of a block is defined by its states and parameters. Consequently the instance variables of a block object have to carry state variables and parameters of the block. State variables and derivatives are provided as numerical classes.

Discrete and digital systems
The object approach is not restricted to typical simulator data types like float and integer, but supports all possible types of data that can be implemented in a class. For example fractional numbers to implement the arithmetic of modern micro controllers etc. For these types all operators and functions are implemented to formulate the state and output equations of the block. So far the considerations were restricted to continuous time systems. But this is not necessary, because state space description also helps to represent the behaviour of discrete time systems like sampling or even discrete event systems.

state equation $\qquad\qquad x_{k+1} = f_{(x_k,u_k,t_k,p_k)}$ $\qquad\qquad$ (3)

output equation $\qquad\qquad y_k = g_{(x_k,u_k,t_k,p_k)}$ $\qquad\qquad$ (4)

The meaning of the variables is the same as for continuous time. The only difference lies in the fact that they are only defined at discrete points in time t_k, t_{k+1} etc., which can be distributed over time in an arbitrary but strictly increasing order.

Implicit systems

The mathematical abstract state space formulation offers the advantage that components and systems from all disciplines can be treated in an uniform manner. The description can be started on component level. The components can easily be combined connecting inputs to outputs to form more complex systems. There exists a correspondence between mathematical and physical blocks. Unfortunately the precondition of an explicit formulation turns out to be too restrictive for some applications dealing with complex mechanical multibody or electric systems, where kinematic or resistive loops couple together several components. In this cases implicit formulations or formulations including coupled differential and algebraic equations have to be managed.

state equation $\qquad$ $\dot{x}_{(t)} = f_{(x,u,t,p)}$ $\qquad$ (5)

output equation $\qquad$ $y_{(t)} = g_{(x,u,t,p)}$ $\qquad$ (6)

algebraic coupling condition $\quad z_{(y,t,p)} = 0$ $\qquad$ (7)

The algebraic conditions $z_{(y,t,p)}$ guarantee for example that in mechanical multibody systems both parts connected to a joint really move together or that in electric systems the sum of currents flowing into a junction results to zero. To conserve the physical structure of the system it has to be formulated in terms of its underlying components and their interconnections.

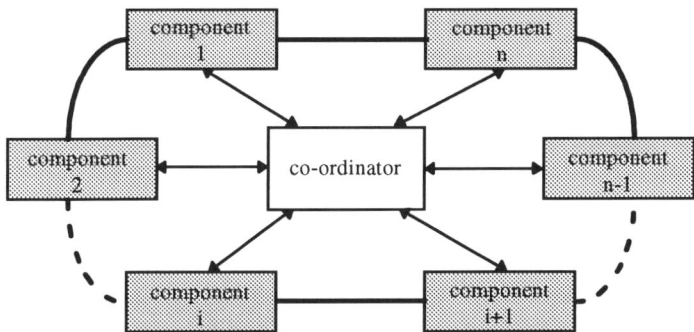

Fig. 3: Co-ordinator solving kinematic loops of mechanical
multibody systems

Fig. 3 shows the example of a kinematic loop from a mechanical multibody system. A co-ordinator is used to collect all kinematic information from the components (e.g. position and velocity of joints) and uses them to calculate the forces acting to the components [Kekc 93].

For special kinematic structures like chains or trees the co-ordinator can be broken up and distributed according to the joints. Fig. 4 shows an example of a mechanical multibody system forming a chain. The kinematic information like inertia and external forces are transported in a first step back to the roots of the chain. During a second step forces and accelerations are calculated for each component of the chain beginning from the root.

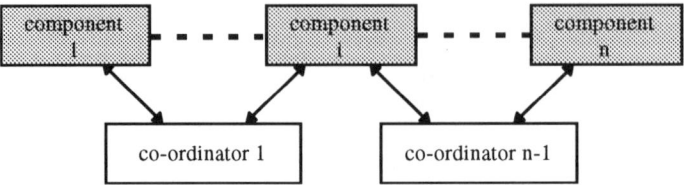

Fig. 4: Distributed co-ordinator solving kinematic chains of mechanical multi body systems

Interfaces

Real physical components are connected together via mechanical joints or electric junctions or connectors for example. Therefore models of these components should look alike. Taking a block diagram as an example, the signal lines between different blocks can be seen as a representation of the physical connections. Signal lines simply transmit data from one block output to one or several block inputs. But in a block diagram data has no semantic information about its meaning. As a consequence you can connect a voltage output for example to a force input. Moreover only a restricted set of data types like float and integer are available as scalars or vectors in a block diagram. Modelling both components and connections as objects these restrictions can be avoided. Connection objects transmit arbitrary objects representing for example physical quantities like voltages, currents, forces and so on. Thus, connecting different components will be safer. Further it was shown that some modelling techniques (for example to solve kinematic loops in mechanical systems) require a very complex interface between components. These interfaces have to transmit physical data plus kinematic information. Connection objects offer the flexibility to implement every modelling method direct with little effort and guarantees a safe use.

Integrating existing tools

There is no need to implement all modelling concepts mentioned above once again in an object oriented environment. Objects offer the flexibility to integrate existing tools, if these only provide some simple interfaces. Block diagram based tools like MATLAB/SIMULINK [Math 92] and ASCET [Eppi 90] can produce C language description for models represented as a block diagram. These C based models can be handled like objects; they implement the model's states and parameters and offer functions to transmit inputs, change parameters, calculate derivatives and outputs. Tools supporting the modelling of multibody systems like Mechanica or Neweul can generate C or FORTRAN code which can be adapted to the behave like objects. A very flexible solution to the modelling problem is offered by tools like Dymola [Dyna 94] or Mathematica [Wolf 91]. Their application is not restricted to a certain domain. Based on symbolic computation they produce equation based models for mechanical, electric or hydraulic components or systems. The symbolic approach makes it easy to adapt the interfaces of the generated models to the object paradigm.

4 Simulation process

Simulating objects
Following a traditionally simulation scheme the simulator uses the system model only to calculate all derivatives of a state vector representing the complete model. The state vector will be integrated to advance to the next time step, then it will be played back to the model. This means the integrator plays the active part. The physical structure of the system represented by the structure of objects cannot be used during simulation. The result are large models consisting of thousands of equations that only can be simulated slowly and never debugged. Therefore it's very convenient to preserve the structure of the objects even during simulation. For all explicit integration methods, the central integrator can easily be broken up and distributed to the objects. This means, each object controls the integration of its own state variables. This allows for a completely distributed integration scheme. Fig. 5 and Fig. 6 shows the sequence necessary to move an object implementing a state space model from time step t_k to time step t_{k+1}. Two cases are important. If the output equations don't depend on the input, they can be calculated independently. When the input is available, the derivatives can be determined and the state variables can be integrated. On the other hand, if the output equations depend on the input, their calculation can take place only after the input is available.

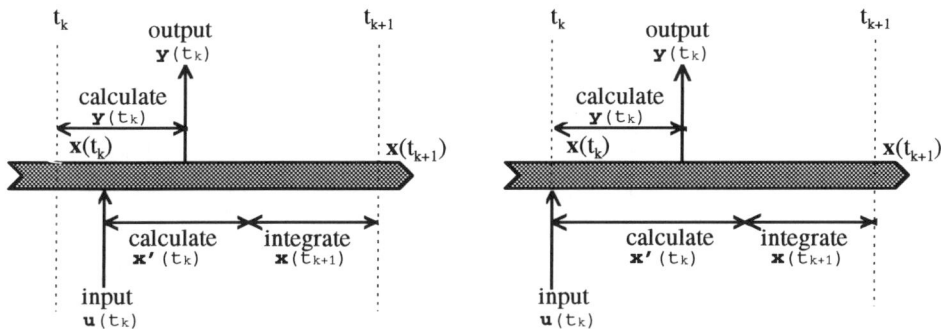

Fig. 5: Objectoriented simulation sequence for state space models with output **y** independent from input **u**

Fig. 6: Objectoriented simulation sequence for state space models with output **y** dependent from input **u**

This behaviour of objects during simulation can very easily be implemented using software processes. Calculation of the output and state equations as well as numerical integration of the state variables are embedded in one or several processes. Inputs and outputs are implemented as communication objects that transmit interface objects between object interfaces. A process will be synchronised by receiving and sending of interface objects. Using the flexibility of processes it is possible to implement even very complicated sequential dependencies arising during simulation for example the solution of kinematic loops as mentioned above. The base concept of independent communicating objects allows to implement all kind of communication schemes and thus supports very different styles of modelling. Communicating objects offer another advantage that only should be mentioned here. On the level of processes the complete model can be distributed on a multi processor system and simulated in parallel. This is of great importance if very large models have to be simulated or real-time conditions have to be fulfilled [Eppi 90].

Continuous time and discrete events
All time varying interface objects carry a time stamp t_k and a time interval ΔT where they are valid, e.g. $u_{(t_k, \Delta T)}$. In a traditional integration scheme with fixed integration step size all inputs received by an object have identical time stamps and intervals. The integration step size depends on the integration method. An Adams-Bashforth 3. order integrator will produce the following equidistant time sequence of inputs, states and outputs:

$$u_{(t_k, \Delta T)} \rightarrow y_{(t_k, \Delta T)}, x_{(t_{k+1})} \tag{8}$$

$$u_{(t_{k+1}, \Delta T)} \rightarrow y_{(t_{k+1}, \Delta T)}, x_{(t_{k+2})} \tag{9}$$

and so on, with $\qquad t_{k+1} = t_k + \Delta T \tag{10}$

The 4. order Runge-Kutta integrator however delivers the following not equidistant sequence:

$$u_{(t_k, \Delta T)} \rightarrow y_{(t_k, \Delta T)}, x_{(t_{k+1/2})} \tag{11}$$

$$u_{(t_{k+1/2}, \Delta T)} \rightarrow y_{(t_{k+1/2}, \Delta T)}, x_{(t_{k+1/2})} \tag{12}$$

$$u_{(t_{k+1/2}, \Delta T)} \rightarrow y_{(t_{k+1/2}, \Delta T)}, x_{(t_{k+1})} \tag{13}$$

and so on, with $\qquad t_{k+1} = t_k + \Delta T \tag{14}$

$$t_{k+1/2} = t_k + \frac{\Delta T}{2} \tag{15}$$

Different simulation step sizes (often called multi rate simulation) can be used for different blocks to optimise calculation effort and numerical accuracy. An object receiving inputs with different time stamps and intervals has to decide what to do. Either force the local integrator to run with the smallest received step size or to accumulate the faster inputs for a slower integration step. Even distributed step size control is possible, if each object supports a roll back mechanism that allows to drop the actual state if one of the inputs has been declared invalid and to go back to stored state of a former point in time.

Sampling data and discrete time systems, very often found in digitally controlled mechatronic products, obviously can be treated as a special case of the communication and simulation concepts presented above, if sampling and delay times are constant and known at least one simulation step in advance. To allow sampling intervals independent from the continuous simulation time steps, however it becomes necessary to interpolate the continuous data according to the used integration method. Sampled data feed back to a continuous time object delivers the time stamp and interval from which the integration strategy can be deduced similar to the strategies proposed above for the case of multi-rate simulation.

To model and simulate digital electronic or measurement and control software systems more accurately, their timing aspects have to be considered more precisely. Digital controllers for example often support effects like interrupts or time and data dependent sampling intervals. Modelling of such phenomena leads to discrete event systems, which means inputs and outputs of these objects can change at arbitrary future times. There exist powerful discrete event simulators to solve that problem. These simulators can be embedded in event driven objects that can be integrated into the communication concept of the simulation of objects.

5 Configuration and parameterization

Representing system hierarchy
So far the system description was based on a flat model. To deal with large systems however, it is necessary to provide a hierarchical concept that allows to built up more complex systems from simpler ones. From this point of view hierarchy is a suitable approach to manage complexity by dividing. Several objects and their interconnections can be packed together and treated as a unit called hierarchy object. Consequently the instance variables of a hierarchy object keep the objects and connections from which the object is built up. The resulting object only represents a structure. The behaviour is completely defined by the underlying base objects. To allow hierarchy objects to be used as building elements of new hierarchies, they have to provide suitable interfaces to interact with other base or hierarchy objects.

Design parameters for abstract models
Model parameters play a very important role describing the behaviour of real physical components. On one hand, they represent real physical data like mass, resistance, stiffness etc. They help to formulate abstract models that can be used to simulate the behaviour of a group or range of components. Taking for example one model of a hydraulic cylinder. Varying parameters like length, piston diameter etc., it is possible to represent the complete product range of a certain manufacturer. Obviously it is very important to choose the best parameters to model a component! There is no mathematical proof, but it has been proven in many applications that it is a good style to work with material and geometrical data directly and use it to determine the desired quantities. On the other hand the designer associates a recommended behaviour of a component. He will characterise it by means of time constants, bandwidth etc.

Representing design knowledge
Lets take an analogue low-pass filter as an example. This circuit may be build up from two operational amplifiers and some passive components like resistors and capacitors. Dependent from their actual values the complete circuit implements a transfer function, specified by abstract filter parameters like ripple, bandwidth etc. For this example the mathematical dependency between the physical component data and the abstract filter data can be formulated without effort. Using this fact the question of model parameterization can be posed in two ways:

1. Specify the parameters of the passive components (for example to validate the function of an existing design). The filter parameters will be calculated.

2. Specify the filter parameters (for example to design a new filter). The parameters of the passive components will be calculated in an optimisation procedure, because only discrete values of resistance and capacitance are available.

For many design problems the dependencies between parameters and system function leads to very complicated mathematical formulations. In some cases software tools are necessary to solve the problems. Both cases can be handled by using objects to represent the parameter dependencies. Even complicated mathematical formulations can be well expressed using objects. Existing tools can be embedded into parameter objects to assist the task of parameterization. This approach leads to a very powerful and easy to use method to represent design knowledge, make it available for simulation and feeds back simulation results to the design process. It can be used by the system engineer to validate that his design idea can be realised with physical components. The mechanical or electronic designer can simulate his designs easily and avoid failures.

Representing parameters in hierarchy
Combining such parameterised objects to a new system on the next level of hierarchy consequently needs to solve all parameter dependencies occurring on that level. If, for example, an analogue band-pass filter has to be build up from a low pass and a high pass filter, the filter parameter of the band-pass, like central frequency, bandwidth etc., have to be expressed in terms of the filter parameters of the underlying low- and high-pass filters. On the other hand the parameters of all passive components must be available on the band-pass level. This step of parameterization is essential for each level of hierarchy. Design knowledge covering the integration problems, arising when several sub-components are put together to form a new function, can be represented using the hierarchical parameter dependencies.

Importance of libraries and OO-databases
Ideally the construction of a new model simply involves the combination of prefabricated sub-models. Modelling with objects strongly supports this methodology. It has been already mentioned that abstract models available through flexible parameterization methods help to simulate complete families of components. Consider the example of a PID controller implemented as a state space model. At the time the controller object is instanciated it can be seen as an ideal abstract PID controller. None of its instance

variables (state variables, parameters) is yet initialised. Initialisation can be done with standard 32 bit floating point number objects representing the values of the desired controller parameters and initial conditions of state variables. The behaviour of this PID controller will approximate very well an ideal one. A second possibility is to initialise the controller object with 24 bit floating point number objects to simulate the behaviour of a PID controller implemented by a floating point signal processor. Last not least representations of fixed point numbers can be used for initialisation to simulate the numerical effects occurring in micro controller using integer arithmetic. As a result of this huge flexibility there is to maintain only one class, but a variety of instances of parameter objects, necessary to initialise the controller object. This game will be repeated on each level of the hierarchy. Take the model of a vehicle powertrain as an example. The model will be represented by a hierarchy object that keeps the components, the vehicle powertrain is build up from engine, clutch, gear etc. This model is highly abstract, because it resembles the structure of the powertrain of nearly every vehicle. Again it will get specific only after initialisation, when the instance variables carry the specific engine, the specific clutch and the specific gear objects. As we see, a large number of different vehicle powertrain models can be represented by one powertrain class, some engine, clutch and gear classes, some more classes for further sub-components, and a very large number of instances necessary to initialise the models correctly. To manage this mass of instances today powerful object-oriented database systems are available [Ahme 91]. These tools are highly optimised to store and retrieve a large number of very different instances of arbitrary classes. Object-oriented databases play a key role concerning the maintainability and the speed of large object based models.

6 Graphical representation and virtual components

Today graphics play a very important role for all types of man-computer interfaces. It's a proven fact that objects are widely used and offer great advantages to build these graphical interfaces. The situation for modelling and simulation tools however is somewhat more specific. Today the engineer will use graphics usually for specification of technical systems, for example a CAD drawing of a mechanical component, the schematic of an electronic circuit or a block diagram of a controller. Originally this form of graphical representation first of all served to support the human communication and to archive designs. The behaviour of the represented physical systems was developed in the brain of the engineer with the help of a sheet of paper and a pencil. Then step by step models were used to simulate the behaviour of the systems. The graphical representation remained unchanged. To visualise the simulated behaviour virtual instruments like voltmeters or oscilloscopes were offered for block diagrams and schematics. In the case of an electronic component this approach simulates the usual work of the engineer. For mechanical systems however in many cases the engineer can observe the behaviour he wants to study. Mechanical bodies and links will move, design relevant positions and velocities of different bodies can be seen at one glance. Consequently simulation results were stored in files and can then be used to animate a 2- or 3-dimensional graphical model of the mechanical system. In the sum 3 different models for CAD, animation and simulation have to be built up and kept consistent.

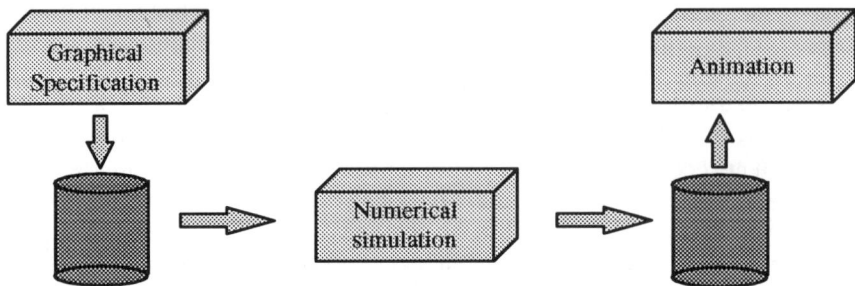

Fig. 7: Traditional graphical support of specification and animation

The flexibility of objects can avoid this overhead by integrating the different models into one. It was explained how communicating objects can simulate the behaviour of real mechatronic systems. To do this, things like internal state, external interfaces or system parameters were used. The internal state in many cases represents the position and velocity of a mechanical system. The external interfaces of the behavioural model exactly correspond to the interfaces really existent and visible. The parameters of the behavioural model directly reflect real quantities like length, distance, diameter etc. Using this dependencies, the models used for behavioural simulation and animation can be coupled together.

Changing the diameter of the piston in a hydraulic cylinder graphically for example will automatically change the value of the associated parameter. Having defined a dependency between the diameter of the cylinder and the diameter of the piston, the dependent parameter will also be changed and visualised. After activating the object and starting the simulation the position of the piston will move according to the oil pressure in both chambers. As a state variable the value of the position will be connected to the geometrical position of the graphical representation. The result is a very direct animation of the model. Fixing a virtual handle at the piston-rod it can be pointed with the mouse and pulled interactively.

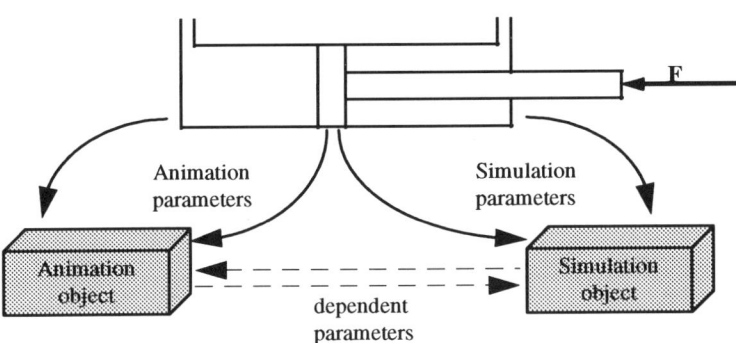

Fig. 8:   Virtual object integrating simulation and animation objects

The objects implementing this integration of behaviour and animation are called virtual objects. Of course they are mainly built up from two parts. One part implements the system behaviour on the basis of communicating objects. The second part represents the systems outlook with an animation object. But it's essential that both parts are completely integrated, which means dependencies exist and are implemented in both directions. Behaviour can influence animation and vice versa.

7 Conclusions

Modelling real mechatronic components and systems as communicating objects including an attractive 2- or 3-dimensional graphical representation pushes simulation technology to new dimensions. Ideally these objects can be recognised and handled by a user like their real world counterparts. Complex tasks like design, manufacturing, assembling, installation, field tests etc., no longer have to be done with expensive and difficult to handle real components. They may be replaced by virtual objects that simulate their outlook and behaviour, but can be manufactured, redesigned, assembled etc., by software with much less effort. The complete life cycle of complex products, consisting of a large number of sub-components, can be simulated with respect to all technical and economical aspects, before only one prototype has to be manufactured.

The last years have shown a dramatically increase in computing power, computer memory, computer graphics and communication technologies. These technologies correspond exactly to the resources that are necessary to model and simulate with virtual objects. All technological previews promise that this process will proceed for the next years. This permanent improvement of the resources will stimulate the application of virtual objects and remove technical limitations existing today.

8 References

[Ahme 91] S. Ahmed et. al., A Comparison of Object-Oriented Database Management Systems for Engineering Applications, Massachusetts Institute of Technology, Research Report R91-12, 1991

[Berg 94] J.-M. Bergé. Analogue VHDL. The VHDL Newsletter # 12, January 1994, S. 7-9.

[Digi 93] Digitalk Inc., Smalltalk/V User's Guide, 1993

[Dyna 94] Dynasim AB, Dymola - Dynamic Modeling Language User's Manual, 1994

[Eppi 90] A. Eppinger, R. Kasper and H.M. Heinkel. Hardware-in-the-Loop design techniques with ASCET. Esprit CIM, CIM-Europe workshop on computer integrated design of controlled industrial systems. Paris 26.4.-27.4.90

[Kekc 93] A. Kekcskeméthy. Objektorientierte Modellierung der Dynamik von Mehrkörpersystemen mit Hilfe von Übertragungselementen, VDI-Verlag, Düsseldorf, 1993, Reihe 20, Nr. 88

[Math 92] The MathWorks Inc., MATLAB Reference Guide, 1992

[Micr 94] Microsoft Corporation, Introducing Visual C++, 1994

[Obje 94] Objectshare Systems Inc., WindowBuilder Pro/V User's Guide, 1994

[Wolf 91] Stephen Wolfram, Mathematica: A System for Doing Mathematics by Computer, Addison-Wesley, 1991

Prof. Dr. Roland Kasper
Otto-von-Guericke-Universität
Institute of Machine Systems and Drive Technology
Universitätsplatz 2
39 106 Magdeburg
Germany
Phone (+49) 391 - 55 92 - 36 06
Fax (+49) 391 - 55 92 - 26 56
E-Mail roland.kasper@masch-bau.uni-magdeburg.de

Dipl.-Ing. Walter Koch
Otto-von-Guericke-Universität
Institute of Machine Systems and Drive Technology
Universitätsplatz 2
39 106 Magdeburg
Germany
Phone (+49) 391 - 55 92 - 28 23
Fax (+49) 391 - 55 92 - 26 56
E-Mail walter.koch@masch-bau.uni-magdeburg.de

Analog Hardware Description Languages for Modeling and Simulation of Microsystems and Mechatronics

Jürgen Bielefeld*, Georg Pelz*, Günter Zimmer*†

*Gerhard-Mercator-University-GH Duisburg,
Dept. Electron Devices and Circuits, Duisburg, Germany
† Fraunhofer Institute of Microelectronic Circuits
and Systems, Duisburg, Germany

Abstract: Modeling and simulation of microsystems and mechatronics is a challenging task because different components like information technology, nonlinear dynamics and control, multi-sensing and -actuating are integrated into systems. Analog hardware description languages are well suited for the needs of these applications. They support modeling techniques for electrical analog/digital circuits, hierarchical design, behavioral modeling, nonlinear dynamics and rigid mechanical constraints.

1 Introduction

Microsystems as well as mechatronic systems require similar methods to model and simulate the system itself and the interaction for instance between mechanics and electronics. In such systems the microelectronics is an integral part which enables intelligent functionalities for sensing, actuating, and controlling [1]. The circuitry for these purposes is by nature a combination of analog and digital components (mixed-circuits). A variety of vendors developed tools for modeling, simulation, and synthesis of mixed-circuits. The exploding circuit complexity of millions of transistors triggered a great demand for system level descriptions, modeling just the analog or digital behavior (function). For this purpose, **hardware description languages** (HDL) have been developed for description of digital or analog circuits in the last few years. In the digital world VHDL (Very high speed integrated circuits HDL) has become the IEEE[1]-standard 1076 in 1987. A common standard language for analog and digital hardware descriptions will be expected at the end of 1996. The formulation of analog behavior with analog hardware description languages (AHDL) is **not** restricted to electrical/electronical circuits. Especially sets of coupled differential equations can be formulated to fullfill nearly all modeling problems in mechatronics or microsystem technology. The aim of our current and future investigations is to integrate the modeling techniques for electronics and mechanics with no loss of information for entire system simulation which improves the knowledge about the interaction of the components. In this contribution it is shown how

---

[1]Institute of Electrical and Electronics Engineers

mechanics can be handled with the AHDL-approach. Examples include coupled nonlinear differential equations, geometrical nonlinearities and flexible plates.

2 Capabilities of Analog Hardware Description Languages

HDLs support the top-down-design methodology, i.e. designing from system level through the hierarchy down to component level. This powerful technique is used in digital HDLs as well as AHDLs. In AHDLs there are some constructs to describe digital behavior. They are necessary for connecting the analog part with the digital part of an electronic circuit to enable analog/digital-simulation. In addition, the full analog electrical models, e.g. complicated transistor models, are preserved.

As said above, AHDLs allow the modeling of arbitrary systems with electrical, mechanical, thermal, and other components. This is accomplished by including mathematical operators $+, -, *, /$, power, and most of the mathematical functions $(\sin, \cos, \exp, \ln, \text{sign}, \text{abs}, \ldots)$. AHDLs have operators for time differential and time integral to write differential equations and they support explicit and/or implicit formulation. Furthermore, they provide capabilities for conditional statements and loop statements as in other procedural programming languages (C, Pascal). C-interfaces are also available to use existing code of special models. With that, it is possible to compute sets of coupled nonlinear differential equations which are typical for multibody systems, provided that the equations of motion are formulated in symbolic representation.

The generation of the symbolic equations of motion can be done manually or automatically e.g. by the modeling tool NEWEUL [2]. For the future it is necessary to integrate state of the art modeling tools for multibody systems [3] enhancing the capabilities of AHDLs.

Each system component has a defined environment with bidirectional terminals (input and output at each terminal) for connecting different components. The connections transmit electrical, mechanical, or other quantities.

An open problem for AHDLs is the direct modeling of **partial differential equations** PDEs for structure dynamics. Hitherto exists a model-transformation approach [4,5] which translates PDEs into equivalent circuit networks for SPICE3[2]. By applying the method of finite differences PDEs are carrid over into sets of coupled differential equations. These sets are implemented easily in equivalent electrical circuits. In this way, arbitrary time dependent, one-dimensional PDEs with spatial derivatives up to the forth order can be computed with electronic circuit simulators. Following Simpson's rule integrations over the spatial dimension are also available, the integrand may be an arbitrary function. These transformations are automated by the CAD-tool MEXEL [4,5]. Now we extend the transformation procedure to the generation of AHDL-code. Modeling PDEs by AHDLs enables simulations for a whole class of problems in microsytem technology and mechatronics. In addition to that the model-transformation approach is not limited to the simulation of mechatronics; electrothermal applications were investigated succesfully [6].

---

[2]SPICE3 is a quasi standard electronic network simulator, which has been developed from the University of California, Berkeley

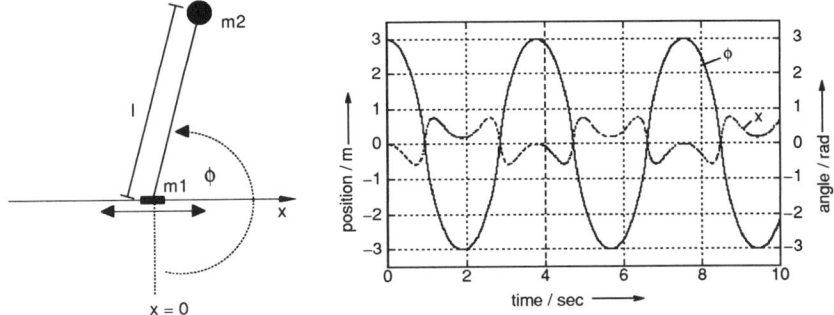

Figure 1: Rigid pendulum with movable bearing

3 Sample Description

To show how mechanics can be formulated by AHDL consider the inverted pendulum with movable bearing depicted in Fig. 1. We use HDL-A™ [7] for this example. The description starts with an ENTITY statement. It defines the model including input/output terminals (PIN), model parameters and initial conditions (GENERIC). The following ARCHITECTURE block contains the model equations. Several ARCHITECTUREs blocks can be formulated for different levels of complexity or simulation purposes. On the one hand, levels of the pendulum could be the nonlinear differential equations to study the motion or the linearized equations for developing a controller. On the other hand, assuming, e.g. temperature or reliability aspects, leads to different views and other kind of knowledge of the same system. Hence, hierarchical design strategy is supported, because specific models of the system can be used for certain system conditions.

The PROCEDURAL statement identifies explicit and the EQUATION statement implicit formulations. With that, the state variables x, x_t, x_tt, phi, phi_t, phi_tt of the pendulum declared by STATE are computed. Matched models can be described for each analysis type, e.g. dc- (operating point), ac- (frequency response), tran- (transient) -analysis.

In summary, the shown description gives good insight into the modeling techniques of AHDLs. But as stated in the previous section, much more is possible.

4 Examples

To underline the potentials of AHDLs we present three examples which were formulated with HDL-A™ [7] and performed with the simulator ELDO® [8][3].

The first example is the inverted rigid pendulum with a frictionless movable bearing from Fig. 1, whose description was given in Fig. 2. The variable ϕ is the angle of the pendulum, x is the horizontal position of the bearing. The masses are assumed to $m_1 = 1\text{kg}$ and $m_2 = 2\text{kg}$, the length l is given with 1m. In the simulation the initial value for ϕ was set to 3rad and x to 0m. In Fig. 1 the typical behavior of the

---

[3]HDL-A and ELDO are both products of Anacad

```
ENTITY hdla_pen IS                        -- model environment
  GENERIC(x_0, phi_0, x_t_0, phi_t_0, l, m1, m2 : real);
  PIN(y1, y2 : mechanical2);
END hdla_pen;

ARCHITECTURE difdg OF hdla_pen IS        -- model level/view
  VARIABLE g: real;
  STATE x,x_t,x_tt,phi,phi_t,phi_tt:analog;
    -- x,   x_t,   x_tt:   position x, tr.  veloc., tr.  accel.
    -- phi, phi_t, phi_tt: angle  phi, rot.   ''  , rot.  ''
BEGIN
  RELATION
    PROCEDURAL FOR init =>
      x_0     := 0.0;     phi_0   := 3.0; -- initial values
      x_t_0   := 0.0;     phi_t_0:= 0.0;
      l       := 1.0;                     -- length
      m1      := 1.0;     m2      := 2.0; -- masses
      x       := x_0;     phi     := phi_0;
      x_t     := x_t_0;   phi_t   := phi_t_0;
      x_tt    := 0.0;     phi_tt  := 0.0;
      g       := 9.81;                    -- gravity
    PROCEDURAL FOR dc, ac, transient =>
      y1.d %= phi;                        -- output terminals
      y2.d %= x;
    EQUATION (x,phi,x_t,phi_t,x_tt,phi_tt) FOR dc =>
      x       == x_0;     phi     == phi_0;  -- static analysis
      x_t     == 0.0;     phi_t   == 0.0;
      x_tt    == 0.0;     phi_tt  == 0.0;
    EQUATION (x,phi,x_t,phi_t,x_tt,phi_tt) FOR ac, transient =>
      x_t             == ddt(x);         -- dynamic analysis and
      x_tt            == ddt(x_t);       -- formulation of second
      phi_t           == ddt(phi);       -- order diff. equations
      phi_tt          == ddt(phi_t);     -- of the pendulum
      (m1+m2)*x_tt    ==   m2*phi_t*phi_t*sin(phi)
                         - m2*phi_tt*cos(phi);
      l*phi_tt        == -x_tt*cos(phi) - g*sin(phi);
  END RELATION;
END;
```

Figure 2: HDL-A model of rigid pendulum

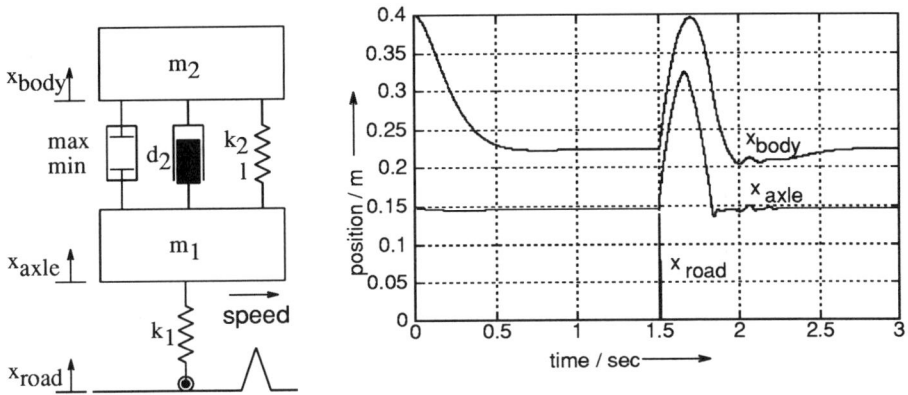

Figure 3: Car suspension with dynamic mechanical constraints

pendulum can easily be observed.

The next example is a simple car suspension system on a rough road with obstacles, see Fig. 3. The model includes dynamic binding of the wheel in relation to the road and a limited range of motion for the suspension system. The wheel is modeled as a simple spring. Its reaction force will be equal to zero, if the wheel flys. The limiter ensures a specific distance between car body and axle. We modeled two rigid mechanical constraints by **ideal inelastic** stops for a maximum and minimum distance. The car has a speed of 25m/s and the obstacle is triangular shaped with a height of 0.14m and a width of 0.3m. For the initial time point the suspension system is not under load except the speed of the car. The Fig. 3 displays a transient response of the suspension due to the gravity. If the car reaches the obstacle, the distance between axle and car body position becomes too small and the limiter enforces a certain difference. Caused by the high speed, the wheel begins to fly. After that we can recognize how the wheel lands on the road.

The first two examples shows the principal potentials of AHDLs for mechanics. The last example is a real-life application from microsystem technology: a micromachined Deformable-Mirror-Device-System [5,9,10], see Fig. 4. The system contains an electrostatically excitated deformable mirror and a controlling circuit. DMDs can be deformed separately due to the applied excitation voltage. In this way, incoming light is modulated in amplitude and phase caused by the deflection of the mirror [10]. The modulated light can be used to create a pixel image on a screen. Arranged in arrays of thousands of them DMDs are developed for projection displays (HDTV: high-definition television).

Using the model transformation approach the upper plate of the square-shaped DMD can be modeled as the 1D-PDE of deflection, for details see [5]. The electrostatic forces depend directly on the deflection of the plate. The controlling circuit is used to address the DMD and to set the electrode's voltage. Fig. 5 depicts the results for the excitation voltage and the reaction of the plate at several positions. The applied

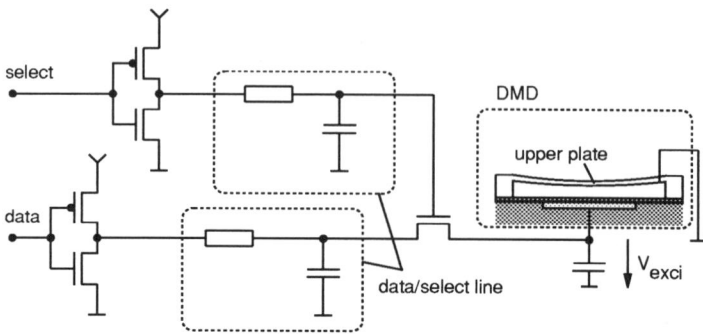

Figure 4: DMD system

input voltages at the data- and select-terminal are 25V. Against that, the amplitude of the excitation voltage is only 23.6V. This reduction of 1.4V is directly due to the non-ideal controlling circuit and leads to smaller deflections as one would expect, see also [5]. Several coupling effects are considered between electronics and mechanics. First, the deflecting plate causes a positive feedback, since it reduces the air gap. As a consequence the electrostatic force increases which causes the deflection. Secondly, the DMD itself is an electrical time-dependent capacitance due to the deflection of the upper plate. This capacitance influences the dynamic behavior of the controlling circuit. The following supplementary effects of the DMD-system can be investigated:

- shape of the excitation voltage,

- dynamic response of the deflectable mirror including damping effects,

- losses caused by the data/select line,

- non-idealities of the electrical signals due to the transistors,

- and shape of the bending line.

As shown, the strong interaction between mechanics and electronics need system approaches for valuable results, which can be fullfilled by AHDLs.

5 Conclusions

Analog hardware description languages have significant advantages for entire system simulation of microsystems and mechatronics. Especially the capabilities for mechanical modeling – namely the formulation of nonlinear differential equations and rigid mechanical constraints – in conjunction with the preservation of the electrical/electronical world deliver the necessary requirements to combine mechanics and information technology. Supporting the hierarchical design strategy AHDLs can help to develop new competetive products in shorter time. With the model transformation approach even partial differential equations can be handled. Modeling of partial differential equations opens new abilities for physically based descriptions by

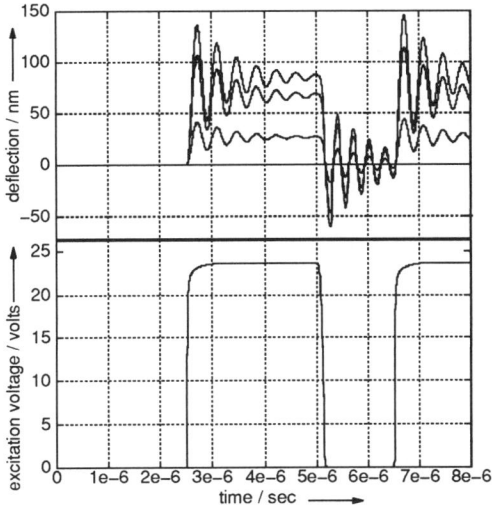

Figure 5: Dynamic simulation results of the DMD system

AHDLs of transport or propagation phenomena within system simulation. The three applications in this contribution – pendulum with movable bearing, car suspension system, micro-electromechanical DMD system – show the generality and suitability of the presented methodology. The coupling effects in the presented DMD system enforce an unified model for the study of sytem dynamics. Current and future investigations focus on the integration of efficient modeling techniques for multibody systems.

6 References

[1] B. Hosticka: Potential of Microelectronics for Mechatronics. Proc. 2nd Conf. on Mechatronics and Robotics '93, Duisburg/Moers, Germany, Sept. 27-29, 1993, pp. 556-565.

[2] E. Kreuzer, W. Schiehlen: NEWEUL – Software for the Generation of Symbolical Equations of Motion. Multibody Systems Handbook, W. Schiehlen, editor, 1990, pp. 181-202. Berlin: Springer.

[3] A. Kecskeméthy: Objektorientierte Modellierung der Dynamik von Mehrkörpersystemen mit Hilfe von Übertragungselementen. Phd thesis, University-GH Duisburg, Germany, 1993. Fortschrittberichte VDI, 88, Düsseldorf: VDI.

[4] G. Pelz, J. Bielefeld, F.-J. Zappe, G. Zimmer: MEXEL: Simulation of Microsystems in a Circuit Simulator Using Automatic Electromechanical Modeling. Micro System Technologies '94, H. Reichl, A. Heuberger, editors, 1994, pp. 651-657. Berlin: vde.

[5] G. Pelz, J. Bielefeld, F.-J. Zappe, G. Zimmer: Simulating Micro-Electro-mechanical Systems. IEEE Circuits and Devices Magazine, Simulation and Modeling, F. Najm, K. Mayaram, editors, (2)1995, pp. 10-13.

[6] J. Bielefeld, G. Pelz, H. B. Abel, G. Zimmer: An SOI MOSFET Model for Circuit Simulators Considering Nonlinear Dynamic Self-Heating. Proc. 20th IEEE Int. SOI Conf., Nantucket, USA, Oct. 1994., pp. 9-10.

[7] HDL-A Language Reference Manual, version 1.0, Anacad Electrical Engineering Software, June 1994.

[8] ELDO Users Manual, version 4.3.x, Anacad Electrical Engineering Software, June 1994.

[9] J. M. Younse: Mirrors on a chip. IEEE Spectrum, (11)1993, pp. 27-31.

[10] L. J. Hornbeck: Deformable-Mirror Spatial Light Modulators. SPIE Vol. 1150, Critical Reviews of Optical Science and Technology, 1989, pp. 86-102.

Dipl.-Ing. Jürgen Bielefeld
Gerhard-Mercator-University-GH Duisburg
Dept. of Electron Devices and Circuits
Finkenstr. 61, 47057 Duisburg, Germany
Tel.: +49-203-3783-151
Fax: +49-203-3783-266
Email: takayama@ims.fhg.de

Dr. rer. nat. Georg Pelz
Organisation, address, telephone, fax: see above
Email: pelz@ims.fhg.de

Prof. Dr. rer. nat. Günter Zimmer
Organisation, address: see above or
Fraunhofer Institute of Microelectronic Circuits and Systems
Finkenstr. 61, 47057 Duisburg, Germany
Tel.: +49-203-3783-111
Fax: +49-203-3783-119

III. Drive Systems

K. Nishibori, S. Okuma, H. Obata
Driving Characteristics of Robot Hand with Fingers using Langevin-Type Ultrasonic Motors

C. Bergmann, M. Gautier
Experimental Identification of DC Electric Drives

R. Scheidl, D. Schindler, G. Riha, W. Leitner
Basics of the Energy-Efficient Control of Hydraulic Drives by Switching Techniques

P. Van den Braembussche, J. Swevers, H. Van Brussel, P. Vanherck
Motion Control and Identification Techniques for Machine Tool Axes

Driving Characteristics of Robot Hand with Fingers using Langevin-Type Ultrasonic Motors

Kenji NISHIBORI

Daido Institute of Technology, Nagoya, JAPAN

Shigeru OKUMA

Nagoya University, Nagoya, JAPAN

Hirohisa OBATA

Aisan Industry Co., Obu, JAPAN

Abstract: An ultrasonic motor is a promising actuator for robots since it has a simple construction, high response, and high torque at low speeds. This paper proposes a robot hand using Langevin-type ultrasonic motors as the fingers instead of moving elements. This robot hand with three fingers can simultaneously grasp and rotate a cylindrical body. The driving characteristics of the robot hand were examined experimentally under various conditions. It was confirmed that high torque was obtained when the optimum contact angle and contact force of the ultrasonic motors were used. The reduction of the output torque is small even when the diameter of the rotational body deviates from the designed value.

1 Introduction

An ultrasonic motor is a promising actuator for robots since it has a simple construction without coils, high response, and high torque at low speeds. Besides the traveling wave-type ultrasonic motor [1]~[5], the Langevin-type ultrasonic motor [6]~[9] takes out the deformation energy with high efficiency as the mechanical vibration from the piezoelectric material. When using this ultrasonic motor for the fingers of the robot hand, instead of the conventional electromagnetic motors, there is a possibility of realizing a high performance in the robot hand. We propose a robot hand using the Langevin-type ultrasonic motors as the fingers without moving elements. The aim is to simultaneously grasp a cylindrical body and rotate it with the fingers. It has many applications, such as fitting axial parts and composing truss structures and pipes in space.

In this paper we made a trial robot hand with three fingers using Langevin-type ultrasonic motors and examined experimentally the basic driving characteristics when grasping a cylindrical body and rotating it under various conditions. The influences of the diameter of the cylindrical body on the contact angles and the contact forces were also considered.

2 Principle of Ultrasonic Motor

Figure 1 shows the structure of a vibration-type ultrasonic motor that is called a Langevin-type ultrasonic motor. Alternative voltage is applied to the piezoelectric material hold between elastic materials, and a longitudinal wave is excited in the elastic solid.

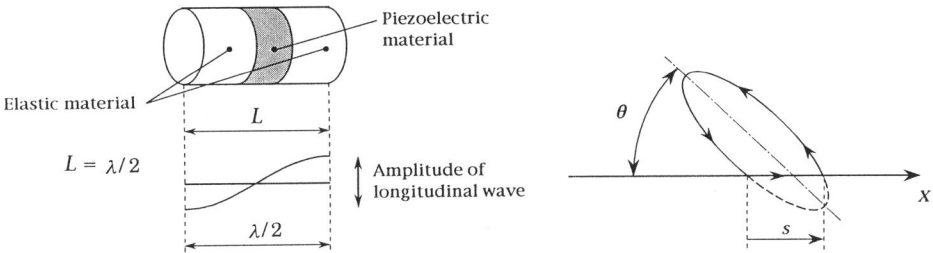

Fig. 1 Structure of ultrasonic motor

Fig. 2 Model of driving force of ultrasonic motor

When a half wavelength $\lambda/2$ coincides with the length L of the ultrasonic vibrator, the vibrating displacement on both tips of the elastic solid gets the maximum value [8]. The resonance frequency f is expressed as:

$$f = C/\lambda = C/(2L) \tag{1}$$

where C is a propagating velocity that is the speed of sound and $C = 5 \times 10^3$ m/s in an aluminum material. If the length of the vibrator is $L = 66$ mm, the resonance frequency of the vibrator is $f = 38$ kHz by equation (1).

The Langevin-type ultrasonic vibrator works as an ultrasonic motor when it is pressed against the surface of an object at the contact angle θ as shown in Fig. 2. Since the vibration is restricted on the surface of the object, the tip of the ultrasonic motor moves on a half elliptical curve [6], [8]. This locus consists of the contact motion pushing the object and the shrink motion of the free vibration as shown by the solid line. Consequently, an ultrasonic motor can rotate a cylindrical body by the friction between the surface of the object and the ultrasonic motor.

3 Experimental Apparatus and Method

Figure 3 shows the experimental apparatus to measure the driving characteristics of a single ultrasonic motor. The ultrasonic motor is pushed to a cylindrical body at the angle of $\theta = 15° \sim 75°$, and the contact force N is changed within the range of 0.98 N $\leq N \leq$ 2.94 N by a weight P. The cylindrical body is made of vinyl chloride that has a diameter of $d = 64$ mm. The characteristics of the motor with the torque load are measured by reeling up weights W. The rotational speed of body, n, is measured by a photo sensor and a slit disk installed on the cylindrical body. The ultrasonic motor has a diameter of 15 mm, the length of 66 mm, and resonance frequency of $f = 38.2$ kHz.

Figure 4 shows the configuration of the robot hand having three parts of an ultrasonic motor as the fingers. This robot hand can rotate a cylindrical body by grasping it without moving elements. Three ultrasonic motors are arranged at equal distance around the cylindrical body. Here the diameter of cylindrical body is defined as a designed value of d_0. The trial robot hand can open the gripper by $B = 55$ mm, and the distance from the base of the gripper to the center of cylindrical body is $A = 128$ mm.

Figure 5 illustrates a robot manipulator having the trial robot hand. Experiments were

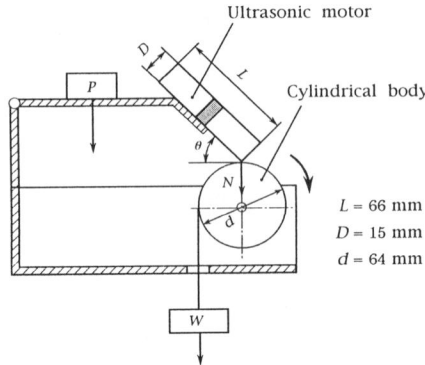

Fig. 3 Schematic outline of experimental apparatus for single ultrasonic motor

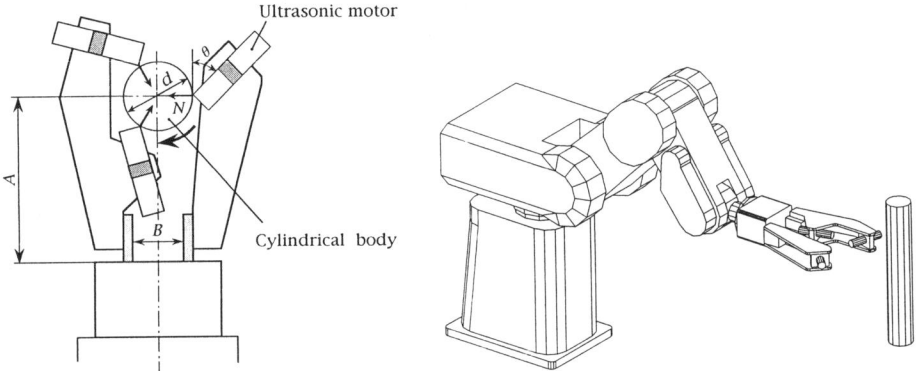

Fig. 4 Configuration of robot hand
having three ultrasonic motors
as the fingers

Fig. 5 Robot manipulator having trial
robot hand

carried out to measure the driving characteristics of the robot hand for the range of the grasping (contact) force, $1.47\,\text{N} \leqq N \leqq 5.39$ N.

4 Experimental Results and Discussions

4.1 Torque Characteristics of Single Ultrasonic Motor

Figures 6 (a) and (b) show the torque characteristics of a single ultrasonic motor for the contact angle θ when the contact force N remains $N = 0.98$ N and 2.45N, respectively. In the case where $N = 0.98$ N as shown in Fig. 6 (a), the rotational speed n, with low torque of $T \doteqdot 0$ N·cm, gets maximum value at the smallest angle of $\theta = 15°$. Beyond 15° the value of n decreases with an increase in θ. This means that the amplitude s in the tangential direction increases as the contact angle θ reduces, as shown in Fig. 2. If the torque T increases, however, the rotational speed n gets the maximum value at the contact angle of $\theta = 45°$. From the fact that the maximum torque is not obtained at the small contact angle of $\theta = 15°$, it is considered that the slip occurs between the ultrasonic motor and the rotational body.

Figure 6 (b) exhibits the result when the contact force is increased to $N = 2.45\,\mathrm{N}$. As the contact angle θ increases from 15°, the rotational speed n increases with high torque T and obtains the maximum at $\theta = 45°$. The obtained torque is greater in comparison with the case where $N = 0.98$ N as shown in Fig. 6 (a). The result shows a drooping characteristic; the rotational speed n decreases linearly with the increase in the torque T. When θ increases beyond 45°, however, the characteristic curves approach the origin, moving parallel.

Figure 7 shows the results when the contact force N is changed. The contact angle remains constant $\theta = 45°$. As the contact force N increases from 0.98 N, the rotational speed n increases and gets the maximum value at $N = 2.45\,\mathrm{N}$. For the further increase in N, however, it is seen that the value of n reduces except the region of high torque.

This result is explained in a model of the driving force for the contact force as shown in

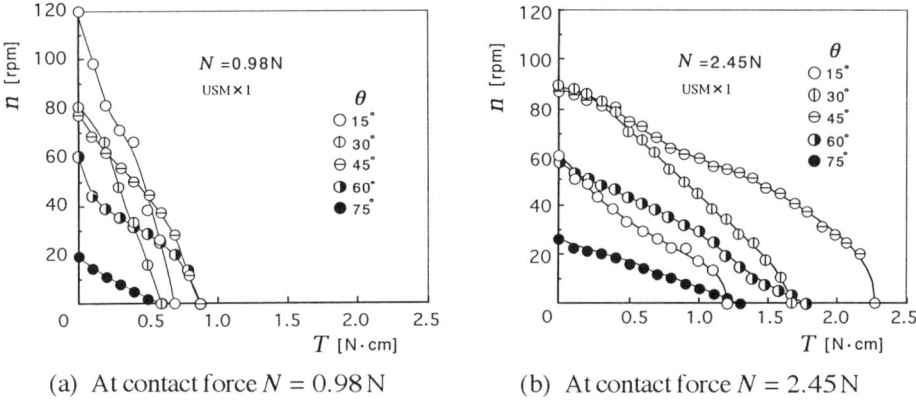

(a) At contact force $N = 0.98\,\mathrm{N}$ (b) At contact force $N = 2.45\,\mathrm{N}$

Fig. 6 Torque characteristics of single ultrasonic motor for the contact angle θ

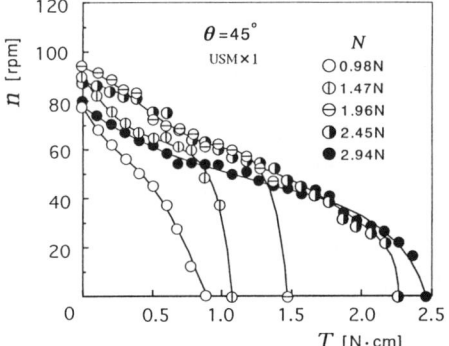

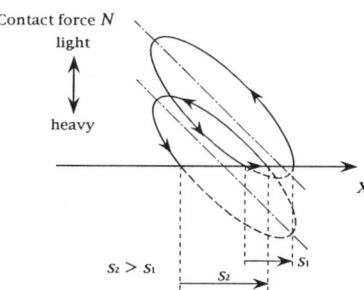

Fig. 7 Torque characteristics of single ultrasonic motor for the contact force N

Fig. 8 Model of driving force for
contact force N

Fig. 8. The surface of the rotational body is assumed plane. The locus of the tip of the ultrasonic motor indicates a half elliptical curve. This locus consists of the contact motion pushing the object and the shrink motion of the free vibration as shown by the solid line. The half elliptical locus is considered to move against the surface of the body according to the value of the contact force N. When the center of the elliptical curve coincides with the surface of the object, the contact length s takes the maximum of s_2 and the rotational speed n consequently gets the maximum. This is the reason the value of n takes the maximum at the contact force of $N = 2.45$ N. For the further increase in the contact force N, the center of the elliptical curve sinks under the surface of the body and the rotational speed n is supposed to decrease. When the contact force N is small in the range of $N \leq$ 1.96 N, as shown in Fig. 7, the rotational speed n decreases abruptly with an increase in the torque T. It is considered that this is due to the slip between the cylindrical body and the ultrasonic motor.

From the above results, it becomes evident that high torque is obtained when the contact angle is chosen to be $\theta = 45°$ and the optimum contact force of $N = 2.45$ N is used for this ultrasonic motor. On this condition, the rotational speed gets $n = 88$ rpm without a load and the maximum torque of $T = 2.25$ N·cm is obtained.

4.2 Torque Characteristics of Robot Hand with Three Ultrasonic Motors

Figure 9 shows the torque characteristics of the robot hand with three ultrasonic motors, when the grasping (contact) force N is changed. When the grasping force N increases from 1.47 N to 2.45 N, the rotational speed n increases in the region of high torque and the robot hand can rotate the cylindrical body even at the range of $T > 5$ N·cm. For a further increase in the grasping force of $N = 4.41$ N, however, the rotational speed n decreases with the same torque. It is confirmed that the grasping force is suitable in the range of 2.45 N $\leq N \leq 3.43$ N. The rotational speed n decreases linearly with the increase in the torque T, which indicates the drooping characteristics of the motor. Compared with the result of a single ultrasonic motor, as shown by a broken line, the maximum torque driven by three ultrasonic motors becomes about 2.3 times as much.

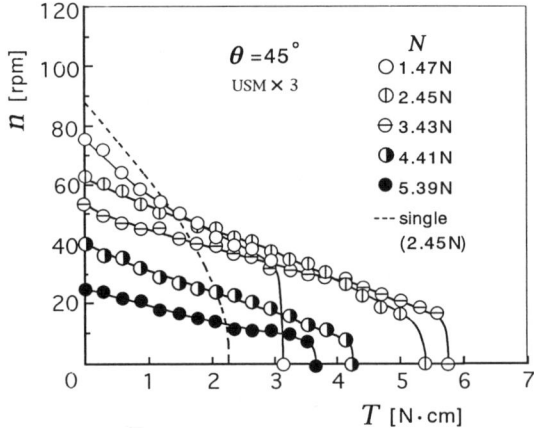

Fig. 9 Torque characteristics of robot hand with three ultrasonic motors

4.3 Torque Characteristics for Various Diameters of Cylindri-cal Body

Figure 10 (a) shows the relation between the torque T and the rotational speed n when the diameter of the cylindrical body is changed to d = 48 mm, 56 mm, 64 mm and 80 mm. The contact angle is θ = 45° and the contact force is N = 2.45 N. When the torque T is low, the smaller the diameter of the cylindrical body, the higher the rotational speed n becomes. As the torque T increases, however, the rotational speed n decreases greatly for the slender cylindrical body. The cylindrical body with the diameter d = 64 mm gets the maximum speed, and beyond this diameter the rotational speed n decreases.

Figure 10 (b) exhibits the peripheral velocity V of the cylindrical body instead of the rotational speed n. It is seen that the peripheral velocity V at the low torque takes approximately the same value, regardless of the diameter d. As the torque increases, however, the peripheral velocity of the slender cylindrical body drops greatly. Then the maximum torque also reduces. This phenomenon means that the slip between the body and the ultrasonic motor increases with the decrease in the diameter d of the cylindrical body. It is seen that the peripheral velocity curve is almost similar in the range of $d \geqq 64$ mm.

4.4 Influence of Diameter of Cylindrical Body Different from Designed Value

4.4.1 Changes in contact angles of three ultrasonic motors

The diameter of the cylindrical body is defined as the designed value d_0, when the robot hand grasps the cylindrical body with three ultrasonic motors at the contact angle of θ = 45°, respectively as shown in Fig. 4. If the robot hand grasps the cylindrical body that has a diameter different from the designed value d_0 without changing the configuration of the ultrasonic motors, the contact angles θ may deviate from 45 degrees.

Figure 11 shows the contact angles of three ultrasonic motors when the diameter d $(= 2\,a)$ of the cylindrical body differs from the designed value d_0. Defining the contact angles of USMs 1, 2 and 3 as θ_1, θ_2 and θ_3, respectively, the following equation is obtained on a triangle of $\triangle OAB$.

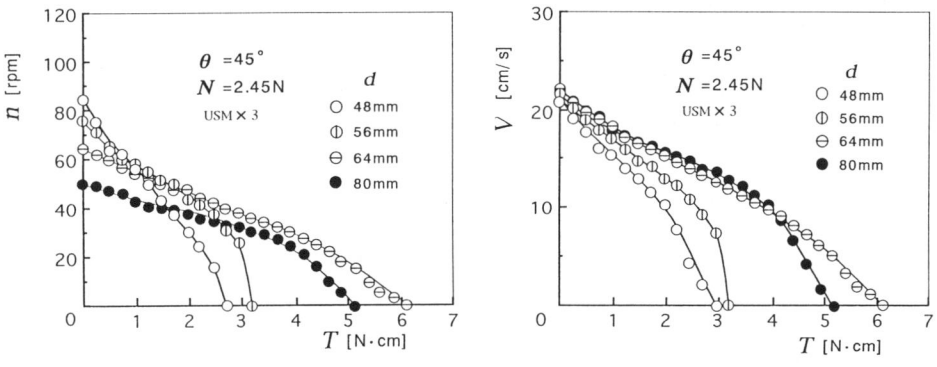

| (a) Changes in rotational speed n | (b) Changes in peripheral velocity V |

Fig. 10 Torque characteristics of robot hand for the diameter d of cylindrical body

$$\theta_2 + \varphi + \gamma = 90° \qquad (2)$$

where the angles of φ and γ are expressed as $\varphi = \cos^{-1}(b/a)$ and $\gamma = 15°$, respectively. Consequently, equation (2) is rewritten as follows:

$$\theta_2 = 75° - \cos^{-1}(b/a) \qquad (3)$$

From the geometric relation, the angle θ_3 is expressed as:

$$\theta_3 = 90° - \theta_2 \qquad (4)$$

The angle θ_1 remains constant of $\theta_1 = 45°$, regardless of the value of the diameter d.

Figure 12 shows the calculated values of the contact angles of three ultrasonic motors when a robot hand grasps a cylindrical body that has a diameter different from the designed value of $d_0 = 48$ mm. It is seen that the contact angles of θ_1 and θ_2 deviate greatly from 45 degrees as the diameter d decreases from the designed value d_0. The maximum length to open the gripper is 55 mm, and two broken lines denote the range of the diameter, 43 mm $\leq d \leq$ 68 mm, which the robot hand can grasp. Then the contact angles of three ultrasonic motors are $\theta_1 = 45°$, $23° \leq \theta_2 \leq 60°$, and $30° \leq \theta_3 \leq 67°$, respectively.

4.4.2 Changes in contact forces of three ultrasonic motors

The contact forces of three ultrasonic motors are equal in strength when the robot hand grasps the cylindrical body that has the same diameter as the designed value. They lose equality, however, when the diameter d of the cylindrical body differs from the designed value d_0. Figure 13 shows the relationship between the diameter of a cylindrical body and the contact forces. According to the balance of three contact forces, N_1, N_2 and N_3, the following equations are obtained.

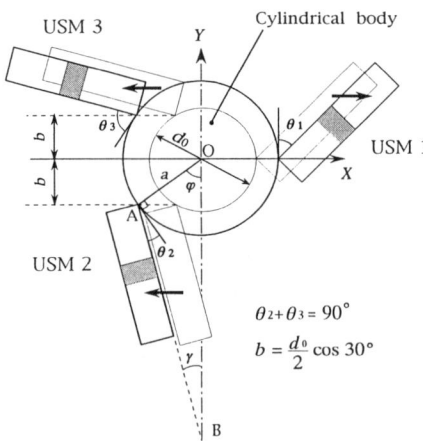

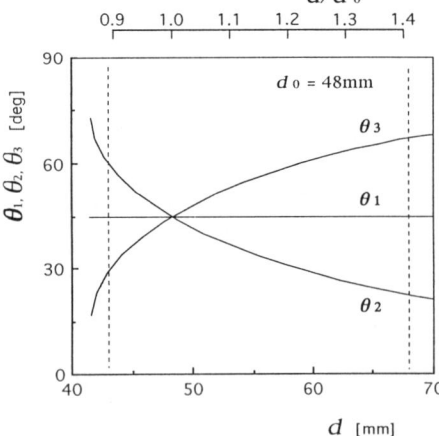

Fig. 11 Contact angles of ultrasonic motors for the diameter of cylindrical body

Fig. 12 Changes in contact angles for deviation from designed value of diameter

$$N_2 \cos \alpha + N_3 \cos \alpha = N_1 \qquad (5)$$

$$N_2 = N_3 \qquad (6)$$

From these equations,

$$N_2 / N_1 = \frac{1}{2 \sin \alpha} \qquad (7)$$

is obtained. The angle α is expressed as follows:

$$\alpha = 90° - \varphi$$
$$= 90° - \cos^{-1}(b/a) \qquad (8)$$

Substituting equation (8) to equation (7), the following relation is obtained.

$$N_2 / N_1 = N_3 / N_1$$

$$= \frac{1}{2 \cos \left\{ 90° - \cos^{-1}(b/a) \right\}} \qquad (9)$$

Figure 14 exhibits the calculated result of the force ratio N_2 / N_1 when the robot hand grasps a cylindrical body that has a diameter d different from the designed value of $d_0 = 48$ mm. As the diameter of the cylindrical body increases from the designed value of $d_0 = 48$ mm, the contact force N_2 of USMs 2 and 3 decreases in comparison with the force N_1 of USM 1. On the other hand, as the diameter of the cylindrical body decreases from the designed value d_0, the ratio of N_2/N_1 increases greatly. In the range of 43 mm $\leq d \leq$ 68 mm, the force ratio is in the range of $1.95 \geq N_2/N_1 \geq 0.63$.

From the above results, it can be considered that the influence of the larger diameter on the driving characteristics is slight, even when the diameter of cylindrical body is different from the designed value d_0.

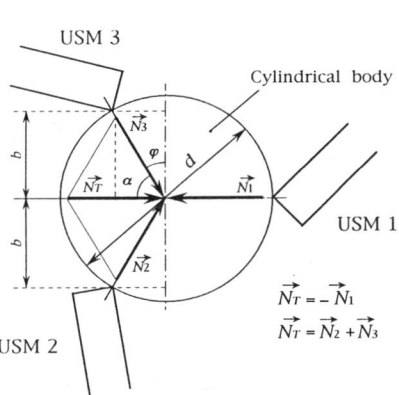

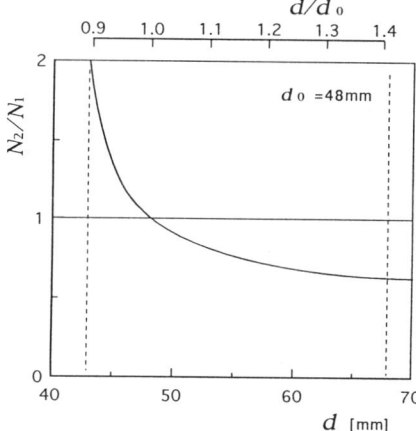

Fig. 13 Relationship between diameter d of cylindrical body and three contact forces

Fig. 14 Change in force ratio N_2/N_1 for diameter d of cylindrical body

4.4.3 Changes in torque characteristics

Figure 15 shows the change in the torque characteristics when the diameter d of the cylindrical body, that the robot hand grasps, differs from the designed value d_0. Experiments were carried out when the robot hand with the designed values of $d_0 = 64$ mm and 48 mm, respectively, grasped the cylindrical body of $d = 64$ mm. In the case where $d_0 = 64$ mm, the robot hand holds the cylindrical body of the same size as the designed value. The contact angles, θ_1, θ_2 and θ_3 of three ultrasonic motors, coincide with 45 degrees, and the contact forces N_1, N_2 and N_3 are equal. In the case where $d_0 = 48$ mm, the contact angles are $\theta_1 = 45°$, $\theta_2 = 25.5°$ and $\theta_3 = 64.5°$, respectively, and the ratio of contact forces is $N_2 / N_1 = 0.66$. In this case the rotational speed n with the low torque almost coincides with the value in the case of $d_0 = 64$ mm. As the torque increases, however, the difference between both of the rotational speeds n increases. This is due to the inequality of the contact angles and the contact forces of three ultrasonic motors. However, it is considered that the reduction of the output torque is small, even when the diameter of the rotational body deviates from the designed value.

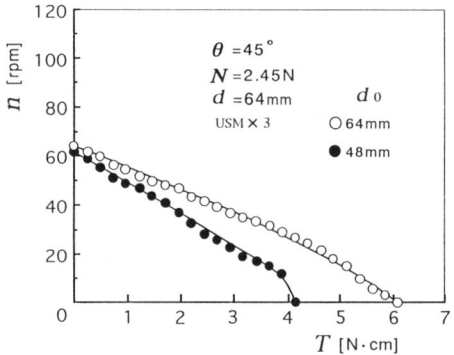

Fig. 15 Change in torque characteristics for different designed values d_0 of diameter

5 Conclusions

A trial robot hand with three fingers using Langevin-type ultrasonic motors was designed and its validity was examined experimentally. The results obtained are summarized as follows:

(1) This robot hand can rotate a cylindrical body by grasping it without moving elements.

(2) High torque is obtained when the contact angle is chosen to be $\theta = 45°$, and when an optimum contact force is used for the ultrasonic motor.

(3) The peripheral velocity of the rotational body is almost constant for the various diameters if torque is low. As the torque load increases, however, the peripheral speed of the slender cylindrical body reduces due to the slip between the ultrasonic motors and the surface of the rotational body.

(4) When the diameter of the cylindrical body differs from a designed value, the rotational speed n decreases with the high torque. This is due to the inequality of the contact angles and the contact forces of the ultrasonic motors. However, the reduction of the output

torque is small, even when the diameter of the rotational body deviates from the designed value.

(5) When the robot hand grasps a cylindrical body that has a larger diameter than a designed value, the influence on the contact angles and the contact forces of three ultrasonic motors is slight.

Acknowledgments

Part of this work was done in cooperation with Prof. R. Kopp of the Institute for Metal Forming in the Rheinisch-Westfaelische Technische Hochschule, Aachen, and Prof. P. Drews and Dipl. F. Quartier in the European Center for Mechatronics, Aachen, Germany when the first author stayed there. Acknowledged are financial supports by the Japanese International Scientific Research Program: University to University Cooperative Research (between Daido Institute of Technology and RWTH) and the Hori Information Science Promotion Foundation. The authors thank Mr. T. Miyamoto of Honda Electronics Co. Ltd. for supplying the ultrasonic vibrators, and a graduate student of DIT, Mr. S. Kondo, for his assistance.

References

[1] M. Tsuboi: Nikkei Mechanical, 135-5 (1983), pp.44-49 (in Japanese).

[2] M. Kuribayashi, S. Ueha and E. Mori: Excitation Conditions of Flexural Traveling Waves for a Reversible Ultrasonic Linear Motor, Journal of Acoustical Society of America, 77-4 (1985), pp. 1431-1435.

[3] S. Iwamatsu, S. Ueha, M. Kuribayashi and E. Mori: Rotary Ultrasonic Motor using Extensional Vibration of a Ring, Japanese Journal of Applied Physics, 25, Supplement 25-1 (1986), pp. 174-176.

[4] K. Nishibori, S. Okuma, Y. Eryu and S. Sakai: Velocity Control of Ultrasonic Motors for Robot Arms by Pulse Width Modulation, Proceedings of the 3rd IEEE International Workshop on Advanced Motion Control, Berkeley, California, U.S.A., Mar. 1994, pp. 823-829.

[5] K. Nishibori, S. Okuma, Y. Eryu and S. Sakai: Position Control of Robot Manipulators with Ultrasonic Motors using Pulse Width Modulation and Fuzzy Controllers, Transactions of the Japan Society of Mechanical Engineers, 60-574, C (1994), pp. 2052-2056 (in Japanese).

[6] M. Tsuboi: Nikkei Mechanical, 89-5 (1981), pp. 100-104 (in Japanese).

[7] K. Uchino: Piezoelectric / Electrostrictive Actuator, Morikita Publishing, 1986, pp. 180-183 (in Japanese).

[8] T. Kenjho and T. Sashida: Introduction to Ultrasonic Motors, Sogo Electronics Publishing, 1991, p. 73 (in Japanese).

[9] M.Kurosawa, K. Nakamura, T. Okamoto and S.Ueha: An Ultrasonic Motor Using Bending Vibrations of a Short Cylinder, IEEE Transactions on Ultrasonics, Ferroelectrics, and Frequency Control, 36-5 (1989), pp. 517-521.

Prof. Dr. Kenji NISHIBORI
Dept. of Mechanical Engineering,
Daido Institute of Technology,
Minami-ku, Nagoya 457, JAPAN
Telephone: +81-52-611-0513
Fax: +81-52-612-5653
Email: nishibo@daido-it.ac.jp

Prof. Dr. Shigeru OKUMA
Dept. of Electrical Engineering,
Nagoya University,
Chikusa-ku, Nagoya 464, JAPAN
Telephone: +81-52-789-2775
Fax: +81-52-789-3140
Email: okuma@okuma.nuee.nagoya-u.ac.jp

Mr. Hirohisa OBATA
Aisan Industry Co. Ltd.
Kyowa-cho, Obu, Aichi 474, JAPAN
Telephone: +81-562-47-1131
Fax: +81-562-48-4013

Experimental Identification of DC Electric Drives

P. Ph. Robet, M. Gautier

Laboratoire d'Automatique de Nantes
U.R.A. C.N.R.S. n° 823
Université & Ecole Centrale de Nantes
ECN, 1, rue de La Noë,
44072 Nantes Cedex 03, France
Tel : (33) 40 37 25 12, Fax : (33) 40 74 74 06
E mail : gautier@lan.ec-nantes.fr

C. Bergmann

GE44- LR2EP
IUT de Nantes
3, rue du Maréchal Joffre
44041 Nantes Cedex 01, France
Tel : (33) 40 30 60 07, Fax : (33) 40 30 60 97

Keywords: Identification, Electrical, Mechanical Parameters, Exciting Trajectories

Abstract : Because of increasing joint speed and acceleration of complex mechanical structures as rigid and flexible robot manipulators, accurate models of joints with electric drives are needed to improve their simulation and their control. We propose a new way to identify the electrical and mechanical parameters of a robot joint driven by a DC motor. The method is based on a closed loop identification of the inverse model which is linear in relation to the parameters. Least Squares techniques and exciting trajectories of current and velocity are used These trajectories minimize a criterion which is a function of the condition number of an observation matrix. An experimental identification of one prismatic joint EMPS300 driven by a DC permanent magnet motor is given.

1. Introduction

The joint drive of robot is generally considered as a constant torque generator and used to get dynamic parameters identification (inertia, first moments, masses, friction) [7]. Because of increasing joint speed and acceleration using advance dynamic control law, accurate models of the drive chain must be taken into account. Electrical and mechanical parameters of the model must be identified because they are unknown or manufacturer's data are not accurate. Since a lot of industrial robots still use DC motor, this paper is focused on *Direct Current motor* drive chain identification.

A method is proposed which comes from important theoretical results and successful experimentation that have been obtained in the area of identification of dynamic parameters of robot manipulators [4,...,7, 13]. A closed loop identification is performed in three main steps. The first one is to obtain an inverse model which is a linear relation to a set of identifiable parameters in order to use simple least squares (LS) algorithm. The second one is to define and generate exciting trajectories which give rich information to get good noise immunity and to decrease bias and variance of the L.S.

estimation. The third step is a classical low-pass filtering data in order to get low noise observation and measurement matrices which ensures the efficiency of the L.S. estimation.

Experimental results of the dynamic identification of a prismatic joint is presented. The different steps starting from the modeling till the validation of the results will be given. Practical issues such that filtering will be adressed to carry out with success such experimentation.

2. Identification model

2.1 Modeling of the drive chain

Let us consider a joint driven by a voltage source amplifier and a DC permanent magnet motor. Electrical and mechanical equations are following :

$$L \frac{dI}{dt} + R I(t) + k_t \dot{q}(t) = v_e(t) \tag{1}$$

$$\Gamma_m = k_t I(t) = J \ddot{q}(t) + F_v \dot{q}(t) + F_s \text{sign}(\dot{q}) \tag{2}$$

where :
Γ_m is the motor torque,
$I(t)$ is the armature current,
$v_e(t)$ is the armature voltage,
R, L, are the armature resistance and inductance respectively,
k_t is the motor torque constant,
J is the total inertia moment of the drive chain referred to the motor axis side,
F_v and F_s are respectively the viscous friction coefficient and the coulomb friction torque of the drive chain.
$\dot{q}$ and $\ddot{q}$ are respectively the motor axis velocity and acceleration.

2.2 The direct state space model

In order to use methods from the Matlab identification toolbox, Eq. 1, 2, must be written in a standard linear state-space form, canceling the non linear coulomb friction effect. The direct model calculates the derivative of the state as a function of the state and control :

$$\dot{x} = A x + B u, \qquad y = C x \tag{3}$$

$x(t) = \begin{bmatrix} \dot{q} \\ I \end{bmatrix}$, is the state vector,

$u(t) = v_e$, is the control input,
$y(t) = x(t)$, is the output vector,

$$A = \begin{bmatrix} \dfrac{-F_v}{J} & \dfrac{k_t}{J} \\ \dfrac{-k_t}{L} & \dfrac{-R}{L} \end{bmatrix}, \text{ is the plant matrix, } B = \begin{bmatrix} 0 \\ 1 \\ \dfrac{1}{L} \end{bmatrix}, \text{ is the control input matrix,}$$

$C=I_{2x2}$, is the output matrix, I_{2x2} is the (2x2) identity matrix.

The predicative errror model (PEM) function has been used to identify matrices A and B [10, 11]. This is difficult because the coefficients of A are very desequilibrated ($\dfrac{f_v}{J} << \dfrac{R}{L}$) [2]. The estimation is very sensitive to the input signal. Unfortunately, this approach doesn't provide method to calculate exciting input signals and it cannot easily take into account non linear state space model. To overcome these drawbacks, we propose to use a method based on the inverse model.

2.3 The inverse model

The inverse model calculates the control input as a function of the state and its derivative. It is naturally given by the equations of the physics (Maxwell, Lagrange or Newton-Euler equations), Eq. (1, 2). These equations are non linear in the state because of sign($\dot{q}$), but they are linear in relation to the parameters to identify and define the identification model as follows :

$$\begin{bmatrix} v_e \\ 0 \end{bmatrix} = \begin{bmatrix} I & \dfrac{dI}{dt} & \dot{q} & 0 & 0 & 0 \\ 0 & 0 & -I & \ddot{q} & \dot{q} & sign(\dot{q}) \end{bmatrix} X, \quad X = \begin{bmatrix} R & L & k_t & J & F_v & F_s \end{bmatrix}^T \tag{4}$$

X is the (Npx1) vector of parameters to identify (Np=6).
They are called standard parameters because they are defined from usual laws of physics.

3. Closed loop identification with inverse model

3.1 Identification method

A common way in robotics is to use multivariable linear least squares (L.S.) techniques to solve an overdetermined linear system of r equations in p unknowns, obtained from a sampling of linear inverse model such as Eq. (4) :

$$Y = W X + \rho \tag{5}$$

W is the (rxp) observation matrix,
p is the number of unknown parameters, p=6,
ρ is the (rx1) vector of errors,
Y is the (rx1) vector of measurements.

Using the inverse model Eq. (4) and samples of $(I, \frac{dI}{dt}, \dot{q}, \ddot{q})$ at different times t_i, $i=1,...,n_e$, with $r=2 \times n_e \geq p$, comes :

$$
Y = \begin{bmatrix} v_e(1) \\ 0 \\ ... \\ v_e(ne) \\ 0 \end{bmatrix}, \quad W = \begin{bmatrix} I(t_1) & \frac{dI}{dt}(t_1) & \dot{q}(t_1) & 0 & 0 & 0 \\ 0 & 0 & -I(t_1) & \ddot{q}(t_1) & \dot{q}(t_1) & sign(\dot{q}(t_1)) \\ ... & ... & ... & ... & ... & ... \\ I(t_{ne}) & \frac{dI}{dt}(t_{ne}) & \dot{q}(t_{ne}) & 0 & 0 & 0 \\ 0 & 0 & -I(t_{ne}) & \ddot{q}(t_{ne}) & \dot{q}(t_{ne}) & sign(\dot{q}(t_{ne})) \end{bmatrix} \quad (6)
$$

The L.S. solution minimizes the 2 norm $\|\rho\|^2$ of the vector of errors.
The unicity of the L.S. solution depends on the rank of the observation matrix W.
W rank deficiency can come from two origins :

– structural rank deficiency which stands for any samples of $(I, \frac{dI}{dt}, \dot{q}, \ddot{q})$ in W. This is the structural parameters identifiability problem which is solved using base parameters in robotics [5]. It the case of drive chain parameters, all the standard parameters are identifiable. This can be checked from QR or SVD factorization of W calculated with random simulated samples of $(I, \frac{dI}{dt}, \dot{q}, \ddot{q})$ [3].

– data rank deficiency due to a bad choice of $(I, \frac{dI}{dt}, \dot{q}, \ddot{q})$ samples in W. This is the problem of the optimal measurement strategy which is solved using closed loop identification to track exciting trajectories. Another reason to use closed loop identification comes from the integrator behaviour of electrical and mechanical systems whose approximated transfer functions are close to 1/Ls and 1/Js respectively (s is the Laplace variable). Then it is difficult to proceed to open loop tests while controlling the current or the velocity and position of the system. This is dangerous and not allowed on industrial robots.

3.2 Exciting trajectories

With a good sampling of $(I, \frac{dI}{dt}, \dot{q}, \ddot{q})$, X can be estimated as the L.S. solution $\hat{X}$ of the full rank linear system (Eq. 5) :

$$
\hat{X} = \underset{X}{Arg.min}\|\rho\|^2 = [W^T W]^{-1} W^T Y \quad (7)
$$

In practical applications, Y and W are random matrices because of noisy measurements. Standard deviation are estimated using classical and simple results from statistics, considering the matrix W to be a deterministic one, and ρ to be a zero mean additive independent noise, with standard deviation σ_ρ such that :

$$C_{\rho\rho} = E[\rho\rho^T] = \sigma_\rho^2 \, I_r$$

where E is the expectation operator and I_r is the (rxr) identity matrix.
The variance-covariance matrix of the estimation error and standard deviations can be calculated by :

$$C_{\hat{X}\hat{X}} = E\left[(X-\hat{X})(X-\hat{X})^T\right] = \sigma_\rho^2 (W^T W)^{-1}$$

$$\sigma_{\hat{X}i}^{\,2} = C_{\hat{X}\hat{X}ii}, \text{ the diagonal coefficient of } C_{\hat{X}\hat{X}} \tag{8}$$

An unbiased estimation of σ_ρ is used and the relative standard deviation $\sigma_{\hat{X}ri}$ is given by the expression :

$$\hat{\sigma}_\rho^2 = \frac{\left\| Y - W\hat{X} \right\|^2}{r - p} \quad , \quad \sigma_{\hat{X}ri} = \frac{\sigma_{\hat{X}i}}{\hat{X}i} \tag{9}$$

When W and ρ are not independent, L.S. estimator is not efficient and bias may occur. Then our strategy to decrease bias and variance is to use good low pass filtering to get noiseless (v_e, $\dfrac{dI}{dt}$, $\dot{q}$, $\ddot{q}$) measurements or estimations and to use exciting trajectories [7].
It is known that the sensitivity of the LS solution of a perturbated linear system (Eq. 5) can be measured by the condition number Cond(W) of the observation matrix W, [9]. In the case of a non perturbated matrix W, comes :

$$\frac{\left\| X - \hat{X} \right\|}{\left\| X \right\|} \leq \text{Cond}(W) \frac{\left\| \rho \right\|}{\left\| Y \right\|} \tag{10}$$

In the case of a little perturbation δW on W ($\left\| \delta W \right\| << \left\| W \right\|$), comes :

$$\frac{\left\| X - \hat{X} \right\|}{\left\| X \right\|} \leq \text{Cond}(W) \frac{\left\| \delta W \right\|}{\left\| W \right\|} = \frac{\left\| \delta W \right\|}{\Sigma_p(W)} \tag{11}$$

The 2-norm condition number is easy to calculate from singular value decomposition (SVD) of W :

$$W = U \begin{bmatrix} \Sigma \\ 0 \end{bmatrix} V^T \; , \; \text{U, V are two orthogonal matrices.}$$

Σ is the pxp diagonal matrix of p singular values Σ_i of W on its diagonal.
$\Sigma = \text{diag}(\Sigma_i(W))$, $\quad \Sigma_1(W) \geq ... \geq \Sigma_p(W)$

The condition number is the ratio of the largest singular value over the smallest one :

$$\text{Cond}(W) = \Sigma_1(W)/\Sigma_p(W) \geq 1 \qquad (12)$$

From (Eq. 10, 11) come that Cond(W) must be taken as close as possible to 1 with large singular values before computing $\hat{X}$. Taking this criterion to define exciting trajectories leads to obtain estimations with equal absolute accuracy for all parameters, then it is not efficient when X is badly scaled.

A second criterion, function of Cond(W diag(Z)), has been proposed to equilibrate the relative standard deviation $\sigma_{\hat{X}ri}$ (Eq. 9) for all parameters. Z is a vector of *a priori* X values, which can be taken as the result of a first identification with Cond(W) close to 1 (<100), or is available from manufacturer's data. This may involve some knowledge about reasonable values of X components as their size or their sign. It has been proved that an exciting trajectory which equilibrates the relative accuracy for all parameters minimizes the following criterion [13] :

$$f(W,Z) = \text{Cond}(\Phi) + \frac{1}{\Sigma_p(\Phi)} \qquad (13)$$

where :

$\Phi = W \text{ diag}(Z) = [\Phi_{:,1} \dots \Phi_{:,p}] = [W_{:,1} Z_1 \dots W_{:,p} Z_p]$

diag(Z) is a pxp diagonal matrix with the absolute value components of Z on its diagonal.

$\Phi_{:,i}$ and $W_{:,i}$ are the ith column of Φ and W, respectively.

$\Phi_{:,i}$ represents the *a priori* contribution of the parameter X_i in the measurement vector $Y_{ap} = W Z$, on the whole trajectory.

$\Sigma_p(\Phi)$ denotes the smallest singular value of Φ.

It can be seen that a trajectory which minimizes Cond(Φ), is a trajectory which gives an equal *a priori* contribution of the Z_i parameter in the *a priori* measurement vector Y_{ap}.

Exciting trajectories are obtained using a sequential quatratic programming (SQP) method (see [8], or the Matlab optimization toolbox[12]), under constraints of the equation of an interpolator and the technological limits of the system (position limits, maximum velocity, maximum current and rated current, ...)[6].

A common way in robotics is to use a bang-bang trajectory generator which is composed of a constant velocity segment with ramp blends (Fig.1). This generator also gives suitable current trajectories using several succeeding interpolators, and can be easily implemented in Simulink using the repeating sequence block.

I_{maxj} : maximum current on the trajectory interpolator j,
ta_j : time of increasing current ramp,
tb_j : time of constant current step.

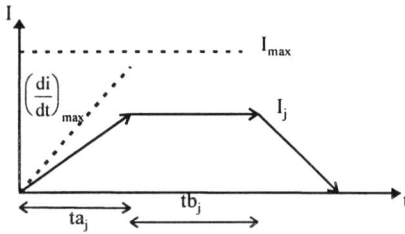

Fig.1 Trajectory bang-bang generator

The equation of one interpolator j is given as following :

Increasing current ramp	Constant current step	Decreasing current ramp
$0 \le t \le t_{a_j}$	$t_{a_j} \le t \le t_{a_j} + t_{b_j}$	$t_{a_j} + t_{b_j} \le t \le 2t_{a_j} + t_{b_j}$
$I_j(t) = \dfrac{I_{maxj}}{t_{aj}} t$	$I_j(t) = I_{maxj}$	$I_j(t) = \dfrac{I_{maxj}}{t_{aj}}(2t_{aj} + t_{bj} - t)$
$\overset{o}{I}_j(t) = \dfrac{I_{maxj}}{t_{aj}}$	$\overset{o}{I}_j(t) = 0$	$\overset{o}{I}_j(t) = \dfrac{-I_{maxj}}{t_{aj}}$

3.3 Essential parameters

When the contribution of some parameters is too small in the measurement vector Y, it is very difficult to get good exciting criterion value. Then they cannot be experimentally identified, in spite of using optimized trajectories. The estimation of the relative standard deviation (Eq. 9) gives a good indicator to detect insignificant parameters to be eliminated in a backward step by step procedure, starting from the largest value of $100 * \sigma_{\hat{X}ri} \ge 10$. This new set of essential parameters defines a simplified model which improves the noise immunity of the identification process and contributes in reducing the computation burden of the model for simulation and control applications.

4. An experimental case study

The linear positionning system EMPS300 connected to the control system from dSPACE™ are used. The control system is based on a TMS 320C31 Texas Instruments™ processor and Matlab-Simulink software, in order to get a high rate numerical control with a big computational capacity. The EMPS300 main components are a DC permanent magnet motor with DC tachometer and current controlled four quadrant PWM chopper, a ball screw drive positionning unit and an incremental encoder [3]. All analog and digital signals are directly accessible between EMPS300 and Simulink using the C code generator RTW Matlab toolbox [12] and the RTI program from dSPACE, to start experiments easily and immediately.
The following practical issues must be observed :
- The armature voltage and current are measured through isolated probes and analog anti-aliasing filters and sampled by AD converters at a high rate (5KHz).

- Low-pass filtering using a decimate filter from MATLAB associated with a difference algorithm provides a digital high-pass or pass-band filter to estimate derivatives at low frequency without derivating high frequency noise.

The rotor velocity q̇ and the current derivative are calculated from the filtered position and current signals respectively using central difference to avoid phase shift. It has been shown that the tachometer signal is contaminated with bias and noise and it is better to get the joint velocities by differentiating the position signal. The rotor acceleration q̈ is calculated from the calculated velocity using central difference algorithm.

- Y and columns of W are both filtered in a process called parallel filtering, using a decimate filter, in order to eliminate high frequency noise and ripple from Y and to keep available the linear system (Eq. 7). The filter bandwidth must be chosen to include the dynamic range in W columns.

All data in figures are given in SI units.

4.1 Global identification

From Eq. 1, 2, it can be seen that both current and velocity tracking cannot be performed with only one control input v_e.

Because it is not possible to easily open the current loop of industrial amplifier (this is also the case for the EMPS300), a current tracking has been chosen using the current controlled PWM servo amplifier. For a current trajectory given by the optimization procedure, Eq.2 is numerically integrated to obtain velocity and acceleration in order to calculate W and the criterion Eq.13. Manufacturer's data are used as a priori Z values. The optimization procedure has given a trajectory fig. 2 with Cond(Φ) =1731 and estimated values are given in table 1.

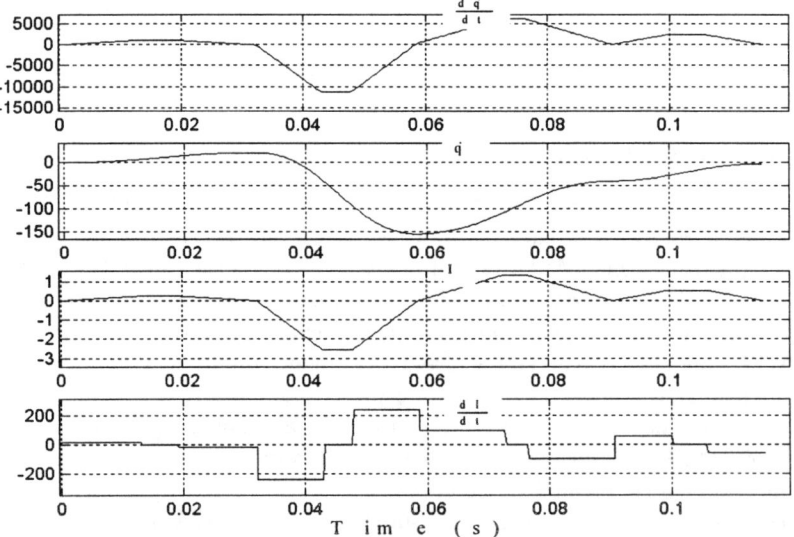

Figure 2 : Exciting trajectory for global identification.

Friction parameters are not significant because of the high value of Cond(Φ) which cannot decrease in spite of the optimization procedure. This is due to the large ratio between electrical and mechanical time constants $(J/Fv \cong 10^4\ L/R)$. It is better to identify them separately by decoupling the electrical and mechanical equations Eq. (1, 2).

4.2 Identification of electrical parameters

As a first step the linear relation between e.m.f. voltage v_{emf} and velocity, assuming zero armature current $(I(t)=0)$ in Eq. 1, is used.

$$v_e = v_{emf} = k_t\ \dot{q} \tag{14}$$

A simple test consists to deconnect the armature motor from the amplifier and to record armature voltage v_e and rotor position while moving the carriage by hand.
k_t is calculated as the mean value of the ratio $(v_e/\dot{q}\ )$, for $\dot{q}$ lying between 10% and 100% of its maximum value :

$k_t = 0.05106 \pm 3e\text{-}5$ (Nm/A)

This is close to the manufacturer's data (0.0525Nm/A) and previous result (Table 1).
As a second step the DC motor resistance and inductance parameters are deduced from Eq.1, locking the rotor to force the speed to zero :

$$v_e = \begin{bmatrix} I & \dfrac{dI}{dt} \end{bmatrix} \begin{bmatrix} R \\ L \end{bmatrix} \tag{15}$$

Eq.15 is sampled on an exciting current trajectory (Fig.3) to obtain a well conditioned overdetermined system as Eq. 5.

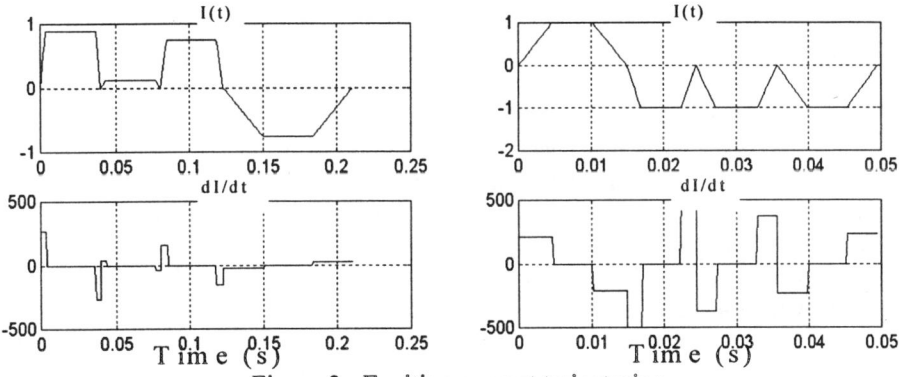

Figure 3 : Exciting current trajectories

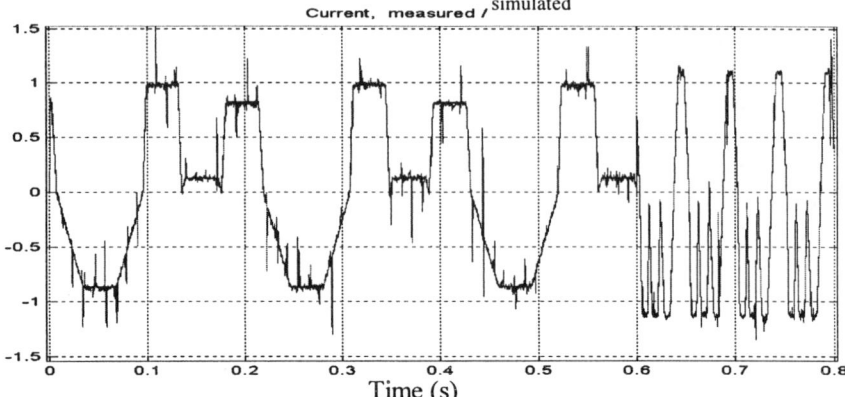

Figure 4 : validation : measured and simulated currents.

The optimization procedure has given Cond (Φ)=31, and identified values are given in Table 1. They are close to those obtained with the global identification.
In order to validate the estimated parameters, the opened loop system is simulated with the identified parameters and the measured v_e input voltage. The simulated current matches the actual current closely as shown on Fig.4.

4.3 Identification of mechanical parameters

The mechanical equation Eq.2 can be rewritten as :

$$\Gamma_m(t) = \begin{bmatrix} \ddot{q}(t) & \dot{q}(t) & \text{sign}(\dot{q}) \end{bmatrix} \begin{bmatrix} J \\ F_v \\ F_s \end{bmatrix} \tag{16}$$

This equation is sampled on an exciting velocity trajectory to obtain a well conditioned overdetermined system as Eq.5.
The motor torque can be measured through the current measurement and the identified value of k_t ($\Gamma_m = k_t$ I), but this needs a current sensor with analog anti-aliasing filter. Another way is to use the digital current reference input V_t calculated by the numerical controller. The current controlled loop can be considered as a static gain G_I in the dynamic range of the mechanics : $I = G_I V_t$.
This has been checked with a Fourier analyzer which has measured a 600 Hz cutt-off frequency with a -45° phase shift and $G_I = 0.27$A/V.
A classical position and velocity PD or PID feedback control law with a sample rate of 1KHz allows to track exciting velocity trajectory obtained with the same bang-bang interpolator and optimization procedure as those used for the current by changing I_{max} to $\dot{q}_{max}$ and $\left(\dfrac{dI}{dt}\right)_{max}$ to $\ddot{q}_{max}$.

The optimization procedure has given Cond (Φ)=8, with Cond(W)=2170, and estimated values are given in table 1.

Table 1 : Identified parameters

Parameters	All parameters Cond(Φ) =1731			Electrical Cond (Φ)=31			Mechanical Cond (Φ)=20		
	$\hat{X}$	$2\,\sigma_{\hat{X}}$	$\%\sigma_{\hat{X}r}$	$\hat{X}$	$2\,\sigma_{\hat{X}}$	$\%\sigma_{\hat{X}r}$	$\hat{X}$	$2\,\sigma_{\hat{X}}$	$\%\sigma_{\hat{X}r}$
L (mH)	0.6	0.1	10	0.45	0.03	3			
R (Ω)	1.78	0.01	0.35	1.725	0.005	0.14			
kt (Nm/A)	0.0515	2e-4	0.2	0.05106	3e-5	0.06			
J (kg m^2)	1.0e-5	3e-6	12				0.94e-5	1e-7	0.4
Fv (Nm/rds^{-1})	5 e-5	2e-4	226				2.2e-5	2e-6	3
Fs (Nm)	6e-3	2e-2	134				6.6e-3	2e-4	1.2

The validation of the estimated values is carried out using the following methods :

— *Identification of a known load*
The parameters are reidentified while the carriage is loaded by a known mass M. In this case the value of J is changed to $J_m=J+M*r_s^2$. We have to check that the calculated value of J_m is the same as that of the identified moment of inertia value with the load. The comparison is given in table 2 with a mass M=4.625Kg and the screw gear ratio $r_s^2=(0.0025/(2*\pi))^2=0.01582e-5(m/rd)^2$, $M*r_s^2=0.073e-5Kgm^2$.

Table 2 : Identified Mechanical parameters with load

Parameters	Unloaded carriage			Loaded, M=4.625Kg			
	$\hat{X}$	$2\,\sigma_{\hat{X}}$	$\%\sigma_{\hat{X}r}$	Calculated	$\hat{X}$	$2\,\sigma_{\hat{X}}$	$\%\sigma_{\hat{X}r}$
J (kg m^2)	0.94e-5	1e-7	0.4	1.02e-5	1.02e-5	1e-7	0.4
Fv (Nm/rds^{-1})	2.2e-5	2e-6	3		2.3e-5	2e-6	3
Fs (Nm)	6.6e-3	2e-4	1.2		6.3e-3	2e-4	1.3

— *validation from closed loop simulation*
The closed loop system is simulated using the direct dynamic model (state space equation) with the identified parameters.
Fig. 5 and Fig.6 show that the current control and position tracking error simulated signals match the experimental signals closely, assuming the same control law and the same tracking trajectory.

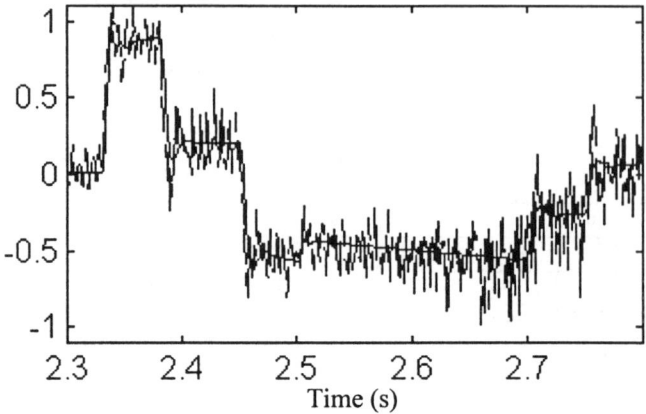

Figure 5 : Current control, —simulated, -- experimental

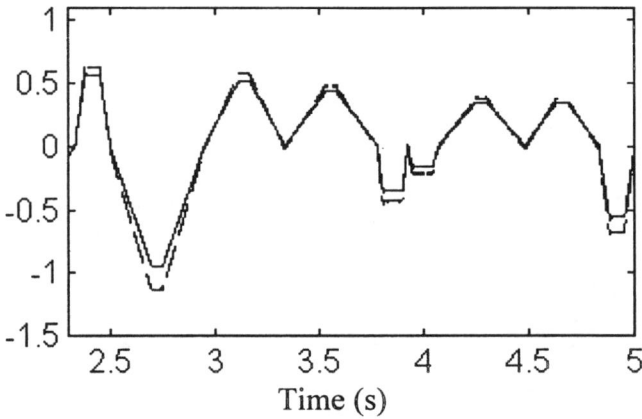

Time (s)

Figure 6 : Tracking error position, —simulated, -- experimental

5. Conclusion

The closed loop identification using two decoupled inverse models allows to take into account state non linearity and to define and track trajectories giving two well conditioned overdetermined linear systems with low variance L.S. solution. More over, the parameters keep their physical meaning which make easier their validation.
Identification of electrical drives with brushless and induction motors is in progress.

6. References

[1] Armstrong, B.: On finding exciting trajectories for identification experiments involving systems with non linear dynamics. *Int. J. of Robotics Research*, Vol. 8, N° 6, 1989, pp. 28-48.

[2] Bergmann, C., Henry, S. : Control and identification of a DC motor in the state space plane, *IMACS*, Atlanta, USA, 1994.

[3] dSPACE[TM] : EMPS300, RTI, DS1102 guides, *dSPACE GmbH*, D 33100, Paderborn, Germany, 1995.

[4] Gautier, M.: Numerical calculation of the base inertial parameters, *Journal of Robotics Systems*, Vol. 8, N° 4, 1991, pp. 485-506.

[5] Gautier M., Khalil W.: Direct calculation of minimum inertial parameters of serial robots, *IEEE Trans. on Robotics and Automation*, Vol. 6, N° 3, 1990, pp. 368-373.

[6] Gautier M., Khalil W.: Exciting Trajectories for the Identification of Base Inertial Parameters of Robots, *The International Journal of Robotic Research*, Vol. 11, N°4, 1992, pp. 362-375.

[7] Gautier M., Khalil W., Restrepo P.P.: Identification of the dynamic parameters of a closed loop robot, *Proc. IEEE Conf. on Robotics and Automation*, 1995, pp. 3045-3050.

[8] Gill, P.E., Murray, W.: User's Guide for NPSOL, Ver. 4.0, Report Sol 86-2. *Stanford University, Software Distribution Center*, 857 Serra Street, Stanford, CA 94305-6225, 1986.

[9] Lawson C.L., Hanson R.J.: Solving Least Squares Problems. *NJ: Englewood Cliffs, Prentice-Hall*, 1974.

[10] Ljung, L. : System Identification Toolbox, User's guide, *The MathWorks*, Inc. Natick, Mass, USA, 1991.

[11] Ljung, L. : System Identification, Theory for the user, *Prentice Hall*, Englewood Cliffs, N. J., 1987

[12] MathWorks, Matlab, Simulink reference guides and Signal, Optimization, RTW toolboxes User's Guides, *The MathWorks*, Inc. Natick, Mass, USA, 1995.

[13] Presse, C., Gautier, M.: New criteria of exciting trajectories for robot identification. *Proc. IEEE ICRA*, 1993, Vol. 3, pp. 907-912.

Basics for the Energy-Efficient Control of Hydraulic Drives by Switching Techniques

R. Scheidl, D. Schindler, G. Riha, and W. Leitner
Department for Technical Mechanics and Foundations of Machine Design
Johannes Kepler University Linz, Austria.

Abstract: Hydraulic drives currently are under strong pressure by the upcoming electric servo-drives. A main reason for this is the poor energy efficiency of many hydraulic drive systems. Switching techniques which are state of the art for electric drives are now also considered for application in hydraulics. This paper reports about a new principle of switching control of hydrostatic drives which is based on periodic wave propagation in a so called resonator. The need for such a resonator is demonstrated by some simple mechanical arguments first. Then a mathematical model in form of a damped wave equation is used to assess its basic performance characteristics and to derive criteria for optimum design. This system turns out to be a pressure converter which controls the output pressure by the pulse-width of the periodic switching between high and low pressure line. Further features to improve the system are described and discussed.

1 Introduction

Any mechatronical device is able to carry out controlled mechanical motion. Among all the engineering planning work to fix the design of such a device the proper selection of the drive(s) is often a major task. For this, like for many other components the mechatronical approach to design should help to find the best solution by utilizing a broad knowledge on available solution principles and by a profound analysis of and judgement on functionality and all the other relevant performance criteria.

Hydrostatic drives (furhter simply called hydraulic drives) have an outstanding position mainly for two reasons: First, they can be controlled very easily and second, they provide high specific power and high specific force both with respect to weight and volume. This favours their application in many mechatronical applications, if a drive is located at a part of a machine which performs fast motions or where very complicated design-geometrical situations do arise. Despite this, hydraulic drives are currently loosing ground mainly due to the upcoming of electrical servo-drives. Classical application fields like forging-press stands [1] or tension-test stands where over decades hydraulic drives have been the first choice are now tending to electro-mechanical drives for two reasons: Their high energy-efficiency and the precision of the motion or positioning in case of sudden load changes, which is accomplished by the high stiffness of mechanical gears (mostly spindles). All the modern electric drive systems make use of switching converters. The basic working-principle common to all these drives is that the drive is connected to the electric supply lines for certain time-periods but fully disconnected the other time. Electric engineers actually use the word "electric valves" as a general designation for these today mostly electronical components which realize this switching. Only small energy losses occur in the converter, as in the switch-on phase the resistance of the valve is extremely low which brings about only small voltage drops and in the switch-off phase no current passes over the valve. Magnetic inductivity in the coils of the drives provides on the one hand storage of the surplus of energy which enters the system during the switch on time and on the other hand filtering of the pulsating support of electrical power due to switching. The former methods of electric power control namely, resistance control and machine-converters have lost their importance at least for higher power ranges.

Hydraulics still holds where electric drives have been two decades ago: Resistance control and variable displacement control (which corresponds to machine conversion for electric drives) are the state of the art. The first method renders excellent dynamical performance possible if fast servo-valves are used but suffers from bad efficiency due to pressure losses at the valve. The second one has a good efficiency and if servo-controlled displacement of the hydraulic-pumps or -motors is used also remarkable high operating frequencies can be achieved, but these devices are expensive, rather voluminous due to a number of accessory elements, and mechanically much more complicated than resistance controlled drives.

Switching control sometimes is considered as a means to replace proportional-valves by switching valves (see for instance [2]). In [3] however, switching control is investigated as a means for energy saving. The method proposed therein simply applies switching valves to a standard hydraulic cylinder with some additional free-running valves. Energy surplus storage is devoted to the load mass inertia. The main shortcoming of this technique is the limited operating frequency, which comes up by the negative effect of the compressibility of the hydraulic fluid (hydraulic capacity), as capacity renders low impedance for high frequencies. The valves and the hydraulic fluid form a R-C element for high frequencies and load inertia then does not come into play and cannot act as an energy storage element.

In this paper we present a new method for an energy efficient switching-type control of hydraulic power. In chapter 2 simple mechanical models will elucidate the basic features for such a type of control. In chapter 3 the key element of our method - a so called resonator - will be discussed. A mechanical model for the fluid flow in this resonator will be established and used for the assessment of the performance of such a system. Comparison with experiments will be shown in chapter 4. Further improvements of the basic concept and some design and control issues will be presented in chapter 5. In the last chapter we will sum up and give an outlook on the further development steps that have to be taken.

2 Fundamental mechanical relations for hydraulic switching control

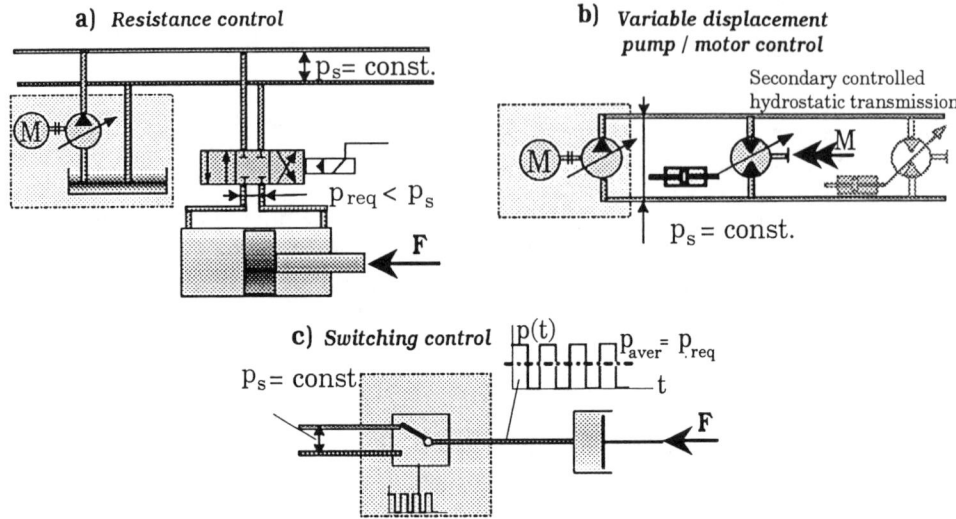

Figure 1: Basic methods for hydraulic fluid power control.

In fig. 1 three basic methods to control a hydraulic drive connected to a constant pressure supply system are shown. Version a) is the classical resistance control and adopts the constant supply pressure p_s to the instantaneously required value p_{req} by inducing pressure losses at the metering edges of the control valve. Variable displacement control in form of the secondary controlled hydrostatic transmission is shown in part b) of fig. 1. An ideal version of switching control according to fig. 1 c) produces the pressure p_{req} only as the time-average value of a rectangularly shaped pressure $p(t)$ at the output. Obviously, energy saving compared to resistance control can only be achieved, if in the switch-off phase hydraulic fluid is entering into the system from the tank line (with low pressure p_0). This elementary form of switching control is realized in the Gall and Senn method [3], the hydraulic circuit of which is shown in fig. 2. The kinetic energy of the load mass stores the energy due to the pressure surplus $p_s - p_{req}$ which enters the system in the switch-on period. The check-valves parallel to the switching valves provide

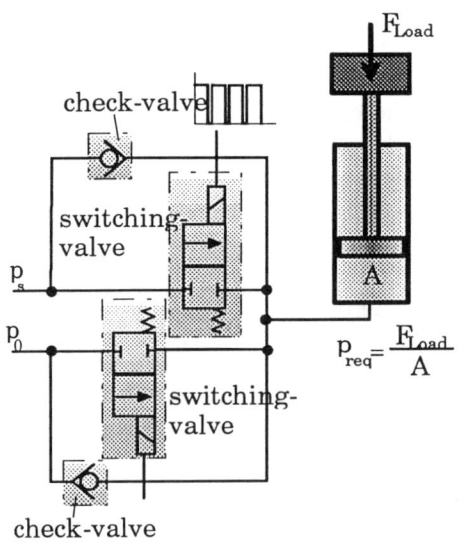

Figure 2: Switching control according to [3].

the aforementioned fluid flow from the tank line into the actuator in the switch-off phase when the drive actuates against the load and from the actuator to the pressure line when lowering the load. In this last case energy is hydraulically recovered. There is an optimal

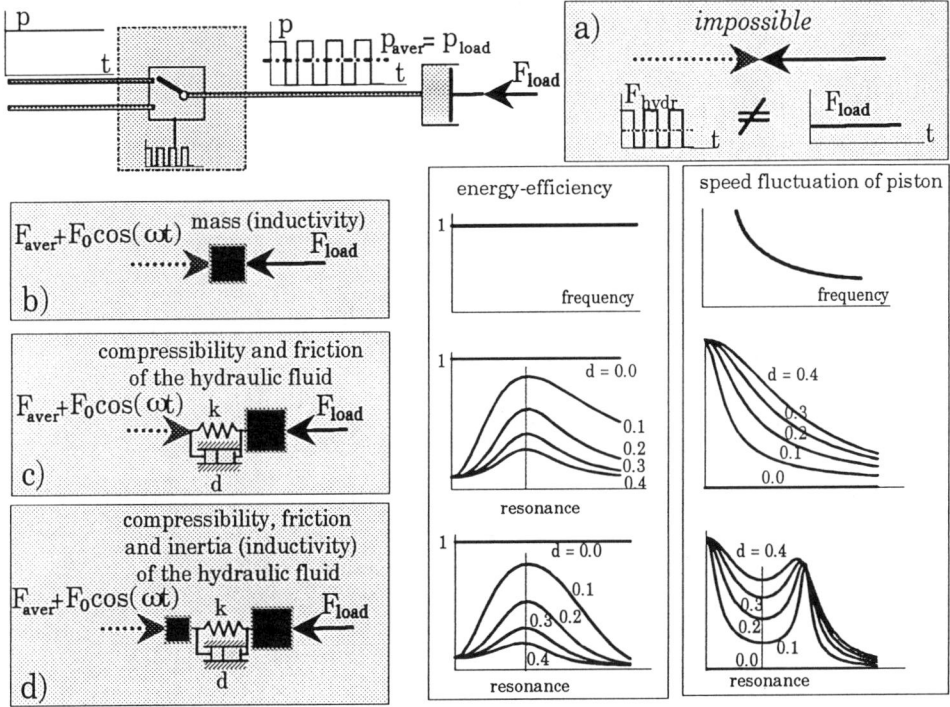

Figure 3: Basic mechanical principles for energy-efficient switching control in hydraulics.

switching frequency with respect to efficiency. In the experiments of Gall and Senn this was about 25 Hz.

Fig. 3 presents a sequence of simple models to make the basic mechanical relations clear. Switching between two pressure levels generates a rectangular hydraulic force. In these simple models this is replaced by the sum of its constant mean value and a sinusoidal force (e.g. its first order harmonic). To compensate for the actual force difference F_{hydr} - F_{load} some additional force is required. Ideally this would be an inertia force F_{inert} generated e.g. by the load mass. Losses are zero, hence efficiency is 100% and we can limit the speed fluctuations of the load by a high switching-frequency (see fig. 3 b)). Unfortunately, fluid compressibility (capacity) makes this model invalid for high frequencies. In fig. 3 c) compressibility is accounted for by a spring. As we cannot avoid some dissipation at the valve and in the hydraulic fluid some dampers are added. Performance of this system now shows an optimal switching frequency which is the resonance frequency of the spring-mass system. Adding an additional mass at the input makes also the fluctuation to show up a local minimum at this resonance frequency as can be seen in fig. 3 d).

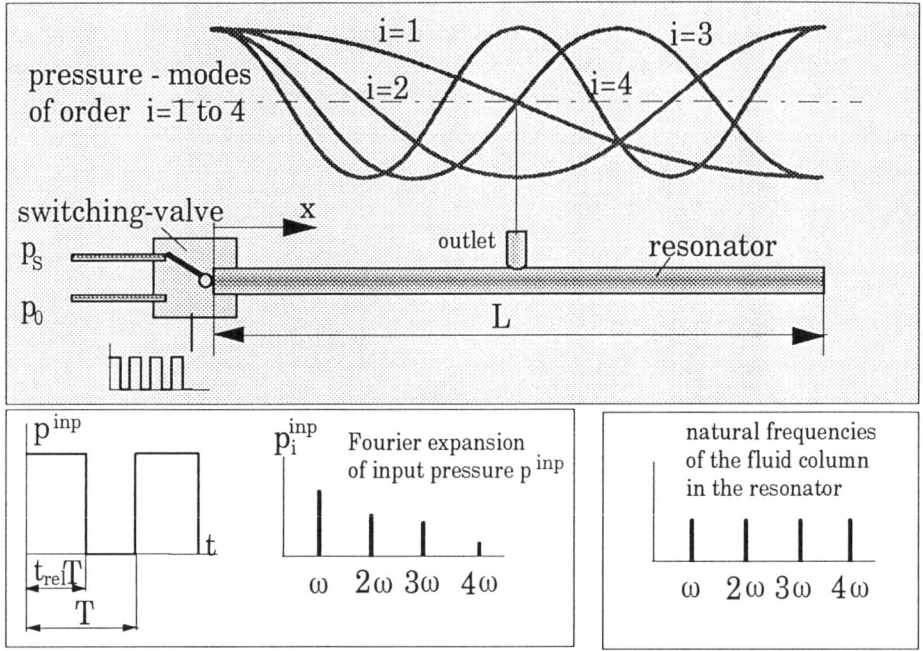

Figure 4: The resonator as an infinite series of oscillators.

So far only harmonic excitation was considered. With a discrete mass-spring system we cannot fulfill optimal resonance conditions in case of rectangularly shaped pressure pulses generated by the periodic switching of the valve. An infinite countable number of oscillators each one optimally adopted to one frequency of the discrete Fourier spectrum would be required. This is given by a homogeneous column of a compressible fluid of

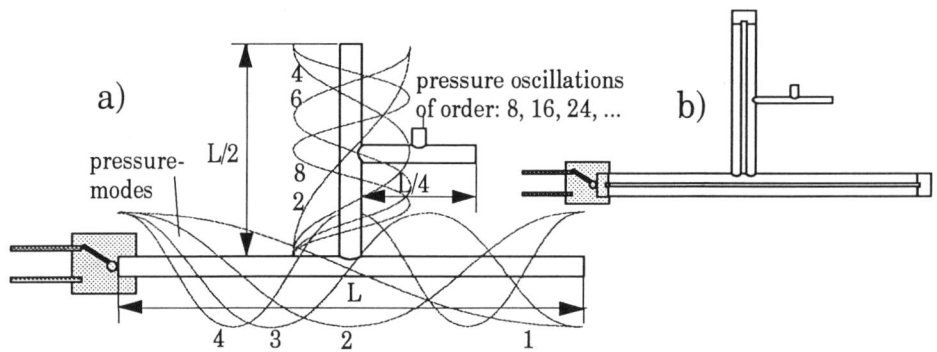

Figure 5: Cascading of n resonators each with half length of its preceeding one filters pressure pulsations up to order 2^n. Only twin-pipe arrangement acc. b) yields the desired result in a stable way.

length $L = c/2f$, where $f = \omega/2\pi$ is the switching frequency and c the speed of sound. Each mode of the longitudinal oscillations represents one discrete oscillator, their natural frequencies form an arithmetic series which means that, if the basic frequency is identical to the switching fequency, the set of the natural frequencies corresponds to the Fourier spectrum of the input pressure. We call that fluid column in the pipe resonator. Clearly, that resonator must have an outlet to apply hydraulic power to the actuator. The first concept we had in mind considered the end of the resonator for that. In a thesis [4] this concept is investigated. In a later proposal which is also the main issue of an application for a patent [5] of the authors the midpoint of the resonator is taken as the outlet, as all the odd order pressure modes have a node there (see fig. 4), thus the pressure fluctuation is drastically reduced there. Utilizing this finding by adding further resonators of half the length of the preceeding ones (see fig. 5) one can filter out pressure oscillations of any order. For stability reasons a system according to fig. 5a) does not work. We will comment on this in the next chapter.

3 A mathematical model for the resonator

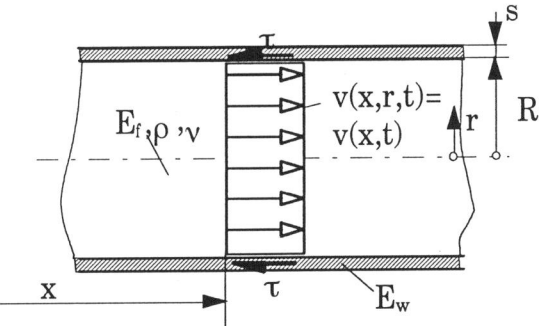

Figure 6: One dimensional flow in a pipe with a plug-type velocity profile.

We work with a plug-type flow model, hence the fluid velocity is constant over the cross-section of the resonator-pipe. Fluid friction is considered by a wall shear stress τ (see fig. 6). As the fluid speed due to wave propagation is of order $O(p_s/E \approx 10^{-2})$ of the wave speed c, and density changes are relatively small, linearized versions of the momentum and the continuity equation are sufficient. These assumptions lead to the following damped one-dimensional wave equation

$$u_{,tt} - c_e^2 u_{,xx} + 2\varepsilon\, u_{,t} = 0 \quad ; \quad (\ )_{,t} = \frac{\partial}{\partial t}(\ ); \quad (\ )_{,x} = \frac{\partial}{\partial x}(\ ) \tag{1}$$

From this displacement potential $u(x,t)$ velocity $\upsilon(x,t)$ and pressure $p(x,t)$ can be derived by the fromulas

$$\upsilon = u_{,t} \quad ; \quad p = -E_e u_{,x}. \tag{2}$$

c_e in (1) is the wave speed considering also the pipe expansion which is related to the pure fluid wave speed c_f

$$c_e = c_f \sqrt{\frac{1}{1 + \frac{E_f}{E_w}\frac{2R}{s}}} \quad ; \quad c_f = \sqrt{\frac{E_f}{\rho}} \quad ; \quad \frac{1}{E_e} = \frac{1}{E_f} + \frac{2R}{E_w s}. \tag{3}$$

E_f and E_w are the fluid modulus of compressibilty and Youngs modulus of the pipe wall respectively. E_e is the effective modulus of elasticity. ρ is the fluid density at zero pressure. R and s are the inner pipe diameter and wall thickness respectively (see also fig. 6). A highly nontrivial problem is friction in the case of nonstationary flow. Analytic solutions of the axisymmetric linear wave propagation in case of laminar behaviour can be found based on series expansions (see for instance [4] and [6]). In many experiments conducted in our hydraulic laboratory strong indication was found that periodic laminar-turbulence transition occurs for realistic hydraulical data. In [7] a method based on some internal state variable for a somewhat improved modelling of such processes is reported, but this is still not fully satisfactory. As we want to proceed here with analytical methods we are limited to rather simple linear friction models. In an actual axisymmetric flow (not bound by our plug-type cross sectional velocity profile) the wall shear stress is proportional to the velocity gradient at the wall. If the qualitative shape of the velocity distribution versus the cross section is assumed to be constant we end up with a relation of type

$$\varepsilon = \frac{k_{frict}\nu}{R^2} \tag{4}$$

for the friction coefficient ε in (1). ν is the kinematic viscosity of the hydraulic fluid and k_{frict} accounts for the wall gradient of the velocity profile. For the parabolic case (Hagen-Poiseuille flow) $k_{frict}=4$.

The general solution of (1) in case of periodic excitation at one boundary in the steady state case is given by the sum of a nonperiodic and a periodic part (see [4])

$$u(x,t) = u_0(x,t) + \sum_i u_i(x,t). \tag{5}$$

$$u_0(x,t) = f_{00}(x) + f_{01}(x)t + f_{02}(x)t^2 \tag{6}$$

$$f_{00}(x) = \left(\varepsilon^2 C_2 x^5 + 5\varepsilon^2 C_1 x^4 + 5c_f^2 C_2 x^3 + 5\varepsilon c_f^2 C_4 x^3 + 15c_f^2 C_1 x^2 + \right.$$
$$\left. 15\varepsilon c_f^2 C_3 x^2 + 15c_f^4 C_6 x + 15c_f^4 C_5\right) / \left(15c_f^4\right)$$

$$f_{01}(x) = \left(2\varepsilon C_2 x^3 + 6\varepsilon C_1 x^2 + 3c_f^2 C_4 x + 3c_f^2 C_3\right) / \left(3c_f^2\right) \tag{7}$$

$$f_{02}(x) = \left(C_1 + C_2 x\right)$$

$$u_i(x,t) = \exp\left(i\frac{\omega}{c_e}\alpha_2 x\right)g_i(\eta) + \exp\left(-i\frac{\omega}{c_e}\alpha_2 x\right)f_i(\xi) \tag{8}$$

$$g_i(\eta) = ff_i \cos\left(i\frac{\omega}{c_e}\alpha_1\eta\right) + hh_i \sin\left(i\frac{\omega}{c_e}\alpha_1\eta\right) \tag{9}$$

$$f_i(\xi) = gg_i \cos\left(i\frac{\omega}{c_e}\alpha_1\xi\right) + jj_i \sin\left(i\frac{\omega}{c_e}\alpha_1\xi\right) \tag{10}$$

$$\alpha_1 = \sqrt{\frac{1 + \sqrt{1 + \left(\dfrac{2\varepsilon}{\omega}\right)^2}}{2}} \quad ; \quad \alpha_2 = \sqrt{\frac{-1 + \sqrt{1 + \left(\dfrac{2\varepsilon}{\omega}\right)^2}}{2}} \tag{11}$$

$\xi = x - c_e t$ and $\eta = x + c_e t$ are the characteristic coordinates.

This solution describes the flow in one pipe and has to be completed by boundary conditions. If we have a network of pipes, (5) must be set up for each branch and Kirchhoff-type flow conditions and pressure continuity conditions must be formulated at each interconnecting node to get the necessary number of equations for the coefficients $(C_1^b, C_2^b, C_2^b, C_4^b, C_5^b, C_6^b, ff_i^b, gg_i^b, hh_i^b, jj_i^b)$ for each branch b.

It is easy to check that for one pipe of length L a sequence of standing waves

$$u_i(x,t) = -\frac{L\,p_i^{inp}}{2\,i\pi\,E_e}\left\{\sin\left(i\pi\left(\frac{x}{L} + \frac{t}{c_e L}\right)\right) + \sin\left(i\pi\left(\frac{x}{L} - \frac{t}{c_e L}\right)\right)\right\} \tag{12}$$

$$p_i^{inp} = \frac{2p_s}{i\pi}\sin(i\,t_{rel}\,2\pi) \tag{13}$$

is generated as depicted in fig. 4, if a rectangular pressure variation is enforced at its input ($x = 0$), the other end is closed, friction is zero ($\varepsilon = 0$), and the resonace condition

$$2\pi f = \omega = \omega_{res} = \frac{\pi\,c_e}{L} \tag{14}$$

is preserved. The solution part u_0 corresponds to a constant average pressure

$$p_0(x,t) = p_{aver} = t_{rel}\,p_s \ . \tag{15}$$

The periodic solution part does not generate any fluid flow over the valve, only the steady state part u_0. This has the important consequence that pressure losses at the valve only occur due to the average flow rate. If we now add friction ($\varepsilon \neq 0$) energy losses arise due to pressure drop in the resonator and some fluid flow over the valve related also to the periodic solution part. Taking only the pressure volume work for defining the efficiency factor

$$\eta_{energy} = \frac{A_{out}\int_0^{2\pi/\omega} P_{out}\upsilon_{out}\,dt}{A_{in}\int_0^{2\pi/\omega} P_{in}\upsilon_{in}\,dt} \tag{16}$$

(kinetic energy portions are relatively small in hydrostatics) we find for a system according fig. 4 which for the sake of simplicity has as the "hydraulic load" only an ideal flow control valve the following formula

$$\eta_{energy} = \frac{p_s t_{rel} Q_{aver} \eta_{press}}{p_s Q_{aver} t_{rel} + p_s Q_{loss}} = \eta_{press} \frac{1}{1 + \dfrac{\upsilon_{Jouk} A}{2 Q_{aver}} q_{in}(\omega_{rel}, t_{rel}, \dfrac{\varepsilon L}{c_e})} = \eta_{press} \eta_{volum} \qquad (17)$$

Q_{aver} is the average flow-rate at any point in the system which is identical to the value adjusted by the flow control valve. η_{press} is the pressure loss from the inlet to the outlet of the resonator due to the average flow rate which for laminar flow is

$$\eta_{press} = 1 - \frac{\varepsilon L \rho Q_{aver}}{p_s t_{rel} A} \quad . \qquad (18)$$

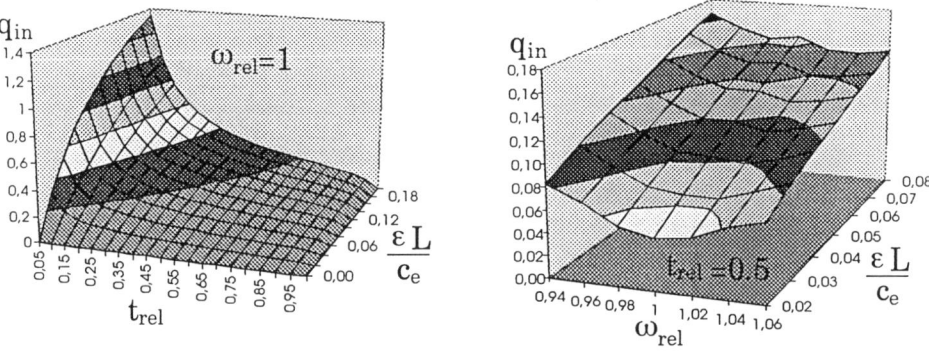

Figure 7: q_{in} versus t_{rel}, ω_{rel}, and $\dfrac{\varepsilon L}{c_e}$.

The remaining factor can be addressed as a volumetric efficiency factor η_{volum} (a common characteristic value in hydrostatic machines). It depends on the average flow rate Q_{aver} and the flow rate produced by the Joukowsky-speed $\upsilon_{Jouk} = c_e p_s / E_e$, which is the fluid particle speed due to a pressure wave, and a dimensionless factor q_{in} which is depicted in figure 7. ω_{rel} is the relative switching frequency scaled by the resonance value ω_{res} as given in (14). Figure 7 clearly indicates that optimal results occur for resonance and that the value of the relative friction coefficient $\varepsilon L / c_e$ has to be small in order to achieve high efficiency, as for other technical reasons Q_{aver} can hardly be extended beyond the Joukowsky flow rate. Reduction of the pipe length improves the efficiency by lowering the relative friction $\varepsilon L / c_e$. The pipe diameter can only be optimized for a given flow rate Q_{aver}, as it determines the cross sectional area A and influences the friction coefficient ε.

Formula (17) gives an immediate insight into the efficiency performance and how it can be tested by experiments. The second term in the denominator Q_{loss} is the relative flow rate of the periodic solution parts which enters the system at the inlet in the switch-on phase $(0 < t < t_{rel}T)$ and clearly has to leave the system at the switch-off phase by passing through the outlet fitting to the tank-line. If $Q_{aver} = 0$ (experimentally by closing the outlet) the absolute value is simply the fluid consumption by the resonator system.

An arrangement of resonator pipes according to fig. 5a) has a solution as described in the previous chapter, namely pressure waves of order $n2^{k-1}$ (n=1,2,3, ...) in the k-th pipe, only in the case $\omega = \omega_{rel}$ and $\varepsilon = 0$. This solution is a structurally unstable one and breaks down for any deviation from this aforementioned unrealistic system state. A modified arrangement as shown in fig. 5b) with a twin-pipe in parallel to each subresonator yields the desired solution in a stable way.

4. Experiments

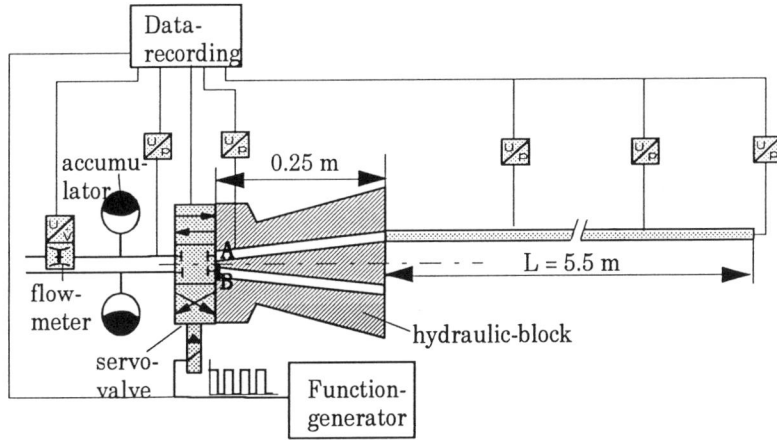

Figure 8: Experimental setup to check validity of the mathematical model.

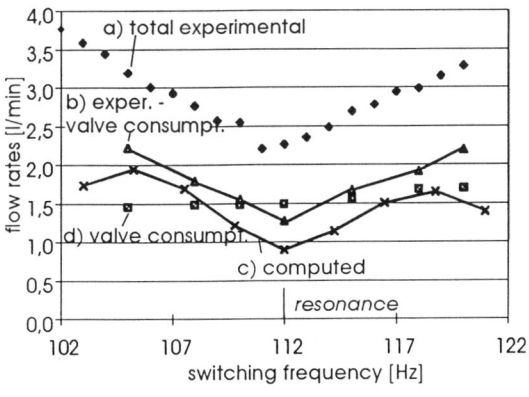

Figure 9: Results of experiments according to fig. 8.

Fig. 8 shows an experimental setup for testing the mathematical model of the previous chapter. A pipeline of approximately 6 m length with one end closed was connected to a fast servo-valve, which was operated as a switching valve. In order to avoid deviations from the ideal resonator shape (a straight uniformal pipe starting immediately behind the metering edges of the servo-valve) as much as possible a special hydraulic block was designed for fitting the pipeline to the valve. Pressure-gauges and flow metering was realized according to fig. 8. The agreement between measured and computed pressures generally was very good (see [4]). Some discrepancies occured for the flow rates as shown in fig. 9. The main reasons were found to be the consumption of the valve, partly caused by the capacity of the flow channels to port B also. Closing the outlet A of the valve (see fig. 8) causes an average flow rate as indicated by curve d) in fig. 9. Subtracting those parts of these values which are devoted to the pilot stage of the valve and capacity of port B from the originally measured values (curve a)) gives much better results.

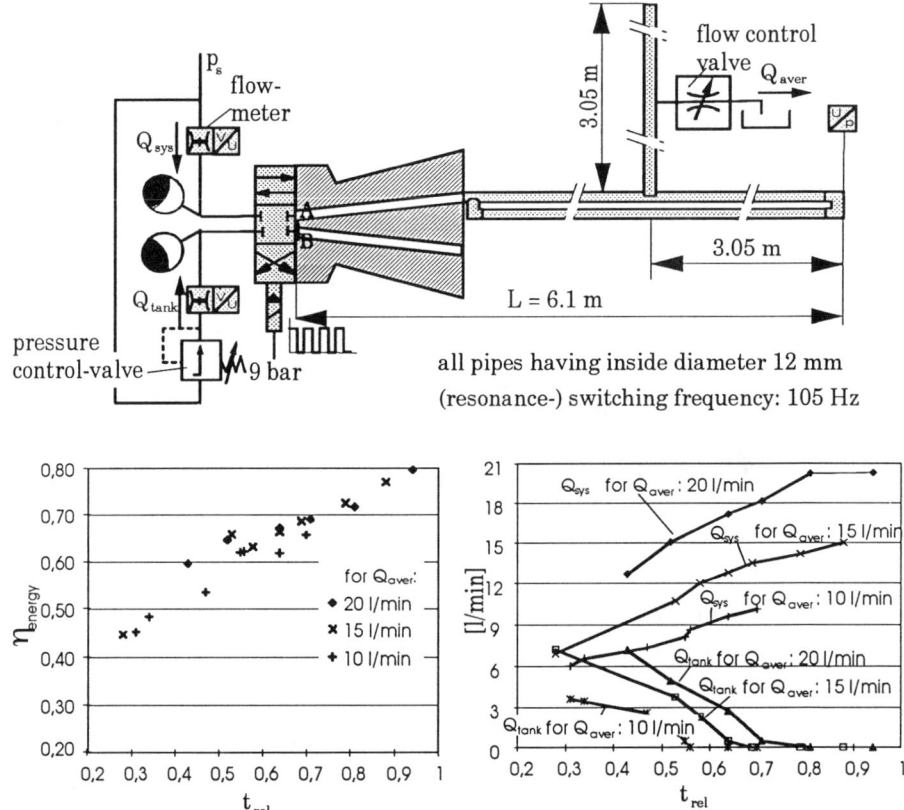

Figure 10: Experimental setup and results for a system with one subresonator. The diagrams show efficiency η_{energy} and flow rates versus pulse-width t_{rel}.

Results of experiments for an arrangement with one sub-resonator are shown in fig. 10. The tank line pressure was lifted to 9 bar in order to reduce cavitation. The results confirm that the main loss adheres to the fluid consumption of the system itself, which is a value rather independent of the net flow rate Q_{aver}.

5 Features for improving the system perfomance

The resonance condition (14) links the resonator length to the switching frequency. To get a practically reasonable size switching frequencies of 1 kHz are required (reducing L to 0.65 m). Reduced length should also contribute to an increased efficiency as outlined in the previous chapter. Realization of such high frequencies by conventional actuation of a translatory sliding valve does not seem to be promising. However, a rotary valve as depicted in fig. 11 can easily generate high frequencies without requiring any acceleration forces. The pulse-width (t_{rel}) is adjusted by the relative angle of the slitted control-jacket. The resonator starts immediately behind the metering edges and the arrangement of the ports in the jackets relative to the inlet- and outlet-chamber respectively provide minimum capacity induced leakage flow.

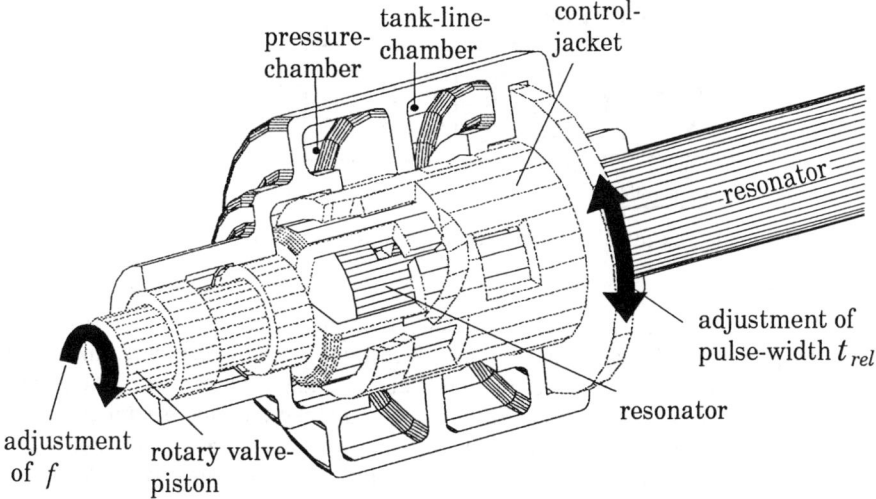

Figure 11: Rotary valve to realize high-frequent pulse-width control.

The angular speed of the rotary valve piston ω must be controlled accurately for two reasons: First, to obtain optimal efficiency and second, to get rid of pressure fluctuations at the outlet. Pressure nodes at the resonator pipes' midpoints only occur if this resonance condition is preserved. This fact can be used to form a control loop as qualitatively depicted in fig. 12. As all the odd order modes do not show up any amplitude at the main resonator's midpoint a convolution of the pressure signal there with a rectangular, symmetric trigger signal derived e.g. from the driving motor of the valve yields zero only in the resonance case and can thus be used as a control signal for a PI-control of the motor speed.

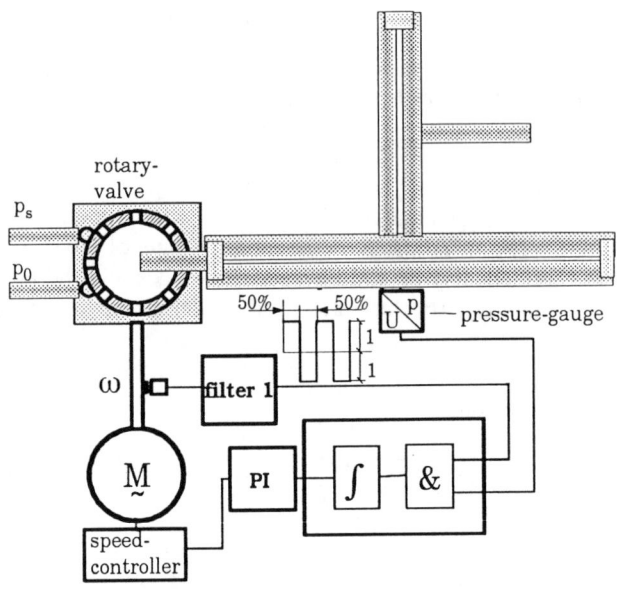

Figure 12: Controller for preserving the resonance condition.

Reduction of losses besides adjusting optimally switching frequency and resonator diameter is possible by minimizing fluid friction. A resonator pipe which performs nearly the same longitudinal motion like the fluid avoids any boundary layers and consequently friction shear stresses. In principle there are two ways to make the pipe behaving in such a manner. The first is to adopt its longitudinal wave speed to that of the fluid. The second requires a material with an extreme Poisson ratio for the longitudinal dilatation induced by circumferential expansion. If this effect is adjusted in such a way that the volume contraction of the fluid related to the local pressure is identical to the volume reduction of the corresponding piece of pipe no relative motion between the fluid and the resonator pipe for the oscillatory part of the flow occurs. A severe problem arises at the connections of consecutive resonator pipes as this connection must be flexible on the one hand and leakproof on the other hand.

6 Summary and outlook

The presented method of fluid power control by means of a pulse width controlled actuation of standing waves in a resonator or resonator-system constitutes a hydraulic pressure converter. The output pressure equals nearly the input pressure times the relative pulse-width. This system aims at realizing a simple, robust, fast, and energy-efficient control. Losses occur mainly due to the internal power consumption to maintain the standing waves against fluid friction. This consumption leads to a flow from the pressure line to the tank line which also shows up in the power of idle motion. Switching frequencies in the order of 1 kHz are necessary to reduce the dimensions of the resonator to a technically reasonable value. A rotary valve is shown which is able to generate such high frequencies and a simple way to control the pulse-width. Strong reduction of the friction effect can be achieved by resonator pipes performing the same longitudinal motion like the fluid. Arranging several resonator-pipes in a proper way provides filtering of pressure oscillations at the outlet of the system.

Currently the authors are looking for industrial partners to built a prototype of such a system, in particular the rotary-type switching valve, and to carry out the further development steps.

7 References

[1] M. Mitze / W. Pasch: Hydraulik ade: AC-Servotechnik im Pressenbau. antriebstechnik, Heft 4, 1995, pp 66-76.

[2] B. Lühmann: Digital gesteuerte Hydraulikventile und ihre Anwendung, Doct. thesis, TU-Braunschweig, 1983.

[3] H. Gall / K. Senn: Freilaufventile - Ansteuerungskonzept zur Energieeinsparung bei hydraulischen Linearantriebem, Ölhydraulik und Pneumatik, **38**(1994) Nr. 1-2

[4] A.S. Abo El -Lail: Investigation of a Switching Method for the Energy-Efficient Control of Hydraulic Drive Systems, Doct. thesis, Univ. of Linz, 1995.

[5] R. Scheidl: Vorrichtung zum Steuern eines hydrostatischen Antriebes, Austrian patent application, February, 1995.

[6] H. Theissen: Die Berücksichtigung instationärer Rohrströmungen bei der Simulation hydraulischer Anlagen, Doct. thesis, RWTH-Aachen, 1983.

[7] R. Scheidl / D. Schindler: Modelling of strongly pulsating fluid flow in pipelines of hydraulic drives, Proc. 13th IASTED Conf. on Modelling, Identification and Control, Grindelwald, 1994.

Prof. Dr. Rudolf Scheidl, Dipl.Ing. D. Schindler, Dipl.Ing. G. Riha, and Dipl.Ing. W. Leitner
Department for Foundations of Machine Design, University of Linz
Altenbergerstraße 69, A-4040 LINZ, Austria

Tel.:   (++43) 732 2468 9745

FAX: (++43) 732 651086

Email: scheidl@tmech1.mechatronik.uni-linz.ac.at

Motion control and identification techniques for machine tool axes

P. Van den Braembussche, J. Swevers[*], H. Van Brussel, P. Vanherck
Mechanical Engineering Department, Division PMA, K.U.Leuven,
Heverlee, Belgium

Abstract: The traditional indirect drive design for machine tool axes, a rotary motor and a ball-screw, is well known, but has some important drawbacks (like limited speed, acceleration, accuracy and lifetime). The use of linear motors can overcome these problems, but it also has its inherent drawbacks. In this direct drive design, all stiffness must result from the servo controller. This makes high demands upon the controller design. This paper presents some control techniques to achieve a good tracking behaviour for both types of axis design. The linear motor ripple results in significant tracking errors for the indirect drive design. A compensation technique is proposed based on an experimentally identified ripple model. The proposed technique is tested on three experimental set-ups: a rotary motor axis with ball-screw, a linear motor axis supported by air bearings and a linear motor axis supported by rolling element slideways.

1. Introduction

The evolution in production techniques (e.g. high-speed milling) and the required shortening of the production cycles, impose more stringent requirements on the velocity and acceleration of the machine tool axes. Traditional axis design for high speed and accuracy machine tools consists of a rotary motor with a ball-screw transmission to the slide. This is called an indirect drive design and has limitations of speed, acceleration, lifetime and accuracy. Recently, several applications of linear motors in machine tools have appeared on the market [6]. A linear motor axis can overcome these drawbacks of the indirect drive design.

Along with the higher speeds, also the requirements on machining accuracy are getting more increasingly stringent. As an example, finishing steps are not allowed in near-net-shape manufacturing and hence more accurate and smoother cutting are desired. At high working speeds (e.g. 30m/min feedrate for high speed milling), the desired accuracy cannot be attained by traditional control techniques. This paper describes an advanced state feedback with feedforward controller for the direct as well as the indirect drive design, and discusses its performance and limitations on the basis of experiments. The salient features of the direct drive design are discussed in Section 2. Section 3 of this

---

[*] Senior Research Assistant with the N.F.W.O. (Belgian National Fund for Scientific Research)

paper presents the identification procedure and section 4 presents the model-based controller.

Pritschow and Philipp [5] have shown that very high tracking accuracy can be achieved with a linear asynchronous motor by designing a good feedforward filter. An important difficulty in the design of an accurate controller for a direct drive linear synchronous motor, is the ripple of the motor thrust force (cogging), which is directly transmitted to the load This ripple forces causes significant tracking errors. A good knowledge of the ripple force is essential to compensate the effect of this disturbance. Sections 5 and 6 discuss the identification and compensation of the ripple effects and show the resulting possible reduction of the tracking error by experiments on two test set-ups.

2. Direct drive versus indirect drive design

The traditional indirect drive design (rotary motor and ball-screw) has some important drawbacks. The transmission unit gives rise to several disturbances, like friction, backlash and flexibilities, and limits the machine speed, acceleration, accuracy, stroke and lifetime. For instance, for a long-stroke machine the low critical speed of the screw limits the maximum speed of the load.

The direct drive design for a machine tool axis does not exhibit these disadvantages. Generally, it has the following advantages over its rotary-motor counterpart:
- no backlash and less friction, resulting in very high accuracy,
- no mechanical limitations on acceleration and velocity: the velocity is only limited by the encoder bandwidth or by the power electronics,
- higher closed-loop bandwidth possible: dynamics are only limited by encoder dynamics, measurement noise, calculation time and frame stiffness,
- mechanical simplicity, higher reliability and longer lifetime,
- unlimited length of movement, without limitations on the maximum velocity.

Despite these advantages, the use of a linear motor axis has some important drawbacks, related to the control design. The stiffness of the axis in the motion direction is entirely determined by the servo controller. This makes high demands upon the controller design of the linear motor axis. In the linear motor design, the motor is very sensitive to load variations and external disturbances and the motor ripple is directly transmitted to the load, resulting in significant tracking errors. This effect plays also a role in the reaction on frame motion [8].

Besides these control design implications, the use of a linear motor axis influences the mechanical design:
- high attraction force between slide and frame, perpendicular to the motion,
- a good sealing is necessary to prevent attraction of the chips by the magnets, which could cause motor damage,
- large contact surface between motor and machine tool body: water cooling is always necessary
- when no absolute position measurement device is available, motor phasing at start for the synchronous motor type might be difficult.

- over-sized motors for the high-force, low-velocity applications: with rotary motors the motor size can be optimised by the use of a mechanical reduction and a fast rotating servo motor.

3. Experimental identification

The first test bench is a linear torque-controlled synchronous motor driving a slide supported by three air bearings (figure 1). The mean attraction force is 2700 N. Three other bearings guide the carriage laterally to withstand the transverse inertia forces that occur when the axis is used in an XY-configuration. The position of the load is measured by an optical linear scale with an incremental measurement system.

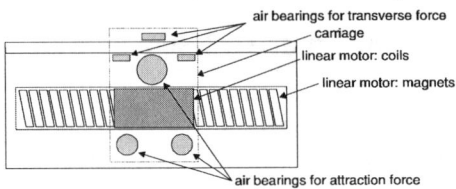

Figure 1 Top view of the linear motor axis on air bearings

A second test bench consists of a rotary velocity-controlled synchronous motor with ball-screw transmission and rolling element slideways. A velocity-controlled motor is a motor enclosed in a high gain analog velocity feedback loop, which in the ideal case behaves as a velocity source. In order to reach the high velocities, a large lead ball-screw is used (50mm per revolution). The position of the load is measured by an magnetic linear scale on the slideways with an incremental measurement system. The controllers of the axes are implemented on a digital signal processor board.

The modelling of the test benches corresponds to the identification of the dynamic relation between the input voltage U to the motor power supply, and the position X of the carriage. This dynamic relation is described by means of a discrete time transfer function H(z). A discrete time representation of the model (using the z-transform) is preferred to a continuous time representation (using the Laplace transform), because of the discrete implementation of the controller, which design is described in section 4.

The identification of the transfer function H(z) is done with a weighted non-linear least squares frequency domain identification method. The minimisation criterion is the total squared relative difference between the measured frequency response function of the test bench and the frequency characteristic of the transfer function. The criterion is non-linear in the model parameters, and therefore, its minimisation requires an iterative search algorithm such as Gauss-Newton's or Levenberg-Marquardt's method [9].

The measurement of the frequency response function is based on time domain measurements, which are transformed to the frequency domain using the discrete Fourier transform (DFT), calculated with the fast Fourier transform (FFT). In order to avoid leakage a multisine excitation, which is a broad band excitation signal, is used:

$$x(t) = \sum_{k=1}^{N} A_k \cos(2\pi l_k f_0 t + \phi_k),$$

with A_k the user definable amplitude of the component at frequency $l_k f_0 \in$ IN. The phases ϕ_k are chosen [9] such that the peak value of the signal is as small as possible, so that it is possible to make measurements with a maximum signal to noise ratio. Broad band excitation signals with minimum peak value allow to reduce the measurement time significantly. Several periods are measured. Then, the Fourrier coefficients are calculated [10].

In the motor power supply of the linear motor axis on air bearings, the input voltage U is linearly transformed into the motor current, i.e. the motor force. If motor friction is negligible (see section 2), this dynamic relation corresponds to a double integration.

The parametric open loop model of the linear motor axis on air bearings, is:

$$H(z) = \frac{X(z)}{U(z)} = \frac{-2.6682}{(z-1)(z-1)},$$

where $U(z)$ is the motor input command, expressed in Volt and $X(z)$ the position of the carriage, in μm. The model contains a double pole at z=1, which corresponds to a double integration. This observation and the good correspondence between the measured frequency response function and the frequency response of the identified model (see figure 2), show that motor friction is indeed negligible.

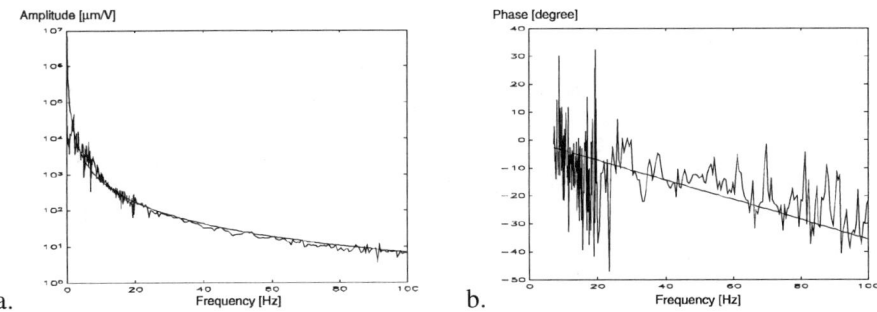

Figure 2   Measured frequency response function and the frequency response
of the model for the linear motor axis: amplitude (a) and phase (b)

For the rotary motor axis, a similar identification procedure results in the following parametric model:

$$H(z) = \frac{X(z)}{U(z)} = \frac{1.808z^2 - 3.791z + 3.276}{z^4 - 3.682z^3 + 5.321z^2 - 3.588z + 0.949}$$

where U(z) is the motor input command, expressed in Volt, and X(z) the position of the carriage, in μm. Figure 3 shows that the measured frequency response function and the frequency response of the model correspond well until 220 Hz.

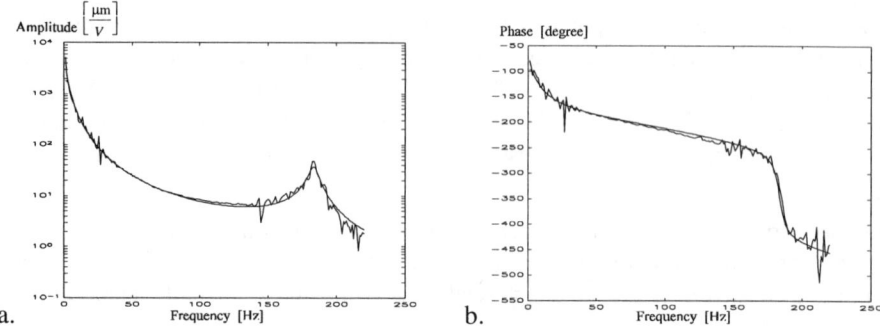

Figure 3 Measured frequency response function and the frequency response
of the model for the rotary motor axis: amplitude (a) and phase (b)

The theoretical model order is 3: the velocity control loop of the motor results in one differentiation (disturbed by the friction around zero Hz) and the flexibility of the screw results in a double pole. The order of the identified model is 4. This is due to the limited analog velocity-feedback gain of the motor, which makes that the motor velocity is not exactly proportional to the input command to velocity loop control of the motor.

4. State space control

Figure 4 shows the controller scheme, used for tracking in the different set-ups. The state observer estimates the system states: $\hat{x}_k$. These are compared with the reference states $x_{m,k}$, which are the desired states calculated in the reference state generator. The difference between the estimated system states and the reference states is fed back to the system with gain K. The reference state generator has an observer-like structure, in order to compensate the computation errors in the feedforward calculation:

$$\begin{cases} x_{m,k+1} = \hat{\Phi} x_{ref,k} + \hat{\Gamma} u_{f,k} \\ x_{ref,k} = x_{m,k} + L_{ref}(y_{ref,k} - \hat{C} x_{m,k}), \end{cases}$$

where $\hat{\Phi}, \hat{\Gamma}$ and $\hat{C}$ are the state space matrices of the system model, $y_{ref,k}$ is the feedforward signal and $x_{ref,k}$ is the reference state at moment k.

To compensate the steady-state errors, the integral of the tracking error is fed back with gain K_i. The controller gains (K and K_i) and observer gains (L and L_{ref}) are defined by pole placement. For the linear motor axis on air bearings, the state feedback poles are set at 40 Hz and ζ=0.7, the integration feedback pole at 10 Hz. The observer poles are set at 100 Hz. The location of the feedback poles is limited by a vibration mode of the frame. The observer gain L_{ref} is selected such that the poles of the reference state generator are located around the closed-loop bandwidth, in order to avoid generating states difficult to track. Instead of the integration feedback, a disturbance observer can be used [15].

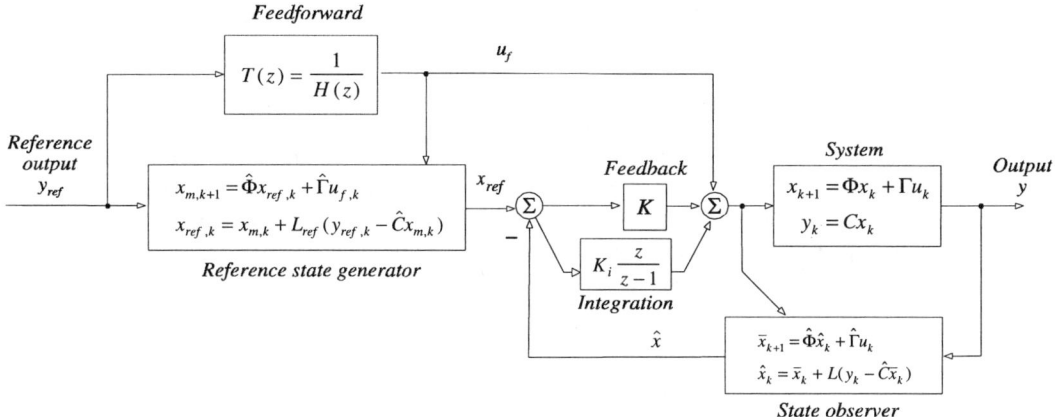

Figure 4 Control scheme

The feedforward command $u_{f,k}$ is based on the inverse of the open-loop transfer function $H(z)$, in order to compensate the dynamic effect of command changes: $T(z)=1/H(z)$. If the system has no zeros outside the unit circle, the exact inverse of the transfer function is stable and can be used to calculate the feedforward command. If the system has zeros outside the unit circle, the inverse of the transfer function must be approximated by the ZPETC-method [12] or its refinements ([2], [3], [5], [13] and [15]). Since the linear motor axis on air bearings has no zeros outside the unit circle, the exact inverse can be taken.

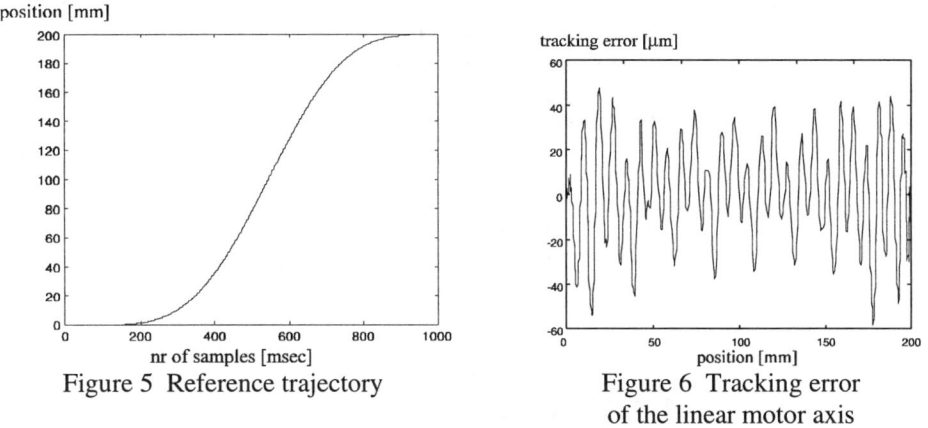

Figure 5 Reference trajectory

Figure 6 Tracking error
of the linear motor axis

Figure 5 shows the reference trajectory, a 9th-order polynomial ([11], p.187), used to validate the performance of the controller. The maximum velocity is 30m/min. Figure 6 shows the tracking error, when the described controller is applied on the linear motor axis to track this reference trajectory. The maximum tracking error is about 50μm. The oscillation appearing in the tracking error is due to the ripple of the linear motor. This error can be removed by adding a non-linear feedforward to the controller, based on a model of the ripple forces. Sections 5 ad 6 describe this ripple model identification and feedforward.

For the rotary motor axis, the state feedback poles are set at 30 Hz and $\zeta=0.7$, the integration feedback pole is set at 10 Hz. The observer poles are set at 150 Hz. Since the axis behaves like a second order system until 100 Hz, a second order model of the axis can be used for the controller design. This simplifies the design and implementation of the feedforward and the feedback without any loss of accuracy. Figure 7 shows the tracking error of the rotary motor axis, when tracking the trajectory of figure 5.

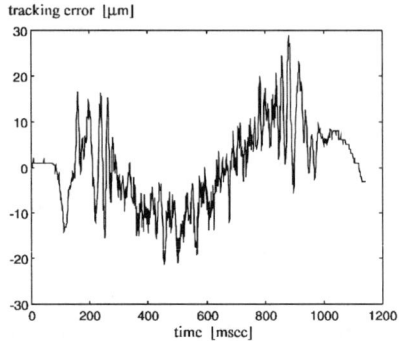

Figure 7 Tracking error of the rotary motor axis

The maximum tracking error is 30μm. While the tracking error of the linear motor axis results mainly from the motor ripple, the tracking error of the rotary motor axis is mainly caused by the friction and by the flexibilities in the system, limiting the closed loop bandwidth. The motor ripple also contributes to the tracking error, but its relative contribution is smaller than in the case of the linear motor axis.

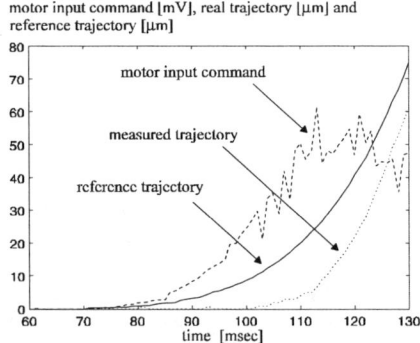

Figure 8 Reference trajectory [μm], measured trajectory [μm]
and motor input command [mV] at start for the rotary motor axis

Figure 8 shows the measured and the reference trajectory (in μm) and the motor input command (in mV) at the start of the trajectory. Although the reference trajectory increases around 70 msec after start, the load does not move initially because of friction. The resulting tracking error is integrated in the feedback loop, which increases the motor input command rapidly until it reaches the value that compensates the friction force and the load starts to move. At the end of the trajectory, the difference between the friction at standstill and during movement of the load, causes the load to overshoot the end position. So that the latter reaches its desired end position in a oscillatory manner, typical for stick-slip motion. This technique is described in more detail in e.g. [1], [7] and [14], which also describe some friction compensation techniques.

5. Identification of the linear motor ripple

In order to increase the tracking error of the linear motor axis, two types of ripple have to be considered: the *position ripple* and the *force ripple*. These ripple effects depend on the position of the motor carriage, more precisely the relative position of the magnets with respect to the coils of the linear motor.

Since the linear motor axis on air bearings is free of friction, the ripple forces can be measured very accurately. Figure 9 shows the experimental set-up: the linear motor carriage is connected to the fixed environment through a force cell. A screw-threaded cylinder between the environment and the force cell makes it possible to measure the force at different positions.

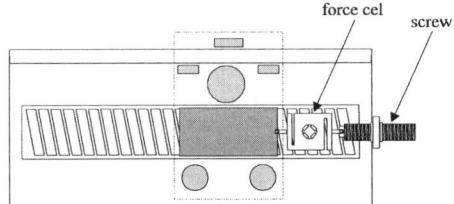

Figure 9 Measurement set-up for the ripple forces

At various positions, the force is measured, when increasing the input gradually from 0 to 10 Volt. Figure 10 shows the measured motor force as a function of the motor input at a certain position. The figure shows a linear relation between the input to the motor current loop and the force, in the range between 0 and 8.5 Volt. Between 8.5 and 9.1 Volt, the force decreases to zero due to the current limitation in the amplifier.

Figure 10 shows the position ripple and the force ripple. The *position ripple* is a magnetic disturbance force. It is the force necessary to keep the carriage at a certain position, with zero motor input current. It depends only on the relative position of the motor coils with respect to the magnets. This force is always present, even when there is no current flowing in the motor coils.

The *force ripple* is an electromagnetic effect caused by the variation of the motor constant (the linear relation between the motor current and the thrust force) with the position. In the paper, the difference between the thrust force at a certain position and the desired thrust force (the mean thrust force over all positions) for a certain motor input, is called the force ripple for that motor input value. The force ripple occurs only if the motor is required to produce a certain force, the value of this force ripple depends on the required force, as well as on the position of the motor.

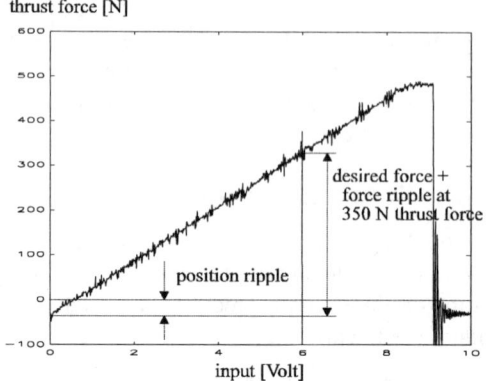

Figure 10 Thrust force versus motor input command at a certain position

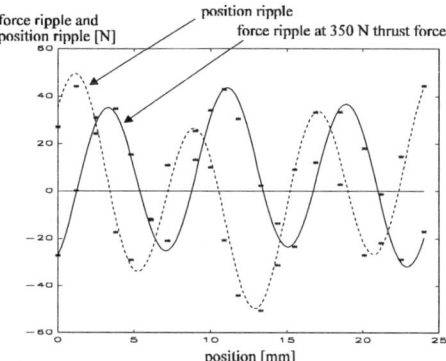

Figure 11 Force and position ripple: measurements and model

Figure 11 shows the force ripple versus position over a stroke of 25 mm. The measurements are shown by an asterisk (*). The solid line is the identified model of the force ripple at 350 N desired force and the dashed line represents the position ripple. Both ripples have the same shape: sinusoidal (with a period equal to the pitch of the magnets) and its third harmonic, but they are shifted in position.

The force F executed by the motor is the sum of the desired force, the position ripple and the force ripple:

$$F = F_{desired} + F_{pos} + F_{force}.$$

The identified model of the position ripple, expressed in Newton, is:

$$F_{pos} = 16.1 \sin(\omega_x x) + 35.9 \sin(3\omega_x x + 0.09\pi)$$

The identified model of the force ripple is:

$$F_{force} = F_{desired}*[\, 0.016 \sin(\omega_x (x + 5)) + 0.0879 \sin(3\omega_x (x + 5)+0.09\pi)]$$

with $\omega_x = 0.266$ rad/mm and x the position in mm.

6. Ripple model feedforward

The position ripple and the force ripple model of section 5 can be used as an extra feedforward command signal in the linear motor axis controller. Figure 12 shows the tracking error, when this ripple model feedforward is applied to the linear motor axis on air bearings. The maximum tracking error is reduced to 15μm, which is a reduction by a factor of three, compared to the tracking error obtained with the controller without ripple model feedforward (figure 6).

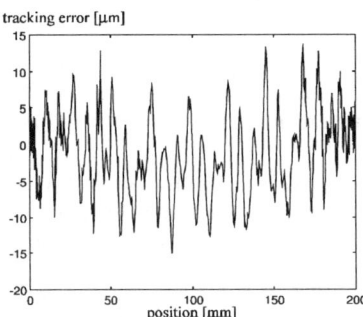

Figure 12 Tracking error of the linear motor axis on air bearings
with ripple model feedforward

Not all magnets of the linear motor are identical. This produces changes in the position and force ripple over the stroke. Therefore ,the ripple model based on measurements over a part of the stroke, is not valid over the whole stroke. The ripple model feedforward, as presented here, is very sensitive to these ripple model errors. Due to these small differences in the magnets, it will never be possible to compensate the ripple effect completely with this method. Figure 11 has already shown that the ripple behaviour is not perfectly repetitive, which makes it impossible to fit a perfect model through the measurements.

The flexibility of the plate guiding the transverse air bearings, is excited by the high frequency content of the ripple model feedforward signal. This gives rise to vibrations, which are another limiting factor for the reduction of the tracking error.

The ripple model feedforward has also been applied to a second linear motor axis, which is supported on rolling element slideways, but contains the same linear motor type as the axis on air bearings. Figure 13 shows the tracking error, when the control scheme of figure 4 is used. The trajectory is a 5-th order polynome with 30 m/min maximum velocity and 20m/sec^2 maximum acceleration.

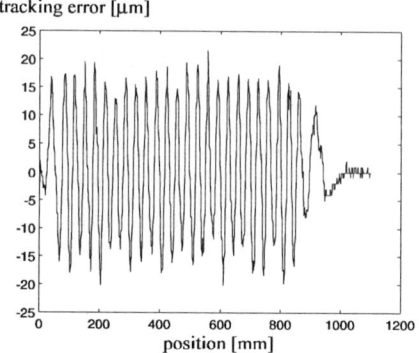

tracking error [μm]

Figure 13 Tracking error of the linear controller on the second axis

Since this second linear motor axis has a higher structural stiffness than the linear motor axis on air bearings, it is possible to use a higher feedback gain, which decreases further the tracking error. The maximum tracking error is now 20μm.

Figure 14 shows the tracking error, when ripple model feedforward is applied. The ripple model has not been measured again, but is the one identified in section 5, based on the measurements of the first linear motor axis (which is the same motor type). The maximum tracking error is reduced from 20μm to 7μm.

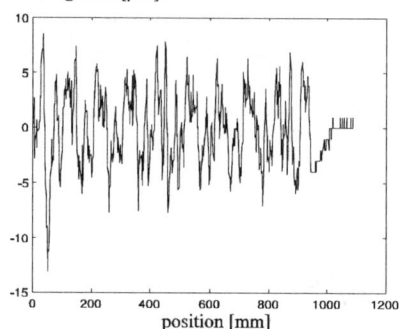

tracking error [μm]

Figure 14 Tracking error of the controller
with ripple model feedforward on the second axis

7. Conclusion

The use of linear motors in machine tool axes is very promising, but has several disadvantages from mechanical as well as control design point of view.

With the traditional design, it is possible to reach good tracking behaviour, when a well-designed controller is used. The tracking error is mainly caused by the friction and the flexibilities in the system, limiting the closed loop bandwidth.

Compared to the traditional design, the stiffness of the axes and thus disturbance rejection, must result from the servo controller and hence more effort must be spent on the controller design. A good controller must be based on an accurate knowledge of the disturbance forces.

Ripple forces of the linear motor cause large tracking errors of too high frequency content, to be compensated by the state feedback. Position ripple and force ripple model feedforward can compensate these effects and decrease the tracking errors significantly.

Acknowledgement

This research is carried out in the framework of BRE2-programme of the European Union, under the project KERNEL, BRE2-5621. This text also presents research results of the Belgian Programme on Interuniversity Poles of Attraction initiated by the Belgian State, Prime Minister's Office, Science Policy Programming. The scientific responsibility is assumed by its authors.

References

[1] Armstrong-Hélouvry B.: "Control of Machines with Friction", Kluwer Academic Publishers, London, (1991)

[2] Chen C.-H.: "Integrated Design Methods for Motion Control Systems", PhD thesis 93D3 ,ISBN 90-72802-30-10, Mechanical Engineering Department, Katholieke Universiteit Leuven, Belgium (1993)

[3] Haack and Tomizuka: "Zero Phase Error Tracking Algorithm for Digital Control", Journal of Dynamic systems, Measurement and Control, Vol 113 (1991), pp. 6-10

[4] Koren Y. and Lo Ch.-Ch.: "Advanced controllers for feed drives", Annals of the CIRP, Vol. 41/1 (1992), pp. 437-441

[5] Pritschow G. and Philipp W.: "Research on the Efficiency of Feedforward Controllers in Direct Drives", Annals of the CIRP Vol.41/1 (1992), 411-415

[6] Pritschow G.: "Increasing Accuracy Through Utilisation of Linear Direct Drives", Proceedings of the 27th CIRP International Seminar on Manufacturing Systems, Ann Arbor, USA, May 21-23, 1995, pp. 135-141

[7] Prajogo T., Al-Bender F. and Van Brussel H.: "Identification of pre-rolling friction dynamics of rolling element bearings: modelling and application to precise positioning systems", Proceedings of the 8th International Precision Engineering Seminar, Compiegne, France, May 15-19, 1995, pp. 229-232

[8] Rankers A.M. and van Eijk J.: "The influence of reaction forces on the behaviour of high performance motion systems", Proceedings of the 2nd International Conference on Motion and Vibration Control, Yokohama, Japan, August 30-September 3, 1994, pp.

[9] Schoukens J. and Pintelon R.: "Identification of Linear Systems", Pergamon Press, Oxford, UK (1991)

[10] Schoukens J., Pintelon R. and Van hamme H.: "Identification of Linear Dynamic Systems Using Piecewise Constant Excitations: Use, Misuse and Alternatives", Automatica, Vol.30, No.7 (1994), pp. 1153-1169

[11] Swevers J.: "Linear Identification and Control of Flexible Robots", PhD thesis 92D1, ISBN 90-73802-08, Mechanical Engineering Department, Katholieke Universiteit Leuven, Belgium (1992)

[12] Tomizuka M.: "Zero Phase Error Tracking Algorithm for Digital Control", Journal of Dynamic systems, Measurement and Control, Vol 109 (1987), pp. 65-68.

[13] Torfs D., De Schutter J. and Swevers J.: "Extended bandwidth zero phase error tracking control of nonminimal phase systems", Transactions of the ASME, Journal of Dynamic Systems, Measurement and Control, Vol.114 (1992), pp. 347-351.

[14] Tung E.D., Anwar G. and Tomizuka M.: "Low velocity Friction Compensation and Feedforward Solution Based on Repetitive Control", Transactions of the ASME, Journal of Dynamic Systems, Measurement and Control, Vol.115 (1993), pp. 279-284

[15] Van Brussel H., Chen C.-H. and Swevers J.: "Accurate Motion Controller Design Based on an Extended Pole Placement Method and Disturbance Observer", Annals of the CIRP, Vol.43/1 (1994), pp. 367-372

ir. Pieter Van den Braembussche
Mechanical Engineering Department, Division PMA, K.U.Leuven,
Celestijnenlaan 300 B
3001 Heverlee, Belgium
tel: +32-16-32 25 36; fax: +32-16-32 29 87
email: pieter.vandenbraembussche@mech.kuleuven.ac.be

Dr. ir. Jan Swevers
Mechanical Engineering Department, Division PMA, K.U.Leuven,
Celestijnenlaan 300 B
3001 Heverlee, Belgium
tel: +32-16-32 25 40; fax: +32-16-32 29 87
email: jan.swevers@mech.kuleuven.ac.be

Prof. Dr. Dr.h.c. ir. Hendrik Van Brussel
Mechanical Engineering Department, Division PMA, K.U.Leuven,
Celestijnenlaan 300 B
3001 Heverlee, Belgium
tel: +32-16-32 26 47; fax: +32-16-32 29 87
email: hendrik.vanbrussel@mech.kuleuven.ac.be

Prof. Paul Vanherck
Mechanical Engineering Department, Division PMA, K.U.Leuven,
Celestijnenlaan 300 B
3001 Heverlee, Belgium
tel: +32-16-32 24 80; fax: +32-16-32 29 87
email: paul.vanherck@mech.kuleuven.ac.be

IV. Robot Control

B. Siciliano
*A Unified Framework for the Design of Interaction Control
Schemes for Robot Manipulators*

P. Fraisse, F. Pierrot, P. Dauchez
*Robust Control of a Two-Arm Robot: an Efficient
Implementation
in a DSP-based Controller*

R. Neumann, W. Moritz
Robot Path Control with a Decentral Structure

G.-W. van der Linden
*Design and Implementation of a Model-Based Nonlinear
Controller for an Experimental Hydraulic Robot*

A Unified Framework for Design of Interaction Control Schemes for Robot Manipulators

Bruno Siciliano

Dipartimento di Informatica e Sistemistica
Università degli Studi di Napoli Federico II
Via Claudio 21, 80125 Napoli, Italy

Abstract: A unified framework for design of control schemes for robot manipulators interacting with compliant environments is presented in this paper. Compliance and impedance control schemes are logically derived from operational space motion control schemes, with integration of contact force measurements. Force control schemes are then presented which are obtained by closing an outer force feedback loop around an inner position or velocity feedback loop. Numerical examples are illustrated.

1 Introduction

One of the fundamental requirements for the success of a manipulation task is the capability to handle *interaction* between the robot manipulator and the environment. The quantity that describes the state of interaction more effectively is the *contact force* at the manipulator's end effector. High values of contact force are generally undesirable since they may stress both the manipulator and the manipulated object [1].

This paper is aimed at establishing a unified framework for design of interaction controllers where contact forces can be controlled with two different strategies; namely, an *indirect* strategy via a suitable use of motion control laws or a *direct* strategy via true force control laws [2].

The performance of operational space motion control schemes [3] is analyzed first. The concepts of mechanical *compliance* [4] and *impedance* [5] are presented, with special regard to the problem of integrating contact force measurements into the control strategy [6]. An interesting feature of compliance and impedance control schemes is that the interaction force can be indirectly controlled by acting on the reference position of manipulator motion control [7].

On the other hand, if it is wished to accurately control the contact force, it is necessary to devise control schemes that allow directly specifying the desired interaction force. The realization of a *force control* scheme can be entrusted to the closure of an outer force regulation feedback loop generating the control input for the position control scheme the manipulator is usually endowed with [8]. Three different schemes are presented; namely, a force control scheme with inner position loop, a force control scheme with inner velocity loop, and a parallel force/position control scheme [9,10].

Numerical examples are presented throughout the paper to illustrate the performance of the various interaction control schemes for a simple planar arm in contact with an elastically compliant plane.

2   Compliance Control

For a detailed analysis of interaction between the manipulator and environment it is worth considering the behavior of the system under a position control scheme when contact forces arise. Since these are naturally described in the operational space, it is convenient to refer to *operational space control* schemes.

Consider the manipulator dynamic model which can be written as

$$B(q)\ddot{q} + C(q, \dot{q})\dot{q} + F\dot{q} + g(q) = u - J^T(q)h, \tag{1}$$

where q is the $(n \times 1)$ vector of joint variables, B is the $(n \times n)$ inertia matrix, $C(q, \dot{q})\dot{q}$ is the $(n \times 1)$ vector of Coriolis and centrifugal terms, F is the $(n \times n)$ matrix of joint viscous friction, g is the $(n \times 1)$ vector of gravity terms, u is the $(n \times 1)$ vector of input torques, $J(q)$ is the end-effector geometric Jacobian, and h is the vector of contact forces exerted by the manipulator's end effector on the environment.

Let x denote the $(m \times 1)$ vector of end-effector location. Let also $\tilde{x}$ be the operational space error between a *constant* desired end-effector location x_d and x. Choose the input torque as a PD operational space control with joint space gravity compensation of the type

$$u = J_A^T(q)(K_P\tilde{x} - K_D\dot{x}) + g(q) \tag{2}$$

where J_A indicates the analytical Jacobian; this is related to the above geometric Jacobian by

$$J = T_A(x)J_A \tag{3}$$

where $T_A(x)$ is a transformation matrix which depends on the set of Euler angles chosen to describe end-effector orientation.

The control law (2), in the absence of contact forces, ensures global asymptotic stability of the desired posture x_d as long as the analytical Jacobian is nonsingular (no kinematic or representation singularities). On the other hand, in the case $h \neq 0$, the control law (2) no longer ensures that the end effector reaches x_d. In fact, at the equilibrium it is

$$J_A^T(q)K_P\tilde{x} = J^T(q)h. \tag{4}$$

On the assumption of a full-rank Jacobian, one has

$$\tilde{x} = K_P^{-1}T_A^T(x)h = K_P^{-1}h_A, \tag{5}$$

where h_A is the vector of equivalent generalized forces that can be related to h by

$$T_A^T(x)h = h_A. \tag{6}$$

Eq. (5) shows that at the equilibrium the manipulator, under a position control action, behaves as a generalized spring in the operational space with *compliance* K_P^{-1} in respect of force h_A. Since the transformation matrix T_A is nondiagonal while matrix K_P is typically diagonal, it can be recognized that linear compliance (due to force components) is independent of the posture while torsional compliance (due to moment components) does depend on the current manipulator configuration through the matrix T_A.

On the other hand, if $h \in \mathcal{N}(J^T)$, one has $\tilde{x} = 0$ with $h \neq 0$, i.e. contact forces are completely balanced by the manipulator mechanical structure; for instance, an anthropomorphic manipulator at a shoulder singularity does not react to any force orthogonal to the plane of the structure.

A detailed description of the contact is demanding from a modeling viewpoint. To point out the fundamental aspects of interaction control, it is convenient to resort to a simple but significant model of contact. To this purpose, a decoupled *elastically compliant environment* is considered which is described by the model

$$h = \begin{bmatrix} f \\ \mu \end{bmatrix} = \begin{bmatrix} K_f & O \\ O & K_m \end{bmatrix} \begin{bmatrix} dp \\ \omega dt \end{bmatrix} = K \begin{bmatrix} dp \\ \omega dt \end{bmatrix}, \tag{7}$$

where dp is the vector of translation along the reference frame axes and ωdt is the vector of small rotation about the axes of such frame. Hence the vector $[\, dp^T \quad \omega^T dt\,]^T$ describes a generalized displacement from the environment rest position. The *stiffness matrix* K is typically *positive semi-definite*. In fact, the environment does not generate reaction forces along those directions where unconstrained end-effector motion is allowed.

In view of (3), Eq. (7) can be written in terms of operational space variables as

$$h = KT_A(x)dx \tag{8}$$

where dx denotes the operational space generalized displacement with respect to the undeformed environment rest position x_e

$$dx = x - x_e. \tag{9}$$

Resorting to (6),(9) gives

$$h_A = T_A^T(x)KT_A(x)dx = K_A(x)(x - x_e) \tag{10}$$

that allows relating the equivalent forces on the manipulator with the environment deformation through the matrix K_A, i.e. the environment stiffness matrix. The matrix K_A^{-1}, if it can be defined, is the environment *compliance* matrix. It represents a *passive* compliance since it describes an inherent property of the environment in the operational space chosen to express manipulator end-effector position and orientation. By observing that

K_A is only positive semi-definite, the concept of compliance cannot be globally defined in all operational space but only along those directions, spanning $\mathcal{R}(K_A)$, along which end-effector motion is constrained by the environment.

On the other hand, notice that the matrix K_P^{-1} in (5) represents an *active compliance* since it is performed on the manipulator by a suitable position control action. With the environment model (10), Eq. (5) becomes

$$\tilde{x} = K_P^{-1} K_A(x)(x - x_e); \tag{11}$$

at the equilibrium, the end-effector location is given by

$$x_\infty = \left(I + K_P^{-1} K_A(x)\right)^{-1} (x_d + K_P^{-1} K_A x_e), \tag{12}$$

while the contact force can be shown to be

$$h_{A\infty} = \left(I + K_A(x) K_P^{-1}\right)^{-1} K_A(x)(x_d - x_e). \tag{13}$$

Analysis of (12) shows that the equilibrium position depends on the environment rest position as well as on the desired position imposed by the control system to the manipulator. The interaction of the two systems (environment and manipulator) is influenced by the mutual weight of the respective compliance features. It is then possible to increase the active compliance so that the manipulator dominates the environment and vice versa. Such a dominance can be specified with reference to the single directions of the operational space. For a given environment stiffness, according to the given interaction task, one may choose large values of the elements of K_P for those directions along which the environment has to comply and small values of the elements of K_P for those directions along which the manipulator has to comply.

Expression (13) gives the value of the contact force at the equilibrium which reveals that it may be appropriate to tune manipulator compliance with environment compliance along certain directions of the operational space. In fact, along a direction with high environment stiffness, it is better to have a compliant manipulator so that it can taper the intensity of interaction through a suitable choice of the desired position. In this case the end-effector equilibrium position x_∞ practically coincides with the environment rest position x_e, and the manipulator generates an interaction force, depending on the corresponding element of K_P, that is determined by the choice of the component of $(x_d - x_e)$ along the relative direction.

In the dual case of high environment compliance, if the manipulator is made stiff, the end-effector equilibrium position x_∞ is very close to the desired position x_d, and it is the environment to generate the elastic force along the constrained directions of interest.

In certain cases, it is possible to employ mechanical devices interposed between the manipulator's end effector and the environment so as to change passive compliance along particular directions of the operational space. For instance, in a peg-in-hole insertion task, the gripper is provided with a device ensuring high stiffness along the insertion direction and high compliance along the other directions (*remote center of compliance*). Therefore,

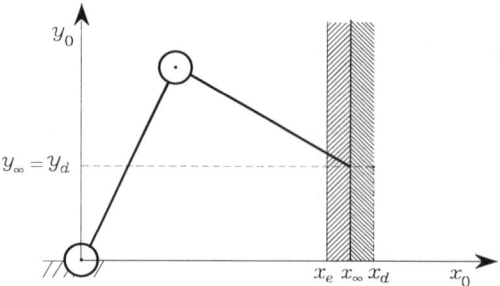

Figure 1: Two-link planar arm in contact with an elastically compliant plane.

in the presence of unavoidable position displacements from the planned insertion, trajectory contact forces and moments arise which modify the peg position so as to facilitate insertion.

The inconvenience of such devices is their low versatility to different operating conditions and generic interaction tasks, i.e. whenever a modification of the compliant mechanical hardware is required. On the other hand, with active compliant actions the control software can be easily modified so as to satisfy the requirements of different interaction tasks.

Example 1. Consider the two-link planar arm whose tip is in contact with a purely frictionless elastic plane; let x_e be the equilibrium position of the plane, which is assumed to be orthogonal to axis x (Fig. 1). The environment stiffness matrix is

$$K_A = K_f = \text{diag}\{k_x, 0\},$$

corresponding to the absence of interaction forces along the vertical direction ($f_y = 0$). Let $p_d = [\,x_d \quad y_d\,]^T$ be the desired tip position, which is located beyond the contact plane. The proportional control action on the arm is characterized by

$$K_P = \text{diag}\{k_{Px}, k_{Py}\}.$$

The equilibrium equations for position and force (12),(13) give

$$p_\infty = \left[\begin{array}{c} \dfrac{k_{Px} x_d + k_x x_e}{k_{Px} + k_x} \\ y_d \end{array}\right] \qquad f_\infty = \left[\begin{array}{c} \dfrac{k_{Px} k_x}{k_{Px} + k_x}(x_d - x_e) \\ 0 \end{array}\right].$$

With reference to positioning accuracy, the arm tip reaches the vertical coordinate y_d since the vertical motion direction is not constrained. As for the horizontal direction, the presence of the elastic plane imposes that the arm can move as far as it reaches the coordinate x_∞. The value of the horizontal contact force at the equilibrium is related to the difference between x_e and x_d by an equivalent stiffness coefficient which is given by the parallel composition of the stiffness coefficients of the two interacting systems.

Hence, the arm stiffness and environment stiffness influence the resulting equilibrium configuration. In the case when

$$k_{Px}/k_x \gg 1,$$

it is

$$x_\infty \approx x_d \qquad f_{x\infty} \approx k_x(x_d - x_e)$$

and thus the arm prevails over the environment, in that the plane complies almost up to x_d and the elastic force is essentially generated by the environment (passive compliance). In the opposite case

$$k_{Px}/k_x \ll 1,$$

it is

$$x_\infty \approx x_e \qquad f_{x\infty} \approx k_{Px}(x_d - x_e)$$

and thus the environment prevails over the arm which complies up to the equilibrium x_e, and the elastic force is generated by the arm (active compliance).

3 Impedance Control

It is now desired to analyze the interaction of manipulator with environment under the action of an inverse dynamics control in the operational space. With reference to model (1), consider the inverse dynamics control law

$$u = B(q)y + n(q, \dot{q}), \tag{14}$$

with $n = C\dot{q} + F\dot{q} + g$. In the presence of end-effector forces, the controlled manipulator is described by

$$\ddot{q} = y - B^{-1}(q)J^T(q)h \tag{15}$$

that reveals the existence of a nonlinear coupling term due to contact forces. Choose y as

$$y = J_A^{-1}(q)M_d^{-1}\left(M_d\ddot{x}_d + K_D\dot{\tilde{x}} + K_P\tilde{x} - M_d\dot{J}_A(q, \dot{q})\dot{q}\right), \tag{16}$$

where M_d is a positive definite diagonal matrix. Substituting (16) into (15) and accounting for second-order differential kinematics in the form

$$\ddot{x} = J_A(q)\ddot{q} + \dot{J}_A(q, \dot{q})\dot{q} \tag{17}$$

yields

$$M_d\ddot{\tilde{x}} + K_D\dot{\tilde{x}} + K_P\tilde{x} = M_dB_A^{-1}(q)h_A, \tag{18}$$

where

$$B_A(q) = J_A^{-T}(q)B(q)J_A^{-1}(q) \tag{19}$$

is the inertia matrix of the manipulator in the operational space; this matrix is configuration-dependent and is positive definite if J_A has full rank.

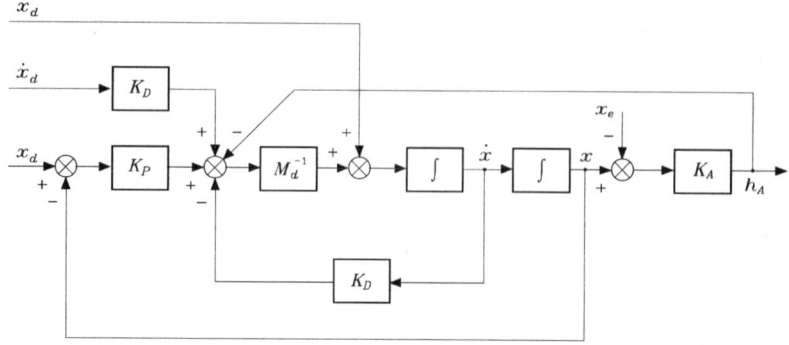

Figure 2: Equivalent block scheme of a manipulator in contact with an elastic environment under impedance control.

Eq. (18) establishes a relationship through a generalized *mechanical impedance* between the vector of resulting forces $M_d B_A^{-1} h_A$ and the vector of displacements $\tilde{x}$ in the operational space. This impedance can be attributed to a mechanical system characterized by a mass matrix M_d, a damping matrix K_D and a stiffness matrix K_P which allow specifying the dynamic behavior along the operational space directions.

The presence of B_A^{-1} makes the system coupled. If it is wished to keep linearity and decoupling during interaction with the environment, it is then necessary to *measure* the generalized contact *force*; this can be achieved by means of appropriate force sensors which are usually mounted on the manipulator wrist. Choosing

$$u = B(q)y + n(q, \dot{q}) + J^T(q)h \tag{20}$$

with

$$y = J_A^{-1}(q) M_d^{-1} \left(M_d \ddot{x}_d + K_D \dot{\tilde{x}} + K_P \tilde{x} - M_d \dot{J}_A(q, \dot{q})\dot{q} - h_A \right), \tag{21}$$

on the assumption of error-free force measurements, yields

$$M_d \ddot{\tilde{x}} + K_D \dot{\tilde{x}} + K_P \tilde{x} = h_A. \tag{22}$$

It is worth noticing that the addition of the term $J^T h$ in (20) exactly compensates the contact forces and then it renders the manipulator infinitely stiff with respect to external stress. In order to confer a compliant behavior to the manipulator, the term $-J_A^{-1} M_d^{-1} h_A$ has been introduced in (21) which allows characterizing the manipulator as a *linear impedance* with regard to the equivalent forces h_A, as shown in (22). The resulting block scheme of a manipulator in contact with an elastic environment under impedance control is illustrated in Fig. 2.

The behavior of system (22) at the equilibrium is analogous to that described by Eq. (4); nonetheless, compared to a compliance control specified by K_P, Eq. (22) allows a complete characterization of system dynamics through an *active impedance* specified by matrices M_d, K_D, K_P. These matrices are usually taken as diagonal; also in this case it is not difficult to recognize that impedance is configuration-independent as regards the force components, while it depends on the current manipulator configuration as regards the moment components through the matrix T_A.

Furthermore, similarly to active and passive compliance, the concept of *passive impedance* can be introduced if the interaction force h_A is generated at the contact with an environment of proper mass, damping and stiffness. In this case the system of manipulator with environment can be regarded as a mechanical system constituted by the parallel of the two impedances, and then its dynamic behavior is conditioned by the relative weight between them. As pointed out above, one may think about constructing a mechanical device with proper passive impedance that allows the manipulator to better cope with the given interaction task.

Example 2. Consider the planar arm in contact with an elastically compliant plane of the previous example. Apply the impedance control with force measurement (20),(21) characterized by:

$$M_d = \mathrm{diag}\{m_{dx}, m_{dy}\} \qquad K_D = \mathrm{diag}\{k_{Dx}, k_{Dy}\} \qquad K_P = \mathrm{diag}\{k_{Px}, k_{Py}\}.$$

If x_d is constant, the dynamics of the manipulator and environment system along the two directions of the operational space is described by

$$m_{dx}\ddot{x} + k_{Dx}\dot{x} + (k_{Px} + k_x)x = k_x x_e + k_{Px} x_d$$
$$m_{dy}\ddot{y} + k_{Dy}\dot{y} + k_{Py}y = k_{Py} y_d.$$

Along the vertical direction one has an unconstrained motion whose time behavior is determined by the following natural frequency and damping factor:

$$\omega_{ny} = \sqrt{\frac{k_{Py}}{m_{dy}}} \qquad \zeta_y = \frac{k_{Dy}}{2\sqrt{m_{dy}k_{Py}}},$$

while along the horizontal direction the behavior of the contact force $f_x = k_x(x - x_e)$ is determined by

$$\omega_{nx} = \sqrt{\frac{k_{Px} + k_x}{m_{dx}}} \qquad \zeta_x = \frac{k_{Dx}}{2\sqrt{m_{dx}(k_{Px} + k_x)}}.$$

Below the dynamic behavior of the system is analyzed for two different values of environment compliance: $k_x = 10^3$ N/m and $k_x = 10^4$ N/m. The arm is characterized by the following data:

$$a_1 = a_2 = 1\,\mathrm{m} \quad \ell_1 = \ell_2 = 0.5\,\mathrm{m} \quad m_{\ell_1} = m_{\ell_2} = 50\,\mathrm{kg} \quad I_{\ell_1} = I_{\ell_2} = 10\,\mathrm{kg\cdot m^2}$$

where a_i is the link length, ℓ_i is the distance of center of mass from joint axis, m_{ℓ_i} is the link mass, I_{ℓ_i} is the link moment of inertia about joint axis, for $i = 1, 2$, and

$$k_{r1} = k_{r2} = 100 \quad m_{m_1} = m_{m_2} = 5\,\mathrm{kg}$$

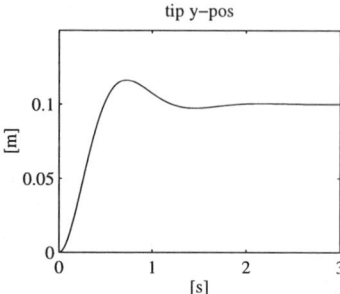
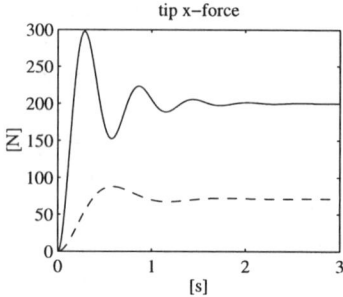

Figure 3: Time history of the tip position along vertical direction and of the contact force along horizontal direction with impedance control scheme for environments of different compliance.

$$I_{m_1} = I_{m_2} = 0.01 \, \text{kg} \cdot \text{m}^2 \quad F_{m_1} = F_{m_2} = 0.01 \, \text{N} \cdot \text{m} \cdot \text{s/rad}$$

where k_{ri} is the gear reduction, m_{m_i} is the mass of joint motor, I_{m_i} is the rotor moment of inertia about joint axis, F_{m_i} is the viscous friction, for $i = 1, 2$; motors are located at the joint axes.

The impedance control with force measurements (20),(21) is chosen as

$$m_{dx} = m_{dy} = 100 \quad k_{Dx} = k_{Dy} = 500 \quad k_{Px} = k_{Py} = 2500.$$

For these values it is

$$\omega_{ny} = 5 \, \text{rad/s} \quad \zeta_y = 0.5.$$

Then, for the more compliant environment it is

$$\omega_{nx} \approx 5.9 \, \text{rad/s} \quad \zeta_x \approx 0.42,$$

while for the less compliant environment it is

$$\omega_{nx} \approx 11.2 \, \text{rad/s} \quad \zeta_x \approx 0.22.$$

Let the arm tip be in contact with the environment at position $p = \begin{bmatrix} 1 & 0 \end{bmatrix}^T$; it is desired to take it to position $p_d = \begin{bmatrix} 1.1 & 0.1 \end{bmatrix}^T$.

The results in Fig. 3 show that motion dynamics along the vertical direction is the same in the two cases. As regards the contact force along the horizontal direction, for the more compliant environment (*dashed line*) a well-damped behavior is obtained, while for the less compliant environment (*solid line*) the resulting behavior is less damped. Further, at the equilibrium, in the first case a displacement of about 7.1 cm with a contact force of about 71.4 N are observed, while in the second case a displacement of 2 cm with a contact force of 200 N are observed.

4 Force Control

In the above schemes the interaction force could be indirectly controlled by acting on the reference value x_d of the manipulator motion control system. Interaction between manipulator and environment is anyhow directly influenced by compliance of the environment and by either compliance or impedance of the manipulator.

If it is desired to accurately control the contact force, it is necessary to devise control schemes that allow directly specifying the desired interaction force. The development of a *force control* system, in analogy to a motion control system, would require the adoption of a stabilizing PD control action on the force error besides the usual nonlinear compensation actions. Force measurements are typically corrupted by noise and then a derivative action cannot be implemented in practice. The stabilizing action is then to be provided by suitable damping of velocity terms. As a consequence, a force control system typically features a control law based not only on force measurements but also on velocity measurements, and eventually position measurements too.

The realization of a force control scheme can be entrusted to the closure of an *outer force regulation feedback loop* generating the control input for the position control scheme the manipulator is usually endowed with. Therefore, force control schemes are presented below which are based on the use of an inverse dynamics position control. Nevertheless, notice that a force control strategy is meaningful only for those directions of the operational space along which interaction forces between manipulator and environment may arise.

4.1 Force Control with Inner Position Loop

With reference to the inverse dynamics law with force measurement (20), choose in lieu of (21) the control

$$y = J_A^{-1}(q)M_d^{-1}\left(-K_D\dot{x} + K_P(x_F - x) - M_d\dot{J}_A(q,\dot{q})\dot{q}\right) \tag{23}$$

where x_F is a suitable reference to be related to a force error. Notice that the control law (23) does not foresee the adoption of compensating actions relative to $\dot{x}_F$ and $\ddot{x}_F$. Substituting (23) into (20) leads, after similar algebraic manipulation, to the system described by

$$M_d\ddot{x} + K_D\dot{x} + K_P x = K_P x_F, \tag{24}$$

which shows how Eqs. (20),(23) perform a position control taking x to x_F with a dynamics specified by the choice of matrices M_d, K_D, K_P.

Let h_{Ad} denote the desired *constant* force reference; the relation between x_F and the force error can be symbolically expressed as

$$x_F = C_F(h_{Ad} - h_A), \tag{25}$$

where C_F is a diagonal matrix whose elements give the control actions to perform along the operational space directions of interest. Eqs. (24),(25) reveal that force control is developed on the basis of a preexisting position control loop.

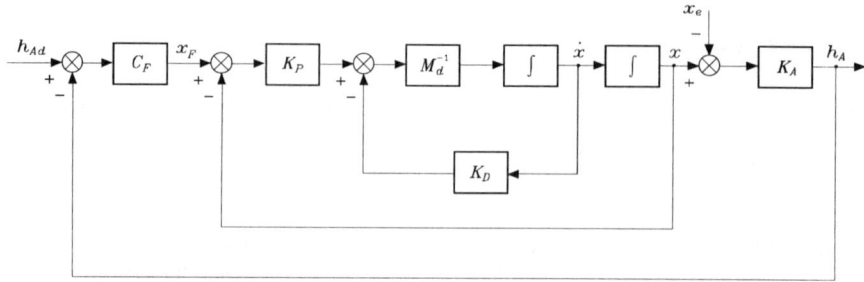

Figure 4: Block scheme of force control with inner position loop.

On the assumption of the elastically compliant environment described by (10), Eq. (24) with (25) becomes

$$M_d\ddot{x} + K_D\dot{x} + K_P(I + C_F K_A)x = K_P C_F(K_A x_e + h_{Ad}).$$ (26)

To decide about the kind of control action to specify with C_F it is worth representing Eqs. (10),(24),(25) in terms of the block scheme in Fig. 4, which is logically derived from the scheme in Fig. 2. This scheme suggests that if C_F has a purely proportional control action, then h_A cannot reach h_{Ad} and x_e influences the interaction force also at steady state.

If C_F has also an integral control action on the components of generalized force, then it is possible to achieve $h_A = h_{Ad}$ at steady state and at the same time to reject the effect of x_e on h_A. Hence, a convenient choice for C_F is a *proportional-integral* (PI) *action*

$$C_F = K_F + K_I \int_0^t (\cdot)\, d\varsigma.$$ (27)

The dynamic system resulting from (26),(27) is of third order, and then it is necessary to adequately choose the matrices K_D, K_P, K_F, K_I in respect of the characteristics of the environment. Since the values of environment stiffness are typically high, the weight of the proportional and integral actions shall be contained; the choice of K_F and K_I influences the stability margins and the bandwidth of the system under force control. On the assumption that a stable equilibrium is reached, it is $h_{A,\infty} = h_{Ad}$ and then

$$K_A x_\infty = K_A x_e + h_{Ad}.$$ (28)

4.2 Force Control with Inner Velocity Loop

From the block scheme of Fig. 4 it can be observed that, if the position feedback loop is opened, x_F represents a velocity reference and then an integration relationship exists between x_F and x. This leads to recognize that in this case the interaction force with

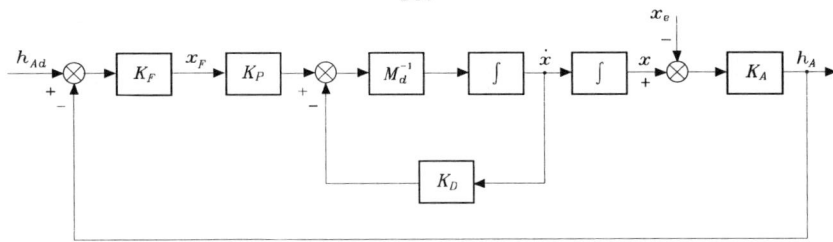

Figure 5: Block scheme of force control with inner velocity loop.

the environment coincides with the desired value at steady state, even with a proportional force controller C_F. In fact, choosing

$$y = J_A^{-1}(q)M_d^{-1}\left(-K_D\dot{x} + K_P x_F - M_d \dot{J}_A(q, \dot{q})\dot{q}\right) \tag{29}$$

with a purely proportional control structure ($C_F = K_F$) on the force error yields

$$x_F = K_F(h_{Ad} - h_A), \tag{30}$$

and then system dynamics is described by

$$M_d\ddot{x} + K_D\dot{x} + K_P K_F K_A x = K_P K_F (K_A x_e + h_{Ad}). \tag{31}$$

The relationship between position and contact force at the equilibrium is given by (28). The corresponding block scheme is reported in Fig. 5. It is worth emphasizing that control design is simplified, since the resulting system now is of second order.

4.3 Parallel Force/Position Control

The presented force control schemes require the force reference to be consistent with the geometrical features of the environment. In fact, if h_{Ad} has components on $\mathcal{R}(K_A)$, both Eq. (26) (in case of an integral action in C_F) and Eq. (31) show that, along the corresponding operational space directions, the components of h_{Ad} are interpreted as velocity references which cause a drift of the end-effector position. If h_{Ad} is correctly planned along the directions outside $\mathcal{R}(K_A)$, the resulting motion governed by the position control action tends to take the end-effector position to zero in the case of (26), and the end-effector velocity to zero in the case of (31). Hence, the above control schemes do not allow position control even along the admissible motion directions.

If it is desired to specify a desired end-effector location x_d like in pure position control schemes, the scheme of Fig. 4 can be modified by adding the reference x_d to the input where positions are summed. This corresponds to choosing

$$y = J_A^{-1}(q)M_d^{-1}\left(-K_D\dot{x} + K_P(\tilde{x} + x_F) - M_d \dot{J}_A(q, \dot{q})\dot{q}\right) \tag{32}$$

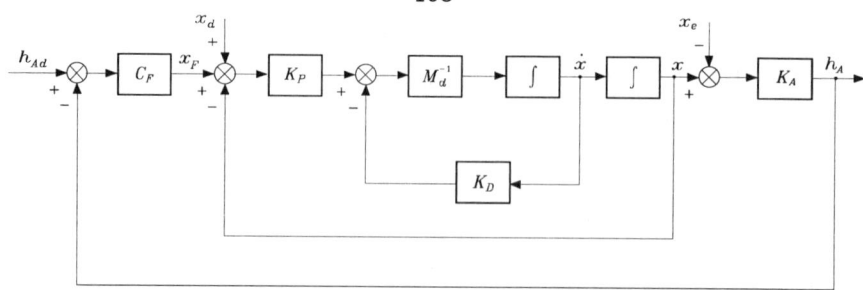

Figure 6: Block scheme of parallel force/position control.

where $\widetilde{x} = x_d - x$. The resulting scheme (Fig. 6) is termed *parallel force/position control* in view of the presence of a position control action $K_P\widetilde{x}$ in parallel to a force control action $K_P C_F(h_{Ad} - h_A)$. It is easy to verify that in this case the equilibrium position satisfies the equation

$$x_\infty = x_d + C_F\left(K_A(x_e - x_\infty) + h_{Ad}\right). \tag{33}$$

Therefore, along those directions outside $\mathcal{R}(K_A)$ where motion is unconstrained, the position reference x_d is reached by x. Vice versa, along those directions in $\mathcal{R}(K_A)$ where motion is constrained, x_d is treated as an additional disturbance; the adoption of an integral action in C_F as for the scheme of Fig. 4 ensures that the force reference h_{Ad} is reached at steady state at the expense of a position error on x depending on environment compliance.

Example 3. Consider again the planar arm in contact with the elastically compliant plane of the above examples; let the initial contact position be the same as that of Example 2. Performance of the various force control schemes is analyzed; as in Example 2, a more compliant ($k_x = 10^3$ N/m) and a less compliant ($k_x = 10^4$ N/m) environment are considered. The position control actions M_d, K_D, K_P are chosen as in Example 2; a force control action is added along the horizontal direction, i.e.

$$C_F = \text{diag}\{c_{Fx}, 0\}.$$

The reference for the contact force is chosen as $h_{Ad} = [\,10 \quad 0\,]^T$; the position reference —meaningful only for the parallel control— is taken as $p_d = [\,1.015 \quad 0.1\,]^T$.

With regard to the scheme with inner position loop of Fig. 4, a PI control action c_{Fx} is chosen with parameters:

$$k_{Fx} = 0.00064 \qquad k_{Ix} = 0.0016.$$

This confers two complex poles $(-1.96, \pm j5.74)$, a real pole (-1.09) and a real zero (-2.5) to the overall system, for the more compliant environment.

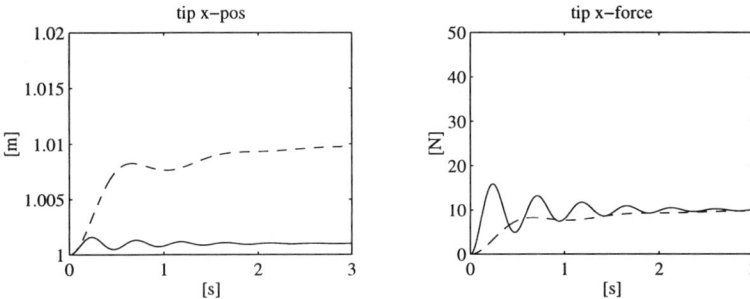

Figure 7: Time history of the tip position and of the contact force along horizontal direction with force control scheme with inner position loop for two environments of different compliance.

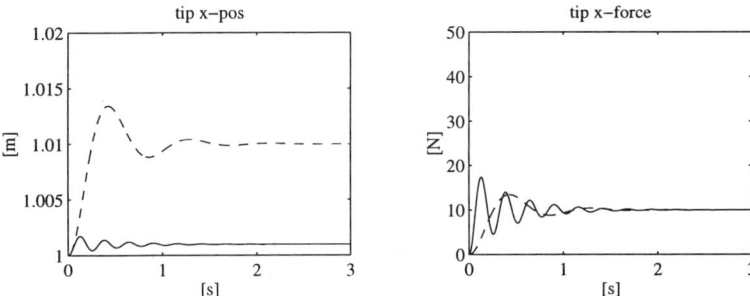

Figure 8: Time history of the tip position and of the contact force along horizontal direction with force control scheme with inner velocity loop for two environments of different compliance.

With regard to the scheme with inner velocity loop of Fig. 5, the proportional control action in c_{Fx} is

$$k_{Fx} = 0.0024$$

so that the overall system, for the more compliant environment, has two complex poles $(-2.5, \pm j7.34)$.

With regard to the parallel control scheme of Fig. 6, the PI control action c_{Fx} is chosen with the same parameters as for the first control scheme.

Figs. 7,8,9 report the time history of the tip position and contact force along axis x for the three considered schemes. A comparison between the various cases shows what follows.

- All control laws guarantee a steady-state value of contact forces equal to the desired one for both the more compliant (*dashed line*) and the less compliant (*continuous line*) environment.

- For given position control actions (M_d, K_D, K_P), the force control with inner velocity

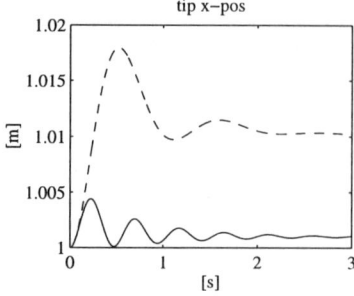
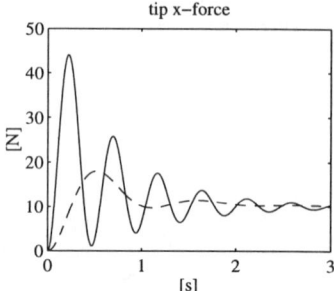

Figure 9: Time history of tip position and of the contact force along horizontal direction with parallel force/position control scheme for two environments of different compliance.

loop presents a faster dynamics than that of the force control with inner position loop.

- The dynamic response with the parallel force/position control shows how the imposition of a position reference along the horizontal direction degrades the transient behavior, even though it does not influence the steady-state contact force. This effect can be justified by observing that a step position input is equivalent to a properly filtered impulse force input.

- The reference position along axis y is obviously reached by the arm tip according to dynamics of position control; the relative time history is not reported.

5 Conclusions

The problem of interaction between manipulator and environment has been tackled by resorting to control schemes which make use of contact force measurements. Force can be controlled either indirectly through an active compliance or impedance, or directly through an outer force feedback loop which generates the reference to the inner motion feedback loop. The performance of the various schemes has been illustrated by means of numerical examples for an arm in contact with an elastic environment. It is understood that the interaction control schemes presented can be integrated in the hybrid control framework which is typically used to control force along constrained motion directions and position along unconstrained motion directions.

6 Acknowledgements

This work was supported by *Ministero dell'Università e della Ricerca Scientifica e Tecnologica*.

7 References

[1] D.E. Whitney, "Historical perspective and state of the art in robot force control," *Int.*

J. Robotics Research, vol. 6, no. 1, pp. 3–14, 1987.

[2] L. Sciavicco and B. Siciliano, *Modeling and Control of Robot Manipulators*, McGraw-Hill, New York, 1996.

[3] O. Khatib, "A unified approach for motion and force control of robot manipulators: The operational space formulation," *IEEE J. Robotics and Automation*, vol. 3, pp. 43–53, 1987.

[4] J.K. Salisbury, "Active stiffness control of a manipulator in Cartesian coordinates," *Proc. 19th IEEE Conf. Decision and Control*, pp. 95–100, Albuquerque, New Mex., 1980.

[5] N. Hogan, "Impedance control: an approach to manipulation: Part I — Theory," *ASME J. Dynamic Systems, Measurement, and Control*, vol. 107, pp. 1–7, 1985.

[6] R.J. Anderson and M.W. Spong, "Hybrid impedance control of robotic manipulators," *IEEE J. Robotics and Automation*, vol. 4, pp. 549–556, 1988.

[7] S.D. Eppinger and W.P. Seering, "Introduction to dynamic models for robot force control," *IEEE Control Systems Mag.*, vol. 7, no. 2, pp. 48–52, 1987.

[8] J. De Schutter and H. Van Brussel, "Compliant robot motion II. A control approach based on external control loops," *Int. J. Robotics Research*, vol. 7, no. 4, pp. 18–33, 1988.

[9] S. Chiaverini and L. Sciavicco, "The parallel approach to force/position control of robotic manipulators," *IEEE Trans. Robotics and Automation*, vol. 4, pp. 361–373, 1993.

[10] S. Chiaverini, B. Siciliano, and L. Villani, "Force/position regulation of compliant robot manipulators," *IEEE Trans. Automatic Control*, vol. 39, pp. 647–652, 1994.

Prof. Bruno Siciliano
Dipartimento di Informatica e Sistemistica
Università degli Studi di Napoli Federico II
Via Claudio 21, 80125 Napoli, Italy
Tel: +39 81 768-3179
Fax: +39 81 768-3186
E-mail: siciliano@na.infn.it

Robust Control of a Two-arm Robot: An efficient implementation in a DSP-based controller.

P. Fraisse, F. Pierrot and P. Dauchez
L.I.R.M.M., Montpellier, France

Abstract: In order to perform complex coordinated tasks with a two-arm robot, we have designed and implemented a robust hybrid position/force control scheme. Previous papers have been dedicated to precise presentation of the method: the main goal of this paper is to present the practical point of view. Of course, we first recall the description of our original control scheme applied to a two-arm robot manipulating a single object: this control scheme has been defined after series of tests which have involved many different control schemes. We discuss also the reason why we have chosen a DSP-based controller to implement such a scheme, we show the ways the control software and the operator-robot interface are implemented on this single-processor architecture. Finally, experimental results are presented which prove both the robustness of the control strategies and the efficiency of a DSP-based hardware for advanced robot control.

1 Introduction

Two-arm robots are supposed to work in various environments and to perform various tasks. In fact, multi-arm robots are often presented as a solution for manipulation tasks in ill-structured environments (under-sea, space, out-door) where they could be asked to carry unknown objects, or to be in contact with unknown surfaces. Unfortunately, it is clear that the dynamics of the closed-loop system consisting of two robot arms and the object, leads to higly non-linear and complex equations and behavior. Moreover, even for an industrial single-arm robot setup, it could be quite impossible to model the environment on which the robot may have to exert forces. Consequently, due to the complexity of tasks where two-arm robots are involved, robustness is definitely requested: in our mind, this means that a two-arm system must have a behavior as constant as possible when payload or environment changes. The general philosophy we follow here is to select a complete control strategy (this includes the task description, the global architecture of the control scheme, the type of controllers, but also the values of the controllers gains) and to keep it for all the tasks the two-arm robot should have to perform.

Basically, we strongly believe that the use of force control is indispensable for a two-arm robot system. The solutions studied in this paper are based on one of the possible approaches for using force information in robot control: the Hybrid Force/Position control method. As a matter of fact we consider basically a Symmetric Hybrid Force/Position task description according to Dauchez [1] rather than any other hybrid description related to a Master/Slave approach as in [2] [3] [4]: this choice is not related to robustness (and actually the solutions presented in this paper to improve robustness can be useful for other task descriptions also) but we consider this approach as one of the most suitable for two-arm robot control. In this paper we focus on two control techniques that we recommend for two-arm robot: according to theoretical and simulation results previously presented [5], we consider in different ways position control and force control. As far as position control is concerned, Sliding Mode control seems to be a good answer to the robustness problem: as it will be shown later, we have suggested an improvement of this general technique in order to implement it on robots. Regarding force control, we have shown that Sliding Mode control is not enough to obtain good robustness, and then proposed a new concept: the Virtual Environment Concept [6]. Then we show how the two techniques we propose can be implemented in a realistic way in a DSP-based control system. We explain also why we move from a multi-processor architecture to a single-DSP architecture; we explain as well some of the various tools we had implemented in order to build a complete setup. Results obtained with a real two-arm robot are finally presented that definitely show the robustness of the control scheme we came up with.

2 Robust force/position control for two-arm robots

We believe that it is of interest to control two-arm robots *via* force information, for instance because:
- it is a convenient way to overcome kinematic modelling errors which can be particularly inconvenient when two arms manipulate a single object,
- it is a very good solution for complex tasks involving contact with the environment (insertion "in-space", surface following).

The first point is then to define the general architecture we work with.

2.1 General architecture

The task description is a symmetric task description where the reference is the object itself, rather than one of the two arms. We control two spaces: one is composed of relative displacements of end-effectors and internal forces into the object, the other one is composed of absolute displacement of the object and external forces acting on the object. In each space, a direction is either controlled in position or controlled in force. This task description is based on the hybrid force/position method proposed in [7] (**Fig. 1**).

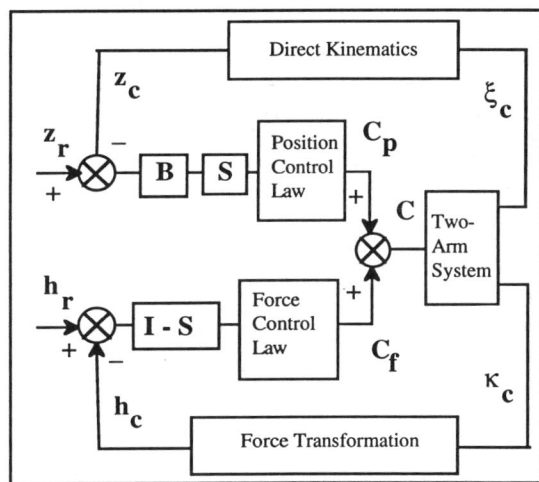

Fig. 1 Symmetric hybrid control scheme

The subscript r represents reference values and the subscript c represents current values. The input vectors **z** and **h** are defined by:

$$\mathbf{z} = \begin{pmatrix} \mathbf{Pa} \\ \mathbf{Pr} \end{pmatrix} \qquad \text{and} \qquad \mathbf{h} = \begin{pmatrix} \mathbf{fa} \\ \mathbf{fr} \end{pmatrix},$$

where:
Pa represents the absolute position/orientation of the object,
Pr represents the relative position/orientation of the arms,
fa corresponds to the external force and moment which produce the object displacement,
fr to the internal force and moment which produce its deformation.

The complete mathematical derivation of **z** and **h** can be found in [1]. The vector ξ represents the joint angles of the two arms. The vector κ represents the forces and moments measured by the force sensors at the wrists of the arms. From ξ and κ, it is straightforward to calculate **z** and **h**. **B** is a matrix which transforms the orientation error into a rotation vector. **S** is a diagonal matrix (12x12) which selects position or force control for such or such component. Its elements are respectively 1's or 0's depending on whether position or force control is chosen. **I** is the identity matrix with the same dimension as **S**. **Cp**, **Cf** and **C** are control vectors which result respectively from position servoing, force servoing, and the sum of the two contributions.
In order to obtain a high level of autonomy, we must be able to guarantee a given behavior when the robot is controlled with such an advanced scheme. With this goal in mind, we came up with two control techniques that may be mixed together to obtain a robust hybrid scheme. In the next sections, we will briefly describe the control laws we suggest for the position loop and the force loop.

2.2 Sliding Modes control for two-arm robot

The control law itself is based on a Variable Structure Control (VSC). It is intrinsically a robust control scheme which allows us to avoid the knowledge of the robot's dynamic parameters: in an availability domain, the system is forced to dynamically behave according to an equation defined *a priori*. This equation is represented in the state space by a surface called "sliding surface". Thus the system's behavior is indifferent to a certain extend to alterations of the system's parameters as well as to disturbances. From a mathematical point of view, it's a non linear control law based on a discontinuous term in the control parameters of the process, which forces the system's state to follow an exponentially stable linear differential equation. As far as we are concerned, for a unique actuator controlled in position, the VSC technique can be summarized by:

$$\begin{aligned} u &= +V_m \quad \text{if} \quad S(x) > 0 \\ u &= -V_m \quad \text{if} \quad S(x) < 0, \end{aligned}$$

where u is the control parameter, V_m is the motor maximum input voltage, x is the position of the actuator, S is the sliding surface. Moreover, S depends on the tracking error e as follows:

$$S = \dot{e} + \lambda e, \qquad e = x_t - x_c.$$

where:
x_t is the reference position value,
x_c is the current value.

With this choice of u, we obtain a simple control algorithm for real time implementation, and as V_m is the physical maximum input value of the controlled actuators, the trajectory described in the phase plane will stay on the sliding surface until the maximun capabilities of the actuators are reached.

As far as a two-arm robot is concerned, two cartesian sliding surfaces are defined, $S_z(z)$ and $S_h(h)$ for respectively position and force control, and they depend on the controlled variables, z and h:

$$S_z(z) = \dot{e}_z + \lambda_z\, e_z$$

$$S_h(h) = \dot{e}_h + \lambda_h\, e_h$$

where $e_z = B(z_r - z_c)$ is the tracking error in position and $e_h = h_r - h_c$ is the tracking error in force. Hence the resulting hybrid sliding surface is:

$$S_x = S\, S_z + (I - S)\, S_h$$

Besides, the changeover condition of the control vector u lies not directly on the sliding surface S_x since it's defined in the cartesian space, but on a joint equivalent, J^+S_x where J is the Jacobian matrix of the two-arm robot.

We have shown [5] that it was more efficient to modify the equation which defines the sliding surface, otherwise a chattering phenomenon appears: this is what we called VSC-HF (High-Frequency). We introduced in the sliding surface an additional term which increases the oscillation frequency far over the bandwidth of the controlled system and so it reduces the influence of the oscillations. To do that, we build a feedback loop with a lowpass filter on each relay which commutes a control component. This part of the control scheme is realized with analog circuits because the oscillation frequency must be much higher than the sampling frequency of the whole servo-loop. Simulating the resulting control scheme [5], the results we obtained in position control showed a very good robustness with respect to the variations of the robot dynamics. On the other hand, the VSC-HF is not very robust for force control as far as variations of the environment stiffness are concerned.

2.3 Virtual environment concept

To cope with the problem of robustness in force control, we have designed the concept of Virtual Environment. In order to understand easily this concept, let us just consider here the basic case shown in Fig. 2:

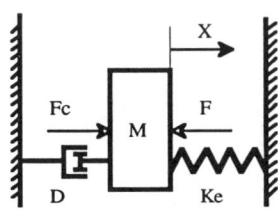

Fig 2. Basic case

where:

- Fc is the actuator force,
- D the damping coefficient,
- M the actuator mass,
- X the actuator position,
- F the force produced by the wall,
- Ke is the wall stiffness.

We have:

$$F(t) = -Ke\ (\ X(t) - X(t{=}0)\ )$$

To simplify the notations, let us consider that:

$$X(t{=}0) = 0,\ X(t) = X\ \text{and}\ F(t) = F$$

Then:

$$F = -Ke\ X.$$

Moreover:

$$M\frac{d^2X}{dt^2} = Fc - F - D\frac{dX}{dt}$$

Considering equation, we obtain, with a simple integral control law, the scheme of Fig. 3:

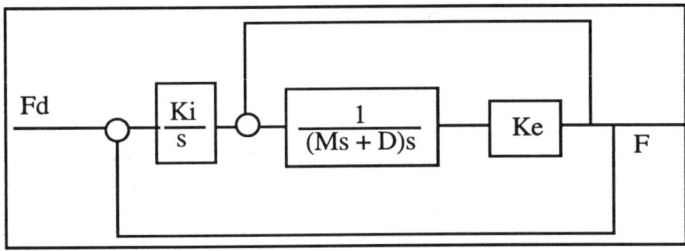

Fig. 3. Integral force control scheme.

From Fig. 3, the following transfer function can be derived:

$$\frac{F}{Fd} = \frac{1}{\dfrac{Ms^3}{KeKi} + \dfrac{Ds^2}{KeKi} + \dfrac{s}{Ki} + 1}$$

The environment influence, namely Ke, is obvious when considering the fact that Ki is multiplied by Ke! We propose to create a virtual environment whose influence is chosen larger than the real one: this situation is schematically shown in Fig. 4, where Kv represents the stiffness of a virtual spring.

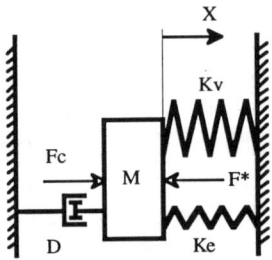

Fig. 4. Basic virtual environment.

In a simple case, if the stiffness of the virtual spring (Kv) is greater than the real environment stiffness, we have:

$$F = - (Ke + Kv) \, X \approx - \, Kv \, X$$

Introducing the virtual environment leads to create several virtual forces: Fd* is the virtual desired force, F* the virtual output force, and, considering Fig. 5 where the virtual aspect is separated from the real aspect (in the grey area), Fv is the force created by the virtual environment.

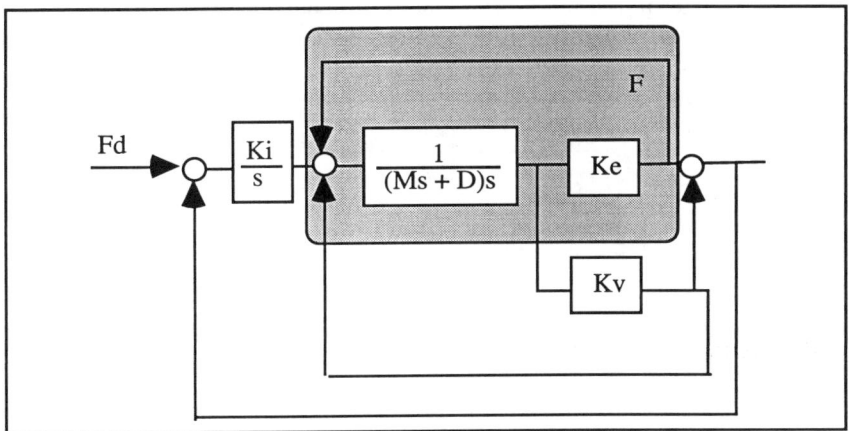

Fig. 5. Virtual and real worlds.

Considering Fig. 5 leads to the following transfer function:

$$\frac{F}{Fd} = \frac{1}{\dfrac{Ms^3}{(Ke+Kv)Ki} + \dfrac{Ds^2}{(Ke+Kv)Ki} + \dfrac{s}{Ki} + 1}$$

Clearly, the influence of Ke decreases: Ki is here multiplied by the sum of Ke and Kv!

2.4 Robust hybrid control for two-arm robot

This concept can be implemented in various control schemes, including of course a hybrid Force/Position control scheme. In that case, the control scheme of Fig. 1 has to be modified as follows. First, the feedback force vector is not h_c but h_c^* :

$$h_c^* = h_c + h_v$$

where h_v is due to the virtual spring and can be computed as follows:

$$h_v = K_v \, B \, (z_c - z_{co})$$

where:

$\mathbf{K_v}$ is the virtual stiffness matrix

$\mathbf{z_{co}}$ represents the initial position, when there is no effort ($\mathbf{h = 0}$).

Second, the contribution of this virtual effort must be substracted from the control vector, just as it was really felt by the robot. We have shown [8] that this modification could take place in the calculation of the cartesian sliding surface. Then it appears a matrix $\mathbf{kI}$ where k overvalues the gain between the sliding surface and the applied force. k depends in particular on the equivalent gain matrix of the non linear servo loop, and on the gain matrix between the torque control vector and the voltage control vector. Then we obtain the control scheme illustrated in **Fig. 6** (on this figure, the "TAS" block represents the "Two-Arm System").

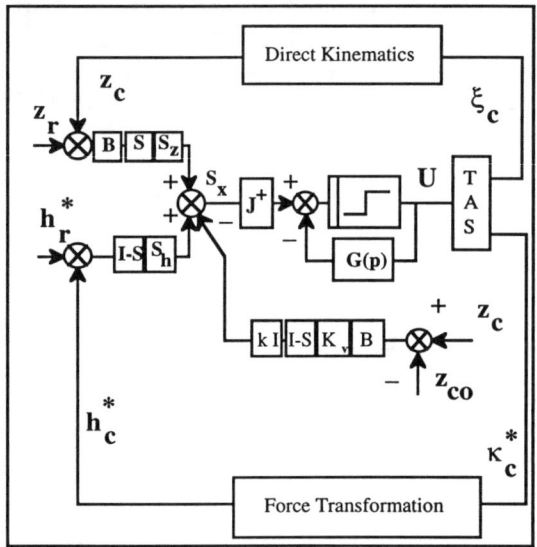

Fig. 6 Robust hybrid control scheme for a two-arm robot

This final control scheme lies on a hybrid symmetric structure, a VSC-HF control law for position servoing, and a VSC-HF with virtual environment control law for force servoing.

3 Implementation on DSP

As a matter of fact, our two-PUMA 560 robot system (Fig. 7), equipped with two 6-axis force sensors, has been working for quite a long time; but it was previously controlled by a multi-processor VME-bus system that gave too little computation power to implement any advanced control laws.

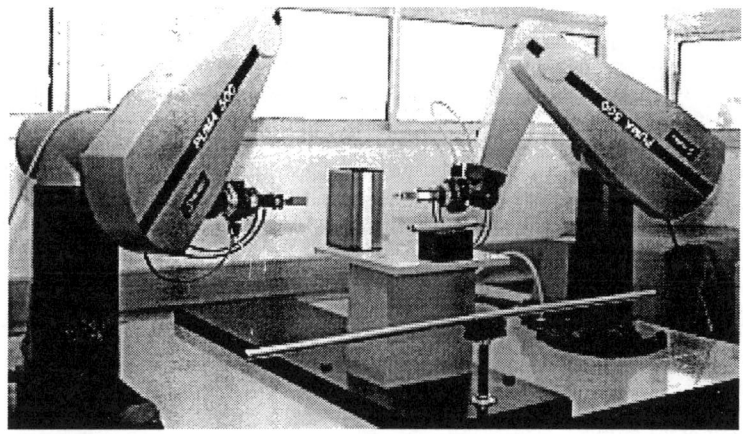

Fig. 7. Two-arm robot setup.

This system was based on five Motorola boards working in parallel on a VME bus: the best sampling period we obtained for the symmetric hybrid control scheme was 51 ms! Obviously, stability problems have been often encountered... At this time, we were looking for another solution that could give:

- computation power,
- portability of our C programs,
- simplicity of programming,
- basic Real-Time facilities,
- possibility of evolution.

Regarding computation power, we had selected a set of programs (PUMA Direct Kinematics, PUMA Jacobian Matrix, 6x6 matrix PseudoInverse), which were run on various processors in order to show their efficiency for typical problems. Those tests gave the following results (Fig. 8):

Processors	Software Package	Computation Time (ms)
Motorola 68020 / 68882 16 MHz	Cosmic C, CESAR	29
Intel 80486 25 MHz	MS C	4
Inmos T 805 25 MHz	Inmos C	4
Sun SPARC 2	CC	0.8
Texas C40 40 MHz (single prec.)	Texas C	0.4

Fig. 8. Comparison of computation times.

Since portability of our own C programs was guaranteed for almost all systems we have tested, we mainly focused on simplicity and evolution capability: we have found two systems that could present these two features:

- Inmos Transputers systems,
- Texas DSP systems.

As a matter of fact, both systems provide very easy and natural ways to work with Real-Time for robot manipulators; basically (this means: without any specific Real-Time environment... and then at no cost!) one can write the "minimum Real-Time loop", plus a "background" task, with only few C lignes: of course, this is just a "minimum set of functionnalities" but, up to now, we did not consider this as a limitation in our work. However, more complex arrangements can be made more easily with Transputers than with Texas DSP (due to multitasking facilities build in the Transputers).
Compiling and loading are very simple tasks in both cases too.
Evolution capability is guaranteed on both systems since the processors provide high-speed links.

Then, the first reason to choose a DSP system is computation power (see Fig. 8). The second reason is coming from the following constraint; research in robotics applications required a lot of interactions between developpers and the control softwares, for many purposes:

- to tune control gains,
- to select a control law or even a control scheme,
- to read parameters,
- to save data,
- ...

Consequently, the availability of a "direct" connection between the Real-Time Control Hardware and the Host Machine has been a key point in the decision process. As a matter of fact, dSPACE hardware and software packages [8] had let us developped a complete setup in a very friendly way.

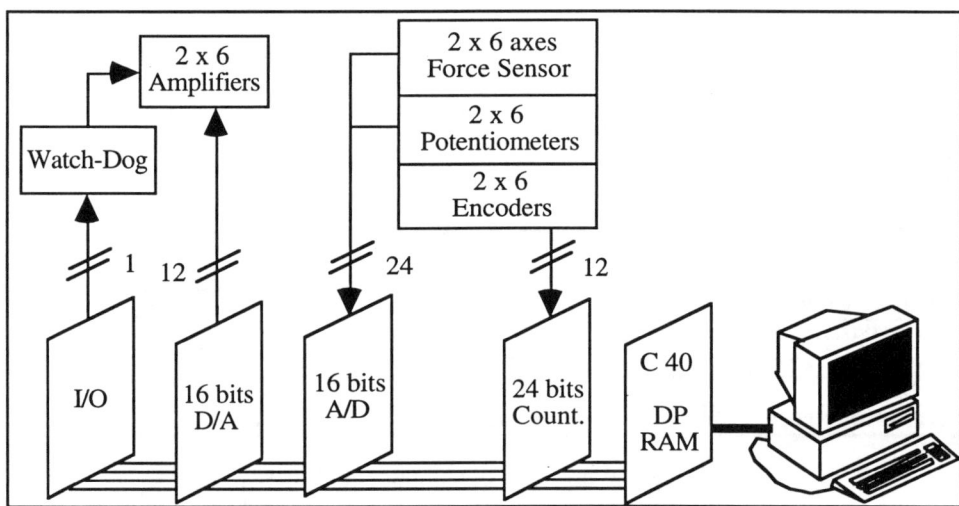

Fig. 9. Connections between robots, controller and host machine.

One of the key-points is data exchange through Dual-Port Memory (DP RAM in Fig. 9), which can be made either by Windows-based softwares (provided by dSPACE) or low-level DOS-based softwares we wrote: we use them when the host machine is an extremely low-cost PC that cannot even support Microsoft Windows, or is equipped with a very small screen (this is sometimes usefull when all the equipment is mounted on some vehicle). As shown on Fig. 9, the system is quite complex, due to the twelve axes (each one is equipped with an encoder and a potentiometer) and the two 6-axis force sensors: this gave 24 Analog Inputs, 12 Analog Outputs and 12 Encoder Inputs. Additionnaly we use one Digital Output to provide the following safety behavior: if the sampling period is greater than a given value, a Watch-Dog board cuts automatically the Power on amplifiers. Obviously, other digital I/O are used for grippers command.

As mentionned earlier, the control software is arranged is a very simple way: a single Real-Time loop is always running and works according Fig. 10.

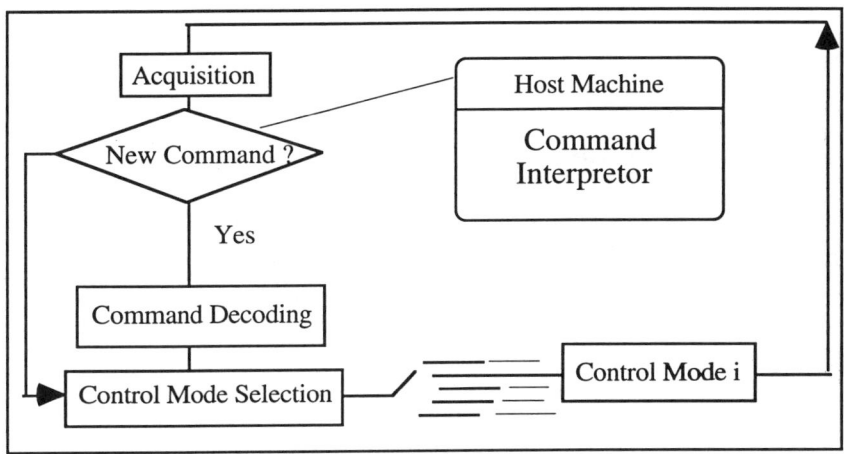

Fig. 10 Control software

The 'Acquisition' block in Fig. 10 actually represents acquisition itself (36 data every sampling period) but also all transformations needed for high-level control schemes: this means for instance that Encoders acquisition is followed by the computation of joints position <u>and</u> cartesian position too (*via* the direct kinematics). Moreover, this block is in charge of storing many data in Dual-Port Memory. All those tasks are done in 0.63 ms.

After acquisition, the software access Dual-Port Memory to check if the operator has made a new interaction; the interactions are done on the PC, *via* a Command Interpretor we developped. This Command Interpretor is able to recognize the type of command, to check if all parameters are present, to verify if all parameters are consistent with the command, and finally to write all information in the Dual-Port Memory.

The software running on the DSP is then able to select the Control Mode and obviously to compute the selected control law. Up to now, we can choose among the following list:

- Joint Space PID, one arm
- Joint Space PID, two arms
- Hybrid One-Arm Control, PID Controllers
- Hybrid One-Arm Control, Robust Controllers
- Symmetric Hybrid Two-Arm Robot Control, PID Controllers
- Symmetric Hybrid Two-Arm Robot Control, Robust Controllers
- Simplified Impedance Control, one arm
- Complete Impedance Control, with Force Feedback, one arm

• Simplified External Hybrid Control, two arms
• Complete External Hybrid Control, with Dynamic Decoupling, two arms

To show the efficiency of this implementation, a single figure could be sufficient: the 51 ms sampling period we had with five boards working in parallel on a VME bus has been divided by about 25: the robust hybrid control scheme is now running with a 2ms sampling period.
However, one limitation remains: we still need (few) analog circuits to realize the High-Frequency oscillations (frequency: 5 kHz) of the VSC-HF control law.

4 Experimental results

4.1 Testing conditions

In order to show the robustness of the techniques proposed in section 2, we have performed various tests involving various payloads, various stiffness coefficients and various velocities. The motions are specified as position steps that are filtered numerically before use (then to modify velocity and acceleration, we just have to tune the filter). Force reference values are simply obtained with force steps.
The error criterion is as follows:

$$I = \sum_{n=0}^{n=n_{final}} (\mathcal{E}(nT_s))^2 T_s$$

where :
T_S is the sampling period,

$\mathcal{E}(nT_S)$ is the error (in force or in position), at time nT_s, in the selected direction.

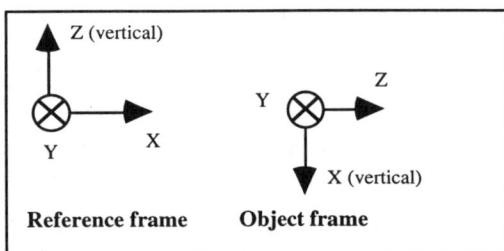

Fig. 11 Definition of the frames

Absolute positions and external forces are expressed in a fixed reference frame. Relative position and internal forces are expressed in a frame attached to the object (**Fig. 11**).

4.2. Robustness of position control

In order to evaluate the robustness provided by the Variable Structure Control with High

Frequency oscillations we proposed, we have implemented a classical PID controller on the same hardware, running at the same sampling period, and we compare here the behavior of both schemes when the two-PUMA robot carries various payloads, at different speeds.

In fact, as shown in Fig. 12a, we made tests with a motion along the Z axis of the reference frame; the robot starts at Z=100cm (the two-arm robot base is located at Z=0), moves down to Z=65cm (low speed) or Z=55cm (high speed) and goes back to Z=100cm immediately. Desired velocities are shown in Fig. 4b : in one set of trials, maximum speed is -40cm/s (low speed), in the other set, maximum speed is -60cm/s.

Actually the high-speed tests were done only with VSC-HF, because PID controller were not able to work at such speed.

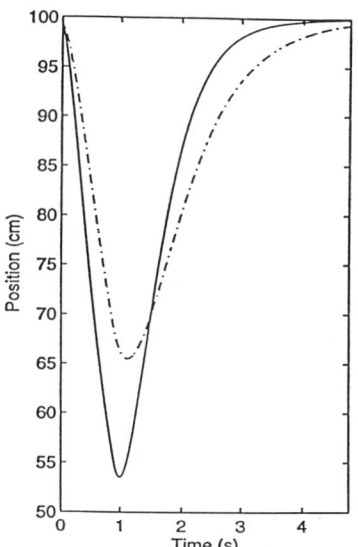

Fig 11a. Desired positions.

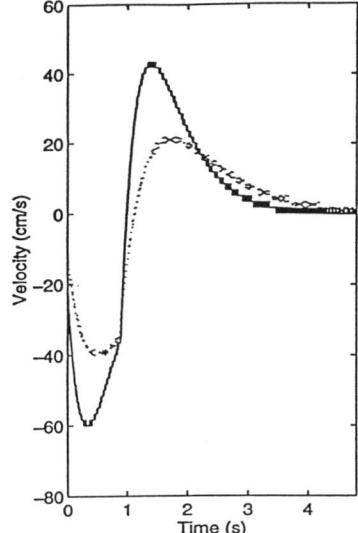

Fig 11b. Desired velocities.

The advantage of VSC-HF versus PID is clearly summarized in Fig. 12 where results for PID (low speed) and VSC-HF (low speed and high speed —dased line—) are plotted for masses from 0kg up to 6 kg.

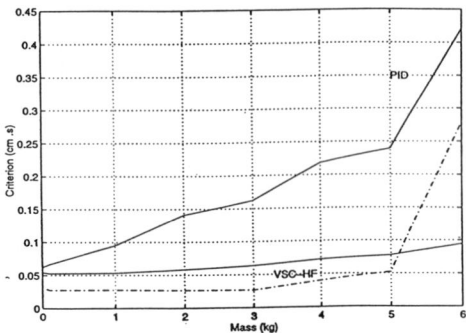

Fig 12. Values of the criterion for PID and VSC-HF.

3.3. Robustness of force control

The task consists in exerting an internal force of 15 N on a deformable object (actually: a spring) along Z axis of the object frame. All directions but Z axis of object frame are position controlled. We compare here two schemes: one involving Virtual Environment Concept (plus VSC-HF), one involving regular force scheme (again with VSC-HF). Unfortunately we had only three different springs whose stiffnesses were: 2 N/cm, 5 N/cm and 10 N/cm. We tuned the gains for the medium value (5 N/cm) and then tried to evaluate robustness with the two other values. A shown in Fig. 13a and 13b, results are quite good in both cases and we were not able to make good evaluation.

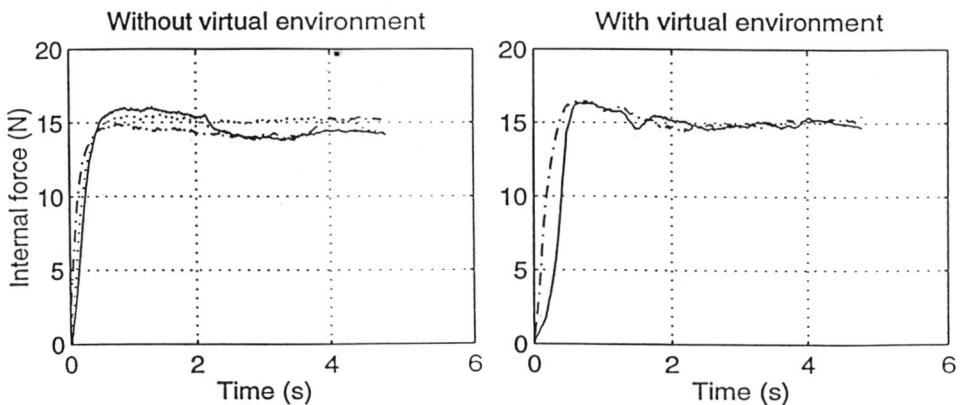

Fig 13a. Force responses.   **Fig 13b.** Force responses.

5 Conclusion

Advanced control techniques have been implemented in a realistic way in a two-arm robot involving 12 DC actuators and 2 six-axis force sensors. Thanks to a DSP-based hardware, a 2 ms sampling period has been obtained that avoids discrete time considerations. We are now close to a 100% software-based solution for VSC implementation: we expect that future evolution of our setup (adding another DSP board for instance) and optimisation of control software will let us obtain a sampling period drammatically smaller.

The robustness, both in position and in force, is quite good. The two-arm robot is now able to move objects whose mass may vary from 0 kg up to 8 kg, at various velocities or accelerations, while keeping almost the same tracking error. The force response is also quite constant when the robot exerts forces on environments whose stiffness vary drastically (namely from 1 to 5): however, the preliminary results presented in this paper are not sufficient to conclude on the robustness of force control. We definitely need more tests regarding that point.

The system described here has also been tested on a prototype of mobile manipulator with success.

6 References

[1] P. Dauchez, "Task Description for the Symmetric Hybrid Control of a Two-Arm Robot Manipulator", Technical Report # 90011, LIRMM, Montpellier, France, 1990.

[2] T. Ishida, "Force Control in Coordination of Two Arms", Proc. of the International Conference on Artificial Intelligence, August 1977, pp. 717-722.

[3] Y.F. Zheng, J.Y.S. Luh, "Joint Torques for Control of Two Coordinated Moving Robots", Proc. IEEE Int. Conf. on Robotics and Automation, San Francisco, April 1986, pp. 1375-1380.

[4] S. Hayati, K. Tso,T. Lee, "Generalized Master/Slave Coordination and Control for a dual Arm Robotic System", Proceedings of Second ISRAM, Albuquerque, November 1988, pp. 421-430.

[5] P. Fraisse, P. Dauchez, F. Pierrot, "Robust Hybrid Control Schemes for a Two-Arm Robot. Performance Analysis", Proc. of IECON '92, San Diego, November 1992, pp. 676-681.

[6] P. Fraisse, X. Delebarre, F. Pierrot, "Virtual environment for Robot Force Control. Experimental results", Proc. ICAM '93, Tokyo, August 1993.

[7] M.H. Raibert, J.J. Craig, "Hybrid Position/Force Control of Manipulators", Trans. ASME, Journal of Dynamic Systems, Measurement, and Control, vol. 103, N 2, 1981, pp. 126-133.

[8] P. Fraisse, "Contribution à la commande robuste position/force des robots manipulateurs à architecture complexe. Application à un robot à deux bras", Thèse de Doctorat, Université Montpellier II, February 17, 1994 (in French).

[9] dSPACE GmbH, TRACE & COCKPIT, 33098 Paderborn, Germany

Dr. Philippe Fraisse, Dr. François Pierrot, Dr. Pierre Dauchez
LIRMM   UMR 9928 CNRS / Université Montpellier II
161, rue Ada
34392 Montpellier Cedex 5   France
Phone: (+33) 67.41.85.56 or (+33) 67.41.85.61
Fax: (+33) 67.41.85.00
E-mail: fraisse@lirmm.fr, pierrot@lirmm.fr, dauchez@lirmm.fr

Robot Path Control with a Decentral Structure

Rüdiger Neumann[1], Wolfgang Moritz[2]
1. University of Ulm, Ulm, Germany
2. University of Paderborn, Paderborn, Germany

Abstract: One of the experiments of the D-2 spacelab mission was the Robot Technology Experiment ROTEX was which designed to test the use of robots in space. The expressly developed joint control concept for the manipulator arm with six joints is described. But first the modelling of a joint shows that the robot has to be modelled with joint elasticity. With the use of decentral disturbance observers for each joint not only the friction torques but also the coupling torques between the axes can be detected and compensated for. Furthermore, the observed states are utilized for a state feedback which allows an active vibration damping. Feedforwards are used to guarantee a high path accuracy whereby with the use of the disturbance observer the feedforward coefficients are independent of variable robot parameters. As the controllers were implemented as a linear discrete state-space system, they could be changed for the ground space during the mission.

1 Introduction

There are robot applications which desire not only a high end-effector accuracy at the end of a point to point movement. There also tasks such as laser welding, water jet cutting or precise effector movements in a restricted environment which have to have a high path accuracy throughout the whole trajectory. An application of this kind was the Robot Technology Experiment ROTEX [1] within the D-2 spacelab mission which was designed to test the use of robots in space. As the manipulator arm with six joints (J1 .. J6) and a multisensor end-effector (EE) were built into a restrained SPACELAB rack, as shown in figure 1, path errors or oscillation of the arm could easily cause collisions.

This was the starting-point for the development of a joint control concept for high path accuracy and oscillation damping for the ROTEX robot, because previous investiga-

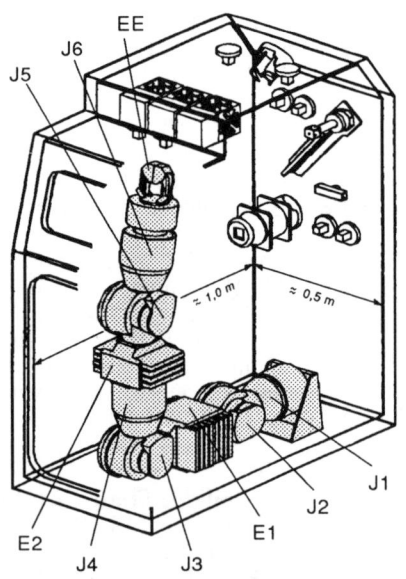

Figure 1: ROTEX robot in the workcell of the spacelab

tions [2] have shown that a conventional cascade controller could not reach the above demands. Therefore, the contribution of the Department of Automatic Control of the University of Paderborn to the ROTEX experiment, which was supervised by the DLR and developed jointly by aerospace companies and research institutions, dealt with the modelling of the robot as well as the structural and the numerical design of a decentral control structure for the control of the joints. Moreover, during the D-2 mission the controller behaviour was subjected to analysis and the controller fine-tuning in consideration of the missing gravity was the first ROTEX experiment.

Although the robot mechanic and the control was especially suited for a space application, the developed control structure is also applicable to industrial robots with geared joints. With it the effects of the following reasons for path errors can be reduced:

- *Friction*, which occurs especially in joints with gears of a high gear ratio, e. g. Harmonic Drive gears.

- *Elasticity*, which has the main orgin in the limited stiffness of the gear and also in the link.

- *Coupling* between the axis of a robot through which the movement of one joint has an undesired influence of the other joints.

- *Variable roboter dynamic* in dependency of the joint positions and the payload.

As a result of the hardware design the ROTEX joint controller structure is purely decentral as shown in figure 2, but it should cope with the above difficulties. Because of the structure and of the limited processor power, there was no way to make use of central algorithms, such as decoupling via an inverse model.

Figure 2 shows that the robot processor calculates the reference position and velocity for each joint controller. They are located in the joint-processors on the robot arms and are getting their reference and measurement values only from the joint that has to be controlled.

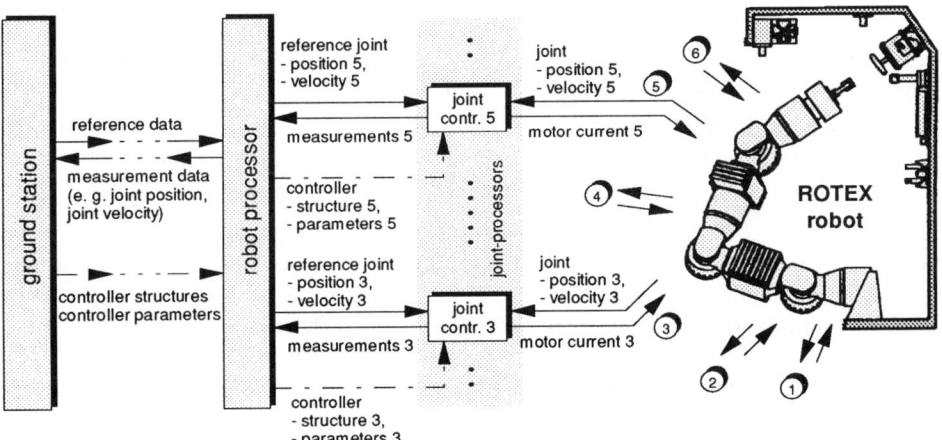

Figure 2: Decentral control structure of the ROTEX robot

This control structure is also typical for industrial robots where the joint controllers are located in a decentral manner on the motor amplifier of the joints. The used control algorithms normally are cascade controllers with an inner PI velocity loop and a superimposed proportional position loop. For the joint control of the ROTEX robot the control algorithms were not fixed to a P-PI cascade, a linear discrete controller of 3rd order was implemented. Altering the four controller matrices of each individual joint controller - a procedure that was effected from ground control during the D-2 mission - it was possible not only to adapt some feedback parameters, but also to generate a different feedback structure. This flexibility had proved its effectiveness especially in the initiation phase before launching the mission. Thus, it was possible to fine-tune the control and modify the joint controllers during the mission.

In comparison to industrial robots were the sampling frequency of the joint controller is about 1 kHz the sampling frequency of the joint controllers was limited to 100 Hz, due to the processors employed. This and also the delay time in computing and measuring had to be taken in consideration during the design process.

2 Modelling

The robot is an aluminium light-weight construction. Unlike conventional robots, this one has all six of his joints built up in the same way, which means that the motor torques of the main axes are unable to set the robot in motion under the influence of gravity. This is why, during the initiation phase on ground, gravity had to be compensated for by counterweights at the deflection pulleys. The high-reduction Harmonic-Drive gears in the joints are operated by brushless DC motors. On the basis of modelling and identification of a single drive performed by means of a test-stand [3], the equations of motion of the entire robot were determined in symbolic form.

2.1 Joint Model

The structure and the parameters of a joint were identified by means of a test-stand that provided measurement of the motor torque and the load-side torque used, in addition to a precise measurement of the motor- and load-side angles.

In order to measure the stiffness, the load-side was blocked and a triangle-shaped torque was applied by the motor. Figure 3, left-hand side, displays the motor-sided difference angle which was transformed to the load-side via the gear ratio traced above the load-side torque.

Measurement reveals the stiffness to be significantly higher and far less amplitude-dependent than indicated in the catalogue. Furthermore, it becomes obvious that there is no backlash in the gear - this being a result of the clutch-free connection between motor and gear as well as of the quality of the gear-teeth without backlash. The slope which indicates the stiffness, figure 3, right-hand side, is obviously reduced for small amplitudes of gear-drive deformation.

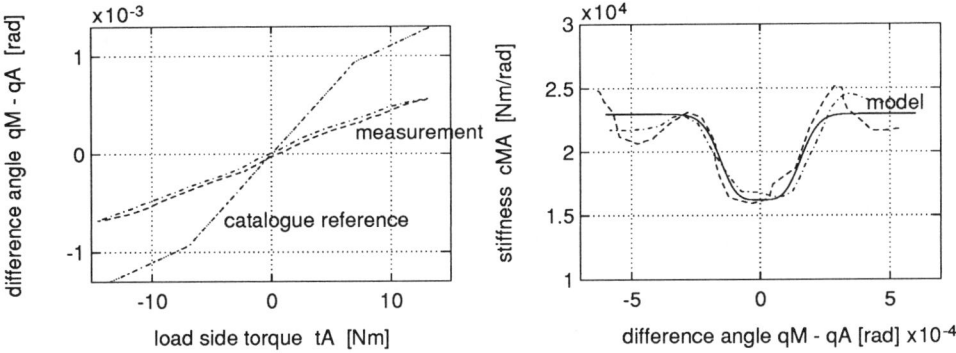

Figure 3: Stiffness of the Harmonic Drive gear type: HDUC-20-160-BLR

Further measurements were performed to determine the friction; they are displayed in figure 4. Measurements at various stages of mounting of the joint led to the origin of the friction being located. The curve b) in figure 4, left-hand side (without circular spline), shows clearly that the velocity-proportional part of the friction which is responsible for the poor efficiency of the gear-drive has its origin in the wave generator, for the friction in the tooth system leads only to an increase in the Coulomb part (cf. c)). What is more, no difference between the static friction and the sliding friction can be detected.

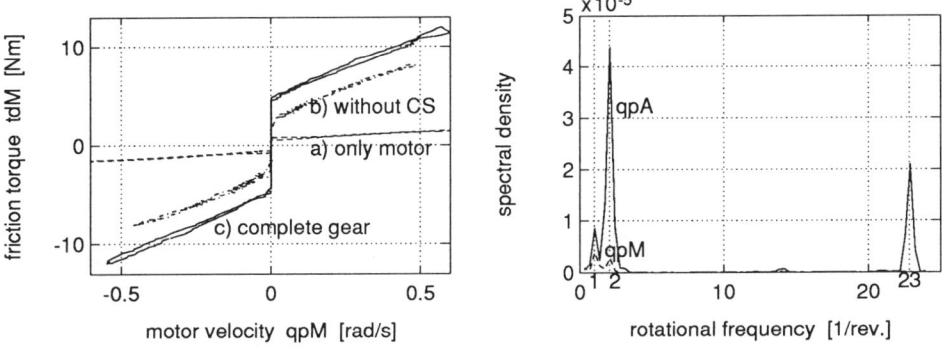

Figure 4: Friction and position dependent disturbances (HDUC-20-160-BLR)

When the gear was operated at constant velocity, a position-dependent disturbance made itself felt by exciting vibration in the gear. Figure 4, right-hand side, shows the power spectral density of the motor-side and the load-side velocities (qp_M resp. qp_A) depending on the number of motor revolutions (on the motor-side). The arising vibrations correlate with one, two, and twenty-three motor revolutions. They result from an inexact alignment of the elliptic wave generator as well as from the 23 balls in the wave-generator bearing and affect mainly the load-side coordinate. Measurements have shown that modelling of a drive by a two-mass oscillating system is possible if the latter is extended by non-linear parts, such as Coulomb friction of the motor side, an amplitude-dependent yet backlash-free gear torsional spring, and a position-dependent load-side disturbance

A comparison of the gear stiffness of the first three axes of the ROTEX robot with that of an industrial robot, such as the "manutec r3", has shown that the 3rd axis of the latter, which has a similar load-side inertia, displays a stiffness ten times higher. Thus, the control is bound to have a particular vibration-damping effect because the drive itself excites vibration. Modelling of the Harmonic Drive gear, especially modelling the effects of elasticity and friction is detailed in [4].

2.2 Entire Model

The results of the modelling of an individual joint were taken into account in the symbolic generation of the equations of motion of the six-axis robot with joint elasticities. With the help of a program written in MAPLE the equations of motion were generated and transformed into state-space form following a symbolic inversion of the mass matrix by a Cholesky decomposition [5]. The non-linear robot model of 24th order, formulated in DSL (Dynamic System Language) [6], was combined with models of friction, amplifier dynamics, sensors, and the discrete joint control to form an entire model. So-called CAMeL tools developed at the institute for the mechatronic design of controls [7], such as simulation, frequency response analysis following automatic linearization, and controller optimization [8], are harmonized to the DSL input description.

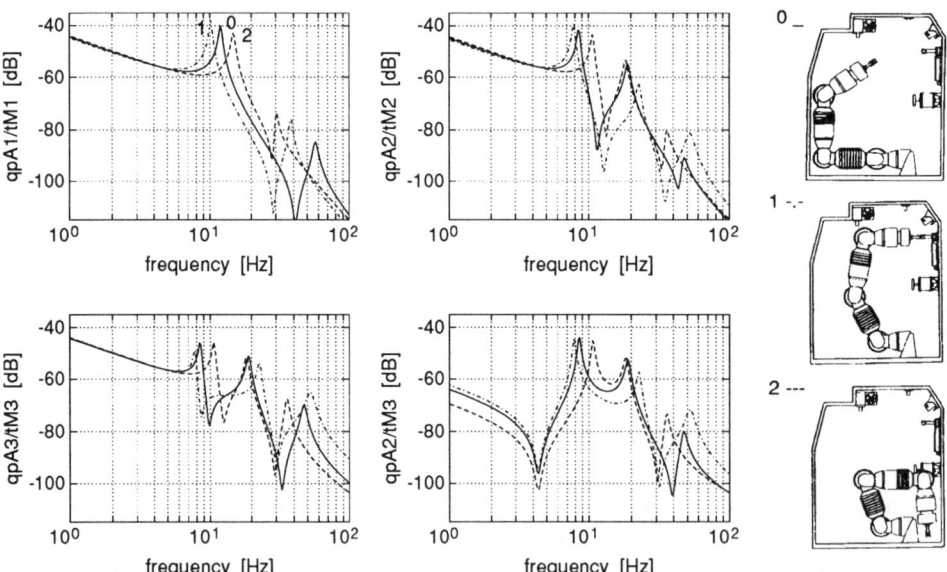

Figure 5: Magnitude of transfer functions in three different robot positions

Figure 5 shows e. g. some magnitudes of the transfer functions from motor torque to load-side velocity, computed from the DSL model in 3 robot positions. The eigenfrequencies vary in accordance with the respective position; due to the interconnection between the axes, the transfer functions do not resemble those of a two-mass oscillating system. In a frequency range near the eigenfrequencies, a distinct coupling between the axes can be detected, as shown by the coupling transfer function in figure 5, bottom right.

3 Control Structure

The modelling of the a single joint and of the manipulator reveals the following demands to the control. The high friction has to be compensated for and the ribble torques have to be suppressed. As the robot has a higher elasticity than industrial robots there is a need of active vibration damping and a reduction of the coupling torques.

For the detection of friction in drives disturbance observers are already used for a long time [9]. But if the drives are coupled as they are in a robot this coupling has to be taken in consideration in the robot model. This would lead to one central observer which has to be adapted to the actual working position. Besides the fact that a central controller structure is not possible for the ROTEX robot such an observer is expensive to realise. An alternative procedure is a decentral disturbance observer structure for each joint which was proposed for a robot model with rigid joint in [10, 11]. The coupling were regarded as external forces and they are observed in the same way as friction. It should be mentioned that the separation principle is not valid anymore as the plant and the observer have a different structure. Therefore the design of the controller and observer parameters not independent. It has to be done simultaneously.

For the ROTEX robot the observer had to be suited to a lot of implementation properties such as a maximum controller order of three, a sample rate of 100 Hz or a velocity measurement with great dead zone. The design of the observer based controllers are described in detail in [12, 13, 14].

If the observer based decentral control is applied to a robot that cannot be treated as rigid as it has a high joint elasticity like the ROTEX robot, there are four components of the control which has to be adjusted to each other. This chapter gives an overview of the components shown in figure 6 of which a detailed description is shown in [4,15].

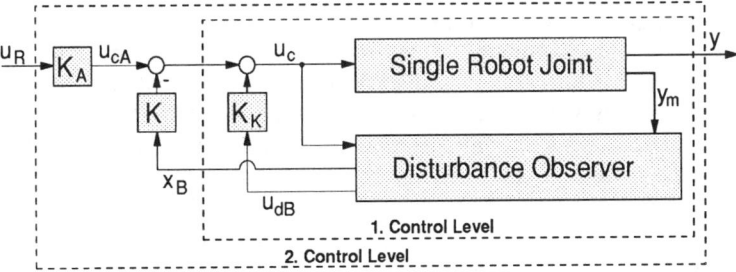

Figure 6: Principle control structure of the observer based decentral control

3.1 Disturbance Observer

As shown in figure 6 each joint needs an independent disturbance observer. For a robot with significant joint elasticity or for a robot application which desire a high path accuracy at high velocity and acceleration, a rigid body observer model cannot be used. The model has to have the structure of a two-mass oscillator. To detect the high motor friction independently from the load side torques and the coupling torques a motor side and a load measurement is needed. For the ROTEX robot the motor velocity and the load side angle were available. The joint model has to be augmented at the motor side and at the load side with an integrator as disturbance model since an integrator is suited for step-shaped disturbances. In figure 7 the observer is in the box marked with 1 and the box with the disturbance model is marked with 2.

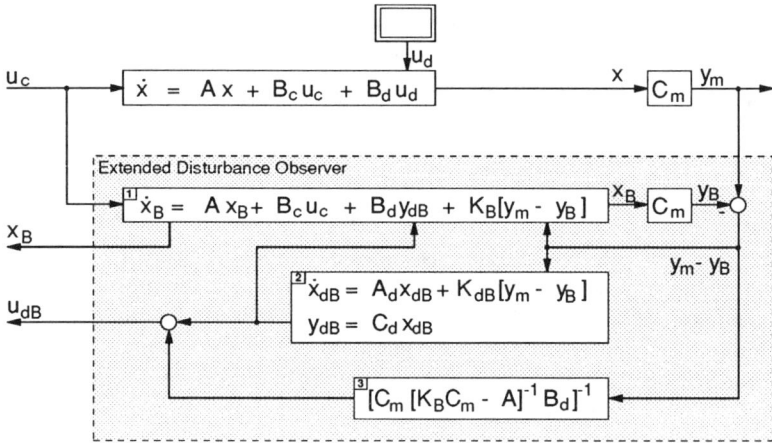

Figure 7: Structure of an observer with extended disturbance model

As shown in the chapter about the joint modelling the load side disturbance is a high frequent ribble torque. To detect such a disturbance an integrator is not fast enough. Therefore, a part which is proportional to the observer error is added to the disturbance model marked with 3 in figure 7. This component is only in action in the transient phase of the disturbance model and the additional output results in a faster disturbances reconstruction.

The proportional gain is calculated in the same way as the disturbance output of an observer without explicit disturbance model proposed in [16]. In combination with a classical observer with an integrator as disturbance model the extended disturbance observer has the following properties. The disturbance reconstruction is faster and it is also possible to reconstruct ramp-shaped disturbances with stationary accuracy. The integral over the difference between the disturbance and the observed disturbance is zero after the end of the transient for step-shaped disturbances. Therefore a reconstruction error during the transient phase does not result in an error in the velocity if the observed disturbance is used for compensation.

3.2 Disturbance Compensation

The observed disturbance has to be added to the motor input in the way that the disturbance has minimal influence on the load side position. If the observer consists only of a rigid body model the compensation gain is 1. But if the load side disturbance observed independently the compensation gain for a static compensation has to be calculated according to [17]. If a dynamic compensation function is desired the inverse transfer function of G1, as shown in figure 8, has to be calculated.

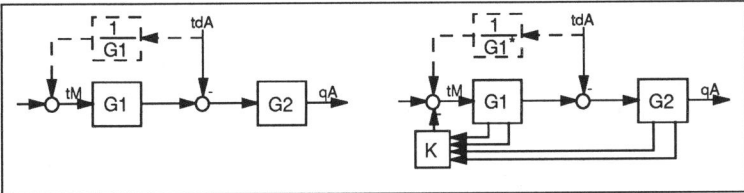

Figure 8: Principle of disturbance compensation of a plant with state feedback

The transfer function G1 is quotient of the command transfer function and the disturbance transfer function. If the plant is controlled with a state feedback the command and disturbance transfer functions have be calculated in consideration of the feedback as shown in figure 8, right hand side.

The disturbance observer and the dynamic disturbance compensation together form the first control level which ensures that joint non-linearities are linearized and that the robot is approximately decoupled [4].

3.3 State Feedback

A state feedback is needed for system stabilization and active vibration damping where as the observer provides the not measured states. Control of interconnected axes whose load-side inertia varies with the robot position will have to be considered by the optimization of control parameters. That is why the individual joint controllers were designed with the help of a parameter optimization with objectives vector [8] in various operating points at the entire model, this procedure being described in more detail by [12,14].

3.4 Feedforward Gains

The last component which insures a position error free path following at a high velocity and acceleration are the feedforward gains. Not only position but also velocity, acceleration and jerk have to be feedforwarded as reference values. The feedforward gains normally depend on plant parameters and feedback gains. As plant parameters can vary like a temperature dependent damping or the load side inertia, the gains ought to be adapted. But if the robot is controlled in a first level with the disturbance observers, the feedforward gains depend only on observer parameters which are exactly known and constant. The described control strategies were not only used in a reduced form at the ROTEX robot they were also implemented at the elastic test robot of Paderborn [4].

4 Controller Optimization during the D-2 Mission

4.1 General Procedures

During seven days of the D-2 mission, many experiments were performed with the robot; in addition to the initial experiments on the control and tracking behaviour, there were also ones of calibration, handling, mounting, and telemanipulation. In all, 6 different control schemes were implemented and their suitability tested. Of these measuring data, only a few segments will be presented here, with the measuring data being contrasted with the data simulated through the model. The reference values for these simulations were the set values that were transmitted simultaneously to the ROTEX robot.

The symbolic entire robot model for the simulation comprised the coupled six-axis robot dynamics including vibrational behaviour with friction and non-linearities in the drives, sensor models including dead time in measuring, the controllers in discrete form,

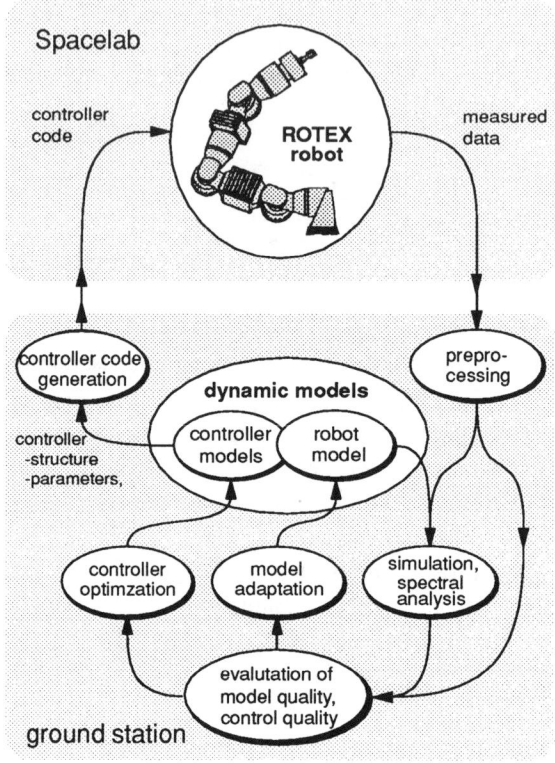

Figure 9: Computer aided control analysis and synthesis

and dead time in computing. The entire model consisted of 29 subsystems described in DSL [6] and interconnected, thus producing an entire model of order 48. Figure 9 displays the structure of the computer-aided analysis and controller synthesis that was used during the D-2 mission to fine-tune the control.

Measurement data (reference and actual link angles and angular velocities) of different tracks were transmitted to ground control. Pre-processing was used to convert the measuring data into the formats and sampling grids required. On the one hand, the quality of the control could now be judged directly, on the other hand spectral analysis was employed to find out if oscillations in the measured variables correlated with structural eigenfrequencies or with the number of motor revolutions.

As the joints were to weak, no final test could be performed under gravity conditions. In space at zero gravity, however, the robot proved to be far more oscillatory than on earth, due to the missing holding device lacking mass and damping. It turned out that vibration damping especially in the first axis was not effective enough. Furthermore, the fourth axis showed a strong, rotation-dependent disturbance that was supposed to be the result of the drive not being mounted in precise alignment.

Thus, in a first step, the robot model had to be adjusted to this fact; the quality of this adjustment was checked up by means of a comparison of the simulated and the measured data. In a second step the control was fine-tuned to the adjusted plant model. After discretization, newly adjusted controllers or controllers with altered structure were formatted and transmitted to the ROTEX robot.

In this way both the fine-tuning of the control and the altered control structures were tested. For analysis and synthesis, tools both of the CAMeL [7] environment and MATLAB were employed; therefore the existing data processing could be used to explain also the causes of events unforeseen that affected experiments performed by other work-groups.

4.2 Measurements Using the Initial Control

The first experiments served as the check-up of the basic functions of the robot and the examination of a control under the influence of zero gravity. Figure 10 displays a small section of the PPE trajectory (Predefined Path Execution) with the control that previous tests on earth had proved to be useful and that was now implemented as the initial control.

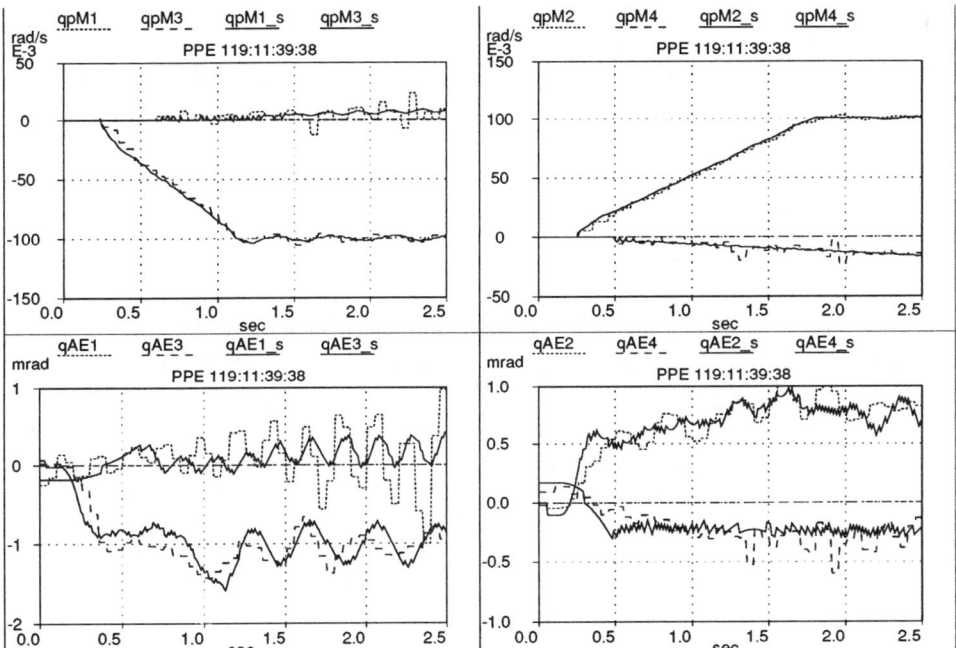

Figure 10: Velocity and control error (initial control)

The two upper partial images display the measured and also the simulated motor velocities of the first 4 axes while the lower partial images show the corresponding control errors for the load side. The control errors make it clear that especially the first axis lacks sufficient damping. Already during simulation the plant simulation model was adjusted to the altered zero-gravity conditions regarding joint damping. Measurement and simulation data reveal a high degree of conformity. It has to be pointed out that contrary to common usage it was not the absolute angles that were compared but the control errors that lie in the millirad range and here reveal only deviations in a few sensor increments. As the real behaviour is so well represented by the model, a re-optimization of the control with the new model adjusted by means of measuring data supplied by the Spacelab will lead to an improvement of the control behaviour.

4.3 Measurements Using the Optimized Control

Between the experiment phase that the above measurement was taken from and the one in which a new controller was to be tested 10 hours had elapsed. During this time the whole analysis and synthesis cycle as represented in figure 9 was performed; it comprised model adaptation and controller design for all six joint controllers with altogether 30 controller and observer parameters to be designed. The first step in the subsequent experiment phase was implementation of the joint control newly adjusted to the special conditions of the robot. During this phase, the PPE trajectory was performed with the course shown in figure 11 representing the same trajectory segment as figure 10.

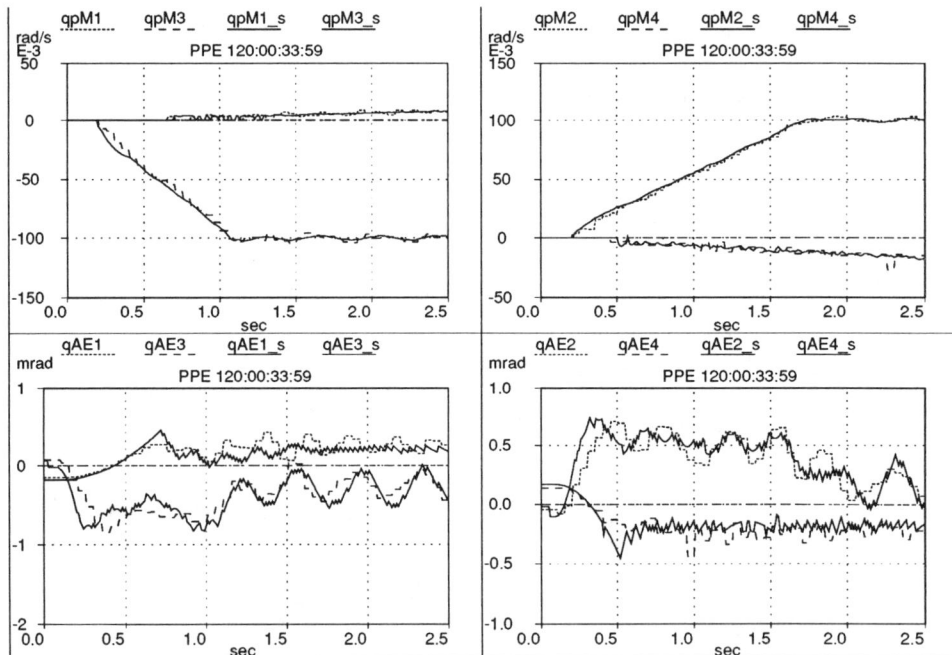

Figure 11: Velocity and control error (optimized control)

The velocity courses show distinctly less vibrations; this can also be seen in the control errors that are also just half as high in the axes with maximum velocity. Due to the lack of acceleration feedforward in the acceleration phase, the control error is higher and will be reduced during the phase of constant velocity. The residual ripple to be seen especially at axis no. 3 is the result of a joint not being mounted in precise alignment and occurs precisely once per motor revolution.

In all, the dynamical control errors at the load side are less than 0.8 mrad (2.7 angular minutes); the stationary errors are 0.2 mrad at the highest (40 angular seconds). The results surpass those of more rigid, conventionally controlled industrial robots which have also far higher sampling frequencies with position control.

This control was further employed in the subsequent experiment phases. Supervision of the control accuracy in the subsequent experiments proved the control to function regularly even when the gripper had contact with the environment or when the robot was operated by superimposed force control.

4.4 Measurements using Modified Controller Structures

As the D-2 mission was extended by one day, there was some time left for additional experiments on altered controller and observer structures. As tests of this kind were not free of risk, the experiments were confined at first to the sixth axis because the consequences of a malfunction would be smallest with the last hand axis.

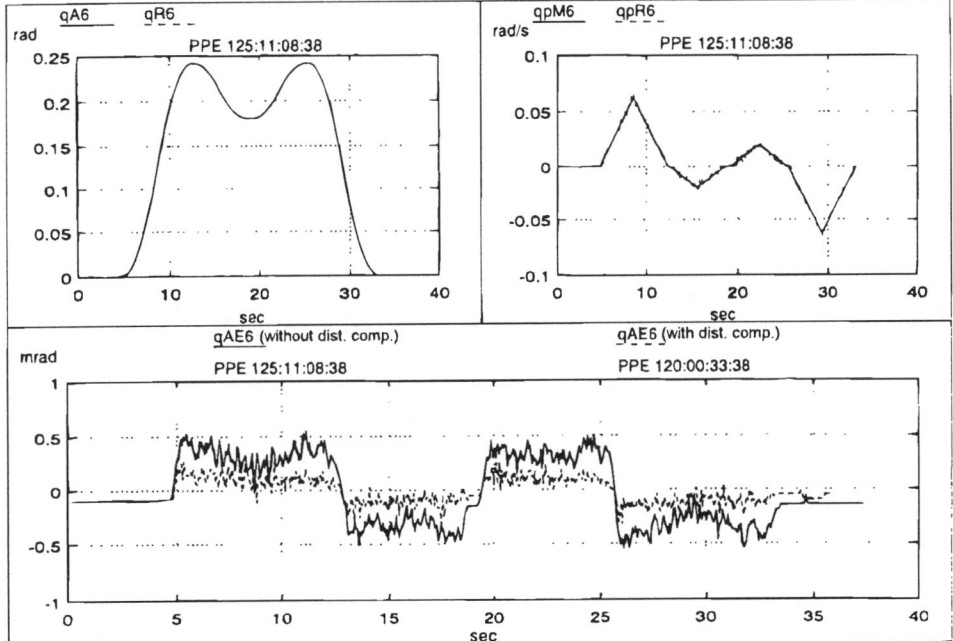

Figure 12: Control of the 6th axis without / with disturbance compensation

In one experiment the observed disturbance was not used for disturbance compensation. The reference path was again the PPE trajectory and allowed comparison with a measurement with disturbance compensation.

Figure 12 shows the entire PPE trajectory of the sixth axis. The upper left partial image shows the load-side actual and reference angles, the upper right one displays the motor-side actual and reference velocities. In this representation no difference between reference and actual values can be detected. Therefore, the lower partial image shows the control errors, on the one hand those of a control without disturbance compensation, on the other hand of a measurement with disturbance compensation; data which are already displayed in figure 11. As disturbance observation and compensation were missing the path accuracy was diminished at a rate of 2 to 3.

5 Summary

Model-supported controller design and its test by computer simulation were the decisive factors of the good tracking behaviour of the ROTEX robot and thus of the success of the different experiments performed during the mission, as no real final test was possible. Every controller structure being transmitted to the Spacelab had to function stable and precisely in order not to endanger the robot.

In spite of the limitation to a decentral structure, it was possible to obtain a decoupling effect for the different axes due to the inner structure of the control. Moreover, the controller parameters as well as the controller structure could be altered from ground control. A control concept of this kind can also be recommended for conventional robots because it is very useful in reaching high path accuracy.

6 References

[1] Hirzinger, G.; Freund, E.; Duelen, G.; Stieler, B.; Lückel, J.: Roboter-Technologie-Experiment ROTEX. Research Program of the Germann Spacelab Mission (ed. Sahm, Keller, Schiewe), WPF, Köln 1993.

[2] Dornier System, Rotex: Advanced Joint Control Analysis, Part IIa: PID Joint Controller Design and Simulation, D2-RX-TN-89025-DS, Friedrichshafen 1989.

[3] Moritz, W.; Neumann, R.: Observer-based Joint Controller Design, Simulation, and Implementation for the D-2 ROTEX Project. Phase C/D Final Report . In: Subsciptors Final Dornier Report, Volume 2 Part B. Automatisierungstechnik, Universität - GH - Paderborn, June 30th, 1992.

[4] Neumann, R.: Beobachtergestützte dezentrale entkoppelnde Regelung von Robotern mit elastischen Gelenken. Dissertation Fachgruppe Automatisierungstechnik, Universität - GH - Paderborn, to appear 1995.

[5] Schütte, H.; Moritz, W.; Neumann, R.; Wittler, G.: Entwicklung eines hochgenauen Roboters in einer mechatronischen Entwurfsumgebung. Beitrag zur GMA-

Fachtagung "Intelligente Steuerung und Regelung von Robotern", Langen, Nov. 9-10,1993, pp. 645-656.

[6] Schröer, J.: A Short Description of a Model Compiler/Interpreter for Supporting Simulation and Optimization of Nonlinear and Linearized Dynamic Systems. CADCS 91. 5th IFAC/IMACS Symposium on Computer Aided Design in Control Systems. Swansea, Wales, July 15-17, 1991.

[7] Jäker, K.-P.; Klingebiel, P.; Lefarth, U.; Lückel, J.; Richert, J.; Rutz, R.: Tool Integration by way of a Computer-Aided Mechatronic Laboratory (CAMeL). CADCS 91. 5th IFAC/IMACS Symposium on Computer Aided Design in Control Systems, Swansea, Wales, July 15-17, 1991.

[8] Kasper, R.; Lückel, J.; Jäker, K.-P.; Schröer, J.: CACE Tool for Multi-Input, Multi-Output Systems Using a New Vector Optimization Method. Int. J. Control 51, 5 (1990), pp. 963-993.

[9] Weihrich, G.: Drehzahlregelung von Gleichstromantrieben unter Verwendung eines Zustands- und Störbeobachters. Regelungstechnik 26 (11, 12), 1978, pp 349-354 and 392-397.

[10] Nakao, M.; Ohishi, K.; Ohnishi, K.; Miyachi, K.: A Robust Decentralized Joint Control Based on Interference Estimation. IEEE Conf. on Robotics and Automation, 1997, pp. 326-331.

[11] Neumann, R.: Untersuchung eines dezentralen Regelungskonzepts unter Berücksichtigung von Reibungs- und Kopplungsmomenten. Diplomarbeit, Fachgruppe Automatisierungstechnik, Universität - GH - Paderborn, Nov. 1986.

[12] Neumann, R.; Moritz, W.: Observer-Based Joint Controller Design of a Robot for Space Operation. Proceedings of the 8th CISM-IFToMM Symposium on Theory and Practice of Robot and Manipulation Ro.man.sy '90. Cracow, Poland, July 2-6,1990, pp. 496-507.

[13] Neumann, R.; Moritz, W.: Gelenkregelung des ROTEX-Roboters bei der D-2 Spacelab-Mission. GMA Fachtagung Intelligente Steuerung und Regelung von Robotern. Langen 1993, pp. 473-482.

[14] Moritz, W.; Neumann, R.; Schütte, H.: Control of Elastic Robots Using Mechatronic Tools. Harmonic Drive International Symposium, Hotaka Nagano, Japan. May 23 - 24 1991, pp 2-1 to 2-23.

[15] Neumann, R.: Observer Based Control of Robots with Elastic Joints. Preprints of IFAC Workshop Motion Control, Munich, October 9-11, 1995.

[16] Bruce-Boye, C.: Entwurf und Analyse eines Störgrößenbeobachters zur Kompensation von Reibkräften am Beispiel eines Roboters mit teleskopartigem Gelenk. Dissertation Universität Bremen, 1991

[17] Müller, P. C.; Lückel, J.: Zur Theorie der Störgrößenaufschaltung in linearen Mehrgrößenregelsystemen, rt-Regelungstechnik, 25(2), 1977, pp. 54-59.

Design and implementation of a model based nonlinear controller for an experimental hydraulic robot

Gert-Wim van der Linden
Mechanical Engineering Systems and Control Group
Delft University of Technology
Delft, The Netherlands

Abstract

To illustrate the capability of todays software and hardware for controller design, the position control of an experimental hydraulic robot is considered. Using model based control concepts it is possible to design and implement a controller in three steps: 1) pressure control, 2) feedback linearising control and 3) linear outerloop control. The total controller has been implemented on the experimental setup using a multiprocessor DSP system, resulting in satisfactory closed loop behaviour. All required steps, including most of the controller implementation, can be performed using standard software.

1   Introduction

To investigate the applicability of modern controller design methods on robotics systems using todays available hardware and software, the control of an experimental hydraulic robot is considered. This robot, see figure 1, has three rotational degrees of freedom and features strong nonlinearities, as well as significant actuator dynamics. This makes this experimental setup an excellent testbed. The system is controlled by means of a multiprocessor DSP (digital signal processor) system, on which complex controllers can be implemented at sufficiently high sampling rates. In this paper a possible solution for high performance position control is sketched, from the modelling phase upuntil the controller implementation. The work as reported here is a continuation of [6, 7].

Control problem
A manipulator typically has to cope with two types of demands: rapid gross motion and cautious fine motion. This paper will deal with the gross motion problem,

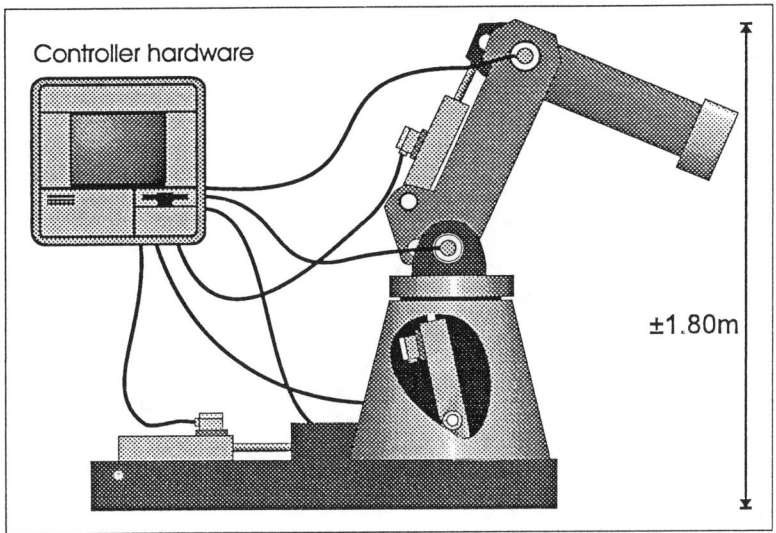

Fig. 1: Experimental hydraulic robot setup

focussing on which solution may be useful, which steps must be taken, and which hardware and software may be of use for this steps.

The key to rapid yet accurate gross motion is obviously high performance position control, taking into account all relevant dynamics, including the nonlinearities. Since the experimental setup is of high quality and behaves quite deterministically, it is possible to obtain an accurate model, especially of the rigid body part. Using standard software, it is possible to do so in a convenient and flexible fashion, allowing the derivation of a well-identifiable model, useful for experiment design, identification, controller design and implementation.

2 Experimental setup

The experimental setup will be treated in two parts: the robot, and the controller hardware.

2.1 Hydraulic robot

The robot is a three axis, direct drive hydraulic robot, which is designed and build at the Delft University of Technology, Faculty of Mechanical Engineering. It features an excellent payload to weight ratio, and can achieve a high bandwidth due to proper components, such as hydraulic servovalves with electrical feedback and the good power to weight ratio of the hydraulic actuators. Because of the direct drive concept and the hydrostatic bearings in the actuators the robot experiences little friction.

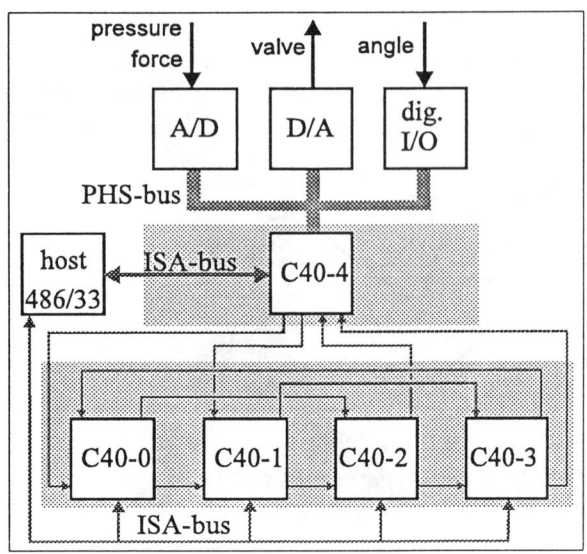

Fig. 2: Controller hardware layout

Available measurements are the joint angles by means of Heidenhain 17-bits encoders, and the actuator pressure differences, by means of Paine analog pressure difference sensors. Control inputs are the (voltage) inputs to the current controllers of the electro-hydraulic servovalves, resulting in a valve spool displacement. For contact tasks the forces on the tip of the robot can also be measured, but that is not used here.

2.2 Controller hardware

The control system is a PC/486 mounted dSPACE multiprocessor TI TMS320C40 system, with dedicated fast I/O, see figure 2. The I/O consists of a 32-channel 16 bits A/D board with 2 multiplexed $5\mu s$ ADC's, a 6-channel parallel 16 bit $3\mu s$ D/A board and a 32-bit digital I/O board to interface the angle encoders. The root DSP (no.4) is connected to the I/O by means of a fast 32-bit bus and is connected by a full grid to the four other DSP's. It is possible to change controller parameters online using the host PC, or to capture data.

3 Control design strategy

Since the experimental setup features quite nonlinear and complex dynamics due to both the actuators and valves and the geometry of the robot, it is currently not possible to perform proper model based controller design in one step. That would require the design of a robust nonlinear optimal dynamic output feedback controller, for which no applicable theory is available yet. Therefore, a 'divide and conquer'

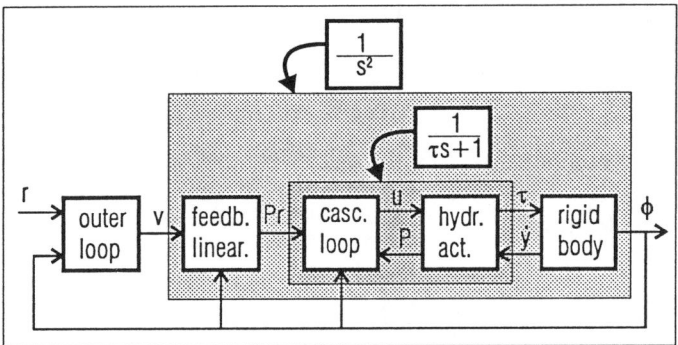

Fig. 3: Controller structure

strategy must be used, such as feedback linearisation plus linear outer loop control. However, feedback linearisation requires an exact model, and usually information about all states is necessary. Because of the quite complex dynamics of the actuators and valves, feedback linearisation cannot be applied directly. Fortunately it is possible to design a slave controller for each actuator, masking its dynamics from the higher level controllers. This leads to the following controller design strategy:

1. Cascade ΔP control of the actuators. This will create a suitable input for a feedback linearisation scheme, since the pressure difference is directly related to the applied joint torques. Furthermore, the effect of nonlinearities is reduced, and the resulting fast dynamics can be neglected in the next controller level. Because of the complex and nonlinear dynamics, the underlying model will be based on nonlinear black-box identification. The current cascade controller does not use the information concerning the nonlinearities, which is subject to current research.

2. Feedback linearisation of the remaining nonlinear mechanical system. This will be based on a theoretical rigid body model structure, which can be derived from first principles, with estimated parameters based on optimally designed experiments [7].

3. Outerloop control of the nearly linear system. This last control layer must realise the desired closed loop behaviour of the controlled robot. Since all available knowlegde concerning the nonlinearities has been used in the previous step, it is most effective to design a linear robust controller, based on a black box model (nominal and uncertainty). The amount of identified uncertainty will be a measure of the efficiency of the previous two layers.

The resulting controller structure is presented in figure 3. The next three sections will discuss the design and implementation of each of the controller parts.

4 Cascade ΔP control

4.1 Actuator model

First of all a basic model of the actuator dynamics is derived. Because of the significant gravity component, a nonlinear model must be used, as the actuator gain is discontinuous around the equilibrium. The most simple nonlinear model for each actuator is the following:

$$\begin{aligned} \tfrac{d}{dt}P &= c_1 u\sqrt{P_s \pm P} - c_2 P + c_3 \dot{y} \\ F &= AP \end{aligned} \tag{1}$$

where u is the valve input, P is the pressure difference across the piston, P_s the supply pressure, $\dot{y}$ the actuator velocity, A the piston area and F the resulting actuator force. The model is parameterised by c_1, c_2, c_3. The actual sign of $\pm$ is opposite to the sign of u.

4.2 Cascade controller design

The aim of the cascade ΔP controller is to control the pressure difference P such that it follows the reference pressure P_r closely. The controller structure follows directly from the actuator model:

$$u = \frac{c_2 P - c_3 \dot{y} - K_p(P - P_r)}{c_1\sqrt{P_s \pm P}} \tag{2}$$

where K_p is the controller gain determining the resulting bandwidth, and P_R the new reference pressure input. The controller parameters c_1, c_2, c_3, K_p can be determined by handtuning the controller on the actuator (for more information, see [5]), reaching a sufficiently high bandwidth of about 100 Hz. The low frequency tracking was however quite bad, mainly due to imprecise knowledge of $c_3\dot{y}$ and valve nonlinearities. Therefore a polynomial NARMAX model [3] was estimated from experimental data, for which a first order loopshaping controller was designed. This indeed improved the low frequency tracking to a sufficient level, as will be shown in the next section.

4.3 Implementation results

The controller was discretized at 5 KHz within MATLAB, and was implemented in C on the root DSP, since the cascade controller has the highest bandwidth requirements, and hence must be placed as close to the I/O as possible. The resulting tracking performance is adequate, as can be seen in figure 4. Future work will include the iterative improvement of actuator model and cascade controller, taking the information about the nonlinearities explicitly into account.

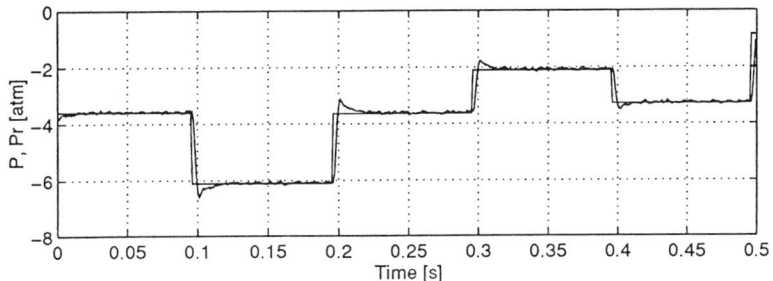

Fig. 4: Random stepresponse cascade controlled actuator

5 Feedback linearisation control

5.1 Rigid body model structure

The software package Autolev [8] is used to generate the symbolic equations of motion of the rigid body mechanics of the robot, based on a system description file. The equations have the standard format

$$\tau = M(\phi)\ddot{\phi} + C(\phi, \dot{\phi}) + G(\phi). \tag{3}$$

A small C-program is used to convert the output of Autolev to a format readable by the symbolic manipulation softwaretool Maple V [2], which only concerns some simple string manipulations. Within Maple, the equations are split into mass, Coriolis and centrifugal, and gravity terms. Symbolic simplifications are performed where possible, since the equations as produced by Autolev are not always optimal. For identification purposes a minimal parameterisation is determined, with the structure

$$\tau = \mathcal{F}(\phi, \dot{\phi}, \ddot{\phi})\Theta \tag{4}$$

with Θ the vector of unknown parameters. All required models can be generated as C-functions using the macroC library for Maple V [1].

5.2 Parameter estimation

Using equation (4) and the procedure as described in [7] the rigid body parameters of the setup can be determined experimentally. This procedure consists of three steps: 1) experiment design, maximizing the condition number of the associated information matrix, 2) data collection and filtering and 3) parameter calculation (stated as a linear least squares problem). The resulting model predicts the joint torques with less than 2% error, which is quite satisfactory. Cross-validation experiments resulted in errors of no more than 3%, demonstrating the accuracy of the resulting rigid body model.

5.3 Feedback linearising innerloop design

The standard solution for the feedback linearisation of a rigid body system is known as the Computed Torque controller. Hence, the feedback linearisation law equals:

$$\tau_r = M(\phi, \hat{\Theta})v + C(\phi, \dot{\phi}, \hat{\Theta}) + G(\phi, \hat{\Theta}). \tag{5}$$

with v the new input and τ_r the desired joint torque. When the feedback linearisation law matches the true system sufficiently close, the resulting system behaves *globally* close to the linear relation $\ddot{\phi} = v$.

5.4 Implementation results

Using the previously mentioned macroC library, the feedback linearisation law is generated as C-function. It is run on a slave DSP at 1 KHz, recalcuting the complete nonlinear model every sample, taking less than 0.4 ms. The effectivity of the feedback linearisation loop will be illustrated in the next section by means of frequency responses of the openloop system.

6 Outerloop control

6.1 Black-box model

By means of a frequency analyser the compensated system can be identified again, such that a suitable model is found for robust controller design. Doing so for multiple operating points, spanning the working area as good as possible, it is possible to derive a nominal model with an uncertainty bound. Using closedloop experiments (using a simple PD controller) in combination with the knowledge on the controller transferfunction allows the calculation of the openloop transferfunctions (from v to ϕ). The resulting magnitude plots are depicted in figure 5. Clearly, for all operating points the entries on the diagonal are close to the theoretical double integrator, but in the high frequency region additional (actuator) dynamics is encountered. The first axis is fully decoupled from the arm, at least below measurement resolution, which indicates that the Coriolis and centrifugal forces have been estimated correctly. However, the decoupling is not achieved for the arm (entry 23 and 32). Therefore, carefull outerloop control is required to take the additional dynamics and the coupling into account. Otherwise, the often proposed PD controller would have sufficed.

The nominal model $G(s)$ is found by fitting the average frequency response by a fourth order model for each axis. The uncertainty bounds are determined by the maximum difference between nominal model and measured openloop responses, which could be modelled simply by an additional uncertain input gain matrix W_{unc}:

$$G_{\text{pert}} = G(s)\,(I + \Delta(s)W_{\text{unc}}) \tag{6}$$

The uncertainty block $\Delta(s)$ is structured (diag($1 \times 1, 2 times 2$)), since the robot arm is decoupled from the first (vertical) axis.

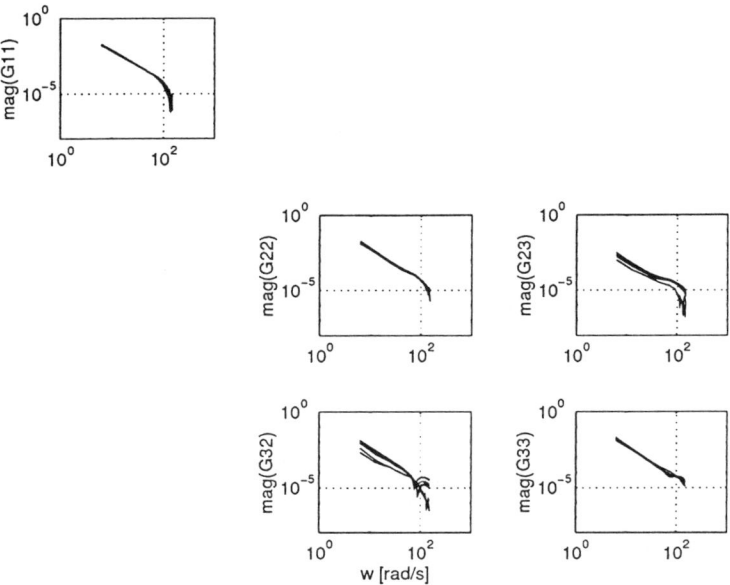

Fig. 5: Magnitude plots of compenated plant

6.2 H_∞/μ controller design

Now a controller can be designed for the identified model to meet the desired trajectory following performance. The most suitable linear robust controller design method is H_∞/μ optimal control, as it allows to accomodate for uncertainties, which indeed occur, as the first two loops were not perfect. The controller is designed using MATLAB and the MUSYN toolbox, maximizing the bandwidth, while maintaining sufficient robustness and disturbance attenuation. The used standard plant layout is depicted in figure 6. For an extensive treatment on H_∞/μ controller design (using

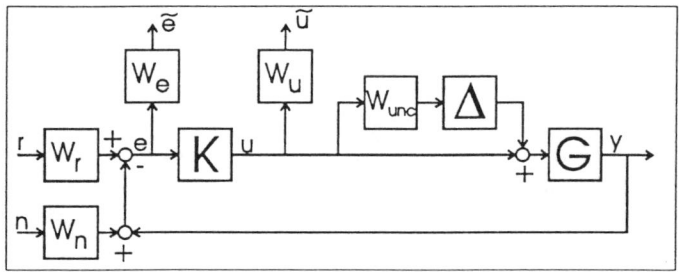

Fig. 6: H_∞/μ design standard plant layout

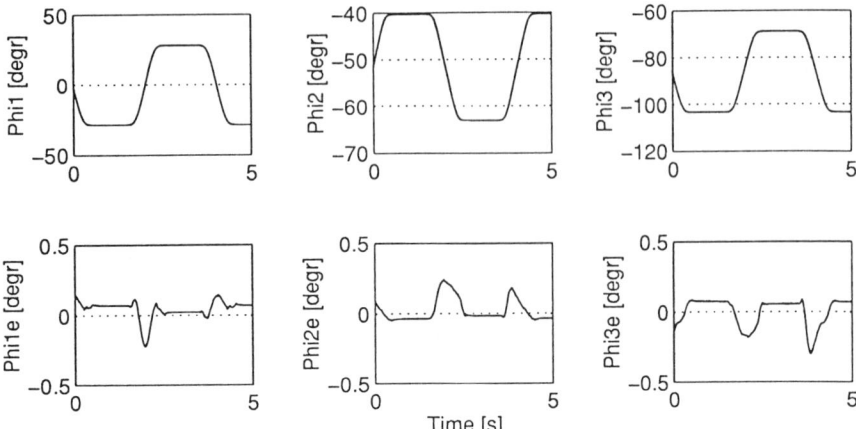

Fig. 7: Trajectory response (upper: angles, lower: errors)

the control of the first axis of the robot for illustration), see [9]. This three axis design has been performed similarly, aimed at a closed loop bandwidth of 7 Hz, a safe value, which may be approved upon in the future.

6.3 Implementation results

The resulting 18th order linear controller (in state-space format) can be coded in C automatically using the softwaretool IMPEX [4]. This tool is also used for discretizing, transforming and scaling of the control law. The control law is run on the second slave DSP at 1 KHz, taking 0.2 ms calculation time. When a smooth but fast reference trajectory is applied to the controlled system the tracking is good, as figure 7 indicates. The tracking error is well below $0.5°$, which is quite good for a trajectory so close to the limits of the robot, in velocity, acceleration and jerk. The frequency responses from reference ϕ_{ref} to output ϕ see figure 8) also show quite satisfactory behaviour: the coupling in the arm is not too strong, while the required bandwidth is indeed reached.

7 Conclusion

A feasable high performance solution for the position control of a robot is presented, exploiting the capabilities of todays software and hardware. The software allows rapid prototyping, since many steps can be performed automatically, not only the controller code generation, but also in the modelling/identification phase. The D-SP hardware allows to implement complex controllers at high sample rates, and therefore puts little restrictions on the control engineer's imagination.

The actual restriction are more due to the limitations of available theory, as in the case of controller design for the experimental hydraulic robot. A one step solution was not possible, but by dividing the controller design into three steps it was possible to achieve satisfactory closed loop behaviour. The use of standard software, such as Autolev, MATLAB and Maple V significantly accelerated the controller design, while the multiprocessor DSP hardware allowed to implement the resulting control laws at a sufficiently high sampling frequency.

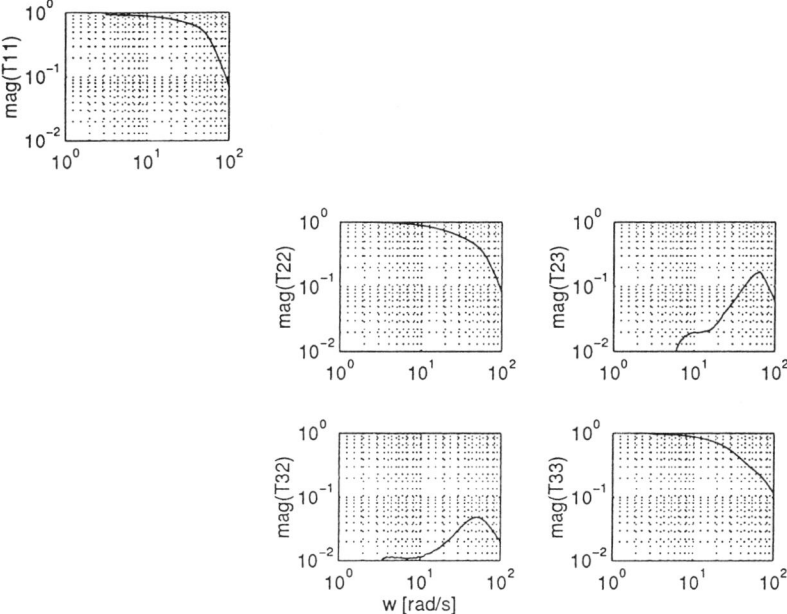

Fig. 8: Closed loop frequency responses (magnitudes)

References

[1] Capolsini P., *MacroC; C code generation within Maple*. INRIA - Université de Nice, FTPsite: ftp.inria.fr, files: /lang/maple/5.2/share/macroc.*, 1992.

[2] Char, B.W. *et al*, *Maple V Language Reference Manual*. Springer Verlag, Berlin, 1991.

[3] Chen S., Billings S.A., "Representations of nonlinear systems: the NARMAX model", *International Journal of Control*, vol.49, no.3, pp.1013-1032, 1989.

[4] dSPACE, *DSP-CITpro IMPEX manual*. dSPACE GmbH. Technologiepark 25. D-33100 Paderborn, Germany, 1989.

[5] Heintze H., Weiden A.J.J. van der, "Inner loop design and analysis for hydraulic actuators, with application to impedance control", in *Proceedings of the 4th IFAC Symposium on Robot Control, Capri, Italy*, pp.401-406, 1994.

[6] Linden G.W. van der, Valk P., "Digital control of an experimental hydraulic manipulator", in *American Control Conference*, pp.2455-2459, 1994.

[7] Linden G.W. van der, Weiden A.J.J. van der, "Practical rigid body parameter estimation", in *Proceedings of the 4th IFAC Symposium on Robot Control, Capri, Italy,* pp.631-638, 1994.

[8] Schaechter D.B., Levinson D.A., Kane T.R., *Autolev User's Manual.* Online Dynamics, Inc., Sunnyvale, CA, USA, 1991.

[9] Linden G.W. van der, Noble R.H.C. le, "H_∞/μ optimal controller design: finding the weights", in *Submitted to the 34th Conference on Decision and Control, New Orleans,* 1995.

Address

Ir. Gert-Wim van der Linden
Mechanical Engineering
Systems and Control Group
Delft University of Technology
Mekelweg 2, 2628 CD Delft, The Netherlands
e-mail: linden@tudw03.tudelft.nl
Fax: (+31)-15784717
Phone: (+31)-15785232

V. Robot Applications

C. Uhrhan, R. Roshardt, G. Schweitzer
User-Oriented Automation of Flexible Sheet Bending

P. Drews, D. Matzner
Smart Welder - A Mechatronic Application for Automated Shipbuilding

W. Roddeck, H.-J. Rehbein
Autonomous Tool-Mover for Laser-Cutting with Industrial Robots

N. Ahlbehrendt, H. Diesing, S. Jakobi
Interactive Robot Master Slave System for Deburring Large Cast Iron

User Oriented Automation of Flexible Sheet Bending

Christoph Uhrhan, René Roshardt, Gerhard Schweitzer
Institute of Robotics and Mechatronics Lab
ETH Zurich, Switzerland

Abstract: The automation concept is based on the integration of the three factors "man, technique, organisation - MTO", developed at the Centre for Integrated Production at the ETH. As a benchmark, a manufacturing cell for flexible bending of metal sheets is under construction. It consists of a laser cutting machine, press brake, conveyor belts, and a robot. These devices will be networked to enable the fully automated programming cycle. The project is jointly pursued by the Institute of Work Psychology, the Institute of Forming Technology, and the Robotics Lab.

In this paper the automated charging of the press brake by a robot is presented in some more details. The robot system should fulfil the requirements of the MTO-concept, meaning, for example, user friendly and application-tuned task level programming, as well as interaction facilities to intervene in decisions on every system level. This automation standard is made possible by the coordinated use of mechatronics hard- and software, in particular a modular robot with sensor integration, and a controller, based on object-oriented real time software.

1 Introduction

The demands on today's production systems are determined by a large number of product variants and low batch sizes. The production has to therefore offer increased flexibility. This requirement can be fulfilled by newly developed production technologies such as stereo lithography. Flexible production can also be guaranteed by the increased use of manufacturing technologies which still offer the demanded flexibility. One such technology is sheet bending at press brakes. Multiformed parts can be produced with one press brake and some once installed standard tools. The flexibility of this technology is not just because of the manual sheet handling at press brakes (Figure 1). Sheet metal bending with press brakes is in fact predestined for low batch sizes and flexible forming of sheet metal parts.

Combined with appropriate automation, bended sheet parts become an economic alternative to welded, cast or cut parts. To automate the mostly manual process of sheet handling, we have developed a robot system which had to be integrated into the material and data flow of a complete sheet metal forming cell [1]. The automation of the flexible manufacturing system was carried out by a concept in which man, technique and organisation - MTO [2], are simultaneously planned and designed. The flexibility of such designed systems is obtained by a complementary function allocation between man and machine.

Figure 1: Manual sheet bending at a press brake

Such a concept accommodates for the fact that the capabilities of humans and machines differ in a qualitative manner, meaning that they cannot replace but only complement each other. For the planning and valuation of our robot system we used a method for complementary design of sociotechnical systems [3, 4]. The method was developed by the Institute of Work Psychology at the ETH and helps us to define which functions should be automated and how they should be automated. It was applied to our application by a team of work psychologists, computer scientists and engineers [5].

Requirements for humane tasks are, for instance: Completeness; planning and decision making requirements; variety of demands; cooperation requirements; learning possibilities and autonomy. As far as the technological aspect is concerned, the design of human-machine systems is very often reduced to the design of optimal (i.e. user friendly) interfaces. To support the requirements for humane tasks the technical system must offer appropriate transparency and the possibility of interaction in tasks and decisions on every system level. To meet this requirement, we developed a robot system consisting of a task oriented programming environment; new interface devices to support efficient robot programming; a highly sensor integrated robot hand and an application adapted manipulator.

2   User Oriented Robot Programming

For robot programming a system was developed, which offers the possibility of user oriented, flexible robot programming. The complete programming system consists of four levels which enable fully automated robot programming with data from a previous planning system; manual editing of application oriented commands and low level motion programming and system configuration [6]. Low level motion programming and system configuration are more system oriented and are useful during the installation phase or during error handling and system optimisation. This level supports fast control adaptation to freely configurable modular robot systems and will not be described here. In the following sections we describe the two upper levels which enable robot programming by describing the product to be manufactured, and programming by application specific commands.

2.1 Programming by Product Description

In our flexible bending cell a closed data flow from CAD to the numerical control of the press brake and the robot controller was aimed for. Due to the data integration, the production data has only to be calculated once inside the cell and is then available for other parts of the production system. The robot program is generated with data from a planning system which calculates the bending sequence from CAD-data of the finished sheet part. This planning system is developed by the Institute of Forming Technologies at the ETH Zurich [7]. The output of the planning level is the description of the flat sheet with additional information about the bending sequence (bending edges, bending angles, bending tools and grip points relative on the workpiece), figure 2.

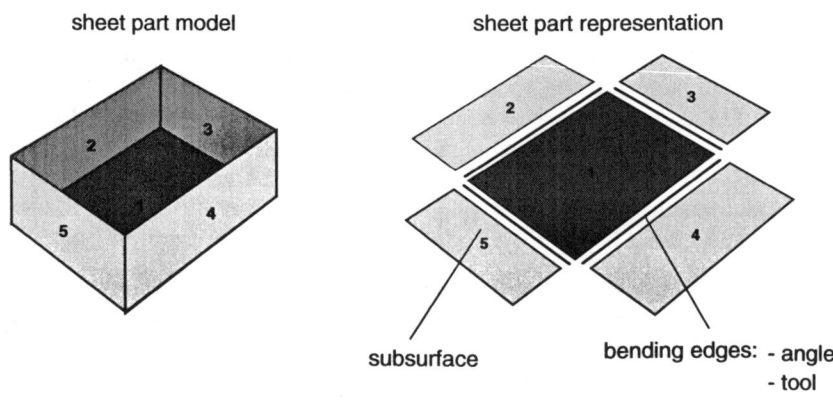

Figure 2: Product data

The resulting database, which is generated from the bending sequence planning system, is used for automated programming of the robot as well as of the press brake. As explained in the introduction, flexibility can only be realized economically by a system concept which takes into account man resources, technique and system organisation. A main issue of automated programming is therefore the possibility of interaction [8], which has to be supported by appropriate system design, interface devices and programming. To make such interaction in automatically generated robot programs possible, transparency of the process is necessary for the operator to react efficiently, fast and safely, to exceptions. This means that the operator must be able to obtain information about actual system data and automatically planned motions. This was realised, firstly, by a graphical representation of the product description which is the basis for the automatic planning system. The data representing the sheet part to be bent can be displayed and edited. This interface also supports direct product editing from technical drawings and makes the system more independent giving greater flexibility, so that the robot can be used without the planning components of the cell. Secondly, for the correction and optimisation of single robot motions, the automatically generated motions are coded in an application oriented way, which is easy to change manually by the operator (see Chapter 2.2). Thirdly, a graphical simulation of the planned robot motions and bending operations for visual control on the shop floor is also used (figure 7).

2.2 Process Oriented Programming

For the optimisation of automatically generated programs, or if no product data is available, a process oriented programming level is offered to the operator. The idea was to program the robot with a minimum of robot specific commands. With such commands an operator used to work on a non automated press brake should be able to program the automated press brake. Therefore, a programming interface has been developed which enables robot programming by describing the task as it would be executed manually by the operator. Generally speaking, four different types of elementary commands for a robot task can be defined:

- gripping
- main manoeuvre
- positioning
- application process

This sequence of events can be cycled until the task is complete. The implementation of the basic commands is application specific and depends on the actual hardware, the robot configuration, available sensors, and the control features. In our application, the gripping of the sheets and the positioning at mechanical backstops of the press brake are carried out automatically by sensor guided motions (see chapter 3.2). The main manoeuvre can be defined by a set of commands which describe the motions of the sheet in an application oriented way, such as "turn sheet x degrees"; "rotate sheet y degrees"; "swivel sheet z degrees"; "put sheet in front of edge A" etc. The application process is, in our case, the bending of the sheet. During the bending process, the robot releases the sheet, or it has to

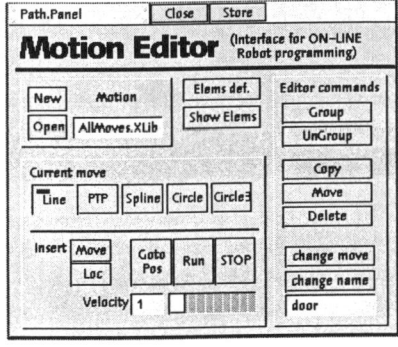

Figure 3: Graphical editor for robot motions

support the sheet during the bending process. With a graphical programming interface, these commands can introduce a sequence with appropriate parameters (figure 3). The command-parameters can either be obtained from data of CAD- (Computer Aided Design) and CAP (Comnputer Aided Planing)- systems, or can be edited manually on the shop floor, or they can be determined by sensor values.

3   Manual Programming Support

The most used, because of its practicality, method of robot programming today is by teaching. Nevertheless exact positioning needs still an experienced operator. To support telemanipulation of our robot, different interface devices are used. Additionally, exact positioning is carried out by the help of sensor informations as shown subsequently.

3.1   Interface Devices

To define the main manoeuvres of the proposed application oriented programming environment, teaching is a very fast method, particularly if a six dimensional joystick such as the Space-Mouse [9] is used. The Space-Mouse offers the possibility of directly steering of the robot in world or tool coordinates in all six dimensions simultaneously. The problem of moving the robot with the Space-Mouse is the exact positioning of the gripper at a defined position. Therefore, other kinds of joysticks, for example the 3D-Mouse [10], appear to be much more practical [11]. Fine positioning, however, requires a great deal of experience with such devices. For the exact positioning of the endeffector we use a so called Jog-Shuttle, which is well known from video recorders, where it is used for the fine-positioning of the video-band. The Jog-Shuttle consists of two controls, an outer and an inner one (figure 4). The twist of the outer one is reset automatically by a spring. Its analog value is used as nominal speed value for the robot. The inner control can be continuously turned, without automatic reset, and is used for the exact positioning of small increments. With the Jog-Shuttle, fine positioning of a joint-, world- or tool controlled, direction is possible.

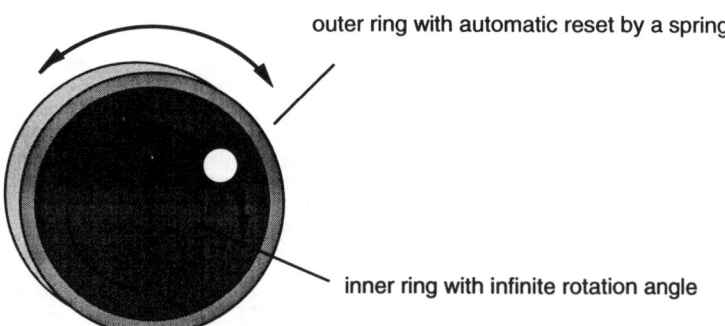

Figure 4: Jog-Shuttle for fine positioning

3.2 Sensor Feedback

Despite the proposed interface devices, exact orientation of the tool centre point is still a problem of manual teaching. For example, if an open gripper has to be positioned the interesting point is virtual, meaning that it is invisible, between the two fingers. Visual perception of the goal position becomes difficult. Further, during the fine positioning phase, the operator normally has to stay close to the robot to obtain a visual feedback, which is always a dangerous situation. To optimise the fine-positioning, which takes much more time than the teaching of manoeuvres, we equipped our gripper system with various sensors [12].

The idea of the sensor gripper is that the teaching of exact positions can be substituted by an automated gripping or positioning phase. At least teaching can be supported by additional sensor information, which would not be available without sensors. Another possibility is that visual feedback of the operator can be substituted completely, so that the operator can stay at a safe distance from the robot during teaching. The sensor feedback to the operator is firstly carried out by displaying the sensor data on a screen graphically. Much more practical than a graphical visualisation is the feedback to other human perception channels. The visual channel is normally used to observe the scene directly. We therefore try to provide an acoustic feedback corresponding to the sensor signals. The varying analog sensor-signals are fedback at different acoustic frequencies, so that the operator can perceive the situation simultaneously with different channels, visually and acoustically.

4   Application Oriented Hardware Design

For the complete application oriented programming system, presented above, appropriate hardware components are necessary. To support user oriented press brake automation, we developed a sensor integrated robot gripper and an application adapted manipulator.

4.1 Sensor Integrated Robot Hand

The developed sheet gripper consists of a vacuum and finger unit (figure 5). This combination enables the robot to grip a wide range of sheet shapes without any change of gripper. For sensor controlled gripping with the vacuum unit, an ultrasonic sensor is used for approximate detection of the work piece in front of the gripper face. To detect the exact deviations, tactile sensors are integrated into the vacuum pads. The sensors give an exact distance between the vacuum pad and the sheet face in a range of 0 to 20 mm. With four such sensors, vacuum pad deviations in position and orientation between the vacuum pads and the sheet face can be determined.

The finger gripper consist of two modules. Similarly to the vacuum unit, an ultrasonic sensor is used for approximate work piece detection. The exact sheet position is measured by three laser- triangulation sensors. The laser-triangulation sensors are mounted in such a way that three points on the sheet plane (one on the upper side and two on the lower one) are measured. With these points, the orientation and position of the plane relative to the gripper is determined. The alignment between the foremost sheet edge and the finger gripper is obtained by determining the time interval between the one and the other detection of the sheet through the laser-sensors.

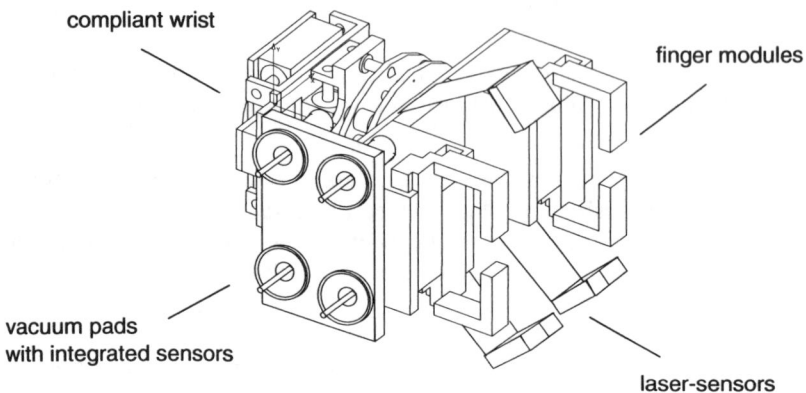

compliant wrist

finger modules

vacuum pads
with integrated sensors

laser-sensors

Figure 5: Sensor gripper for sheet parts

As well as the gripping of the sheets, the positioning of the sheets at mechanical back-stops inside the press brake is supported by a sensor equipped compliant element [13]. The element is installed between the robot and the gripper and was especially designed for the large payloads and torques which occur in the wrist in our application [14]. The compliance offers three prismatic and three rotational degrees of freedom of +/-15 mm and +/-2 degrees. The stiffness of the prismatic degree of freedom can be pneumatically varied. With this compliance, the positioning at the backstops becomes much simpler, as compared to increase the absolute accuracy of the robot or adding force control. The complete compliance can be fixed during large manoeuvres and the gripping process. The deviations of all compliant axes are measured and used for updating the following locations. Similarly to the gripper sensors, the sensed deviations of the compliant element can be used for additional feedback during manual robot teaching.

4.2 Application Adapted Manipulator

The analysis of the man-machine function allocation demands for flexible and automated process execution – with appropriate interaction facilities – as well as manual execution [5]. As an example, the programming concept for the numerically controlled press brake has been developed. For initialisation and adjusting the brake, manual sheet handling can be used as it will be more practical than automated sheet handling by the robot. The possibility of manual press brake charging increases the flexibility of the complete system in such a way that complex parts, which cannot be handled by the robot, can be manufactured manually or, if the robot system fails, the availability of the press brake is still guaranteed.

To fulfil the requirement of manual press brake charging, the robot has to be designed in a way which does not restrict and handicap the manual charging. Therefore, a gantry was used which was extended by an arm with four additional axes (figure 6). This structure covers the minimally necessary work area in front of the press brake, so that no unnecessary workspace is occupied by the robot and that more free and in particular safe areas exist for other components of the bending cell. The robot was built from standard elements. The modular robot design is supported by a configurable control software which can easily be adapted to the actual hardware configuration.

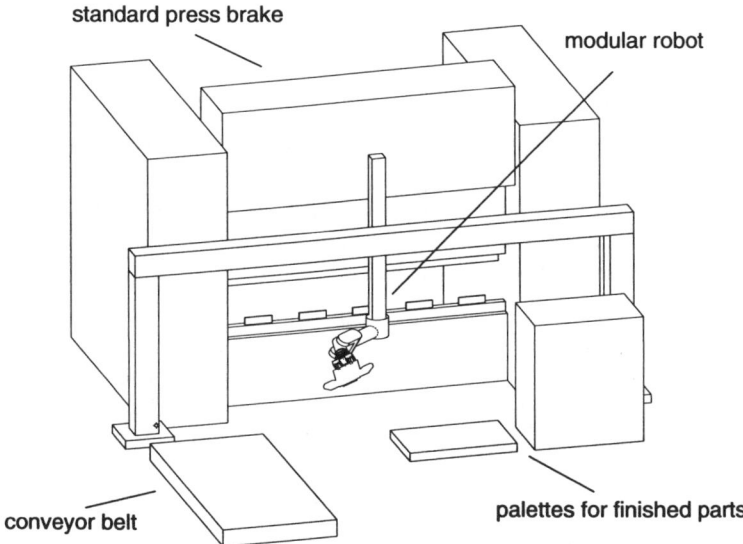

Figure 6: Application adapted robot

5 Realisation

The realisation of the presented robot programming environment was achieved with the Oberon System [15]. The Oberon System is a programming language with object oriented programming features as well as an operating system. A Cross-Compiler, XOberon, for VME- based systems, with additional real time features is available, which enables program development on many platforms, e.g. Macintosh, SPARC computers and PC [16]. The robot control was implemented in an object-oriented fashion. To support modular hardware configurations, a configurable control software was developed which enables fast system adaption to the actual hardware configuration.

The graphical user interfaces are realised with the Gadget system of Oberon [17], which is an experimental graphical user interface management system that allows the construction of user interfaces at runtime. In the next phase robot programming will be executed on a pen-computer as interface, so that the operator becomes more movable than with a computer installed at a fixed location.

The construction of the forming cell was carried out by the Institute of Forming Technology. In cooperation with the Work and Organisational Psychology Unit of the ETH Zurich, we try to implement the global concept, of realising a human oriented automation approach. The complete sheet forming cell including laser cutting machine, conveyer belts, press brake and the robot were installed at the ETH-Technopark. The project is part of the Centre for Integrated Production (ZIP), which is composed of several institutes of the ETH. In the ZIP, the automation concept MTO which plans and designs the human working force and the use of technology in its entirety is applied in different projects.

6 Conclusions

Holistic approaches of automation concepts such as MTO (Man Technique Organisation) can only be realised with the appropriate technical components. In addition to novel products, mechatronics can also support new methods such as a user oriented design of technical systems. This was demonstrated in an application, where a robot system was developed for the sheet bending at a press brake inside a flexible sheet forming cell.

The holistic automation concept demands the simultaneous planning of human work force, technique and organisation. This leads to independent units (fractals, groups, cells) where human beings will have to take more responsibility. Therefore, intervention in decisions and tasks at any time and on every production level has to be supported. To fulfil these requirements an application oriented programming environment was presented, that enables fully automated robot programming with data from previous planning systems, as well as interactive program optimisation and error correction by task level commands on the shop floor.

Efficient programming and interaction was supported by new interface devices and the use of sensor guided motions. A sensor integrated robot gripper was therefore developed. The demanded flexibility was guaranteed because automated sheet handling as well as manual sheet bending is still possible. Manual sheet handling was enabled by a special design of the robot which does not restrict manual bending in any way.

7 References

[1] Uhrhan, Ch.; Automated sheet bending with press brakes, Proc. of the 25th ISIR, Internat. Symposium on Industrial Robots, Hannover, April, 1994, pp. 81-88

[2] Ulich, E.: CIM - eine integrative Gestaltungsaufgabe im Spannungsfeld Mensch, Technik, Organisation. In: Cyranek, G. & Ulich, E. (Hrsg.). CIM - Herausforderung an Mensch, Technik, Organisation. Schriftenreihe Mensch Technik Organisation (Hrsg. E. Ulich), Band 1. Zürich: vdf / Stuttgart: Teubner, 29 - 43, 1993

[3] Grote, G.: A participatory approach to the complementary design of highly automated work systems. In: Bradley, G. & Hendrick, H.W. (Eds.), Human factors in organizational design and management - IV. Amsterdam: Elsevier, 1994

[4]   Weik, S., Grote, G., Zölch, M.: KOMPASS Complementary Analysis and Design of Production Tasks in Sociotechnical Systems, IOS Press, 1993

[5]   Roshardt. R., Uhrhan, C., Wäfler, T., Weik, S.:A Complementary Approach to Flexible Automation, Internat. Conf. on Architecture and Design Methods for Balanced Automation Systems BASYS'95, Vitoria ES, Brasil, 24.-26. July 1995

[6]   Uhrhan, Ch., Roshardt, R.:User Oriented Robot Programming in a Bending Cell, Proc. of the Internat. Conf. on Intelligent Robots and Systems, Munich, 1994, Proc. Vol.2, pp. 1103-1109

[7]   Huwiler B., Reissner J.: Fertigung von Blechbiegeteilen rechnerunterstützt planen, Bänder Bleche Rohre 10, 1992

[8]   Schweitzer, G.; The Intelligent Interactive Robot. In: Mechatronics & Robotics I, IOS Press, 1991

[9]   Space Control: Space Mouse Software Interface/User's Handbook, Malching, Germany

[10] Logitech Ind.: 3D Mouse & Head Tracker, Technical Reference Manual, Fremont (CA), USA, 1992

[11] Weber, P.: Mensch-Maschine-Interface für die Nanorobotik, Diplomarbeit am Institut für Robotik, ETH Zürich, 15. Juli 1994

[12] Keller, S., Mathis, S.: Sensorgestütztes Greifen von Blechbiegeteilen, Diplomarbeit am Institut für Robotik, ETH Zürich, 10. Februar 1995

[13] Uhrhan, C.: Compliant Robot Wrist with Passive and Active Features, 26th Internat. Symp. on Industrial Robots, ISIR'95, Singapore, 4-6 October 1995 (to appear)

[14] Verdan, P.: Bau eines Compliance-Elementes mit Sensorik, Semesterarbeit am Institut für Robotik, ETH Zürich, 25. Februar 1994

[15] Wirt, N., Gutknecht, J.: Projekt Oberon: the design of an operating system and compiler, New York, ACM Press (etc.), 1992

[16] Diez, D., Schweitzer, G.; Realtime Systems for Mechatronics, 2nd Conf. on Mechatronics and Robotics, Duisburg/Moers, Sept. 1993

[17] Gutknecht, J.: Oberon System 3 - Visionen einer Softwaretechnologie der Zukunft, Informatik Nr. 3, Juni 1994

Dipl.-Ing. Christoph Uhrhan, Dipl.-Ing.(ETH) René Roshardt,
Prof. Dr. Gerhard Schweitzer
Institute of Robotics, ETH Zurich
CH-8092 Zurich
Telephone   ++41 1 632 3584
Fax   ++41 1 632 1078
Email   uhrhan@ifr.mavt.ethz.ch, roshardt@ifr.mavt.ethz.ch
               schweitzer@ifr.mavt.ethz.ch

Smart Welder - A Mechatronic Application for Automated Shipbuilding

Paul Drews, Dieter Matzner
APS - European Centre for Mechatronics, Aachen, Germany

representing the CLEOPATRA[1] consortium

Abstract: Within the framework of the ESPRIT programme, the CLEOPATRA project represents a cluster of four industrial real-time applications. All of these applications have a common goal: The introduction of High Performance Computing technologies will boost the market potential of the products and improve their competitiveness. The Smart Welder represents one of these industrial applications. It is designed for the automatic welding of complex sections of a ship hull. For that purpose a CAD system, an off-line programming system, a transputer based robot controller, a vision system and a multi-axis mechanical system including the welding equipment have been combined to form a complex integrated production system. The paper reports on the state of the art of the partners' work on the different subsystems. Main emphasis has been put to the embedded robot control system which uses a transputer based parallel computer architecture for the required time-critical calculations. A flexible algorithm for the on-line motion generation in Cartesian space and some implementation results of the inverse co-ordinate transformation are described in more detail.

1 Introduction

Attempts to robotize the shipbuilding process have already been conducted for several years. However, the presently available technology enables the use of robots only at geometrically simple parts of the ship hull. A substantially larger number of robot installations requires much more sophisticated and intelligent solutions. Several fundamental technological difficulties have to be evaluated before robots can be applied economically to weld any part of a ship hull. Especially in the front and back parts, the geometrical shapes of these sections are very complex. In order to weld these sections automatically, flexible robots with more than six degrees of freedom and an enlarged

---

[1] The work in the CLEOPATRA project is performed within the framework of the ESPRIT programme and partly funded by the Commission of the European Communities. The following companies form the consortium: AEG(D), APS(D), C-VIS(D), DASA(D), DBAG(D), DLR(D), ECSA(F), GRAPHIKON(D), MCS(F), OSS(DK), PAC(UK), PERIMOS(D), THO(F), UBM(D).

working area have to be integrated into the production process. The Smart Welder application represents a typical mechatronic system for such a complex production process. It is an approach to increase the degree of automation in shipbuilding through an intensive use of parallel High Performance Computing (HPC) technologies. By integrating such systems into industry, productivity and competitive power of the European shipyards will be strengthened.

The Smart Welder is composed of a CAD system, an off-line programming system, a transputer based robot controller, a vision system and a multi-axis mechanical system including the welding equipment (see figure 1). The system works according to the following scenario: Firstly a CAD model of the workpiece including the weld lines and process parameters has to be defined. This CAD model is used by the off-line programming system to generate an instruction program for the robot controller. The controller interprets and executes this instruction program. The vision system facilitates the identification of the actual block position, the detection of corners and the supervision of the working space. These tasks are performed by comparing real world images with the CAD model. Finally, the multi-axis manipulator is used to compensate for the detected deviations between the real world and the CAD model.

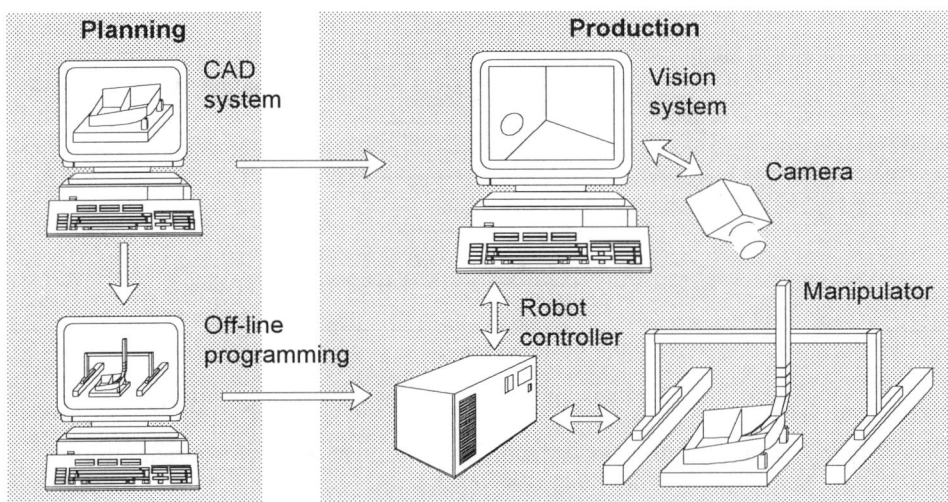

Figure 1: Overall Smart Welder system architecture

The application specified in this way forms a flexible system able to overcome the current problems, as they appear in the use of conventional robot systems for automated welding. This is achieved by the extensive use of advanced information technology and the integration of transputer based HPC within the field of controlling, vision and mechatronics.

2 Transputer based robot controller

2.1 Functional robot controller concept

The transputer based robot controller forms the logical 'heart' of the complete Smart Welder system. It combines various information of the user, the off-line programming system, the vision system and the welding process in order to perform the desired welding task at complex sections of the ship hull. All this input is used to generate output for the system actuators, namely the mechanical system and the welding equipment.

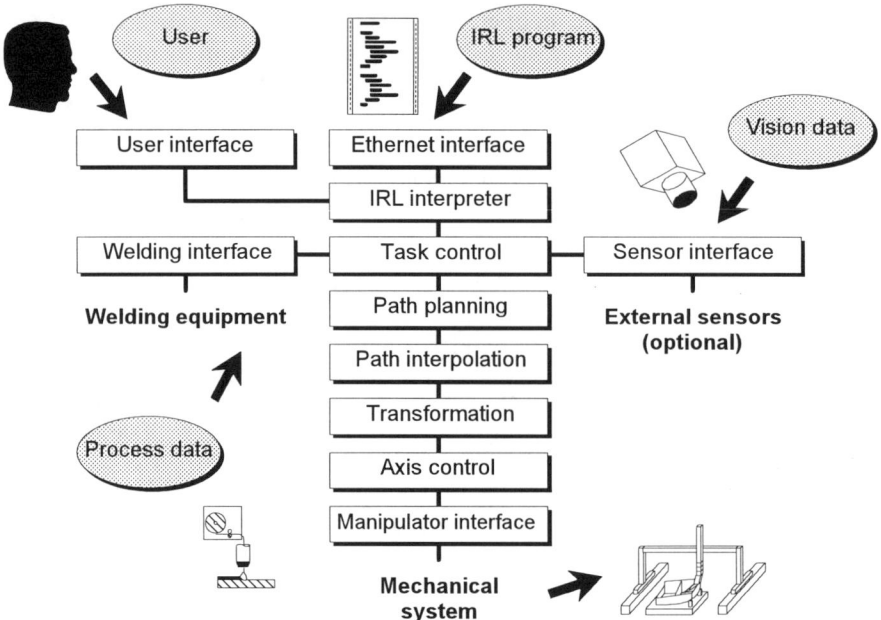

Figure 2: Functional concept of Smart Welder robot controller

Generally the robot controller has to handle four different types of external information: user data, Industrial Robot Language (IRL) instruction data, vision data and process/welding data. These data are integrated into the controller at the user interface, the IRL interpreter, the sensor interface and the welding interface. Figure 2 illustrates the functional concept of the robot controller including the mentioned data types.

2.2 Robot controller hardware architecture

The hardware platform of the Smart Welder robot controller is designed in a way that it provides as much flexibility as possible. The controller system consists of three main components:

- a host system based on a standard 486 PC,

- a standard transputer network for real-time calculations and

- three I/O bus systems to interface the manipulator and the welding equipment.

The 486 PC is used as host system for the underlying transputer network. It provides a graphical user interface for the complete Smart Welder system. In addition, the file management, the Ethernet communication and the interpretation of IRL instruction programs belong to its tasks. The PC is equipped with the OS/2 operating system.

The robot controller uses a transputer based parallel computer architecture for the required time-critical calculations. The hardware platform is composed of six standard T805 transputers. For this real-time transputer network Parsytec MTM-2 transputer modules are used. The transputer network is interfaced to the PC host system via two external links and to the I/O bus systems via four links. The configuration of the network can be taken from figure 3. For the actual project phase, APS has decided to use T805 transputers for the necessary real-time computing. However, the system layout facilitates an easy upgrade to more powerful processors in a future project phase. In this case the transputer network will simply be replaced by another parallel computing system.

Three I/O bus systems provide access to the different input and output signals of the gantry, the robot and the welding source. The information transfer from the bus protocol to the transputer link standard is done by a transputer based adapter card that maps the I/O bus data into the transputer memory. The adapter card is alternatively equipped with a T222 or a T805 transputer. Via three of these adapter cards the transputer network is connected to the three I/O bus systems. The adapter card developed by APS is working according to the following principle: The four links of the transputer are used for communication with the overlying transputer network. Via address decoding and corresponding driver devices a part of the transputer memory is mapped onto the hardware of the I/O bus. A corresponding mapping of system variables performs the direct access to various I/O modules connected to the bus system.

The Smart Welder robot controller interfaces towards the manipulator and the welding equipment via five different types of I/O modules for digital in-/output, encoder input and analog in-/output. All I/O modules conform to the standard Euro format with a 96-pin VG plug at the rear. The transputer network and the three I/O bus systems are each realised as 3U plug-in module. All four modules are housed in one compact 18U rack including the necessary power supply.

2.3 Controller software structure

The real-time software of the Smart Welder robot controller is based on a parallel process concept. Thereby, various parallel processes with local variables communicate via channels. At the same time the communication synchronises the involved processes. This concept has two essential advantages: On the one hand, the software can be enlarged with further parallel processes if a supplement of the hardware is necessary. Thus a good scaleability of the computing power is given. On the other hand, this concept supports the modular structure of the controller software. Each individual process remains clear and the process interfaces are exactly specified by the channels.

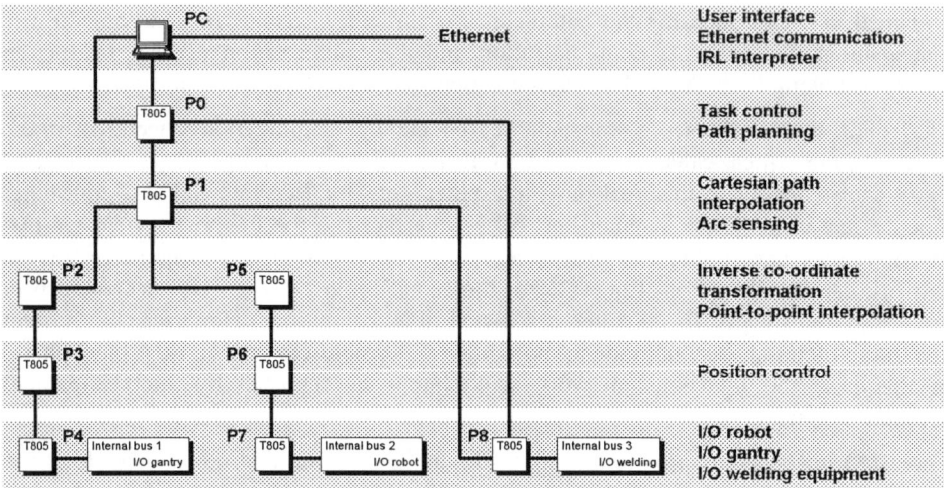

Figure 3: Robot controller hardware configuration and software processes

The software of the transputer network is realised in the programming languages C and occam. As programming tools, the INMOS C Toolset and the INMOS occam 2 Toolset are used. The main controller software processes and their mapping onto the various hardware components can be taken from figure 3. A flexible algorithm for the on-line motion generation in Cartesian space and some implementation results of the inverse co-ordinate transformation are described in more detail in the subsequent sections.

2.4 Flexible motion generation in Cartesian space

Special emphasis has been put to the motion generation in Cartesian space - especially to the path interpolation module. APS has chosen a highly flexible and open structure for the motion generation, also usable for more intelligent and autonomous motion strategies in a future project phase. This is achieved by using algorithms which allow the variation of all motion parameters while the manipulator is moving. Two independent path interpolation modules are implemented: one for the gantry and one for the robot movements. In Cartesian co-ordinates, the geometrical path can be regarded as a desired temporal course of position and orientation of the manipulator end-effector

$$\underline{X}(t) = \left(x(t), y(t), z(t), \phi(t), \theta(t), \psi(t)\right)^T .$$

(1)

For the generation of the desired path, not only the geometrical shape of the trajectory is important, but the time as an additional dimension has to be considered too. The speed profile along the path is defined by the time t. Using the time directly as parameter for the path interpolation will lead to strong limitations for the allowed speed. These limitations have been avoided by introducing an additional path parameter. This parameter facilitates a speed profile along the path that is completely independent of the geometrical shape. For that purpose a suitable parameter is given by the length s of the trajectory from the start to the actual point of the path

$$s(t) = \int_{t_0}^{t} \sqrt{\left(\frac{dx(t)}{dt}\right)^2 + \left(\frac{dy(t)}{dt}\right)^2 + \left(\frac{dz(t)}{dt}\right)^2} \, dt \,. \tag{2}$$

The path parameter s is always positive and increasing monotonously along the path. The first derivation of s to the time represents the path speed and the second one the path acceleration. Following this approach, the generation of the trajectory can be divided into two steps: Firstly the path parameter s is calculated as function of the time t. Its temporal change influences only the speed along the path. Secondly the geometrical interpolation of the trajectory is performed. Thus for every sampling point the actual pose $\underline{X}(t)$ of the end-effector is provided.

First step: $t \Rightarrow s(t)$ $\tag{3}$

Second step: $s(t) \Rightarrow \underline{X}(s) = (x(s), y(s), z(s), \phi(s), \theta(s), \psi(s))^T$ $\tag{4}$

The calculation of the path parameter and the generation of the trajectory have to fulfil some special requirements. For modern industrial manipulators a smooth course of the trajectory is required - abrupt changes of the acceleration are not allowed. For that purpose not only the acceleration itself but also the increase of acceleration, the jerk, has to be limited. In conclusion, the following requirement can be formulated: The Cartesian pose $\underline{X}(t)$ has to be continuous up to the second derivation to the time.

2.4.1 Calculation of path parameter

The calculation of the path parameter represents the first step of the Cartesian path interpolation. The temporal course of the path parameter has to fulfil the following requirements: Path speed and path acceleration have to be continuous. In addition, limitations of jerk, acceleration and speed have to be considered. A flexible algorithm that fits all these requirements works according to the following principle [4]: The actual state variables path acceleration a, path speed v and path parameter s define the jerk r in the next interpolation cycle. The length of the trajectory s_e, the maximal speed v_{max} and the limitation of the acceleration a_{max} also have an influence on the setting of r.

By this method, the jerk is not continuous but temporarily constant in certain intervals. The jerk is integrated three times up to the path parameter. Thus one receives as temporal courses

- for the path acceleration a linear functions,
- for the path speed v parabolas of second order and
- for the path parameter s parabolas of third order.

The used algorithms are real-time capable. They allow the dynamic calculation of the path speed within the cycle time of the path interpolation. By the described method the path acceleration at the beginning and at the end of the trajectory may take any positive value. The temporal path interpolation has to transfer the path parameter s, the path speed v and the path acceleration a from a start value to an end value. During the movement of the Tool Centre Point (TCP) the following conditions have to be kept:

$$|r(t)| \leq r_{max} \tag{5}$$

$$|a(t)| \leq a_{max} \tag{6}$$

$$0 \leq v(t) \leq v_{max} \tag{7}$$

During a time-optimal change of the manipulator pose, the end-effector is moved from the start to the end point in a short as possible time. The required complete time can be divided into up to seven intervals. Inside each interval the jerk takes the positive or negative limitation or the value zero. For each interval $[t_{i-1}, t_i]$ the following equations are valid:

$$t_{i-1} \leq t \leq t_i; \quad i = 1...7$$

$$r(t) = \begin{cases} r_{max} & \text{for } i = 1,7 \\ 0 & \text{for } i = 2,4,6 \\ -r_{max} & \text{for } i = 3,5 \end{cases} \tag{8}$$

$$a(t) = a(t_{i-1}) + r(t)(t - t_{i-1}) \tag{9}$$

$$v(t) = v(t_{i-1}) + a(t_{i-1})(t - t_{i-1}) + r(t)\frac{(t - t_{i-1})^2}{2} \tag{10}$$

$$s(t) = s(t_{i-1}) + v(t_{i-1})(t - t_{i-1}) + a(t_{i-1})\frac{(t - t_{i-1})^2}{2} + r(t)\frac{(t - t_{i-1})^3}{6} \tag{11}$$

The implementation of the algorithm on a digital computer requires temporal discretisation $t = nT$ with the sampling time T and $n = 1, 2, 3,....$ To generate the path parameter in real-time the equations 9 to 11 have to be evaluated for each sampling point. In addition, within each cycle the conditions for the change into another interval have to be checked. A change can be caused by achieving the maximal acceleration, the maximal speed or the start of the brake path.

2.4.2 Generation of geometrical path

The computation of the end-effector pose $\underline{X}(s)$ represents the second step of the path interpolation. The geometrical shape of the trajectory has to fulfil the following requirements: no sharp bends and no abrupt changes in curvature. For the generation of the trajectory, interpolating algorithms are required that are differentiable double continuously to the path parameter s. The Smart Welder application requires interpolation algorithms for

- linear movements and

- path smoothing between separate path segments.

The desired trajectory can be combined of several path segments. The linear interpolation shows the required continuity but passing from one straight line to another, discontinuities in derivation arise. In these cases, the path smoothing algorithm enables a movement along the complete path without stop. The chosen algorithms for the linear interpolation and the path smoothing are described below. During linear interpolations the position and the orientation of the robot end-effector are transferred from a starting point $\underline{X}_0$ to an end point $\underline{X}_e$ using the following algorithm

$$\underline{X}(s) = \underline{X}_0 + \frac{s}{s_e}(\underline{X}_e - \underline{X}_0) \tag{12}$$

where s_e is the complete distance which the end-effector covers during this interpolation. Constrained by the welding process, the passing from one straight line to another must be possible without stop. The pose and the first two derivations to the path parameter have to be continuous in order to avoid abrupt changes or impulses of acceleration. This requirement can be fulfilled by the insertion of an additional smoothing path segment. The continuous smoothing of two path segments is performed inside a defined smoothing radius. This description conforms to the notation of the IRL standard. Inside the specified radius

- the pose is transferred from $\underline{X}_1$ to $\underline{X}_2$,

- the first derivation is transferred from $\underline{X}'_1$ to $\underline{X}'_2$ and

- the second derivation is transferred from $\underline{X}''_1$ to $\underline{X}''_2$.

Thus the smoothing path has to fulfil six boundary conditions that are specified by the actual course of the trajectory at the points of entering $(\underline{X}_1)$ and leaving $(\underline{X}_2)$ the smoothing circle. A polynomial of fifth order

$$\underline{X}(s) = \underline{a} + \underline{b}s + \underline{c}s^2 + \underline{d}s^3 + \underline{e}s^4 + \underline{f}s^5 \tag{13}$$

is able to fulfil these boundary conditions. Its coefficients are determined by the requirements of continuity at the beginning and end of the smoothing path. The coefficients are computed with the actual supporting points and derivations of the connected path segments before each smoothing path.

2.5 Inverse co-ordinate transformation

The inverse co-ordinate transformation transfers the Cartesian pose values in specific joint values for the mechanical system. These joint values are input for the underlying position control. The path interpolation module inputs the Cartesian position and the orientation of the TCP to the inverse co-ordinate transformation. By this transformation routine, a calculation of the joint values is performed which are afterwards outputted to the position control. Two independent transformations running on different transputers exist for the gantry and the robot. Thus a calculation time of less than 1 msec can be guaranteed.

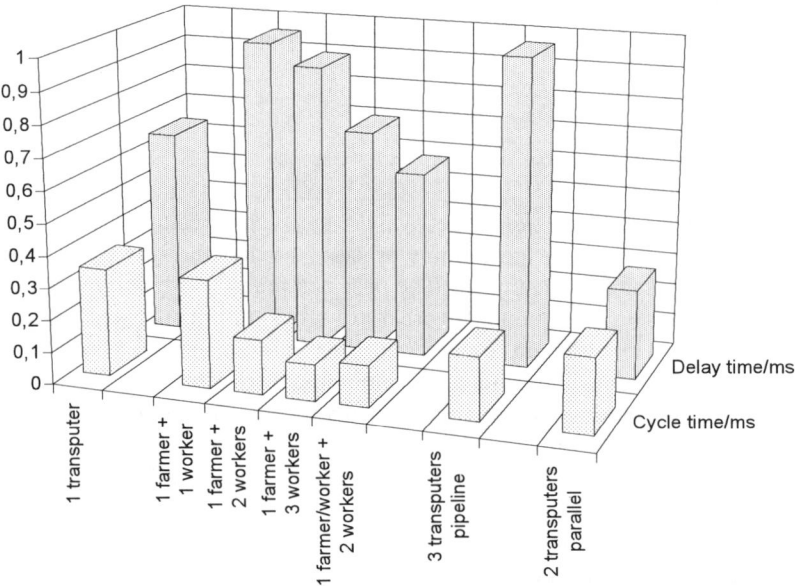

Figure 4: Comparison of several concepts for the inverse co-ordinate transformation

Standard algorithms are used for the inverse co-ordinate transformation. They are based on the robot description scheme of Denavit-Hartenberg. Parameters describe the position of two combined joints. These parameters are specified by the kinematical structure of the mechanics. For each axis n a co-ordinate system with its origin on the joint $n+1$ is introduced. With a transformation matrix $^{n-1}A_n$, combining translations and rotations, the co-ordinate system of axis n can be transformed to the co-ordinate system of axis $n-1$. By that the position and orientation of the TCP (T_6) in the shop-floor co-ordinate system can be calculated by the multiplication of the A matrices of the different axes (e.g. for six axes):

$$T_6 = {^0A_6} = {^0A_1} \, {^1A_2} \, {^2A_3} \, {^3A_4} \, {^4A_5} \, {^5A_6} \tag{14}$$

This calculation is called the co-ordinate transformation. To determine the joints from a given position and orientation of the TCP, the inverse co-ordinate transformation is used. It can be formulated as follows

$$q = g \, ({^0A_6})^{-1} \tag{15}$$

where q describes a vector of joint values and g a pose vector of the TCP. The calculation of the inverse matrix $^0A_6$ is done by using the analytic method of Paul. The inverse co-ordinate transformation results in an output of the joint values. These outputs are sent to the position control.

APS has measured the behaviour of the inverse co-ordinate transformation algorithm of Paul on different transputer networks. Two measurements have been used to make a statement of the real-time abilities:

- the delay time (passed time between data input and output) and
- the cycle time (interval between the receiving of different outputs).

Figure 4 shows the results of some tests of the inverse co-ordinate transformation for a 6-axis industrial robot with different network structures. It could be proved, that the best delay time can be reached using two parallel transputers whereas the best cycle time is reached by using the farmer/worker concept.

3 Multi-axis mechanical system

The complete Smart Welder manipulation system contains eleven degrees of freedom and consists of two independent subsystems: a gantry system and a robot mounted at the gantry. The general concept of the multi-axis mechanical system is shown in figure 5. The different components are

- the gantry system,

- an 8-axis robot mounted at the gantry which gives the system very high flexibility for the welding tasks and

- interfaces between the servo amplifiers/encoders and the transputer network of the robot controller.

The gantry system is based on an xyz-gantry which has been strengthened and equipped with some additional axes. Due to the extra degrees of freedom and the enhanced stretching range, special attention has been paid to the stiffness and accuracy of the separate components. During the design process the system has been analysed using a finite element method in order to optimise the mechanical structure. Furthermore individual components with very fine tolerances have been selected. Thus the maximal total backlash of the manipulator could be limited to 1.2 mm in the worst case. The complete mechanical system has been modelled in RobCAD which is a graphical simulation package for robot installations. RobCAD has been used to estimate the utility of the system concerning working space, agility and simulation of the welding process.

The functions of the gantry are to increase the working space of the robot and to compensate for a translational and rotational offset detected by the block identification module of the vision system. At the same time, the additional axes are used to increase the working range in complex workpiece subsections where it is not possible to position the robot with only three translational gantry axes. The manipulator is able to reach a complete workpiece within its working space of approximately 4 m x 6 m x 2 m.

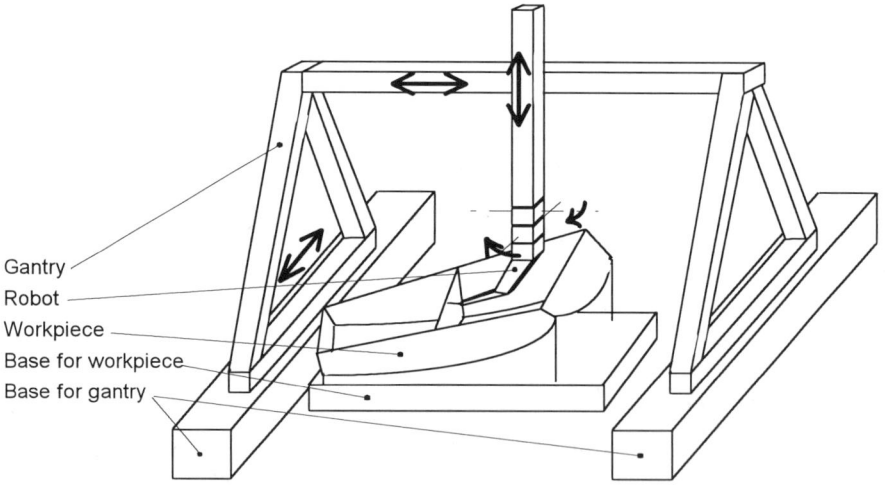

Figure 5: Multi-axis mechanical system in general

The robot is equipped with a welding source for welding defined lines. This includes superimposed weaving and seam tracking. The welding robot is very agile so that it is able to weld in complex sections and confined spaces of a block. It is possible to get actual information about welding current and voltage from the welding equipment. These data are used by the arc sensing module of the robot controller. The arc sensing system determines the required correction values for the robot movement.

Near the robot end-effector, a small video camera is installed. It is used by the vision system to grasp pictures for the identification of the block position and the corner detection. The positions, where images have to be grabbed, are reached by moving the manipulator. The movement is controlled by the Smart Welder robot controller interpreting the off-line generated IRL instruction program.

4 Vision system

The vision system can be divided into three separate sub modules: the block location module, the corner detection module and the safety module. The purpose of the block location module is to compute the transformation matrix between the CAD position of the block and the actual position of the block on the shop-floor. This matrix is computed by comparing special features (e.g. edges of the workpiece) of the CAD model with the real world, represented by a number of pictures. The displacement of the block is determined by reconstruction of the 3D scene from the 2D edges extracted from three images (see figure 6). The area to supervise is approximately 4 m x 6 m.

The corner detection module has to detect the position of corners/butt joints in the welding subsections of the workpiece. By detecting these corners deviations within the workpiece compared to the CAD model of the workpiece can be compensated for. The safety module continuously surveys the area around the manipulator to identify any unexpected objects. If a mismatch between the expected and the realised scene is encountered, it is assumed that an unexpected object is found and an emergency signal is sent to the robot controller.

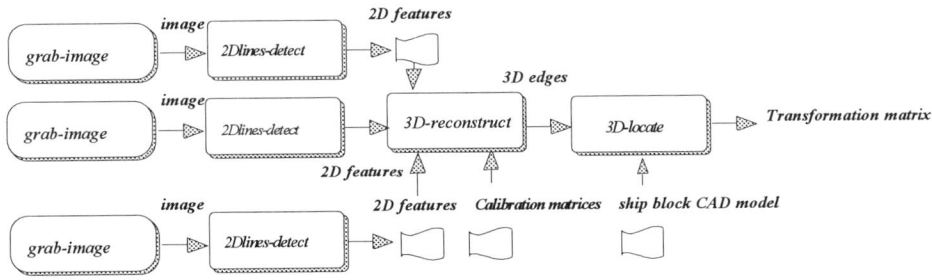

Figure 6: Block positioning module - network of operators

5 Conclusions

The Smart Welder application is designed for the automatic welding of complex sections of the ship hull. For that purpose a CAD system, an off-line programming system, a robot controller, a vision system and a multi-axis mechanical system including welding equipment have been combined to a complex integrated production system. The functional system architecture has been defined in such a way that the Smart Welder can easily be integrated into an overall CIM environment in shipbuilding. The chosen concept can be adapted to related areas in the heavy industry - an aspect which is important for the later exploitation of the project results.

The physical system structure of the Smart Welder is based on standard techniques wherever possible. These techniques are used for the inter-module communication realised by standard Ethernet connection lines as well as for the HPC components used in the different subsystems. The robot controller uses a T805 transputer network for real-time computing, and the vision system architecture is based on the ARVISA vision machine platform for time-critical image processing operations. By integrating these available HPC technologies into the shipbuilding process the involved partners aim to increase the effectiveness of production.

Especially the Smart Welder robot controller takes significant benefits from the embedded High Performance Computing concept. The hardware platform has been specified with the objective to gain as much flexibility as possible. The controller consists of three main components: a PC based host system, a transputer network and several I/O devices. The chosen system layout facilitates an easy upgrade to more powerful HPC processors in a future project phase. The controller software structure follows a parallel process concept. Most of the modules use standard algorithms and methods well known in the area of robotics and control. Special emphasis has been put to the motion generation in Cartesian space. The specified path interpolation module forms the basis for intelligent and autonomous operations in the future. A highly flexible structure has been achieved by using algorithms which allow the variation of all parameters while the manipulator is in motion.

In a future project phase, the autonomous functions of the Smart Welder application will be extended. The autonomy will be increased by the integration of off-line programming functions into the robot controller. This may lead to elegant, time-saving and advanced methods for collision-free motion generation in confined spaces. Furthermore, autonomous manoeuvres based on intelligent algorithms for on-line adaptation of the path will lead to higher productivity.

6 Acknowledgements

The work in the CLEOPATRA project is partly funded by the Commission of the European Communities. Apart from APS - European Centre for Mechatronics, two other companies work on the Smart Welder application: Odense Steel Shipyard Ltd. in

Odense, Denmark and Thomson Broadcast Systems S.A. in Cesson-Sévigné, France. Parts of this paper describe the work that has been performed by these partners.

7 References

[1] N. Ayache: Construction et fusion de représentations visuelles tridimensionnelles - applications à la robotique mobile. Thèse d'état, Université de Paris-Sud, Orsay, May 1988.

[2] J.J. Craig: Introduction to Robotics, Mechanics and Control. Addison-Wesley: 1986.

[3] J.C.S. Jensen: On the Model Errors in Robotics. Control Engineering Institute, Technical University of Denmark, 1993.

[4] J. Olomski: Bahnplanung und Bahnführung von Idustrierobotern. Vieweg: Braunschweig, Wiesbaden, 1989.

[5] P. Puget / T. Skordas: An optimal Solution for Mobile Camera Calibration. Proceedings of the First European Conference on Computer Vision, Antibes, France, April 1990.

[6] C. Venaille: Reconstruction tri-dimensionnelle de réseaux vasculaires en vision trinoculaire. Thèse d'état, Ecole Nationale Supérieure des Télécommunications, 1991.

[7] M. Vukobratovic / M. Stotic: Applied Control of Manipulation Robots. Springer: Berlin, Heidelberg, 1989.

Prof. Dr.-Ing. Paul Drews

Dipl.-Ing. Dieter Matzner

APS - Europäisches Centrum für Mechatronik
Reutershagweg 4, D-52074 Aachen, Germany
Phone   +49-241-8864-0
Fax   +49-241-875715
Email   mechatronik@aps.rwth-aachen.de

Autonomous Tool-Mover for Laser-Cutting with Industrial Robots

Werner Roddeck, Hans-Jürgen Rehbein
Fachhochschule Bochum, Germany

Abstract: In the last few years laser-cutting technology has been spread in a wide area of industrial applications. Moving the laser cutting-head with an industrial robot, 3-dimensional laser-cutting of large sized sheet metals, as they are needed in the automotive sector, can be done with low cost equipment. Specially Nd:YAG-lasers are qualified for laser-cutting systems of this purpose, because the radiation of this type of laser can be transmitted by a glass-fiber. Such a fiber can be easily handled by an industrial robot, so that the cutting-beam can be moved 3-dimensionally in its workspace.
One problem connected with small-shaped contour elements as circular holes is the accuracy of tool positioning by a robot. This problem can be solved by an autonomous tool-mover with two separate numerical axes, which is attached at the end of the kinematic chain of the robot arm.

1 Introduction

Nowadays the production of prototypes in the automotive industry is often done by laser-cutting equipment because at this moment the die, with which the car body is shaped in the series production, is not yet available. If laser equipment of the type of an CNC milling machine tool is used for this working of large sized sheet metal parts one gets an investment of more than 1 million DM. Whereas if the relatively cheap Nd:YAG-laser is combined with a six joint vertical robot arm as the beam guiding system, one gets roughly the half investment combined with higher flexibility in 3-dimensional movement. This is possible because the radiation of the Nd:YAG-laser (wavelength 1.06 μm) can be transmitted by a glass-fiber with reasonable loss. The radiation of a CO_2-laser (wavelength 10.6 μm) however, which is often used in material processing, can only be guided by reflecting mirrors, because it's losses inside a glass-fiber are too high.

The accuracy of an industrial robot in positioning and in motion along a path that has the required load capacity (> 25 kg) is inside that tolerance, which is necessary for sheet metal parts, but normally worse than that of CNC equipment mentioned above.
However the accuracy of movement along a path of a robot becomes worst if small shaped contours as circular holes of diameters between 2 and 30 mm are to be cut.
This is due to the high compliance of the complex kinematic structure of a 6-axis robot, the great demands on the dynamic of the electrical drives and on the interpolation of the

path of movement. These problems result in deformations of circular holes to be cut, as they are shown in fig. 1. In such an application deviation of the path for circular holes of a diameter of 20 mm can be in the range of +/- 0.5 mm although the position error of the robot has the specification of 0.1 mm.

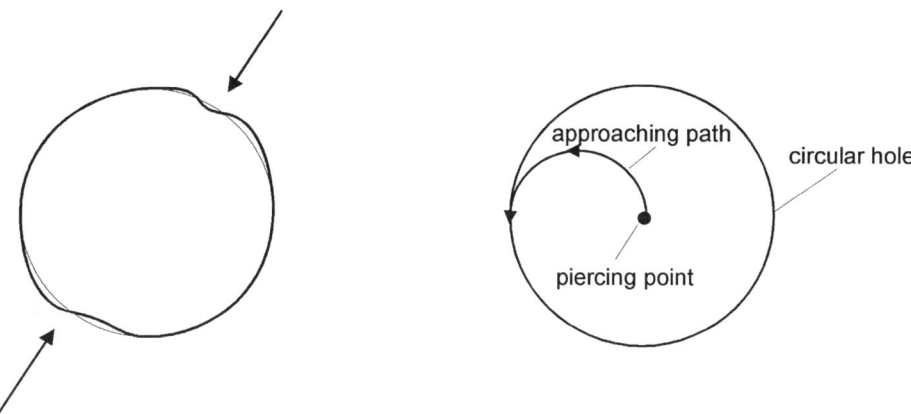

Fig. 1: Typical distortion of a   Fig. 2: Path of movement for cutting a
       circular hole.   circular hole with a laser.

The reasons for these deviations are on the one hand mechanical problems, like low stiffness of components, backlash of gears and hysteresis. On the other hand there are electronically and software problems within the servo system and the interpolation, be-

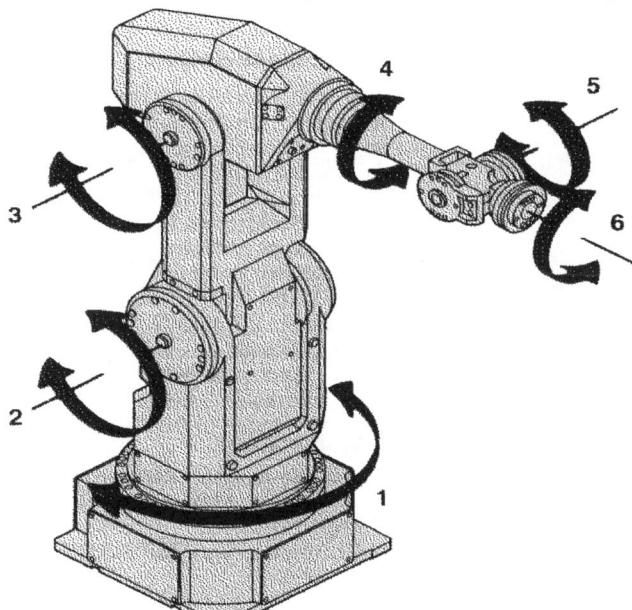

cause in some cases all six joints of the robot arm are involved for 3-dimensional movements. This often happens with great differences in velocity and acceleration of the axes, which share the same movement. Fig. 3 shows the type of industrial robot that should be used for laser-cutting.

As circular holes often occur in body making there are accuracy problems when using robot equipment. Therefore the object of the development described here was to en-

Fig. 3: Industrial robot for laser-cutting

able the cutting of small, circular holes with a tolerance in the diameter of +/- 0.01 mm with an additional appliance at the wrist of the robot as a tool-mover.

The described problem was researched in a developing project sponsored by the ministry of economics of Northrhine-Westphalia [1], [2].

2 Special requirements to the tool-mover

There exist several marginal conditions for the additional robot appliance for laser-cutting, which have great influences on the construction of the tool-mover.

● The tool-mover should be adjustable between diameters of 1 mm up to 30 mm and should give cutting velocities up to 30 mm/s.

● Adjustment of the tool-mover has to be done under program control of the robot and miscellaneous functions as tangential approach to the contour must be possible (see Fig. 2).

● The robot, which should be used with the tool-mover has a nominal load capacity of 30 kg. Therefore the appliance has to be light weighted and compact because the load capacity of the robot is used up to its third part by the weight of the cutting-head, a sensor guided servo system for the focus control (tool-axis Z) and additional forces, which are caused by the movement of the glass-fiber. These masses and forces that are already efficient become even more important when the tool-mover is installed, because it is mounted between the robot's face plate and the Z-servo axis and by this the lever-arms are lengthened. Normally the nominal load capacity of industrial robots is specified with reference to attachment at the face plate so that the problem becomes even more important. Figure 4 shows the locating place of the tool-mover between the face plate and the laser cutting-head.

● When the cutting-head is moved along the circular path it must not rotate around the Z-axis, because otherwise the glass-fiber would be drilled and might be damaged.

● The 2-dimensional movement of the appliance must be possible in all positions of the robot's workspace with any orientation of the wrist. Therefore the torque of the used drives must be great enough to lift the weight of the appliances itself, plus the weight of the Z- servo drive, the cutting-head and the glass-fiber against the gravitation with sufficient acceleration.

It was not possible to include the last rotational axis of the robot in the 2-dimensional movement of the cutting head, because this would have strongly restricted the potential movements of the robot. Therefore the tool-mover had to have two controllable axes which could be moved in a functional dependency.

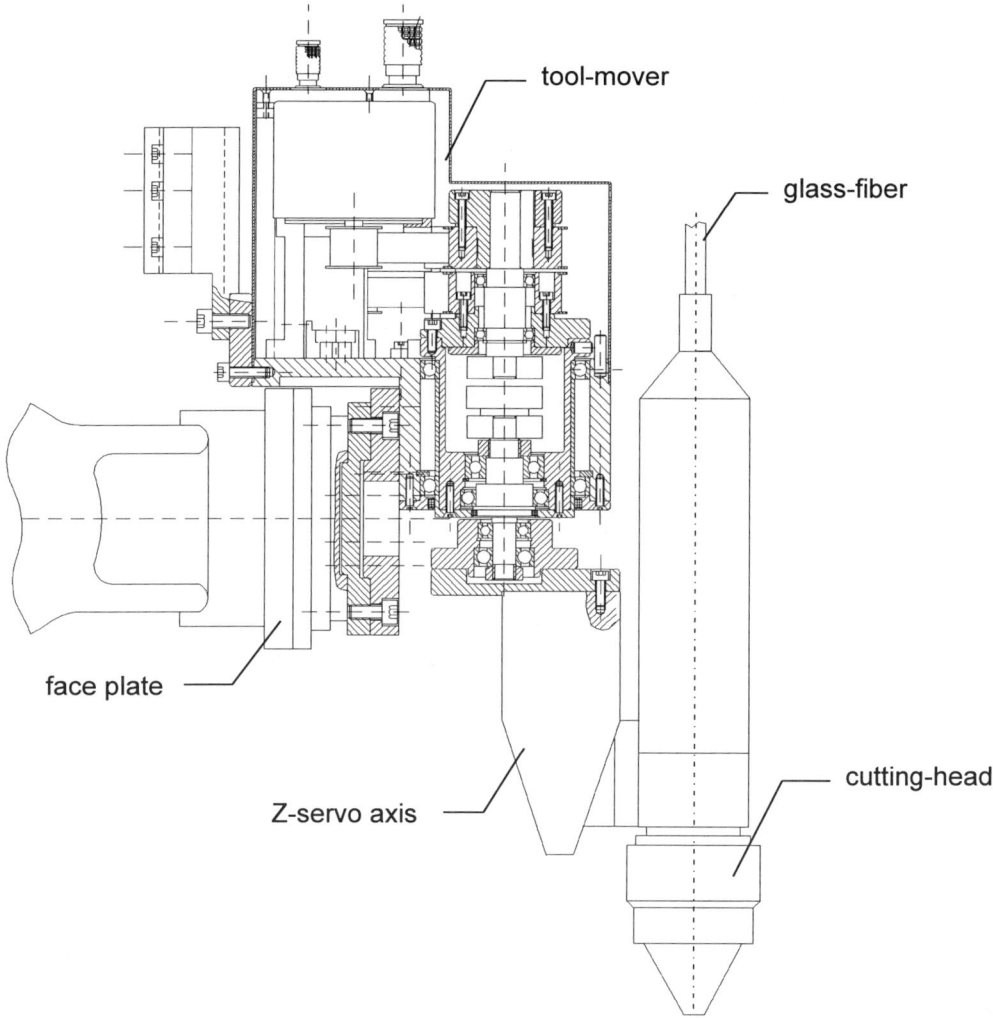

Fig. 4: Tool-mover mounted at the face plate.

3 Design of the kinematic of the tool-mover

The simplest approach to create a circular path of movement with adjustable diameter would be to use a kinematic combined from one linear axis and one rotational axis (fig. 5). But this version can be eliminated at once, as the glass-fiber, connected to the cutting head, would be drilled when the cutting head moves along the circular path.

The next approach for a kinematic to move along a circular path is to use two linear axes. For this version path coordinates must be generated by interpolation which can

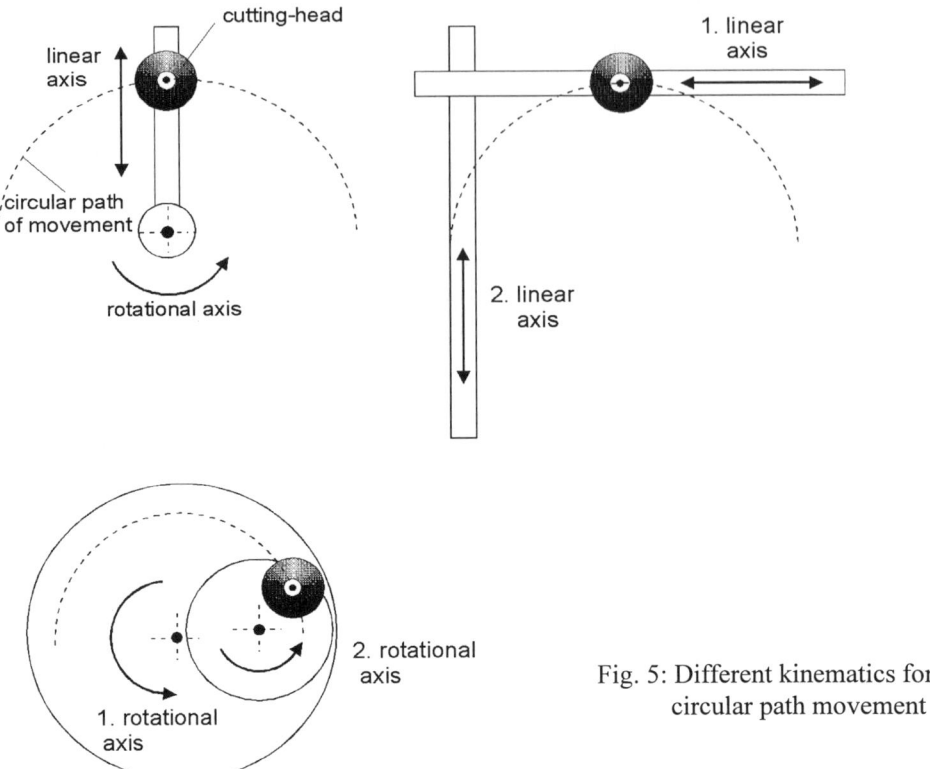

Fig. 5: Different kinematics for
circular path movement

further more give position errors. Such a configuration was also rejected because the volume and weight of two linear axes are disadvantageous for the job of the tool-mover. Moreover the rectangular workspace of two linear axes cannot be used completely when cutting circular holes.

We therefore preferred a combination of two rotational axes (see fig. 5), with which the circular movement for the cutting of the hole can be done very exactly by moving the rotational axis in the center of the kinematic. To avoid drilling the glass-fiber, two assemblies of this type must work on an eccentric appliance, which moves the cutting head on a circular path without a rotation around the Z-axis. Figure 6 shows the kinematic representation of such a *twin eccenter system*.

It consists of two symmetric parts with each partial system containing a rotary beared hollow shaft as the first rotational axis and an eccentric shaft as the second rotational axis that is beared inside the hollow shaft. The corresponding axes in the partial systems are each coupled with a crogged belt, to achieve synchronous movement of the two shafts in each eccentric system. As it is shown in fig. 6, the two adjustable eccentric systems have at the end of the eccentric shaft another eccentric arranged bolt, which are linked by a stiff connecting-rod. The laser cutting-head is fixed to this connecting-

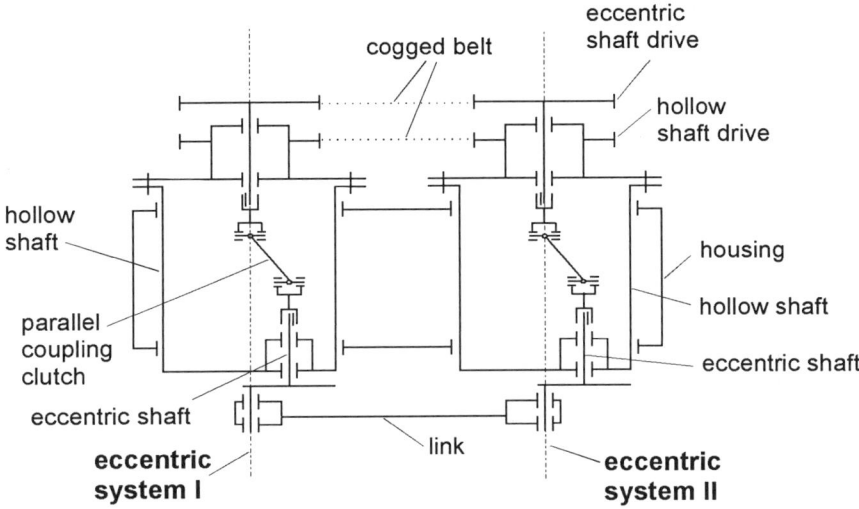

Fig. 6: Twin eccentric system for circular movement

rod and is moved by it when the eccentric systems turn around. To bridge the displacement between the driving side of the eccentric shaft and it's driven end, a parallel coupling clutch is inserted into the kinematic chain.

The bolt at the driven end of the eccentric shaft is located on a hole circle, the radius of which corresponds exactly with the displacement between the driving and driven end of the eccentric shaft. Therefore the eccentric shaft can have a position inside the hollow shaft (see part 0 of fig. 7) which would generate a circular path of movement for the cutting-head with a diameter of zero. When both eccentric shaft and hollow shaft are driven in this *zero position* with the same number of revolutions the cutting head stays in its position without motion. When afterwards both rotation axes are driven from zero position with a different number of revolutions for a short time, a distortion of hollow and eccentric shaft takes place as it is shown in part 1 of figure 7. By this the bolt at the driven end of the eccentric shaft moves from the center of the hollow shaft to it's outside which corresponds to a diameter setting of the tool-mover. If then the hollow and the eccentric shaft are driven with the same number of revolutions again, the cutting head, which is connected to the two eccentric shafts by the connecting-rod, moves on a circular path (see part 2 - 4 in fig. 7) that's radius is equal to the distance of the eccentric bolts from the center of the hollow shaft. By this the cutting head moves along its circular path without turning around it's Z-axis as was required above.

4 Technical realisation of the tool-mover

The driving motor of the system should be simple and light weighted with sufficient torque. Therefore stepper motors with 1/10 step mode were used and to increase the torque a cogged belt was used to gear down the rotation of the two shafts. To get a

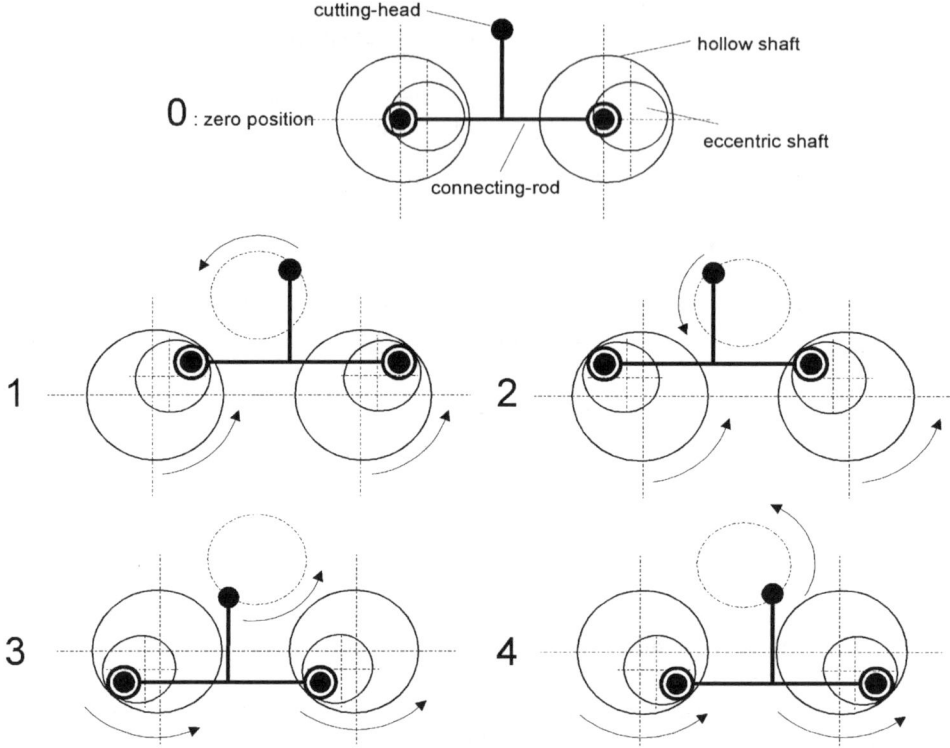

Fig. 7: Circular path movement of the cutting-head by twisting the eccentric shaft
relatively to the hollow shaft.

maximum cutting velocity of 30 mm/s even for small diameters of 5 mm a maximum
stepper frequency of 100 KHz was used; so one gets a resolution of 13 steps/degree
when turning around the rotational axes. The stepper motors are controlled and powered
by a separate stepper unit. The reference position of the hollow and the eccentric shaft
are generated by fine adjustable micro-initiators. The cutting parameters like diameter of
the circle, cutting velocity, maximum turning angel and shape of the path between the
piercing point and the circular contour and the start signal of movement are transmitted
by the robot controller to the stepper controller by binary inputs.
As the two rotational axes can be moved in every functional relation other shapes of
contours than circles can be achieved, e. g. ellipses, long holes or even rectangular
holes.

To get a small weight of the tool-mover all parts of the housing are made from a special
magnesium alloy, which reduces weight down to 70 % of that for aluminium but with
the same solidity. Its mass is therefore about 8 kg. As it is shown in fig. 4 the tool-
mover is constructed in a way that the masses are distributed closely around the tool-
center-point of the robot located in its face plate.

In fig. 8 the complete installation of the autonomous tool-mover at the robot arm is shown. The laser cutting-head with the glass-fiber and the additional Z- servo axis for the focus distance control is mounted to the connecting-rod of the tool-mover.

The order of working is, that the robot positions the laser cutting-head, which is linked to the tool-mover, to the piercing point above the sheet metal, than the laser is switched on and finally the motion command is transmitted to the tool-mover. With the robot in a

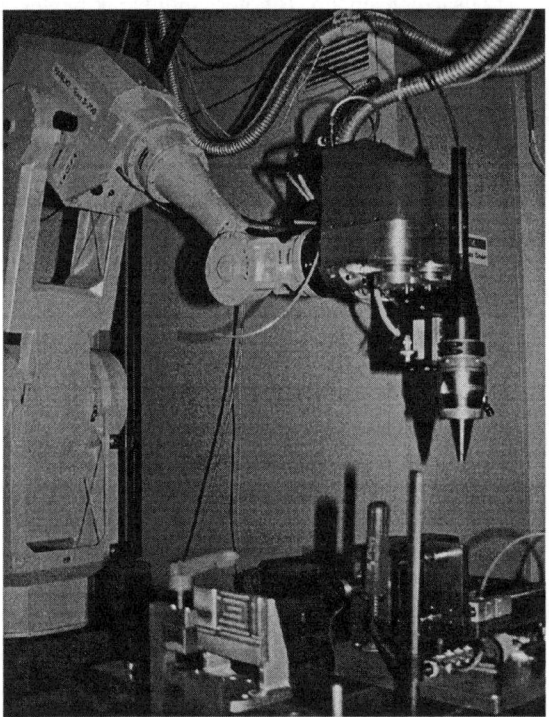

Fig 8: Robot with autonomous tool-mover

clamped position the tool-mover executes the cutting motion for the contour element to be cut. When this is finished the controller of the tool-mover transmits a ready signal to the robot controller so that the robot can execute his next motion command.

5 Conclusion

With the equipment described here circular holes with the required tolerances can now be cut with a laser robot. Also other production technologies that need more accuracy in positioning when used with industrial robots can take advantage of the autonomous tool-mover.

6 References

[1]   Roddeck, W.   Abschlußbericht Forschungsprojekt 3D-Mobil.
      Rehbein, H.-J.   Entwicklungsteilprojekt Formschneidwerkzeug.
                            Gefördert vom Ministerium für Wirtschaft, Mittelstand
                            und Technologie des Landes Nordrhein-Westfalen

[2]   Roddeck, W.   Zusatzachsen am Roboterhandgelenk bewegen
      Rehbein, H.-J.   Laserschneidkopf.
                            Maschinenmarkt, Nr. 23, Juni 1995, S. 42 - 47

Prof. Dr.-Ing. Werner Roddeck
Dipl.-Ing. Hans-Jürgen Rehbein

Fachhochschule Bochum
Lennershofstraße 140
D 44801 Bochum

Tel.: 0234/7007165
      0234/7007227
Fax: 0234/7094222

Interactive Robot Master Slave System for Deburring Large Cast Iron

N. Ahlbehrendt, H. Diesing, S. Jakobi
IpeA GmbH, Berlin, Germany

Abstract: A short survey is given on the newly developed Interactive Robot Master Slave System ("AUTOMANIPULATOR") used in the Gießerei Britz GmbH (Britz Foundry Ltd.). The most essential improvement is the automation-aided manipulation that enables the operator to operate this system continuously and with a constant machining quality. The system consists of several mechatronic components necessary to implement the technical and ergonomic functionality and process reflection. Programming activities are not necessary, even if parts are changed. Conclusions are drawn from the first experience gained under production conditions, and prospects are given for the near future.

1 Introduction

Until now, small- and medium-sized enterprises did not make much of a profit from advanced automation technology. The increased requirements to the flexibility of these enterprises regarding market and customer reaction, innovation of products and small lot sizes with a competitive economy call for a (partial) automation strategy which involves the human being in a conscious and purposeful way.

The implementation of an automation integrating the human being requires a change in the way of thinking (see also /1-4/) from the former goal of full automatisation with all the consequences for the organisation of labour ("Lean production", group work, decentralised, autonomous decision competence, humanisation and attractiveness of workplaces, ...). It is also a challenge to the "mechatronics" as regards the procurement of respectively designed means of automation and, in particular, the interactive means allowing to utilise the natural capabilities (see /5/), experiences and proficiency (ergonomics, interactive interfaces, co-operation of automatic and manual operations, ..).

A typical example for these requirements is the deburring and fettling processes in foundries which are mainly carried out in heavy physical manual work that is injurious to health.

Within the framework of the integrated project initiated by the Federal Ministry for Research and Technology called "Fettling Shop 2000" the Interactive Robot Master Slave System (IRMS) had been developed for and tested in a foundry in deburring processes of internal and external contours at large-sized cast parts. Using the advantages of automatic robot and manipulator technology, this system nearly fulfils the set targets.

2 Tasks of the deburring system

The IRMS had been developed on the basis of precise user requirements to debur the internal and external contours of large cast parts. These parts are individually cast parts. From the economical point of view a fully automatic subsequent treatment cannot be realised, because of its high programming demand and the high qualification requirements to the operating and set-up staff (which is not available) resulting therefrom. Thus, the subsequent treatment must be done directly, or at least supervised by the workman or fettler. In this connection, the same treatment processes are required again and again. Due to the high concentration demands and risks of maloperation, the manipulator technology which is in fact destined for such tasks, can be applied for long-term use in a limited way only (see e.g. /6/).

The most important user requirements are:

- Automation of subprocesses, which can be operated at any time in a manipulative manner

- Automation-aided manipulation in order to reduce the demands to concentration, to minimise user actions, and to exclude maloperations

- No programming work in the treatment of different work pieces

- Protection of the user from dust and noise. Full utilisation of the operating space of the slave arm in a sitting and ergonomic arm position of the user

- Deployment of manual fettlers with a minimum training period

- Low investment costs and short amortisation periods

These user requirements can be generalised to other fields of application. They are mainly fulfilled by the IRMS described in the following paragraph.

3 Description of the system

3.1. Overall system

The innovative heart of the IRMS consists in the utilisation of advantages of the manipulator technology and the automatic robot technology in form of a master slave system based on commercial industrial robotics. It allows the exclusive manipulative, the fully automatic and, in particular, the interactive, i.e. the automation-aided and manipulative, operation.

The IRMS is composed of two industrial robots and interconnected standard controls. One robot is used as the master robot. It is equipped with an information-processing force-moment sensor and a double handle with a permissive switch and a coupling switch attached to the sensor. The slave robot is also equipped with an information-processing force-moment sensor and a compliance system to take up the grinding machines.

3.2.2 Force-moment sensor technology

Both robots are equipped with force-moment sensors consisting of six components. Their measuring information is transmitted to the control information to trigger and co-ordinate the motions of the robot-master-slave system.

Via the double handle of the master robot the user applies forces and moments (manual control forces) which are measured by the master force-moment sensor, and converted in the control into the manipulative part of the motion control.

The force-moment sensor of the slave robot measures the forces and moments (reaction forces) arising at the tool which are used in the control to fulfil different functional subtasks in motion control: overload monitoring, safety functions, force reflection to the master guidance, automatic contour monitoring, and force-adaptive motion control.

Force-moment sensors in industrial use have to meet high demands as regards the transfer rates, interference immunity, degrees of protection, and internal pre-processing, which can only be realised by a compact analogue and digital signal processing integrated in the sensor. The high demands set to the system can be illustrated by the measured values of the pure force-moment information (fig. 4a), measured at a 1.7 kW HF grinding machine with one cutting-off wheel. Measuring signals having an interference signal ratio of 160 to 200 per cent cannot be converted into defined motion commands by the robot control! Therefore, the sensor has to pre-process these measuring signals with much higher measuring frequencies using digital (non-linear) filtering and the redundancy of the force-moment information to obtain information pertinent to control (see fig. 4b). The use of force-moment sensors as mechatronic components with local intelligence is the necessary prerequisite to solve such tasks.

Adapted to the weight of the tool and the machining forces, the S 420 slave robot in this configuration is equipped with the six-dimensional KMS 200 force-moment sensor (measuring ranges 2,000 N and 300 Nm).

The S 5 master robot is equipped with the KMS 10 force-moment sensor (measuring range 100 N and 10 Nm) consisting of six components.

These sensors are designed as fully enclosed deformation parts with resistance strain gauges applied thereon. They were developed and manufactured with the necessary intelligence in the IpeA GmbH. The analogue and digital (single-chip microcomputer) electronics integrated in the sensor allows to fulfil the following functions:

- Analogue processing of measured values

- Monitoring of limits

- EMERGENCY-OFF activation when the threshold value of the resistance strain gauge bridges is exceeded

- Component selection (robot command)

- Digital processing of measured values into information pertinent to control

Especially the last item makes the described field of application suitable for industrial purposes (see also fig. 4a, b).

The following characteristics prove the high interference immunity and integrability of the system:

- Degree of protection: IP 54
- Interface: RS 422 serial interface (adjustable up to 153.6 kbaud)
- Immunity to interference: potential-free optocoupling
- Power: non-stabilised A24-VDC (from robot control)

Other types of force-moment sensors are available for other system configurations.

3.2.3 Compliances

For reasons regarding the theory of control and stability, the tool (in the tool-guided type) in contact processes (tool - work piece) has to be mounted damped elastic at the robot. The required stiffness is mainly determined by the reaction times of the continuous-path controlled robot to obtain the desired travelling speeds. In order to fulfil the most diversified fettling and deburring tasks, this has to be done three-dimensionally. Furthermore, the tool orientation towards the robot hand shall be kept constant. This orientation condition allows an extended machining range excluding possible maloperations and damages (For example, when parting at the riser, a change of the tool orientation would result in a rupture of the cutting-off wheel). At the same time it increases the transparency of the orientation manipulation. For machining tasks with lower demands to the orientation stability of the tool, simpler structures may be used. Straightforward systems on an elastomere basis were tested as well as constructions with rotatory degrees of freedom avoiding the influence of the dead weight of the grinding machines on the excursion of the compliance system.

For the described system a three-dimensional, translatory compliance system with combined balance and leaf springs was developed. The critical damping is executed by pneumatic cylinders. Furthermore, it is equipped with a safety switch for direct EMERGENGY-OFF triggering, and a releasing gear in case the excursion is too high.

4 Control

4.1 Hardware configuration and interfaces

Figure 5 gives a rough survey on the control interface configuration of the overall field-tested system.
Two Fanuc RJ2 standard controls were used. The master control activates the two force-moment sensors, takes up their measured information via RS 422 serial interfaces, and calculates the sequences of motion for the master and the slave robots depending on the operating mode and the actual sensor information. The slave motion commands and the operating mode are transmitted through the RS 422 serial interface to the slave control. The operating modes are conveyed by the keys of the handle (see item 5) and the control desk via parallel interfaces.

A SPC is used to control the manipulator carrier system (one translatory axis/rail) and the tool carrier system (travel axis vertically and horizontally to the travel axis of the manipulator carrier system, one rotatory axis).
Direct EMERGENCY-OFF lines are laid at different points to increase the labour safety (exceeding of power threshold, door safety switch when unauthorised persons enter the room, ...).

4.2. System and function software

The system software of the robot manipulator system is in a hierarchical order. In the first level, both robot controls and both force-moment sensors are initialised; standby or error messages are signalised. When the standby signal is released, the second level is activated where robot programmes can be executed without regard to the sensor (e.g., starting of prepositions), and parameter for master control (master-slave transmission ratio, ratio between expenditure of force and master travel, ...) can be set individually. In the third level the sensor-controlled motion modules are activated for master control. Sensor-aided automatic functions can be superimposed thereto. In the fourth level there are the function modules which - in case of serious damage - make the robot system again ready to work by safety-regulated manipulation.

The sensor-aided motion setpoints in position and orientation (P_M and P_S) are set differentially by the master control in the actual handle coordinate system K_M for the master, and in the actual tool coordinate system K_S for the slave. Further conversion of motions is done by the standard routines of the robot controls.

Essential subfunctions and parameters in this connection are:

Force reflectivity (F_M, F_S)

The master travel is mainly determined by the manual guidance forces F_M applied by the user. The slave machining forces F_S give the user the necessary haptic sense for the machining forces as the most essential precondition for a practical and long-time manageable, interactive work. Additionally, the machining forces F_S are used to realise the automatic function tools for the slave robot.

Online-gravitation compensation (F_M, F_S)

Since the "control-relevant" processed information of both force-moment sensors observe the dead weights of the handle and the grinding machine, their calculation has to be reduced to the pure manual guidance forces (F_M), the machining forces and moments (FS), taking into account the actual orientations of both robots.

Reflectivity of the working area (A_M, A_S)

The operating space together with the robot-oriented axle travel limits and the user-oriented cartesian operating and safety areas is constantly monitored online. The master operating space A_M is mainly used for user protection.

Avoiding maloperations (Mod$_M$)

Due to acceleration limitations hectic and shaky manual guidance is not converted into robot travels. The mode (Mod$_M$) mainly determines geometrical travel restrictions that are automatically observed by both robots. In this connection, force is constantly monitored, so that an EMERGENCY-OFF signal is released in case the threshold is exceeded.

Individual setting (Par$_M$, Ü, ORI)

With Par$_M$ the ratio between the manual forces to be applied and the travel intensity of the master robot can be set. During pure manipulative operation the differential travel settings of the master are transmitted via an adjustable transmission ratio (Ü) to the slave control. Here, the relative orientations of handle and tool always remain constant. For ergonomic reasons these relative orientations are freely selectable during operation in relations of 90 degrees (ORIentation prepositions).

Overlaying of automatic functions (Typ)

A decisive feature is the overlaying of selectable (type), sensor-guided automatic functions that can be manipulated at any time. 'Typ' characterises the type to be selected. The automatic functions that can be accomplished at present are explained in the following paragraph where, in particular, the automatic contour tracing of unknown contours using different tools shall be stressed.

The general basic structure of the online-path generation in a simplified way is as follows (means the cycle time of the robot controls in the cartesian interpolation level):

$$P_M(t+\tau) = P_M(t) + F_M\{F_M , F_S , A_M , A_S / Mod_M , Par_M\} , \qquad P_M \in K_M , \qquad (1)$$

$$P_S(t+\tau) = P_S(t) + Ü* F_M\{.../...\} + F_A\{ F_S , A_S / Typ\}, \qquad P_S \in K_S. \qquad (2)$$

Interaktive = reflektive Manipulation [F$_M$] + automatic [F$_A$]

---

5 Working modes and ergonomics

It should be pointed out that in the development of this comprehensive system the fettlers of the Britz iron foundry had been involved from the very beginning. The experience gained in testing the new system and the recommendations given by the users were taken into consideration during the ergonomic and functional development and design of the entire system. Despite the extended functionality and flexibility of the system, no programming work of the user is necessary. He only has to act.

5.1 System operation

The individual motion units - the work piece carrier system, the manipulator carrier system and the robot-manipulator system - are controllable by an alphanumeric keyboard.

5.2 Ergonomics of the manipulative operation

First, an individual setting of the manipulator guidance is planned. Using Ü (see formula(2)), the transmission ratio of the horizontal travel for master and slave (1:0.25 up to 1:2) to the very precise up to the quick rough manipulation is set. The ratio of the travel intensity to the expenditure of manual guide force can be adjusted individually.

Then, the interactive work is done through manual guidance. Tests showed that the manipulative guidance of all six degrees of freedom with only one hand required too high demands to concentration and, therefore, resulted in symptoms of fatigue so that a double handle was developed and used for manipulation: First, after pressing the permission switch, the master robot can be moved only with the right handle. Only when the user presses the coupling key by his thumb, the slave robot travels synchronously in the set transmission ratio. The orientations of the handle and the tool are kept the same. When the permission key "orientation" is activated additionally by the left hand, a manipulative guidance will be possible in all six degrees of freedom.

When decoupling the slave robot, the operator can always draw the handle - independent on the slave position - into that position that is the most favourable for him. After that he can connect the slave robot again. By pressing the key he can automatically adjust other orientation relations between the slave tool and the master handle. These relations are graded by 90 degree, so that - on the one hand - the orientation transparency of the operator's hand to the tool is always observed, and, on the other hand, orientation modifications at the master handle can only be carried out up to 45 degree which are ergonomically justifiable.

With the

- force reflection of the machining forces,
- operating area reflectivity of master and slave,
- control-related avoidance of hectic manually-guided actions,
- individual adaptability of the master guidance,
- permanent force monitoring for damage protection,

the major ergonomic and functional preconditions for an efficient manipulative work exist. With an individual, ergonomic arm position of the sitting operator, the entire operating area of the slave robot can be exploited.

5.3 Automatic functions and interactive operation

The decisive stress relief of the operator is accomplished by the automatic function, which - at any time - can be interfered in a manipulative way to modify the sequence of motions, to interrupt the process (e.g., by simply lifting the slave tool from the work piece), or to switch to a new machining task at any place within reach, or to switch into another operating mode. The most simple automatic tools are the automatic restriction of the manipulative controllable degrees of freedom to the technologically required number (straight line, plane, constant tool orientation, orientation changes only at constant tool-center-point). Even these tools relieve the user, and avoid maloperations.

At present the following automatic machining tools are implemented:

Surface machining:

Two points of the surface to be machined are approached by the robot tool in a manipulative and orientation-true manner, and then stored by pressing the key. After that, the surface is machined automatically in an oscillating motion between these points using power-controlled propulsion within the selectable lateral or peripheral grinding. In this mode, the tool orientation remains constant; only a translatory, manipulative influence in three dimensions is possible. This automatic orientation observance leads to constant machining results, the manipulative actions are reduced to the necessary ones, and, at the same time, maloperations are nearly avoided.

Separation of the riser:

This automatic function only differs from the surface machining as regards the machining tool (cutting-off wheel), and the additional limitation of the manipulation to the plane given by the orientation of the cutting-off wheel. The automatic observance of these geometric restrictions just allows this (interactive) process and excludes maloperations. Modifications of the orientation or the position in the normal line of the cutting-off wheel would result in jamming or even rupture of the wheel. These conditions cannot be realised practically with a pure manipulative guidance of all six degrees of freedom.

Deburring of straight part:

Activation and execution similar to surface machining.

Automatic deburring of contours:

In this mode, the tool is only approached to the contour for deburring. As soon as the contact between tool and contour is accomplished, the slave robot traces the non-programmed contour in a tool-oriented direction which can be reversed by pressing the key. This powerful automatic tool was developed and tested in peripheral and lateral grinding for shank grinders and rough-grinding wheels.

No programming regarding the application or parts is necessary when using these functions.
The user determines, observes and corrects. The work is done by the slave robot.

6 Final remarks and outlook

The development presented in this paper is only a first step, which lets a number of unrealised wishes open. Besides an extended functionality and reduced control-cycle times in order to involve other fields of application with higher speed requirements, we would like to mention the problem of the user-transparent visualisation of the machining processes in non- or hardly visible areas. Within the framework of the above mentioned project, an experimental study was elaborated by the Fraunhofer Institute IFF/Magdeburg.
From the "mechatronical" point of view, the positive experience gained in this project is proposed for discussion by the following generalising theses:

1. The involvement of the human factor (interaction, ergonomics in hardware and software, capability to keep the process under control, qualification requirements) also constitutes a challenge to the "mechatronics" (which was until now exclusively understood in a technical and technological sense) being a factor of procuring the mechatronic automation components as tools "capable to work automatically" in the most diversified fields of application.

2. Alternative programme technologies have to be developed to reach the broad market segment of small- and medium-sized enterprises with their small and quickly growing lot sizes, which involve the local experience and skills directly: To use experience instead of formalising.

3. The future user (in this case practised with the fettlers of the foundry) has to be involved even in the development and design of such technologies.

The proposed robot-master-slave system was promoted within the framework of the AuT (Labour and Technology) compound project "Fettling Shop 2000". It was developed by the IpeA Gmbh by order of and in close cooperation with the iron foundry Eisengießerei Britz, where both sides.

7 References

/1/ Martin,H.; Rose,H.: Erfahrungen sichern statt ausschalten. ZWF/CIM 88(1993)3, pp. 34-41

/2/ Fuchs,H.: Einheit von Mensch und Automatisierung.. Workshop of the FFT e.V. and the TFH Berlin "Trends und Auswirkungen der menschenintegrierenden Automatisierung", Berlin, 12 Nov. 1992, Vortragbd. pp. 7-20.

/3/ Erbe,H.   Die Qualifikation der Facharbeiter als Basis einer werkstattorientierten Produktionsunterstützung. Workshop of the FFT e.V. and the TFH Berlin "Trends und Auswirkungen der menschenintegrierenden Automatisierung", Berlin, 12 Nov. 1992 Vortragbd. pp. 49-56.

/4/ Autorenkollektiv: Materialband zum Workshop des BMFT Verbundprojekt "Verbesserte, benutzerorientierte Programmierung von Industrierobotern". Aalen 10.03.1993.

/5/ Ahlbehrendt,N; Fuchs,H; Diesing,H;: Anwenderorientierte Programmierung von Robotern - "Erfahrung nutzen statt formalisieren", Int.J. Automation Austria, Jg. 2, (1994), pp. 7-25.

/6/ Verbundprojekt Putzerei 2000, Bericht Nr. 08 (1994), Bericht zur Projektphase A der Eisengießerei Britz (1993).

Prof.Dr.sc.nat. Norbert Ahlbehrendt, Dr.sc.techn.Harald Diesing, Dipl.-Ing.Steffen Jakobi,
Institut für prozeßadaptive und erfaherungsgeleitete Automatisierung-IpeA GmbH, Aninstitut der TFH Berlin und der FHTW Berlin.
Rudower Chaussee 5, IGZ-Gebäude, 12489 Berlin.
Telephon: 030 - 63 92 62 90,   Telefax: 030 - 63 92 63 00

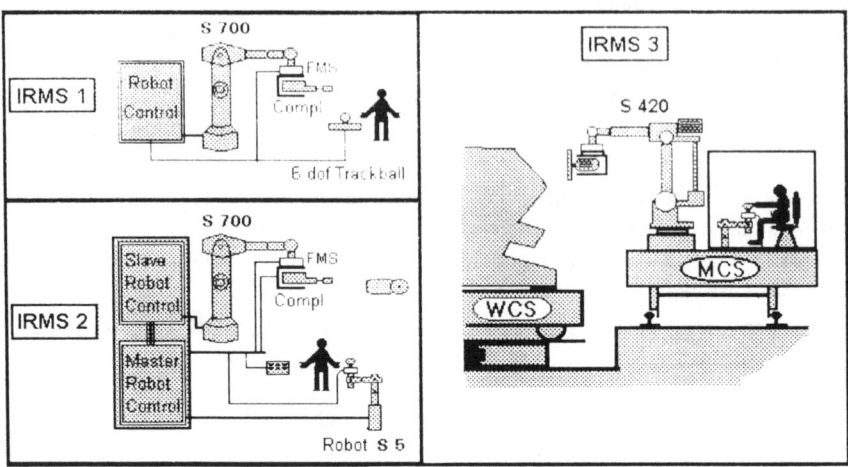

Fig.1: Configurations of the Interaktiv Roboter Manipulator System. (FMS : Force-moment sensor, Compl: Compliance system, MCS: Manipulator Carrier System, WCS: Carrier system for work pieces).

Fig. 2: The IRMS 2 at the GIFA 1994, Düsseldorf.

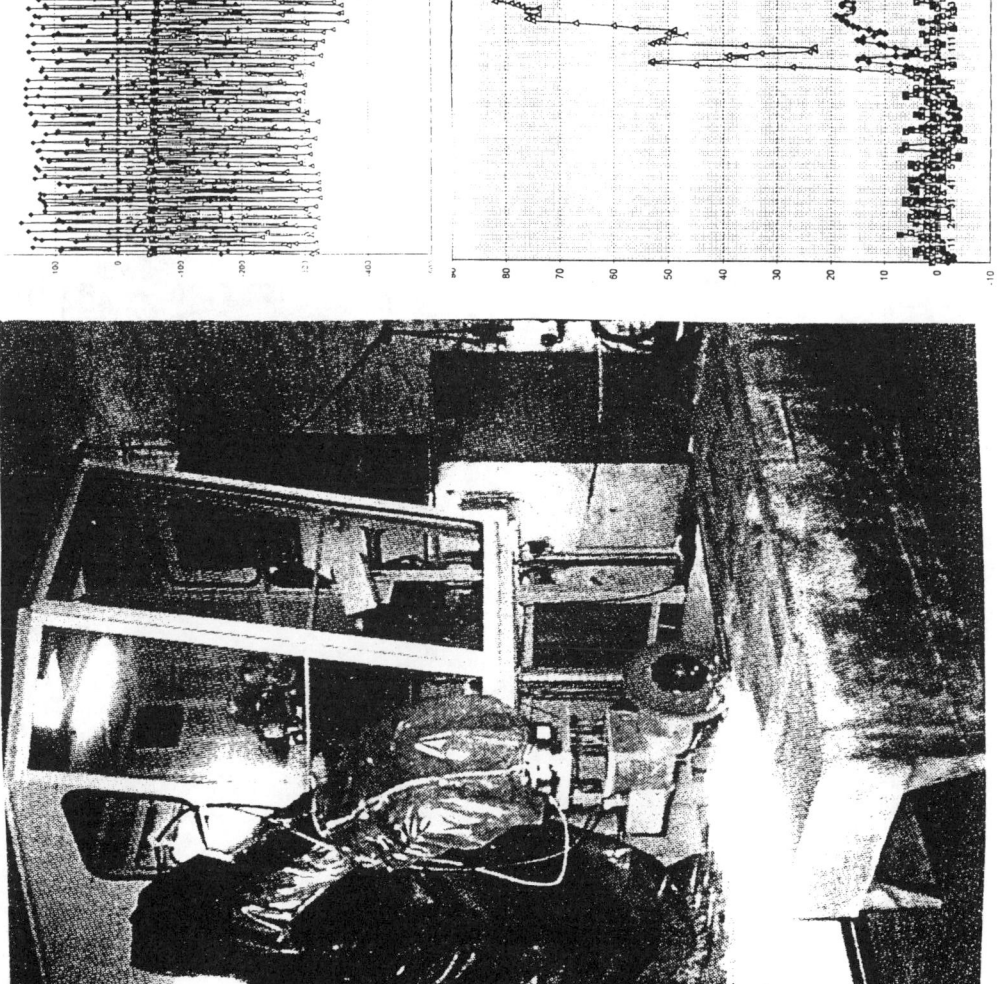

Fig. 4: Measuered values a) of the pure force moment information and b) after digital pre-processing

Fig. 3: The IRMS 3(Eisengießerei Britz, February, 1995)

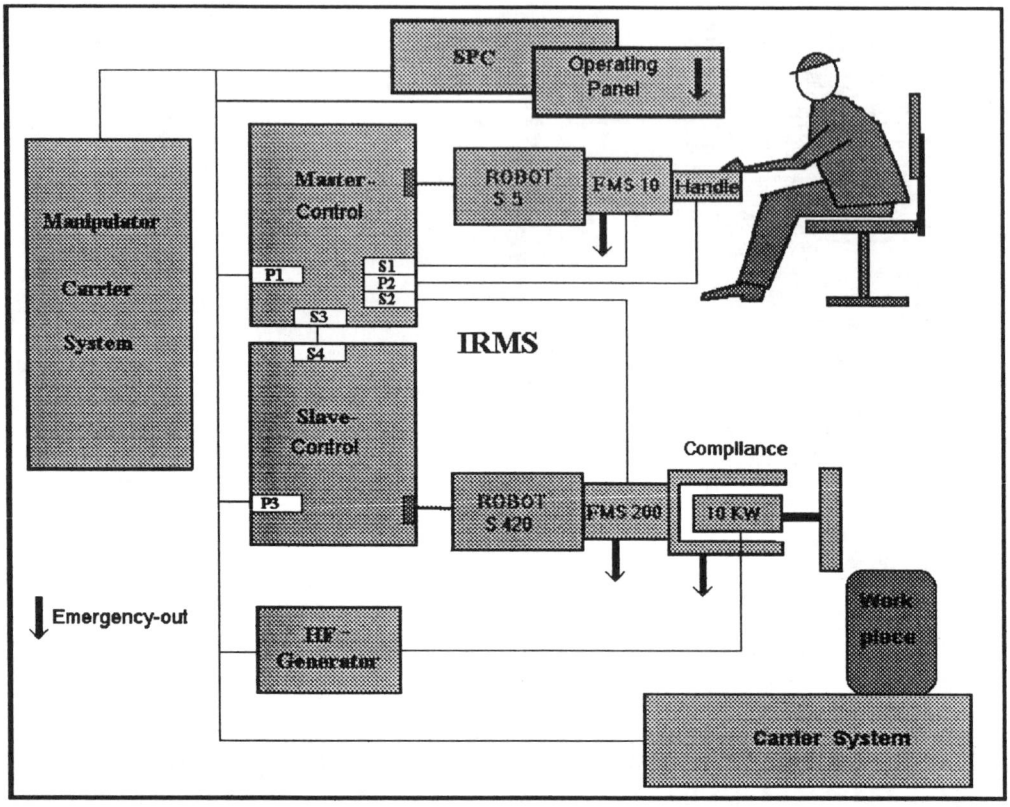

Fig. 5: Scheme of the control interface configuration of the overall system. (Si: RS 422 serial interface, Pi: parallel interface).

VI. Mobile Robots

E. Kreuzer, F.C. Pinto
Remotely Operated Vehicle - A Mechatronic System

P. Alexandre, A. Preumont
A Free Gait Algorithm for Improved Mobility Control of Walking Machines

M. A. Busetti de Paula, M. C. Zanella
Application of a Simulation with Hardware-in-the-Loop Environment in Automated Guided Vehicles

Fr.-W. Bach, J. Seevers, M. Hahn, M. Rachkov
Wall-Climbing Robots for Inspection and Maintenance

Remotely Operated Vehicle
A Mechatronic System

Edwin Kreuzer   and   Fernando C. Pinto

Technical University Hamburg–Harburg
Ocean Engineering Section II
Hamburg, Germany

Abstract: The major problems encountered in the development of control systems for Remotely Operated Underwater Vehicles (ROV) are analyzed. The modeling of the hydro-dynamic effects on the vehicle itself and on the umbilical cable is discussed. A robust control based on "sliding–modes" is used and the sensor systems which can be employed are presented. Experiments with an underwater double pendulum and with a small, laboratory size, vehicle are used to check the simulation results.

1   Introduction

In recent years Remotely Operated Vehicles (ROVs) have been built for various purposes. They found applications ranging from oil-prospection and inspection of underwater oil derricks to exploration of sea resources. Their development has been motivated by the high cost of human divers and the enormous risk to life with working under water. For the most part a ROV consists of a highly manoeuvrable platform on an umbilical cable from the mother ship, fitted with one or more robotic arms.

Such kind of underwater vehicles are complicated active mechanisms operating in hazardous environments [10]. In their design knowledge from different fields of engineering, especially from mechanical engineering, electrical engineering and computer science, play an important role. In the design of ROVs the mechanical engineer must cope with many problems. Among them is dynamic and structural stability and, due to the high pressure and the corrosive medium they need to withstand, also sealing problems are important. On the other hand, the power electronics for the vehicle's propulsion and the signal transmission between the ROV and the mother ship challenge the electrical engineer. The design of control systems in order to facilitate the work of the operator needs also a strong knowledge of the nature of the hydro-dynamic forces acting on the system [1]. Thus, ROVs are characterized by an extensive integration of concepts from a variety of individual disciplines. They are therefore specific examples of mechatronic systems.

2   Mathematical Model

The dynamic behavior of a ROV can be described in the common way [1][2] through a set of second order differential equations of motion. The interaction of the vehicle with the surrounding medium water must be accounted for as external applied forces. The motion of the vehicle gives rise to a pressure distribution over its surfaces. This distribution should be integrated to obtain the hydro-dynamic effect of this interaction. However, the mathematical description of these forces is complicated by the numerous effects which are present. Especially the vortex-shedding and the forces generated by waves require the use of powerful numerical methods.

For low relative velocities, as in the case of the modeling of ROVs, they can, however, be described with sufficient accuracy using the so called *Morison-equation* [13]:

$$F_h = C_d \frac{1}{2} \rho_w A(\mathbf{v}|\mathbf{v}|) + C_m \dot{\mathbf{v}} \tag{1}$$

where F_h is the hydro-dynamic force itself, ρ_w is the water density, A expresses a reference area and $\mathbf{v}$ and $\dot{\mathbf{v}}$ are the vectors of the relative velocity between fluid and vehicle and of the relative acceleration, respectively. The coefficients C_d and C_m depend on the direction of the relative velocity. They are also functions of the frequency of an incident wave. Since the ROVs commonly work at depths where the waves do not have a great influence at all this dependency can be neglected. In practice the coefficients cannot be analytically calculated for general bodies and must, therefore, be determined experimentally. In this case the mean effect of the vortex shedding is also present.

Since the acceleration of the water flow is normally very small, the second term on the right hand side of (1) can be added to the inertia terms of the mass matrix. The coefficient C_m can then be called *hydro-dynamic added mass*. The first term describes the quadratic damping of the water and represents the nonlinear effect of the hydro-dynamic forces.

2.1   Umbilical Cable

In order to include the hydro-dynamic effect of the umbilical cable which connects the vehicle to the mother ship a variety of mechanical models could be used. The description as a one-dimensional continua leads to a very complex mathematical model. A finite element model is not well suited for large displacements and dynamical motion of the cable segments. Another possibility is a multi-body system model for the cable, resulting in a chain of rigid cylindrical elements connected through revolute joints, springs and dampers to account for the stiffness of the cable and the dissipation of energy along it. This model allows a better analysis of the global motion of the umbilical and its influence on the behavior of the vehicle. Nevertheless, the elasticity of the cable segments is only approximated.

In applying (1) we must consider a velocity profile for **v** along the cable segment and then integrate the resulting forces [5]. For constant water current, a linear distribution of the velocity can then be assumed along the rigid segment. The integration along the cable element provides also the resulting hydro-dynamic moments relative to the center of mass of the segments. The coefficients C_m and C_d for slender cylindrical bodies can be found in the literature [13].

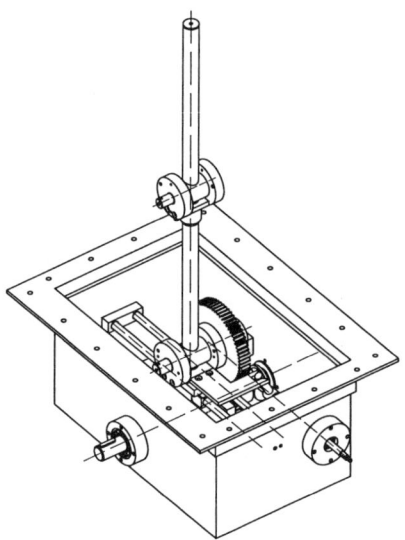

Figure 1: Experimental set-up

An experimental set-up consisting of a submerged inverted double-pendulum, Fig.1, was designed in order to verify the applicability of (1) for the simulation of the dynamic behavior of the umbilical [7]. The double pendulum represents the simplest model of a small portion of the cable. The static equilibrium of the up-right position is guaranteed by the buoyancy forces. The study of the dynamics of the double pendulum should experimentally verify the applicability of Morison's equation. The study of the nonlinear dynamic stability in case of parameters disturbances is also of interest. The lower segment is forced by means of a servo-motor controlled to perform a harmonic angular motion.

Figure 2 shows a comparison between the simulated and the measured behavior of the pendulum for the case of harmonic excitation of 0.4Hz. The figure shows the time history of the relative angle φ_2 and of the relative angular velocity $\dot{\varphi}_2$ between the two segments. The dynamic stability was also investigated through numerical and symbolical methods. For more details concerning the stability analysis refer to [8].

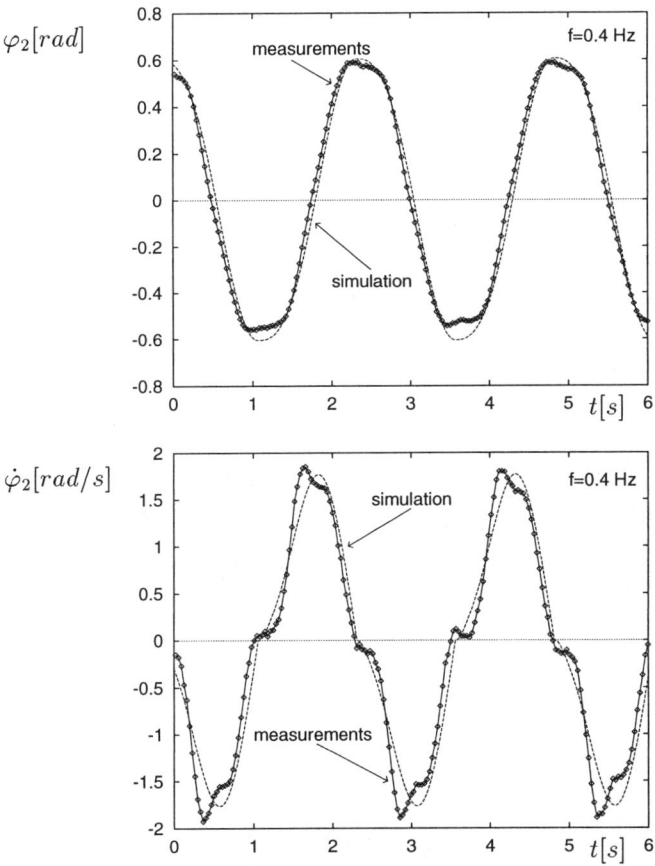

Figure 2: Simulated and measured behavior

The good agreement with the experimental results [7] indicates that the *Morison-equation* can be used successfully for the description of the global motion of the umbilical. Higher order effects such as vortex shedding and structural energy dissipation along the cable are indeed not considered with this approach [3]. Although the vortex shedding could be visually observed by moving the pendulum segment on the water surface, the velocity dependent forces which are associated were not capable of exciting the system.

2.2 Vehicle

The modeling of the hydro-dynamic forces which act on the ROV's hull can be accomplished in the same manner. In this case indeed it is not possible to calculate the necessary coefficients on (1). They must be measured experimentally. To obtain the values of the coefficients C_d a wind tunnel test was performed. The ROV model, which is described in section 5, was fixed to a platform which was capable of measuring the forces and moments acting on the vehicle in the three spatial directions. These values were related to the frontal area of the vehicle furnishing the desired values for the damping coefficients. Figure 3 shows the set-up in the wind tunnel. It was possible to rotate the vehicle relative to the flow in order to measure the dependence of the forces on the angle of incidence of the relative velocity.

Figure 3: ROV in the wind tunnel

The values of the coefficient for the hydro-dynamic damping force which acts longitudinally on the model ROV are shown in Fig. 4. These values are highly dependent on the actual geometry of the ROV. Different configurations of the working tools and packages mounted on the vehicle could lead to different coefficients. The control system must cope with these possible variations, which are to occur even during normal operations of the ROV [15]. A dependence of the coefficients on the value of the relative velocity was not observed in the range of interest.

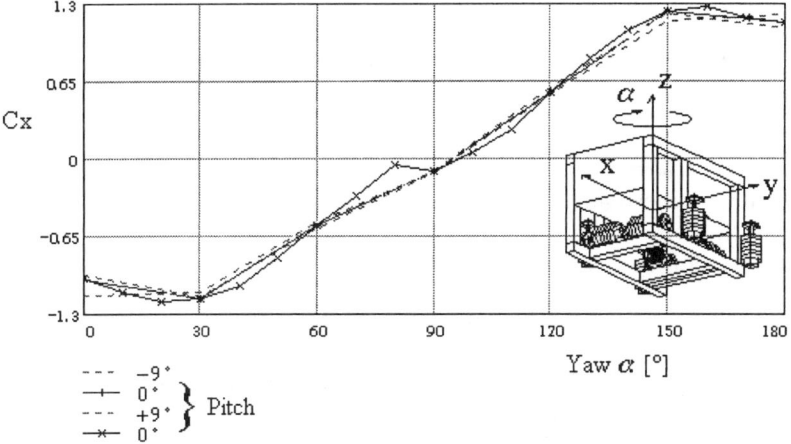

Figure 4: Coefficient C_x for the longitudinal direction

3 Robust Position Control

The pilot of the ROV is responsible for the positioning relative to a structure. This task is difficulted by the lack of good position information under water. However, the effectivity of the operation of a ROV can be improved if its positioning could be assisted by a control system.

The design of the control must take the variations on the system parameters **p** into account. The pilot alone is not able to recognize the changes with the available information. Among other effects the hydro-dynamic forces acting on the vehicle can vary as a consequence from different configurations of the system. The efficiency of the thrusters may change during their life. During the completion of tasks the inertial characteristics of the ROV can be modified by picking tools or parts of the structure which are to be transported. To avoid collisions with the underwater structures it is very important to guarantee that the ROV possesses a similar dynamic behavior in all these situations. This implies a *robust* control system. A way of achieving robustness [14][16] is through the use of sliding-mode [11] or variable structure controllers. The unmodeled dynamics of the umbilical cable can be treated as an external disturbance of the system. If the cable forces are bounded the required robustness may not be affected to a great extent.

For most ROVs the pitch and roll motions are stabilized through the inherent hydrostatic characteristics of the construction itself. The control system shall deal only with the depth z, the cartesian positions x and y and with the yaw-angle α. In general the uncontrolled angles for roll and pitch motions remain small and the depth can be decoupled from the other coordinates. The presented control system concentrates on the motion in the xy-plane for the depth control being achieved with the same approach.

The set of second order equations of motion can be transformed in the usual state-space representation:

$$\dot{\mathbf{X}} = \mathbf{F}(\mathbf{X}, \mathbf{p}) + \mathbf{U}(\mathbf{X}, \mathbf{c}), \tag{2}$$

For the case of a ROV, $\mathbf{F}(\mathbf{X}, \mathbf{p})$ is a non-linear function of the *state-vector* $\mathbf{X}$, representing all mechanical and hydro-dynamical effects on the ROV. The vector $\mathbf{U}(\mathbf{X}, \mathbf{c})$ of the control forces has also a strong nonlinear character due to the control law based on sliding-modes.

The approach of a multi variable sliding-mode controller is to confine the dynamics of the controlled system in a high dimensional manifold. The control parameters $\mathbf{c}$ describe this manifold in the state-space. Their choice must also guarantee the stability of the motion, i.e. once the system achieves the manifold it should converge to the desired position, sliding along it. The simplest definition for the manifold $S = 0$ is a linear one:

$$\begin{aligned}
S_x &= c_x E_x + V_x, \\
S_y &= c_y E_y + V_y, \\
S_a &= c_a E_a + V_a.
\end{aligned} \tag{3}$$

It should be noted that S_x, S_y and S_a must be defined using the coordinates x and y projected in a local coordinate system, fixed to the ROV. E_x, E_y and E_a are the position errors in this local system according to:

$$\begin{aligned}
E_x &= +(x_s - x)\cos(\alpha) + (y_s - y)\sin(\alpha), \\
E_y &= -(x_s - x)\sin(\alpha) + (y_s - y)\cos(\alpha), \\
E_a &= \alpha_s - \alpha.
\end{aligned} \tag{4}$$

The index s indicates the values of the set-points. Assuming a position control instead of a trajectory control the desired velocities upon reaching the final position should all be equal to zero, so that V_x and V_y are the linear velocities, expressed also in the local system, and V_a is the angular velocity of the yaw motion.

The control forces in the vector $\mathbf{U}(\mathbf{x}, \mathbf{c})$ which are needed to obtain the sliding mode on the manifold are of nonlinear nature. A discontinuous control law as in the case of the so-called *bang-bang* controllers is a common choice. The major disadvantage is due to the high frequencies which are necessary on the actuators. This problem can be minimized if the discontinuity is smoothed by the use of a function of the form:

$$F_i = \frac{2 \arctan(p\, S_i)}{\pi}, \tag{5}$$

where the index i represents x, y and a from (3). The discontinuous character of the force-law in the sliding-mode control can then be controlled through the parameter p. With the smoothing effect due to the use of the arctan function, the high frequencies on the control forces are also smoothed, improving the performance of the actuators. The generalized forces determined by (3) and (5) are then transformed through a linear combination in the actual thrusts of the propellers.

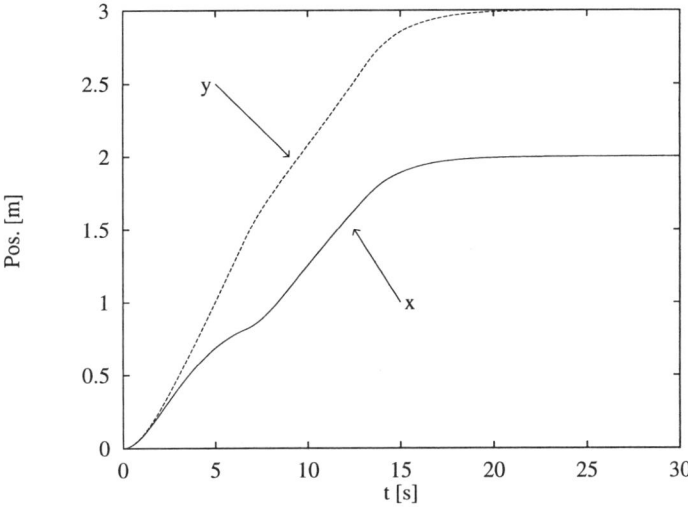

Figure 5: Simulation of the ROV's motion

In order to design the controller the control forces can be substituted into (2) which can be linearized in the vicinity of the set-points for the position. Requiring that the real part of the eigenvalues of the linearized system be negative the domain where a stable motion can be achieved may be determined as a function of the parameters c_x, c_y and c_a and of the characteristics of the thrusters. This leads to the conclusion that all coefficients must be negative. Too small absolute values lead, however, to a very slow system dynamics once in the manifold. On the other side high values for c_x, c_y and c_a increase the necessary power of the actuators in order to achieve the sliding condition. Coefficients on the range -0.3 to -3 are a good compromise.

Figure 5 shows the time history for the motion of the vehicle from the origin to the point with coordinates $x_s = 2$ and $y_s = 3$ with orientation $\alpha_s = \arctan(y_s/x_s)$. The parameters of the controller are $c_i = -0.5$. The ROV has a mass of 40kg and the thrusters, see section 5, are capable of exerting 5N each in the x-direction and 7N in the y-direction. The actual thrust in x is, however, limited by the necessity of simultaneously applying a moment about the z-axis. The robustness of the control system can be seen in Fig. 6 where the actual trajectory on the $x - y$-plane is shown for different values of the mass of the vehicle. The behavior of the ROV is almost the same in the three cases although the mass varies from 20 to 80kg. The dynamics on the manifold is shown in Fig. 7 for the coordinate x. In Fig. 8 the rotation of the vehicle is drawn to show the effect of the control on the angular position.

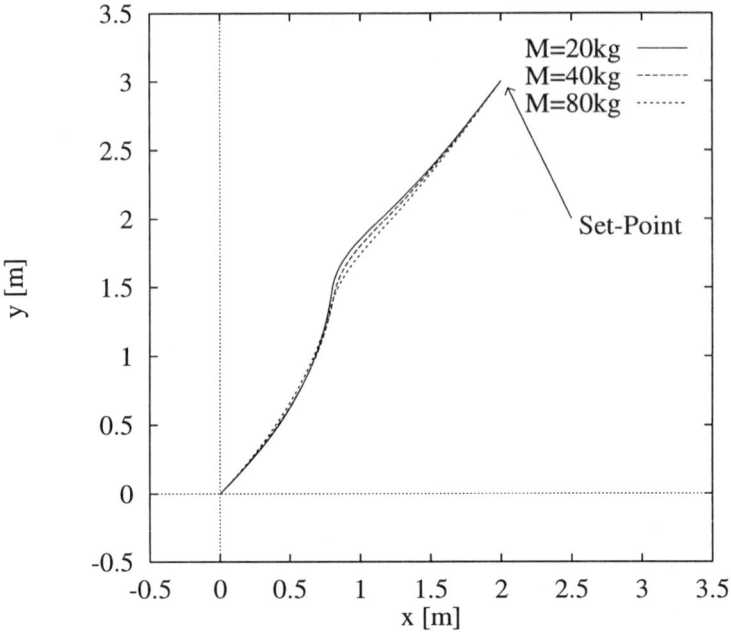

Figure 6: Trajectories in the xy-plane

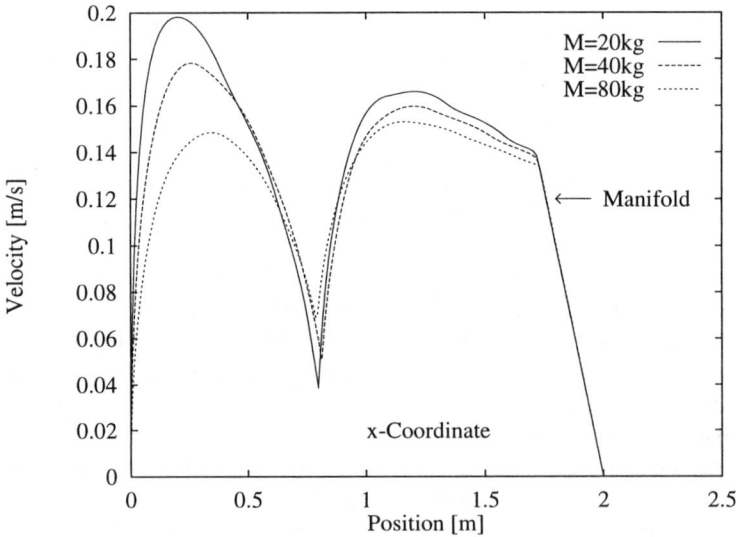

Figure 7: Dynamics of the system in the manifold

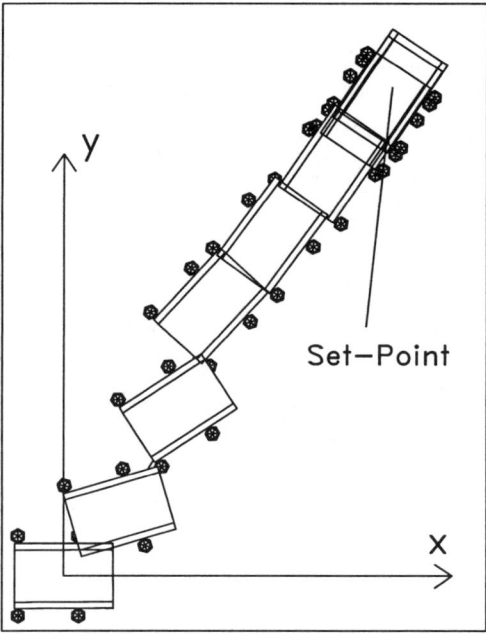

Figure 8: Yaw motion

4   Sensor Systems

One major problem in the automatic control of ROVs is the measurement of the actual position and orientation in a known reference frame [4]. Even for the case of positioning close to a structure, it can be very difficult to determine the relative position of the vehicle. There are many systems [6] ranging from inertial platforms, for the short-time navigation until the structure is reached, to sonar systems and image processing.

Close to the desired position under water, systems which create a mechanical link with the structure can be used with advantages. Especially the use of a mechanical manipulator as a position sensor is an appropriate choice. Once fixed through the end effector to some known point on the workpiece, the relative position and orientation can be determined through the simple direct kinematic of the manipulator. The actuators of the arm may also be used to help positioning.

Figure 9 shows the precision achieved by this system through a slice in its workspace corresponding to the motion of the second and third joint of the manipulator, θ_1 and θ_2. Considering the precision by the measurement of the joint angles as being ± 1 degree and the mathematical equations [9] for determining the position of the end effector the characteristic of the sensor can be determined. Figure 10 shows a

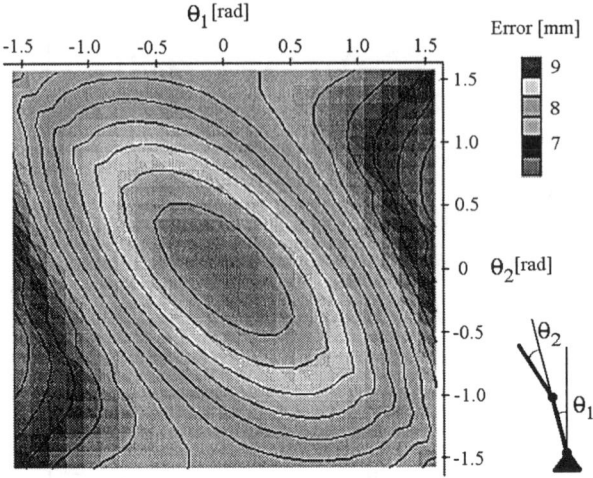

Figure 9: Precision of the measurements with the arm

sketch of the model ROV with two arms to be used in this way.

5 Experimental Model

A small experimental ROV has been developed at the Ocean Engineering Section II of the Technical University Hamburg–Harburg, see Fig.10. This vehicle was designed to operate in the wave channel of the department serving as a test-bed for different kinds of propulsion, sensor and control systems.

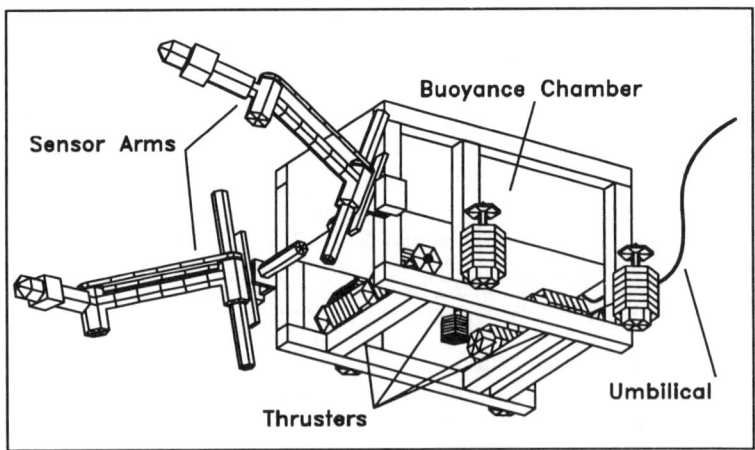

Figure 10: CAD–model of the ROV

The vehicle possesses two mechanical arms which are intended to be used as position sensors in the proximity of a structure. Each arm possesses six degrees of freedom. The joint positions can be measured with special potentiometers. The propulsion is achieved by eight electrical motors with propellers. Two of them are arranged longitudinally and two of them transversally on the ROV, see Fig.10. The other four actuators are directed vertically and are responsible for the depth control. A buoyancy chamber guarantees the static stability of the upward position. Sensors for the angular velocity and for the accelerations of the ROV are placed inside it.

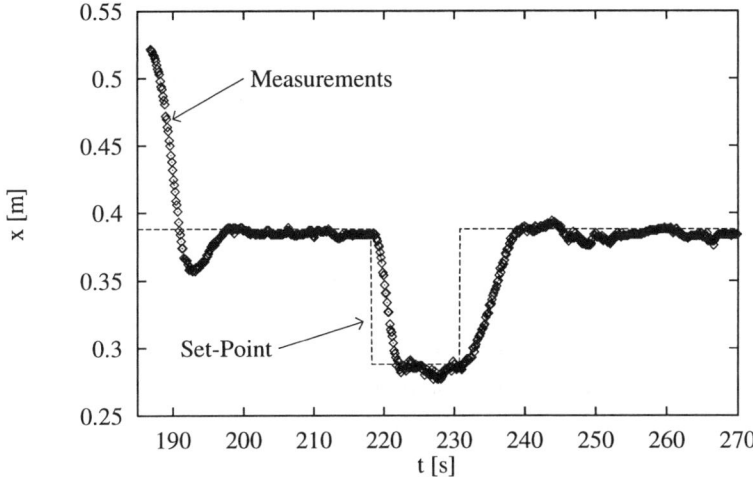

Figure 11: Measured x and y coordinates of the ROV

An umbilical connects the vehicle in the channel with the controlling computer and the power supply for the motors. The measurements are done through an A/D-converter. The power supply is driven by means of a D/A-converter, which has its output transformed and amplified in a PWM-style and sent to the motors. The robust sliding-mode controller is implemented by software. The position and velocity information is given by one of the manipulators which is in turn fixed on a structure in the basin.

Figure 11 shows the time history of the x and y coordinates of the ROV. The ROV is driven by the control system to the desired position, where it stays at rest. The set-point is changed for approximately 15 seconds and is then finally changed back to its original value. The vehicle follows the determined positions.

The effects of the umbilical cable slow down the whole motion. This is due to the relatively high stiffness of the cable. The necessity of transmitting enough power for the electrical motors does not allow the use of a cable with a very small diameter. The precision of positioning is, however, of about 0.05m for each direction. The over-shooting which is not present by the simulations is caused mainly by the unmodeled cable and actuator dynamics.

6 Conclusions

In underwater robotics concepts from mechanical engineering, electrical engineering, fluid dynamics, computer science, control engineering and machine design play an important role. A mathematical description of the hydro-dynamic effects of the water on the system ROV plus umbilical is achieved using the *Morison-equation*. Experiments with an under water double pendulum have confirmed the applicability of the proposed modeling for the analysis of the global motion of the umbilical. The respective coefficients for a model ROV were measured in a wind tunnel. Robust control of the position relative to a structure is achieved with the use of a sliding mode controller. The design takes into account the local and global stability of the system. A manipulator is used as a position sensor in the region close to a structure under water. A model ROV was constructed and used as a test-bed for the sensor and control systems described.

References

[1] Bevilacqua, L.; Kleczka, W.; Kreuzer, E.:
On the Mathematical Modelling of ROV's. In: Proc. of the Symposium on Robot Control, Vienna, pp. 595-598, 1991.

[2] Lewis, D.; Lipscomb, J.; Thompson, P.:
The Simulation of Remotely Operated Underwater Vehicles. ROV '84, Marine Technology Society, 1984.

[3] Every, M.; Davbies, M.:
Predictions on the Drag and Performance of Umbilical Cables. ROV '84, Marine Technology Society, 1984.

[4] Hsu, L.; Costa, R.; Cunha, J.:
Medição de Posição Pela Estratégia "Taut- Wire" Para o Controle de Posição de Um VOR. Relatório interno COPPE/UFRJ, Rio de Janeiro, 1990.

[5] Kleczka, W.; Kreuzer, E.; Pinto, F.:
Analytic-Numeric Study of a Submerged Double Pendulum. ASME International Symposium on Flow-Induced Vibrations & Noise, Anaheim, 1992.

[6] Pinto, F.; Kreuzer, E.:
Uma Comparação entre Sistemas de Sensores Para Um Veículo Submarino. Fórum sobre VOR na Petrobrás, Rio de Janeiro, 1993.

[7] Kleczka, W.; Kreuzer, E.; Pinto, F.:
Experimental and Analytical Investigations of a Submerged Double Pendulum. 1st European Nonlinear Oscillations Conference, Hamburg, 1993.

[8] Kleczka, W.:
Symbolmanipulationsmethoden zur Analyse nichtlinearer dynamischen Systeme am Beispiel Fluid-gekoppelter Strukturen. Fortschritt-Berichte der VDI-Zeitschriften, Reihe 11, Nr. 213, Düsseldorf: VDI-Verlag, 1994.

[9] Kreuzer, E.; Pinto, F.:
Sensing the Position of a Remotely Operated Underwater Vehicle. 10th CISM-IFToMM Symposium on Theory and Practice of Robots and Manipulators, Gdansk, 1994.

[10] Polomsky, S.:
Rechnergestütztes Entwerfen von ferngesteuerten Unterwasserfahrzeugen. VDI Fortschritt-Berichte, Reihe 20, Nr. 48,Düsseldorf: VDI Verlag, 1991.

[11] Utkin, V:
Sliding Modes in Control Optimization. Berlin: Springer-Verlag, 1992.

[12] Fiedler, O.:
Strömungs- und Durchflussmesstechnik. München: R. Oldenburg Verlag, 1992.

[13] Newman, J.N.:
Marine Hydrodynamics. 4th Ed., Massachusetts: MIT-Press, 1982.

[14] Suzuki, H.; Yoshida, K.:
Trajectory Tracking Control of a ROV for Lifting Objects. In: Proc. 1st Int. Offshore and Polar Eng. Conf., Edinburgh, 1991.

[15] Sayer, P.; Miller, C:
The Hydrodynamics of ROVs Carrying Work Packages. In: Proc. 1st Int. Offshore and Polar Eng. Conf., Edinburgh, 1991.

[16] Yamamoto, I.; Nagamatu, T.:
A Control System Design of a Tethered Underwater Vehicle. In: Proc. 4th Int. Offshore and Polar Eng. Conf., Vol. 2, Osaka, 1994.

Prof. Dr.-Ing. Edwin J. Kreuzer
M.Sc. Fernando A.N.C. Pinto

Technical University Hamburg–Harburg
Ocean Engineering Section II
Eissendorfer Str. 42, D–21073 Hamburg, Germany
Tel.: (+49)(40) 7718-3220
Fax: (+49)(40) 7718-2028
email: kreuzer@tu-harburg.d400.de

A Free Gait Algorithm for Improved Mobility Control of Walking Machines

Paul Alexandre & André Preumont
Université Libre de Bruxelles
Department of Mechanical Engineering and Robotics
CP 165, 50 av. F.D.Roosevelt, 1050 Brussels, Belgium

Abstract: This paper reviews the main strategies for the leg coordination of walking machines with emphasis on the *free gait*. A *free gait* algorithm is presented which allows a smooth and stable motion for an arbitrary velocity vector of the vehicle.

1 Introduction

Because of their superior mobility, walking machines have attracted attention for about two decades [1]. A major problem in designing such machines is the high complexity of the joint coordination, which requires a sophisticated control architecture to achieve efficient autonomous movements with minimal intervention from a human operator.

The control architecture of walking machines is often hierarchical, with the following levels: *(i) Level A*: navigation & planning; *(ii) Level B*: gait control (leg coordination) and *(iii) Level C*: leg trajectory. This control approach is also known as *supervisory control*. Level *A* is not significantly different from what it is for a wheeled robot; it will not be discussed here; it is assumed that the velocity vector $(v_x, v_y, \omega_z)^T$ is prescribed by a human operator with a three-dimensional joystick. Level *B* coordinates the motion of the legs; three different strategies can be used:

- central supervision with fixed sequence (e.g. regular gaits);

- leg coordination based on control influences exchanged between the legs, duplicating the behaviour observed on insects (neurobiological coordination [2]);

- rule-based control within a finite set of gait states (free gait [3,4]). Besides its flexibility with respect to changes of direction, the free gait can accommodate a damaged leg.

Level *C* (trajectory and servo control) can typically be distributed at the leg level (decentralized control architecture), but some coupling with level *B* is necessary to account for the tactile information (obstacle avoidance) and for attitude control.

This paper focuses on level *B* and is organized as follows: the regular gaits are shortly reviewed in section 2, the principles of the neurobiological coordination are presented in section 3 and the free gait algorithm is developed in section 4.

2 Regular gaits

The time-space coordination of the motion of the various legs involves a decision regarding what legs should be lifted or placed, in order to achieve the desired velocity while maintaining static stability.

For a given support pattern, each supporting leg moves backward from the anterior extreme position (*AEP*) to the posterior extreme position (*PEP*). Conversely, the projection of the centre of gravity moves forward continuously inside the stability polygon (Fig.1). The minimum of the distance S in the direction of the motion is called *front stability margin*, S^*. It depends on the gait, but the minimum value is always obtained right before a leg is placed.

A regular gait is such that all the legs perform the same cyclic motion, except for a phase difference. A symmetric gait is such that the legs on one side are exactly half a cycle out of phase with respect to the legs symmetrically located on the other side. A regular symmetric gait is entirely defined by the phase lag between successive legs on one side. With the numbering of Fig.1, a *forward wave gait* is such that the phase differences are equal to the duty factor β [5]:

$$\phi_3 - \phi_1 = \phi_5 - \phi_3 = \beta \tag{1}$$

It is the gait with the largest stability margin for any given β. The particular case $\beta=1/2$ is the *alternated tripod*. Observations on insects show that they walk according to an alternate tripod on a regular terrain. Explicit results for the front stability margin of any regular symmetric gait are available in [6].

Regular symmetric gaits are very easy to implement in the supervisory control framework. However, they are too restrictive for non-uniform velocity or irregular terrain, where more freedom is necessary to exploit the full potential of the walking machine. There are two possibilities for achieving a more flexible gait: the first one is based on the duplication of the neurobiological coordination of the stick insect and the second, named free gait, consists of a set of rules aiming at selecting the most stable gait state compatible with the geometric constraints.

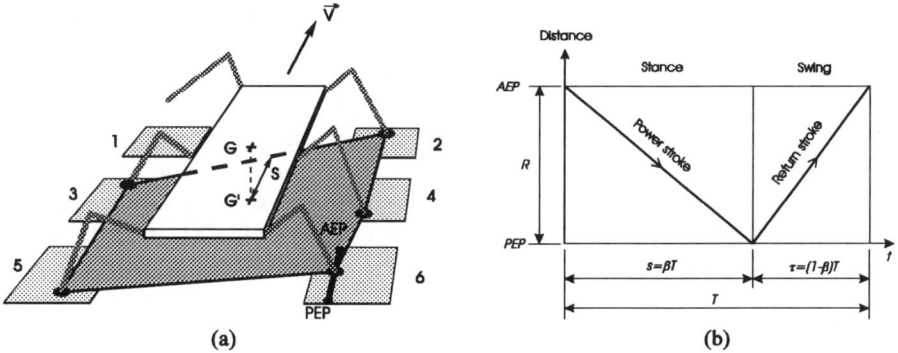

Fig. 1. (a) Leg numbering and definition of the stability polygon.
(b) Movement of one leg during the cycle time (β is the duty factor).

3 Neurobiological coordination

The neurobiological coordination mechanism of the stick insect has attracted a lot of attention from biologists (e.g. [2]) and a model of the coordinating influences between the various legs has been devised [7]. The stick insect is a 7cm long six-legged slowly climbing arthropod. As opposed to the centrally driven hierarchical control which can produce fixed gait patterns, its gait control mechanism is essentially based on a democratic decision mechanism involving coordinating influences from the sense organs of the neighbouring legs. These coordinating influences affect the timing between the pattern generators of individual legs (single leg controllers).

The principle of a single leg controller is illustrated in Fig.2. The transition between stance (support) and swing (return) is achieved by a decision relay producing the target position: a positive output corresponds to a swing (target position AEP) and a negative output to support (target position PEP). The output of the decision relay is then taken as the reference input to a velocity feedback acting as a lower level local controller. The stance is terminated when the sense organs signal that the PEP has been reached; then, the swing phase is initiated and the leg moves towards the AEP.

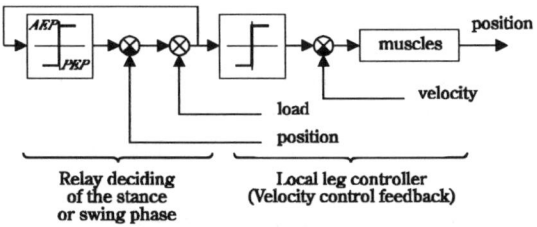

Fig. 2. Principle of the single leg controller.

Each leg is influenced by its neighbours which change the thresholds AEP and PEP according to the coordinating mechanisms of Fig.3. The mechanisms numbered on the figure are the following:

#1. The swing phase of leg L_i inhibits the swing phase of the leg in front of it, L_{i-1} (for obvious stability reasons). The inhibition influence is achieved by shifting backward the PEP threshold in the decision relay of the front leg.

#2. The start of the retraction (support) of L_i excites the start of the swing of L_{i-1} and R_i.

#3. Backward position of leg L_i excites the start of swing of the rear leg L_{i+1}; this is achieved by moving forward the PEP of leg L_{i+1} as leg L_i moves to the rear.

#4. During the swing phase of the middle and backward legs, the AEP is adjusted in such a way that the leg never interferes with the leg in front of it: $(AEP)_{Li} < (PEP)_{Li-1}$. This mechanism is known as "targeting".

#5. Increased resistance produces increased force.

#6. "treading-on-tarsus" reflex: If a leg steps on the foot of the leg in front of it, it is lifted again and placed slightly at the rear. This complement to the targeting is obviously aimed at preventing stumbling.

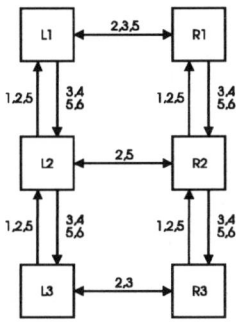

Fig. 3. Summary of the coordinating mechanisms according to Cruse.
The arrows indicate the direction of the coordinating influence.

Computer simulations and hardware implementation [8] have shown that the mechanisms described above can produce step patterns similar to those observed on insects and stable with respect to a wide variety of disturbances, including arbitrary starting configuration.

4 Free Gait

Regular gaits are well adapted when the vehicle moves on an even terrain with a constant velocity. Their fixed stepping sequence, however, is not suited to such situations as turning, walking sideways, or moving on an irregular terrain, for which a walking machine should have a definite advantage on other types of mobile robots. On the other hand, we have seen that, for insects, terrain adaption is achieved through a democratic decision process based on the influence of the *AEP* and *PEP* thresholds of the transition relay by the sense organs of the neighbouring legs.

An alternative to the neurobiological coordination can be obtained with a rule-based central decision process based on the sensory information of the vehicle and any information available regarding possible restrictions of the workspace by obstacles.

The algorithm presented below is inspired by the work of A.Halme et al. [3,4]. In their free gait, the notion of *Gait State* (*GS*) is introduced to simplify the description of the walking process : all the supporting leg configurations of the walking robot are represented as a finite set of Gait States for which the state of each leg is binary (supporting or not). This rough representation leads to a simple classification of the various walking configurations (6 legs) into 3 categories, in relation with their static stability (Fig. 4):

Unstable *GS*: for which 3 or more contiguous legs are in swing at the same time. Those configurations are always unstable whatever the position of the supporting legs.

Marginally stable *GS*: for which there is one pair of contiguous legs lifted. Those configurations can be stable or unstable depending of the position of the supporting legs within their respective workspace.

Absolutely stable *GS*: for which 2 contiguous legs cannot be simultaneously in swing.

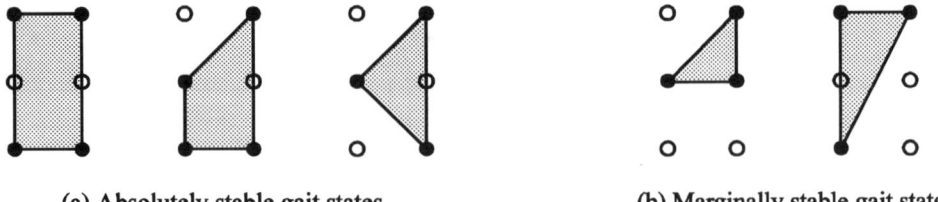

(a) Absolutely stable gait states   (b) Marginally stable gait states

Fig. 4. Examples of gait states.

For most walking machine realisations, those *GS* are stable, independently of the position of the supporting legs within their workspace.

The proposed free gait algorithm produces a sequence of *GS* chosen amongst the absolutely stable ones.

Although the gait control can be seen as a finite state automata (into the finite set of allowed gait states), the gait state information is not sufficient to control smoothly and efficiently the walking process. To achieve that, more information about the leg positions in their respective cycle time is needed. This information can be represented by the six component time-vector called *Leg Phase State* (*LPS*), defined as follows (Halme et al., 1993)(Fig. 5):

A negative component of *LPS* indicates that the leg is lifted: $LPS(i)<0$ represents the *minimum recovery time* at the maximum swing velocity, that is the time necessary for the transfer leg to reach the desired foothold (*AEP*). A zero value indicates that the leg is available for support.

If positive, the component indicates that the leg is in support: $LPS(i)>0$ represents the *maximum predicted duty time* for the current velocity vector of the body, that is the maximum time available for support before hitting the current estimate of the limit of the workspace (*PEP*).

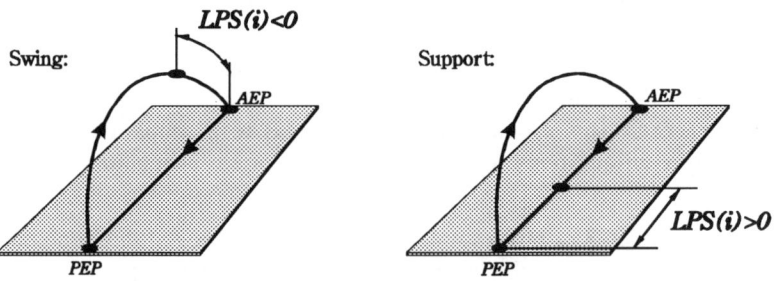

Fig. 5. Leg Phase State definition

In this way, the walking process can be represented by the trajectory of a point in an hyperspace of dimension 6 or, more conveniently, in 6 *LPS*-diagrams of dimension 2 in which the *LPS* of each leg is represented versus the *LPS* of the contiguous leg (Fig. 6). The *LPS* trajectories make an angle of 45 degrees with the axis for uniform velocity since the estimated times to reach the *AEP* in transfer phase or the *PEP* in support phase are decreasing with the same rate for both legs (negatively for legs in transfer phase). For each leg, the maximum value for the *LPS* is equal to the duration of an entire support phase s_i for the current velocity, while the minimum value is the duration of the transfer phase, $-\tau_i$. The discontinuities of the trajectories correspond to the lifting and the lowering of the legs: for example, BB' (Fig. 6) is the lifting of leg i (LPS_i varies from $+0$ to $-\tau_i$) and CC' is the lowering of leg i (C$\rightarrow$C' : LPS_i varies from -0 to $+s_i$).

The set of absolutely stable *GS* is characterised by the property that two contiguous legs cannot be simultaneously in swing (i.e. both *LPS* negative), so that a sequence of absolutely stable *GS* provides trajectories in the 6 *LPS*-diagrams that never enter the part corresponding to both legs in swing (shaded in dark in Fig. 6).

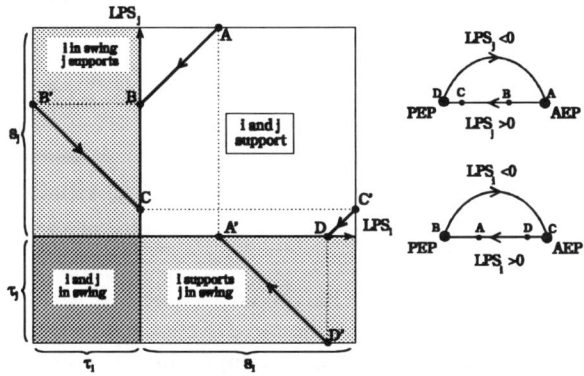

Fig. 6. *LPS* Diagram of a pair of contiguous legs.

The trajectory for a periodic gait is represented.

As an example, consider a *forward wave gait*, characterized by the phase differences given earlier [Equ.(1)]. The corresponding trajectories in the *LPS*-diagram are shown in Fig. 7, for a duty factor $\beta=2/3$. Figure 7.a refers to a couple of successive legs on the same side : the trajectory during one cycle is ABB'CC'A'A; the various segments have the following meaning :

AB   : i and j in support
BB'   : lifting of i
B'C   : i in swing, j in support
CC'   : grounding of i and lifting of j
C'A'   : i in support, j in swing
A'A   : grounding of j

Note that the rear leg *(i)* has a phase advance of *1-β* with respect to the front leg *(j)*. This is the minimum value; any reduction of this phase difference would move point C to the left and the trajectory would enter the shaded area corresponding to i and j being simultaneously in swing.

Figure 7.b illustrates the trajectory for opposing legs (either at the front or at the rear).

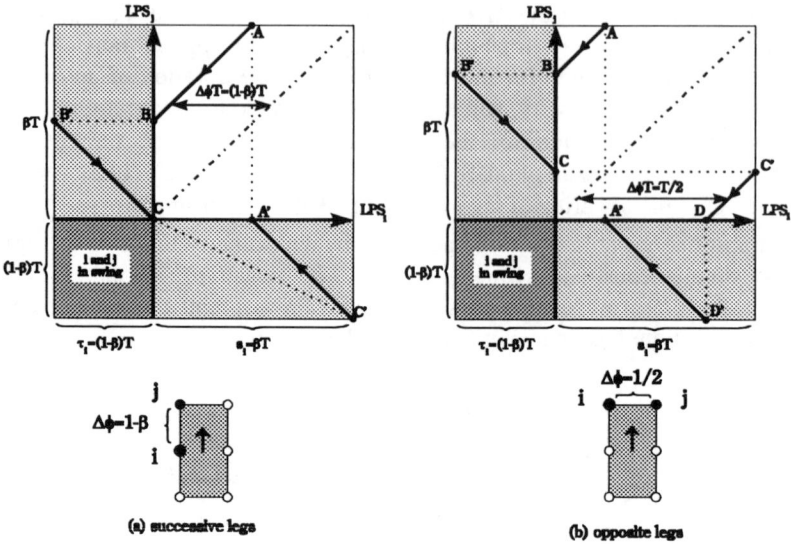

(a) successive legs

(b) opposite legs

Fig. 7. *LPS* diagrams for a forward wave gait (β=2/3).

These legs are half a cycle out of phase and this results in the *LPS* diagram being symmetrical.

If the reference velocity is no longer longitudinal (v_y or $\omega_z \neq 0$), the gait ceases to be regular and the various legs do not have the same cycle time; as a result, the concepts of duty factor and phase difference are no longer applicable. On the contrary, the concept of *LPS* trajectories fully applies and allows to exploit the mobility of the vehicle. The diagrams become rectangular, because $\tau_i + s_i \neq \tau_j + s_j$ (Fig. 8).

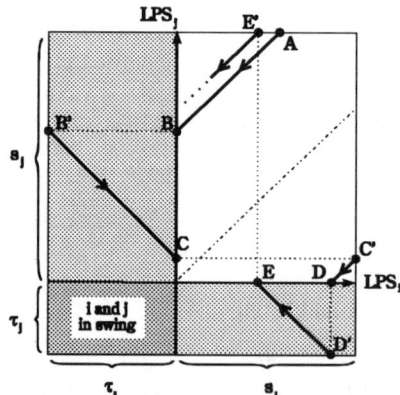

Fig. 8. *LPS* diagram for an non-periodic gait

The idea of the proposed free gait algorithm is to apply geometric rules to the *LPS*-diagrams to control and synchronise the legs movements. The *LPS*-vector is computed every planning cycle time ΔT for the current speed reference (v_x, v_y, ω_z). Next, the algorithm proceeds as follows :

1) Compute of the speed reduction factor of the vehicle: with the *LPS*-diagram, it can be foreseen if some legs needed for support in the next *GS* will not be available in time (the recovery times are too long). The solution is to reduce the speed of the vehicle by some factor $r \leq 1$.

2) Mark the legs to be lifted, either because reaching the end of their workspace, or because two contiguous legs are reaching the end of their workspace within a short interval.

3) Mark the legs to be grounded. There are two cases: if some legs are marked to be lifted, the two contiguous legs have to be grounded if not already supporting; the legs at the end of their swinging phase are also candidate for grounding but in some situations described below it is more suitable to delay the grounding.

The 3 steps described above can be executed with the help of geometric rules related to some critical areas in the *LPS*-diagrams. The areas are highlighted in Fig. 9 and described below :

Area I is the part of the diagram corresponding to both legs in swing. It should never be reached, otherwise the *GS* is no longer absolutely stable.

Area IIa corresponds to leg *i* reaching its *PEP* within ΔT (*planning cycle time*). Therefore, it is time to raise leg *i*.

Area IIIa corresponds to the situation for which leg *i* supports the body ($LPS_i > 0$) while leg *j* swings ($LPS_j < 0$) with $LPS_i < - LPS_j$. That is to say that for the current speed, leg *i* would reach the end of its workspace before leg *j* is ready for support. Therefore, it is necessary either to reduce the speed of the robot with as effect an increase of the *LPS* of the supporting legs or else to choose a closer foothold (i.e. to lower the swinging leg before LPS_j is near zero). The dashed line in area IIIa corresponds to the new limit of the area if it is allowed to ground a leg as soon as 75% of its swinging phase is performed.

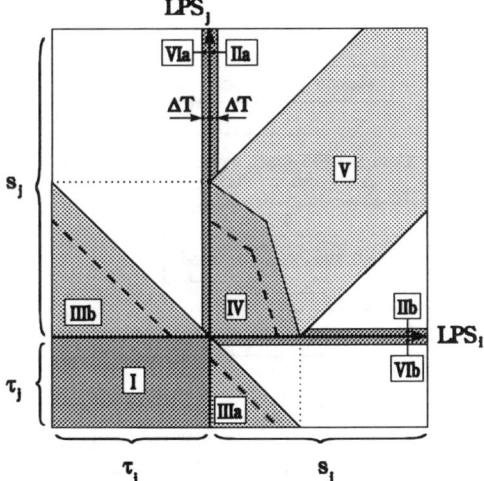

Fig. 9. Critical areas of the *LPS*-diagram

Area IV corresponds to the following situation : both legs will reach the limit of their workspace within a short period, leading the trajectory to an area of type III. To avoid this, it is necessary to raise immediately one leg (taking into account the other *LPS* diagrams) or to reduce the speed if this is not possible. The border of this area corresponds to the border of area III if one leg is lifted (through the discontinuity in the *LPS*-trajectory). The dashed line is the other possible limit if shortening of the swinging phase is considered as above.

Area V informs that the trajectory will cross area IV. It is therefore advisable to avoid grounding a leg with *LPS*-vector in this area.

Area VIa informs that a leg has reached the end of its swinging phase ($LPS > -\Delta T$) and is thus available for support. Two actions are possible in this case: the leg can be lowered immediately or kept in store for future needs of support. The first solution increases the current stability margin of the vehicle but the new *LPS* vector may be located in area V which is not desirable. To avoid this, it is more suitable to delay the grounding. This last possibility is also useful to shift the trajectory in the *LPS* diagram, making possible the transition of the gait towards a forward wave gait as shown later. The new *GS* is of course always statically stable if the preceding one was.

Area IIb, IIIb and VIb are similar to respectively IIa, IIIa and VIa.

Note that for periodic gaits areas I, III, IV and V correspond to the violation of the stability condition: the phase difference between contiguous legs must be at least equal to $1-\beta$.

From these critical areas, the decision steps of the algorithm can now be described (Fig. 10). The algorithm is executed every planning cycle time ΔT.

1. Compute the speed reduction: the factor is obtained by considering the points located

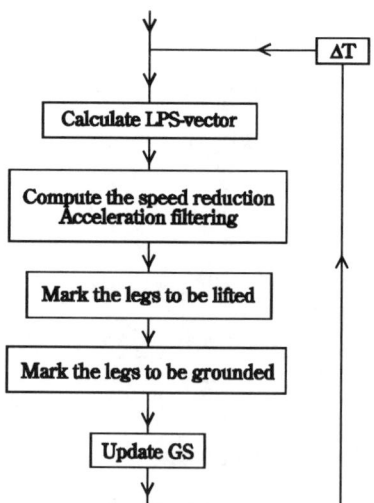

Fig. 10. Flow-chart of the free gait algorithm

in area of type III in the 6 *LPS*-diagram. The speed reduction has the effect of increasing the *LPS* of the supporting leg. Since the *LPS* of the swinging legs remains unchanged, this produces an horizontal translation of the image point in area IIIa (vertical for area IIIb). The speed reduction factor is the largest value ≤ 1 that brings the points located in areas III on the border of their respective areas (Fig. 11).

2. Mark the legs to be lifted: the supporting legs reaching the end of their workspace (areas II in the *LPS*-diagrams) are marked to be lifted. If an area IV is reached in one of the *LPS*-diagrams, it is necessary to lift one of the concerned legs. First, it is checked if the leg with *LPS* minimum can be lifted. If not the other leg is checked. In the particular case where neither of the two legs can be lifted and the point is inside the dashed line limit, a new speed reduction factor is calculated in such a way that the point is brought at least at the limit of the area.

3. Mark the legs to be grounded: mark the legs contiguous to the legs considered at point 2 to be grounded if not already supporting. Moreover, mark the legs at the end of their swinging phase (area VI) to be grounded if the resulting point in the *LPS*-diagram is not located in area V (fig 12).

One can verify that starting from a safe situation (absolutely stable *GS*), the algorithm will always find new suitable *GS* when the vehicle is moving according to the current prescribed velocity vector. Indeed, if the maximum velocity compatible with absolutely stable *GS* is exceeded, it is reduced to an acceptable value. As soon as possible, it is increased gradually (i.e. with an acceleration compatible with the actuator capability) until the reference value or the highest value achievable with the actuators.

 Nevertheless, too frequent and sudden velocity reductions are not desirable because it reduces the efficiency of the motion as well as its smoothness. This is why, it is preferable, instead of reducing the speed, to allow to lower a swinging leg before it reaches *AEP* (i.e. to take a shorter step).

This flexibility in the length of the steps, combined with a filtering of the desired velocity to avoid large accelerations exceeding the actuator capability, improves considerably the smoothness of the motion particularly after changes of direction.

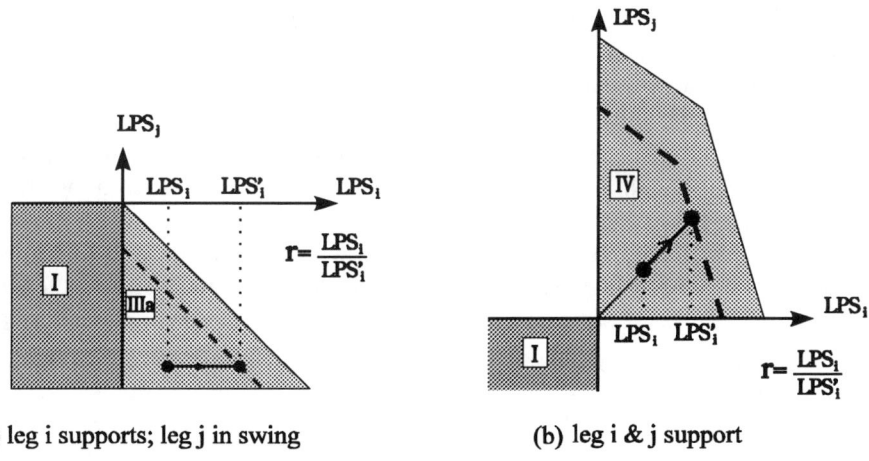

(a) leg i supports; leg j in swing   (b) leg i & j support

Fig 11. Construction for the calculation of the speed reduction factor

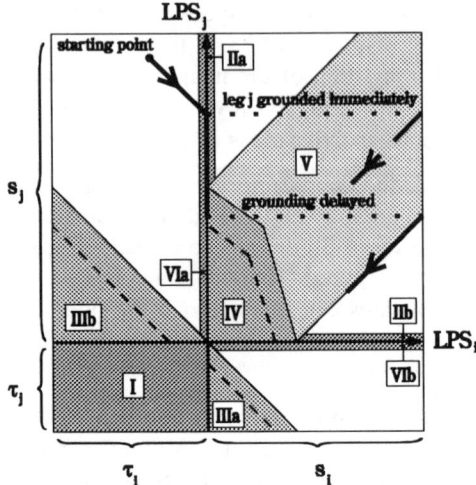

Fig. 12 : Strategy for grounding legs reaching the end of their swinging phase

Although the algorithm works well for difficult situations such as turning, walking sideways; etc..., its stability margin is not optimal for forward motion at constant speed (for $v_y \approx 0$ and $\omega_z \approx 0$) on regular ground. In these circumstances, it is known that the forward wave gait remains the best gait pattern. Hence, a good property of the algorithm would be that it ensures a transition to a forward wave gait as soon as the appropriate conditions are met.

As seen earlier, the characteristics of a forward wave gait are a phase difference of 0.5 for a pair of opposite legs (with respect to the direction of motion) and a phase difference of $1-\beta$ between successive legs on each side. One can verify that if an initial state of the robot corresponding to a forward wave gait is given as input, the present free gait algorithm will keep the gait pattern of a forward wave gait (Fig. 7).

The strategy used to converge from any *LPS*-vector towards a forward wave gait is now described :

First, the phase difference of 0.5 is imposed for the leading pair of legs with respect to the motion and is maintained. This can be done progressively by shortening or lengthening, when appropriate, the swinging phase of the leading legs. The rates of shortening and lengthening have to be limited to avoid too abrupt transitions.

Next, the phase difference of $1-\beta$ between the successive legs is simply obtained by lengthening when possible the swinging phase of the rear legs with as effect a progressive decrease of the phase difference until the minimum allowed value $1-\beta$ is reached. To smooth the transition to the forward wave gait, the delay before lowering the legs to lengthen the swinging phase is also limited to a maximum value of a few ΔT.

After a few steps (depending of the initial condition), the algorithm reaches and maintains the desired forward wave gait with the duty factor β corresponding to the desired velocity (Fig. 13).

The free gait algorithm has been successfully implemented on ULB's walking machine (Fig. 14) whose kinematic is described in [9].

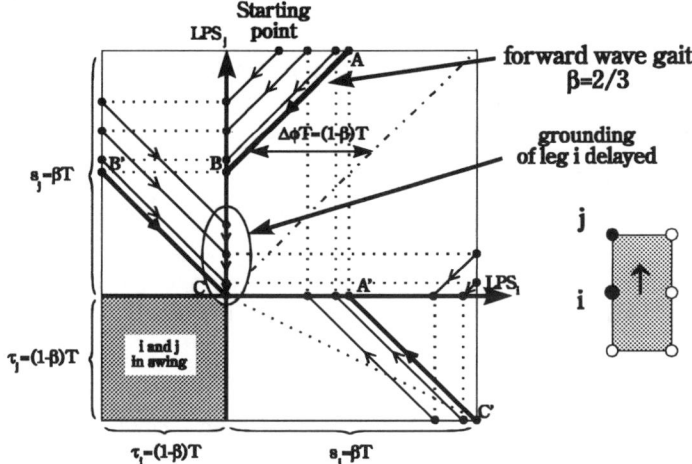

Fig. 13. Convergence of the *LPS*-trajectory towards a forward wave gait for a pair of successive legs

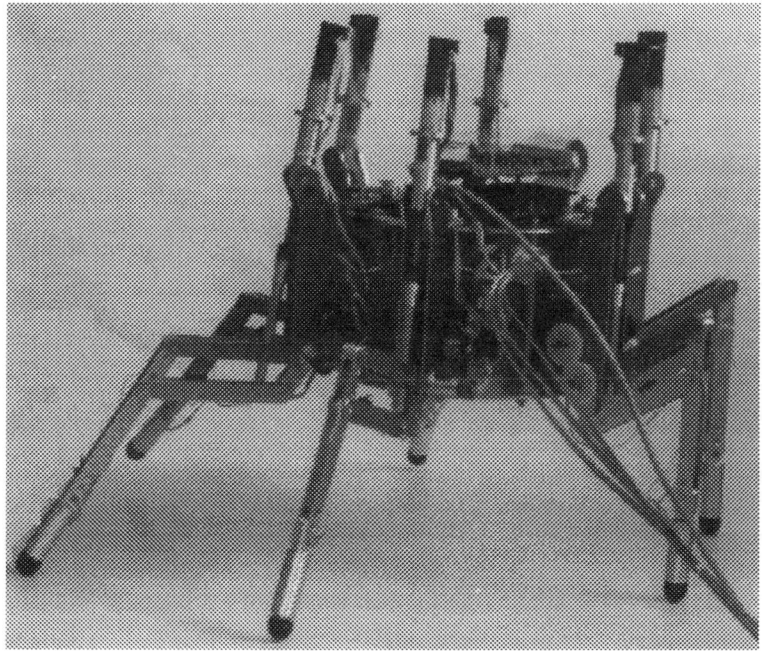

Fig. 14. ULB's walking machine

5 Conclusion

This paper presents a free gait algorithm for coordinating the motion of the legs of a six-legged walking machine, for an arbitrary velocity vector, including a spin of the body about the vertical axis.

The coordination strategy consists of preventing gait states where two neighbouring legs are lifted simultaneously. This is achieved by coordinating influences which are translated into graphical rules into the *leg phase state* diagrams.

Unlike the neurobiological coordination, the coordinating influences are the same for each pair of neighbouring legs, irrespective of the location of the legs with respect to the body.

When the velocity vector is an uniform translation (no spin), the free gait converges toward a forward wave gait, maximizing the static stability of the vehicle.

6 Acknowledgements

This work was done within the ESPRIT Working Group - 8474: Autonomous Legged Mobile Robot (LEGRO).

7 References

[1] A.Preumont & P.Alexandre: *Some Trends in walking robots in Europe*, 2nd Int. Conf. on Motion and Vibration Control *MOVIC*, Yokohama, Japan, 1994.

[2] H.Cruse: What mechanisms coordinate leg movement in walking arthropods?, *Trends in Neurosciences, Vol.13, No1,15-21, 1990*

[3] A.Halme, K.Hartikainen & K.Kärkkäinen: Terrain adaptive motion and free gait of a six-legged walking machine,1st *IFAC* Workshop on Intelligent Autonomous Vehicles, Southampton, UK, 1993.

[4] S.Salmi & A.Halme: Implementing and testing a reasoning based free gait algorithm in the six-legged walking machine "Mecant",2nd *IFAC* Workshop on Intelligent Autonomous Vehicles, Espoo, Finland, 1995.

[5] S.M.Song and K.J.Waldron: *Machine that Walk: The Adaptive Suspension Vehicle*, MIT Press, 1989.

[6] A.Preumont, D.Ghuys & C.Malekian: On the stability of hexapods, 3rd Int. Workshop on Advances in Robot Kinematics, Ferrara, Italy, 126-133, 1992

[7] H.Cruse, J.Dean, U.Müller & J.S.Schmitz: The stick insect as a walking robot, Int. Conf. on Advanced Robotics, Pisa, 936-940, 1991.

[8] H.J.Weidemann, F.Pfeiffer & J.Eltze: The Six-legged *TUM* Walking Robot, International Workshop on Intelligent Robots and Systems *IROS*, 1994.

[9] A.Preumont, An investigation of the kinematic control of a six-legged walking robot, *Mechatronics, Vol. 4, No. 8, 821-829, 1994.*

[10] P.Alexandre & A.Preumont: On the gait control of a six-legged walking machine, 2nd *IFAC* Workshop on Intelligent Autonomous Vehicles, Espoo, Finland, 1995.

Paul ALEXANDRE
Professor André PREUMONT
Department of Mechanical Engineering and Robotics
CP 165, 50 av. F.D.Roosevelt, 1050 Brussels, Belgium
Telephone: (32) 2 650 2687
Fax: (32) 2 650 2710
Email: scmero@ulb.ac.be

Application of a Simulation with Hardware-in-the-Loop Environment in Automated Guided Vehicles

Marco A. Busetti de Paula, Mauro C. Zanella
Centro Federal de Educação Tecnológica do Paraná, Curitiba, Brazil

Abstract: The objective of this paper is the description of an environment of simulation with hardware-in-the-loop for mechatronic systems and will be used as a case study an Automated Guided Vehicle - AGV. The AGV controllers were developed using the design tools environment CAMeL ("Computer Aided Mechatronic Laboratory") [1]. The concept TRANSIENT ("Transputer Based Simulation Environment") [2] is discussed as real-time simulator with hardware-in-the-loop and an implementation of this concept on a VXIbus architecture is presented. The result importance is based on the dynamic study of AGVs when they are developing complex tasks and on the application of a simulation with hardware-in-the-loop environment with characteristics of flexibility and user friendly.

1 Introduction: Mechatronic and AGVs

The automation and mechanization of industrial plants sensibly reflect on the problems of transportation, storage and handling of pieces and products, whose costs are extremely important on the final product. The AGV (Automated Guided Vehicle) is an advanced material-handling system that involves a driverless vehicle which follows a guidepath and is controlled by an on-board computer. The AGVs represent an important task working in the handling field of materials. Due to their controller it is possible to take advantages of their characteristics and applications. The intelligent control level of AGVs is the key of system flexibility. AGV is a typical mechatronic system example. Mechatronic systems are mechanical systems coupled with electrical, hydraulic and information processing systems. These systems are characterized by signal inputs, processing, and output of processed signals. The AGV combines actuators, sensors and information processing in a mechanical system. The typical design methodology in the development of mechatronic systems are: analysis, synthesis and implementation. CAMeL (Computer Aided Mechatronic Laboratory) [1] tools were used to support the design the system. For tests with a real AGV system, it was applied a hardware-in-the-loop simulation environment (TRANSIENT) [2]. When the tests are finished with a successful controller it is possible to export the controller algorithm to industry standard equipment, for example PLCs.

2   The Design Environment

To study the AGV dynamics, we used a methodology based on three main topics: *analysis* - i.e. from a real system one can generate its physical model, parametric mathematics model as well as the identification of the parameters; *synthesis* - generation of the algorithm of controller according to the problem formulation and the system analysis with feedback; *implementation* - that is the realization strategy of the algorithm realized on the second phase [3]. The whole cycle in the design of mechatronic systems (see figure 1), from modeling through analysis, identification , synthesis up to the implementation in the lab and final tests, the CAMeL (Computer Aided Mechatronic Laboratory) [1] tools have to meet the requirements at every step.

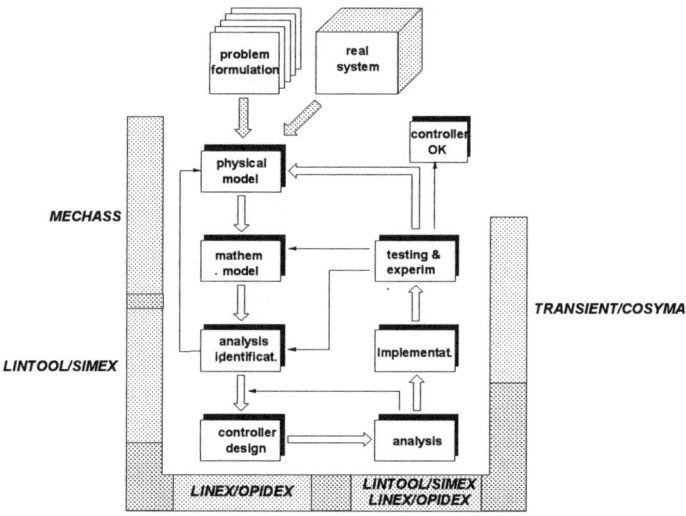

Figure 1: Design development cycle of mechatronic systems

The CAMeL was used as a platform of software for generation of models, controllers and optimization. The CAMeL has a descriptive language of dynamic systems (DSL - Dynamic System Language) [4] and auxiliary tools such as: simulator of nonlinear systems (SIMEX - Simulator Expert); optimizer of parameters of nonlinear systems (OPIDEX - Optimization and Identification Expert); tools for linear systems (LINTOOL - Linear System Tools); real-time simulator of linear systems (COSYMA - Control System for Mechatronic Applications); and a real-time simulator of non-linear systems (TRANSIENT - Transputer Based Simulation Environment). DSL enables a mathematical description of nonlinear systems on the basis of the state space representation

$$x = f(x, u, p, t)$$
$$y = g(x, u, p, t).$$

In this way, a system can be described by indication of its inputs u, outputs y, states x, parameters p and the differential and output equations. Thus simple functional units (denominated as basic systems) can be defined. These basic systems can now be combined in a modular way to form units of higher level (defined as coupled systems), by the formulation of the input/output relation between the basic systems involved. Coupled systems formed in this way can be assembled with other basic or coupled systems to form units of higher level, thus enabling a hierarchical structuring of a system [3].

3 The Simulation with Hardware-in-the-Loop Environment

3.1 The Concept TRANSIENT

We applied the concept of TRANSIENT (Transputer Based Simulation Environment) tool [2] to implement the platform of hardware-in-the-loop simulation for the AGV. This concept is structured in two representations: an experiment oriented and a hardware oriented.

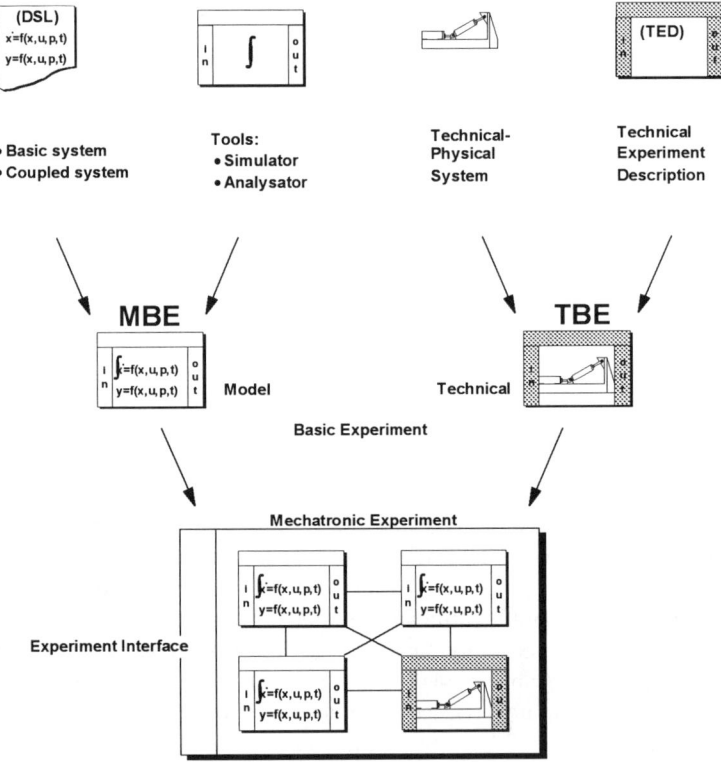

Figure 2: The Experiment Oriented Representation of the TRANSIENT Concept

The experiment oriented representation consists in two kinds of basic experiments: the Model Basic Experiment (MBE), which consists in a description of dynamic system (for example DSL) associated with a tool (for example, a simulator); the Technical Basic Experiment (TBE), composed by the technical-physical system and the description of sensors, actuators and respective interfaces (for example TED - Technical Experiment Description). The figure 2 shows the experiment oriented representation of the TRANSIENT concept.

The hardware oriented representation consists in three kinds of computer units: the Model Computer Unit (MCU), which consists of a processor with memory aggregated and an operational system that enable to run a specific tool; the Technical Computer Unit (TCU) that consists of a processor and a bus. The sensors and actuators of a technical-physical system are connected to the bus through the interface boards; the Network Controller Unit (NCU), which is responsible for the real-time control in the network and to store the data to be visualized by the user. The figure 3 shows the hardware oriented representation of the TRANSIENT concept.

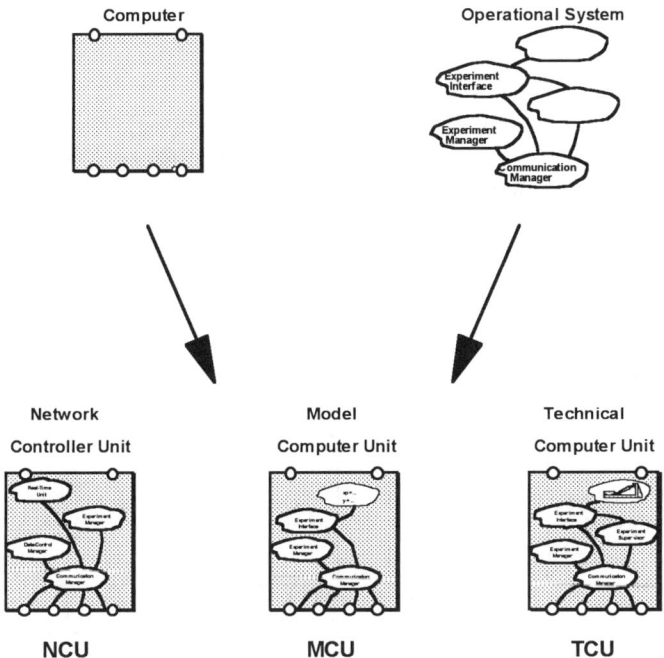

Figure 3: The Hardware Oriented Representation of the TRANSIENT Concept

Computer group is defined as a set of computer units. In this way there are a Model Computer Group as a set of MCUs and a Technical Computer Unit as a set of TCUs. The interface of a computer group has the same interface of a computer unit as showed in figure 4. In this case is used the transputer module hardware interface that consists in four serial links for data transfer and service lines to control of the processing. For more details you can find in next subchapter.

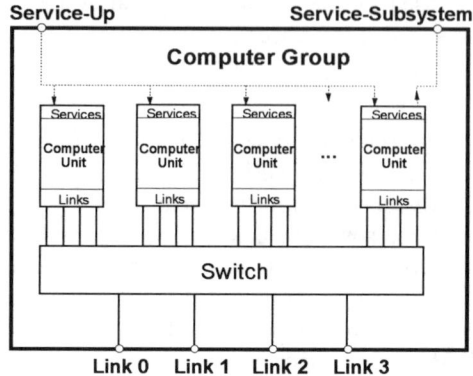

Figure 4: Computer Group

Both computer groups are connected in a network (see figure 5), which has a Network Controller Unit, where it is possible to connect a User Interface Unit. The NCU can also switch the links path between groups and units. The service lines establish a hierarchical connection between units and groups that means, for example, if an error occurs in the Model Computer Group, such error is not propagated to the Technical Computer Group. This is a hardware security for the simulation with hardware-in-the-loop.

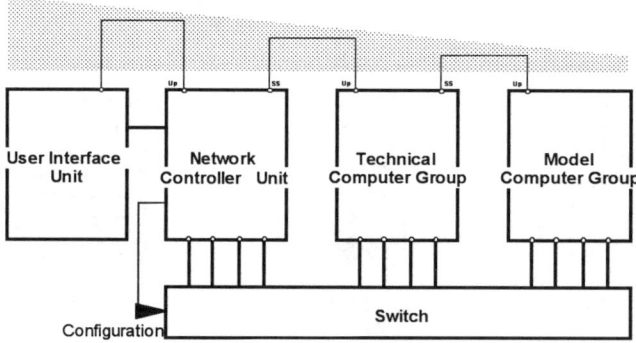

Figure 5: The Network in a Hardware Oriented Representation of TRANSIENT

3.2 The Implementation of TRANSIENT Concept with VXIbus

As platform of hardware to implement the above concept it was used an architecture of RISC parallel processing (transputers) operating in real-time. The Technical Computer Group was implemented with a VXIbus whose controller is a transputer. The Model Computer Group and the Network Controller Unit were implemented as a transputer network located on a board at VXIbus. The VXIbus is an open architecture for modular instruments based on the VMEbus. As an open instrument architecture, the VXIbus satisfies the market need for computer based, down-sized interchangeable and interoperable instrumentation. The hardware in use consists of a certain number of transputers modules (TRAM) by INMOS. Every 10 of them can be assembled on a module motherboard (IMS-B008). Several motherboards can again be linked together. The transputer nodes communicate via 4 links maximum, two of which form a fixed circuit with neighboring nodes. The other links can, via a link switch (IMS-C004) be freely connected with another free links. A host computer (AT) controls the transputer network; it is linked to the first transputer of the network via an interface. The transputer module (TRAM) can be a single transputer with a local memory each or a transputer with a bus (for example, VXIbus) to communicate with sensors and actuators of AGV. On the transputer network runs a simple and dedicated operational system that is responsible by exchange protocol data and codes. The host computer is used as User Interface Unit and has a graphical interface users friendly. Figure 6 shows respectively a example of a implementation of the Technical Computer Unit and of the Model Computer Group. Virtual Transputer Module is defined as a transputer module with the same INMOS logical and electrical interface but not physical compatible.

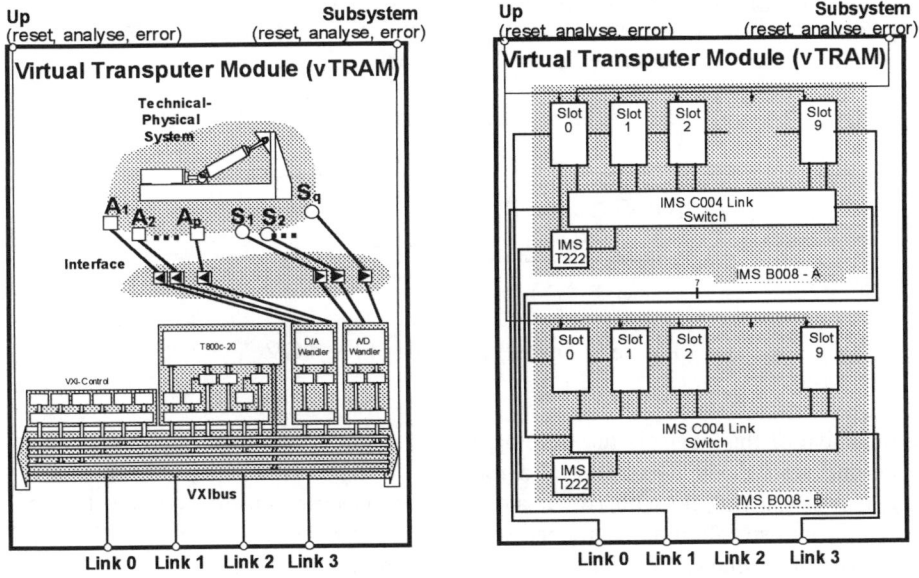

Figure 6: The implementation of some units of TRANSIENT

4   Case Study

4.1 Description of AGV of Case Study

The AGV [5] in study is a single triangle, where the steering motor and the drive motor on the front wheel (see figure 7).

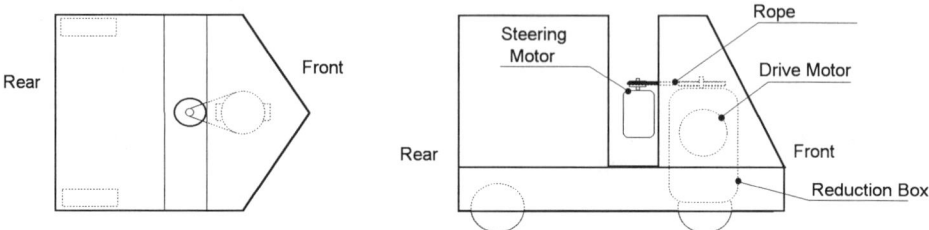

Figure 7: Single triangle AGV

The steering motor is a triphase induction motor (1/2 HP), also the drive motor is a triphase induction motor (1 HP). Both are controlled by a converter with a voltage DC-link circuit for variable-speed AC drives. Automatic load compensation is carried out by the current control system which permits automatic adaptation to the system requirement by means of Flux Current Control (FCC).

The guide path to follow is a passive tracking. It's used optical method because is more simple and is a good prize. Don't required big investments on cables under floor and sophisticated sensors. There are two analogical optical sensors and one lamp in a black box, and the sensors are connected at the control rack (see figure 8).

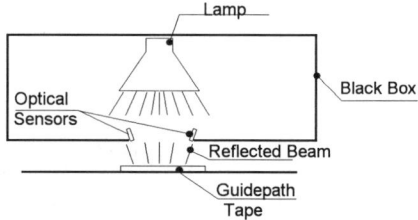

Figure 8: Optical guidance of the AGV

The objective of this paper is the development of a study field of the AGVs dynamics, that permits tests flexibility of control algorithms on them. The AGV, utilized as a case study, is a single tricycle with two analogical photoelectric sensors for detection of a band on the industry floor and equipped with two triphase induction motors - one for the traction and other one for driving - both operating on the front wheel.

4.2 Model of the AGV

The complete system of the AGV is represented in four basic systems: "car", "actuator", "sensor", "control" and "trajectory" [6]. Each basic system is described in DSL and has as interface inputs, outputs and parameters. Basic systems are coupled in a high hierarchical level of representation and denominated "agv", as show figure 9.

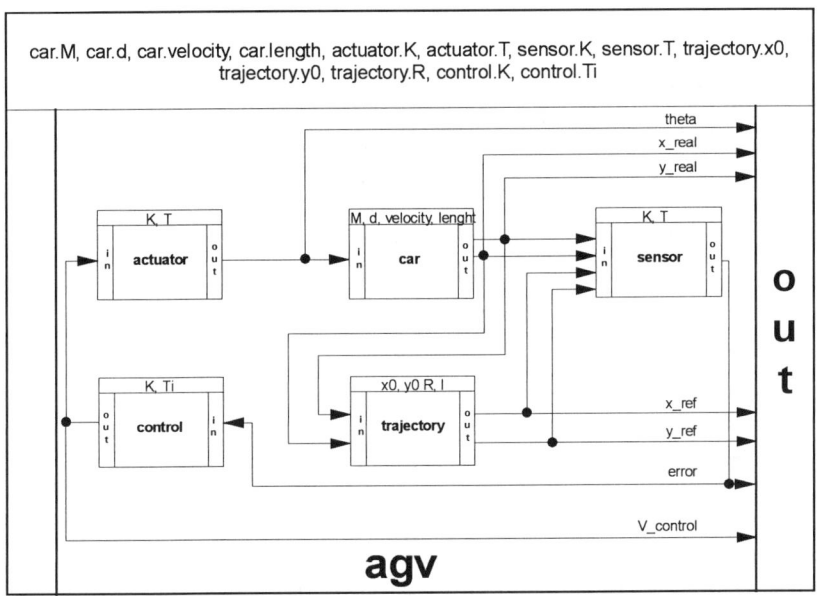

Figure 9: Coupled System "AGV"

4.3 Results of the Simulation

The figure 10 shows a graphic interface example of the developing environment and display the time response of reference and real cartesian position and also the trajectory on the x-y plan.

The model parameter identification was realized off-line through the application of the OPIDEX tool. The applied controller in this example is a simple PID algorithm, which the parameters was optimized with the same tool. The actual state of this work is to improve the control algorithm by simulation with hardware-in-the-loop.

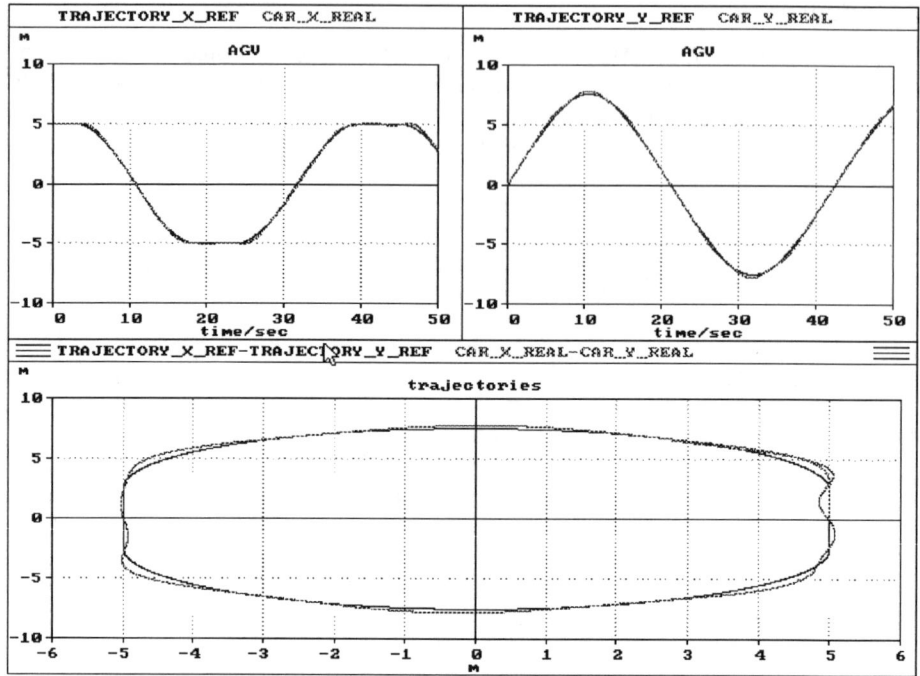

Figure 10: Simulation Results

5   The Results Significance and Conclusions

This paper presented a methodology of developing mechatronic systems using as example an AGV.

The main objective was to present an integrated environment to developing simulation with hardware-in-the-loop. The next step will be an integration of this environment with standard industry packages (equipment and software).

The importance of the results is based on the dynamic study of AGVs, seeking a better conservation of energy on the transportation in the industrial plants, as well as in the operation of these vehicles when they are developing complex tasks. The developed controller should be improved with complex algorithms.

This case study and related environment are also a good integrated system to education and research in mechatronic field.

6   Acknowledgments

The authors gratefully remind that the CAMeL environment was developed by Prof. Dr.-Ing. J. Lückel at the Mechatronic Laboratory Paderborn of University-GH Paderborn,

Germany and the AGV used in the presented work was given by Prof. Dr. Jaime Szajner at the Faculty of Electrical Engineering of University of Campinas, Brazil.

7 References

[1] Jäker, K.; Klingebiel, P.; Lefarth, U.; Lückel, J.; Richert, J.; Rutz, R.: "Tool Integration by Way of a Computer Aided Mechatronic Laboratory (CAMeL)", Preprints of the IFAC Symposium on Computer Aided Design in Control Systems, Swansea, UK, July 15-17 1991.

[2] Engelke, A.; Busetti de Paula, M.: "Transient - Ein Werkzeug zur verteilten Simulation mechatronischer Systeme unter Echtzeitbedingungen", Preprints of the Transputer Anwender Treffen '92, Aachen, pp. 362-379.

[3] Rutz, R.; Richert, J.: "Optimal Mechatronic Design Using CAMeL", Preprints of the International Conference on Machine Automation, Tampere, Finland, February 15-18, 1994.

[4] Schröer, J.: "A Model Compiler/Interpreter for Supporting Simulation and Optization of Nonlinear and Linearized Dynamic Systems", Preprints of the IFAC Symposium on Computer Aided Design in Control Systems, Swansea, UK, July 15-17 1991.

[5] Gouvêa da Costa, S.: "Sistema de Controle de Direção de um Veículo Autoguiado", Proceedings of the 9° Congresso Brasileiro de Automática, vol. 1, Vitória, Brazil, 1992, pp. 390-394.

[6] Zanella, M.; Gouvêa da Costa, S.; Busetti de Paula, M.: "Development System of Automated Guided Vehicles Control", to be published in the preprints of the 11th ISPE/IEE/IFAC International Conference on CAD/CAM, Robotics and Factories of the Future", Pereira, Colombia, August 28-30, 1995.

Address:
M. Sc. Marco Antonio Busetti de Paula
B.Sc. Mauro César Zanella
Centro Federal de Educação Tecnológica do Paraná - CEFET-PR
Núcleo de Pesquisa em Energia e Controle - NuPEC
Av. Sete de Setembro 3165
80230-901 Curitiba - Paraná - Brazil
Phone: +55 -41 2342005
Fax: +55 -41 2245170
E-Mail: busetti@cpgei.cefetpr.br
                zanella@cpgei.cefetpr.br

Wall Climbing Robots for Inspection and Maintenance

Fr.-W. Bach, J. Seevers, M. Hahn, M. Rachkov*

University of Hanover, Hanover, Germany

*Humboldt Research Fellow

Abstract: The climbing robot tele-operated remote systems were developed for inspection and maintenance. The design of the robotics systems provides automation of these processes and economic efficiency in its applications. Two variants of the robots are considered: design of increased load capacity and design of light weight. A sensor-based computer control system is used.

The paper contains an overview of the technical parameters and experimental characteristics of the robots transport module, control system, some design calculations, and the schemes of the robots application.

1 Introduction

Wall Climbing Robots (WCRs) are intended for motion on vertical and sloping surfaces. On board they can have different technological equipment and can perform necessary technological operations during continuous motion or whilst they are stationary [1,2].

The possibility of serving big areas of constructions without scaffolding and the possibilities of using in complex or extreme conditions mean that the WCRs can be employed in an economically efficient and operation safety manner [3,4].

In the programme [5] of research and training for the European Atomic Energy Community in the field of remote handling in hazardous or disordered nuclear environments, the emphasis is on improved nuclear safety by the development of tele-operated remote systems to meet the aims of the nuclear industry by:

- reducing the radiation exposure of human operators in nuclear installations to the lowest reasonably practicable level;

- ensuring the safety of nuclear installations and thus the protection of humans and the environment by the development of remote handling equipment to aid operators to perform inspection, maintenance, repair and replacement tasks more safely, efficiently and with higher quality;

- providing vital assistance in accident and post accident situations and therefore again protecting humans and the environment.

So, it is expedient to apply climbing robots for inspection and maintenance.

There are many designs of WCRs [6,7], but the WCRs for inspection and maintenance of increased load capacity and light weight have the best reliability for these technological operations.

2 Design of the robots

The design Hydra II of the WCR (Figure 1) was completed in order to increase the payload coefficient, which is equal to the ratio of a robot's transport module weight to its nominal payload.

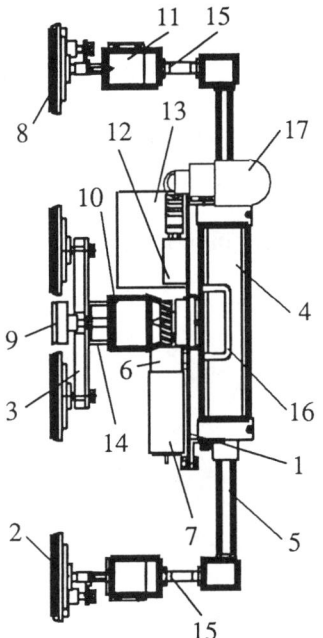 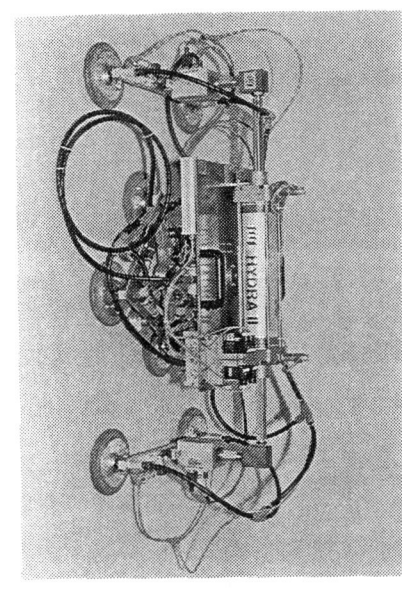

Figure 1: Design (diagram and view) of Hydra II

1 - platform, 2 - external group of pedipulators, 3 - internal group of pedipulators, 4 - transport pneumatic cylinder, 5 - piston-rod, 6 - rotation block, 7 - electromotor, 8 - force gripper, 9 - sealing gripper, 10 - internal lifting cylinder, 11 - external lifting cylinders, 12 - power ejector, 13 - control pneumatic block, 14 - force guides, 15 - adjusting unit, 16 - carrying handles, 17 - video camera.

The transport module of the WCR has a platform and two groups of pedipulators [8]: an external group and an internal group. A transport pneumatic cylinder is mounted on the platform. The external group of pedipulators is fixed on the piston-rod of the transport pneumatic cylinder. The internal group of pedipulators is connected with the platform by means of a rotation block with an electromotor.

A two-staged gripper system has been developed in order to improve the reliability of the robot's motion on rough surfaces. Each group of pedipulators consists of four vacuum force grippers and four sealing grippers.

The internal group of pedipulators has one lifting pneumatic cylinder, and the external one has two lifting pneumatic cylinders. The power ejector and control pneumatic block are attached beneath the platform. A video camera is installed on the platform.

The system is supplied by a compressed air source which is connected with the transport module by an umbilical.

The Hydra II moves as follows: when the internal group of the pedipulators is connected to the motion surface by means of the grippers, the external group has the opportunity to move easily with a piston-rod relative to the platform, or the platform can rotate to change the direction of motion by means of a rotating block. When the external group of pedipulators is fixed on the surface, the internal group can move with the platform, and so on.

A light weight design of the WCR Hydra III was completed in order to apply it for the fragile and fine vertical surfaces (Figure 2).

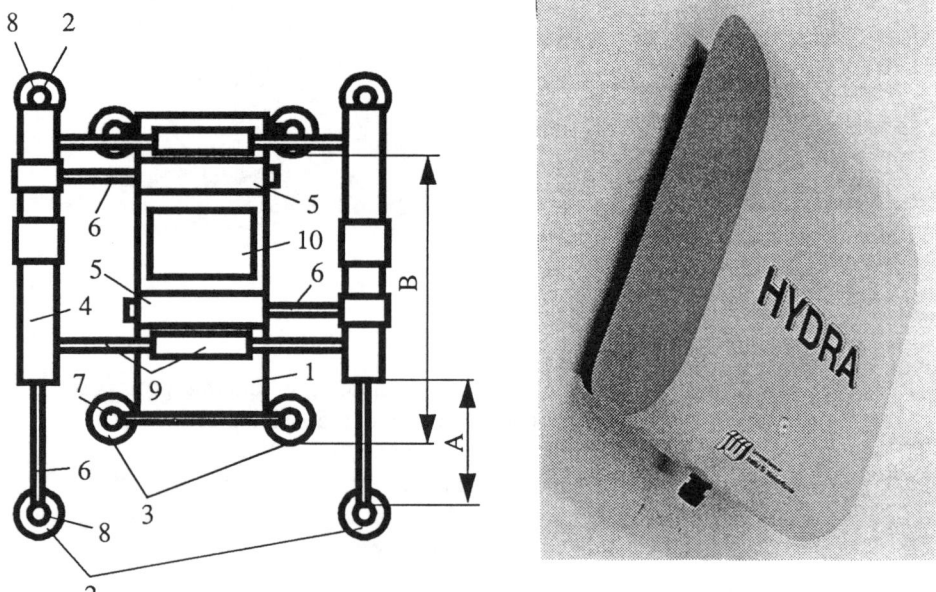

Figure 2:   Design (diagram and view) of Hydra III

1 - platform, 2 - external group of pedipulators, 3 - internal group of pedipulators, 4 - vertical transport pneumatic cylinder, 5 - horizontal transport pneumatic cylinder, 6 - piston-rod, 7 - internal lifting cylinder, 8 - external lifting cylinder, 9 - force guide, 10 - on-board control system; A - cylinder stroke, B - gripper spacing.

Two horizontal transport pneumatic cylinders are mounted on the platform. The external group of pedipulators is fixed on the piston-rods of the vertical transport pneumatic cylinders by means of external lifting cylinders. The internal group of pedipulators is connected with the platform by means of internal lifting cylinders. The vertical transport pneumatic cylinder are fixed on the piston-rods of the horizontal transport pneumatic cylinders.

The Hydra III moves as follows: when the internal group of the pedipulators is connected to the motion surface by means of the grippers, the external group has the opportunity to move easily vertically with the piston-rods of the vertical transport pneumatic cylinders relative to the platform. When the external group of pedipulators is fixed on the surface, the internal group can move horizontally with the platform by means of the horizontal transport pneumatic cylinders, and so on.

The positions of the corresponding vacuum grippers of the external and internal groups of pedipulators are located step by step on the same axis. The cylinder stroke must be equal to the gripper spacing to overcome maximal width of the obstacles during vertical motion. In this case the width of the obstacles can be equal to the cylinder stroke minus the gripper diameter.

This design with the direct connection between the vertical and horizontal transport pneumatic cylinders allows to perform a two-coordinate motion without a rotation unit and to use the plastic pneumatic cylinders, so the weight of the robot is minimal.

Main technical specifications of the WCRs are shown in Table 1:

Specifications		Hydra II	Hydra III
1.	Payload: - nominal transport - stationary	100 kg 250 kg	10 kg 20 kg
2.	Weight	38 kg	3.5 kg
3.	Maxium Stride	300 mm	80 mm
4.	Maximum Stepover Height	25 mm	40 mm
5.	Overall Dimensions: - length - width - height	1120 mm 560 mm 430 mm	540 mm 430 mm 180 mm

Table 1: Main technical specifications of the climbing robots

According to weight and sizes of technological equipment the corresponding design of the WCRs can be chosen.

3 Control system of the robots

The WCRs control system performs transport and technological motions of the robot in programme and manual modes by means of computers.

The controlling components are located in two different places. There is one computer with additional electronics at the host station. This computer is mainly used for the dialogue between the user and the WCR. The second computer is placed on the WCR and will control the safety function of the electronic and mechanical parts. An intelligent supervision and control system directly on the robot helps to increase the security of this handling system by enabling quick and continuous emergency handling. Both computers are connected by a serial data cable. A diagram of the WCRs control system is shown in Figure 3.

The host station includes the host computer and Man-Machine interface. The host computer is based on a Motorola 68000 processor. The peripheral electronics on the board include comfortable data monitoring and a complex interrupt structure. External components can be easily adapted for the computer by using the internal galvanic decoupled bus structure. All of these installations can be easily tied in under the programming language 'C'. A real time operation system can be used, but should not be necessary because time-critical tasks and emergency tasks are executed in their entirety at the board computer on the WCR.

The power supply is controlled by the host computer. Current and voltage supervision are used as an additional information source to monitor the electric components and cables on the WCR.

The on-board control system is designed under the aspect to provide a permanent secure state of the WCR, because every mistake in the software and electronics can be dangerous and can cause damage to the WCR. It has the movement and emergency control blocks to perform point-to-point control and obstacle-avoidance control.

Main task is the dialogue with the WCR. Commands coming from the control panel, the keyboard or from an external data source or data file are interpreted and sent to the WCR. The commands between host and WCR are on a low level, which can be easily and quickly interpreted and dealt with by the WCR. There is no transformation of commands in the WCR, this is completely done at the host station. So there is enough time on the board computer for high speed supervision to reach high level safety.

The interface is directly connected to the I/O ports of the micro controller. These cells provide the contact between controller and actuators, valves and sensors. The cells supervise these components autonomously. In case of unexpected events they create the interrupt messages for the controller. This architecture is used to locate the supervision and reaction 'intelligence' directly at the place where it is needed. Additionally the computer system will not completely break down in case of a failure of one of the control units.

The WCRs control system provides an adaptive control of the gripper system position to avoid an unsafe connection of the vacuum grippers to the motion surface because of a vacuum leakage from the possible surface damages under the grippers.

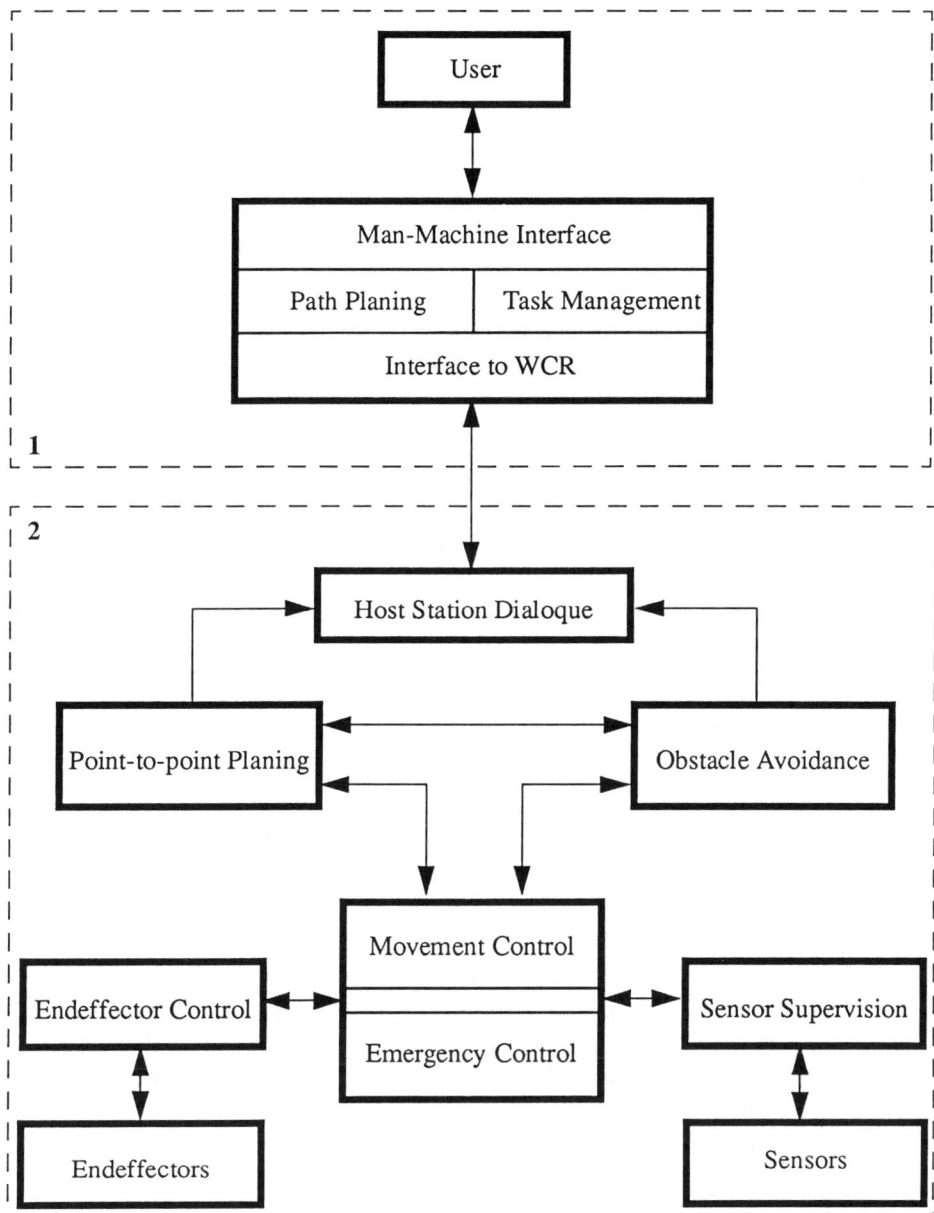

Figure 3: Diagram of the WCRs control system
1 - host station, 2 - on-board control system

The computer interface is connected with the vacuum sensors of the grippers. The vacuum sensors provide two different data, an analogue data representing the vacuum and an adjustable digital data which is used as an interruption signal when the vacuum is falling under a certain level. In this case feedback control provides an automatic change of the vacuum gripper position to attain the reliable connections of each of the actuated grippers to the surface.

The control system has a video unit for remote control of the WCR. The video unit consists of a CCD video camera and three servomotors to obtain three degrees of freedom. There are horizontal and vertical rotatable axes with two corresponding motors in the video camera unit. The third servomotor allows views either from the upper side of the platform for orientation and navigation of the WCR, or under the platform for surface inspection or checking technological equipment during the working phases.

Main technical specifications of the used video camera are shown in Table 2:

Specifications		Value
1.	Resolution	297984 Pixels (512 H x 582 V)
2.	Lens	f = 4.3 mm
3.	Sensitivity	1 Lux
4.	Video signal	CCIR
5.	Voltage	12 V =
6.	Diaphragm adjustment	automatic

Table 2: Main technical specifications of the video camera

The control system performs a special gait of the robots to provide a reliable motion across the rough and uneven surfaces by sealing the grippers to the surface. The sealing phase of the WCRs motion is shown in Figure 4.

At the initial stage of the robot's step all of the lifting cylinders have a direction of the piston force towards to motion surface. The internal grippers are connected to the surface from the previous step by vacuum. If there is some roughness between the external gripper and the surface in final piston-rod position, the vacuum sensors give a signal to the control system to fulfil a sealing stage of the gait.

During the sealing stage the internal cylinder piston force acts to press the platform with the external gripper to the surface. As result maximum value R of the roughness is sealed by means of an elastic working surface of the grippers. After that the external grippers are fixed to the motion surface by vacuum.

At the final stage the volumes inside the internal grippers are connected with atmosphere and the robot can perform the next step.

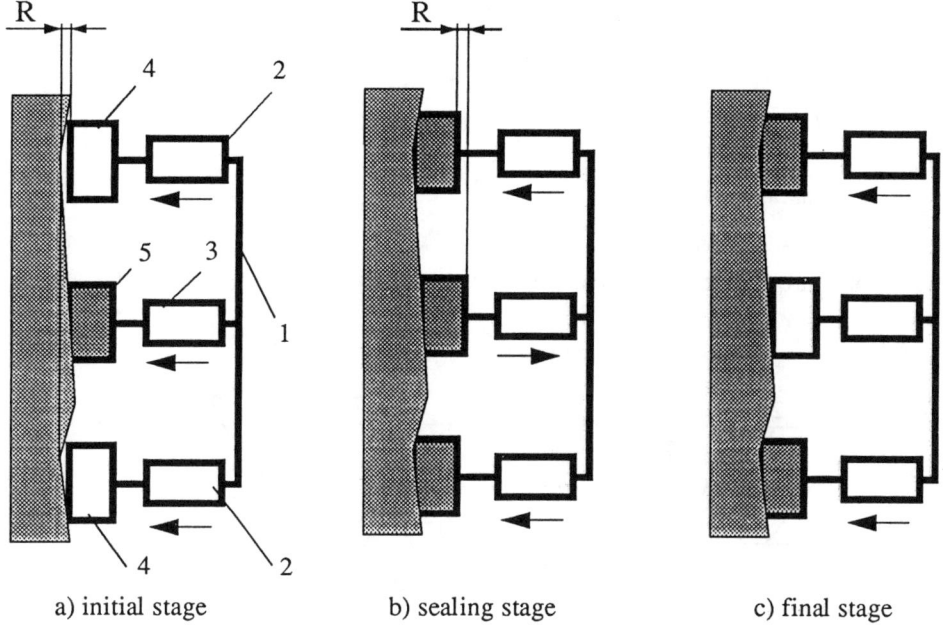

a) initial stage   b) sealing stage   c) final stage

Figure 4:   Sealing phase of the robots gait across the rough and uneven surfaces

1 - WCR's platform, 2 - external lifting cylinders, 3 - internal lifting cylinder, 4 - external grippers, 5 - internal gripper, R - maximum value of roughness

■ - vacuum ;   □ - atmosphere

━━━▶ - direction of the pressure force on the piston

4   Design calculation of vacuum grippers

Figure 5 shows the method for the calculation of the numbers and diameters of vacuum grippers for the different friction coefficients of motion surface. The theoretical characteristics for the diagram are obtained from the formulae (1,2) for the normal detaching force and the diameter of one gripper. The normal detaching force is equal to

$$F_n \geq \frac{F_t}{\mu} + F_e \quad , \text{where} \tag{1}$$

F_t - tangential detaching force,

F_e - normal detaching force from technological equipment,

μ - friction coefficient.

The diameter of one gripper can be found as

$$d = \sqrt{\frac{4 \cdot F_n}{n \cdot \pi \cdot \Delta p}} \ , \text{where} \tag{2}$$

F_n - normal detaching force,

n   - number of grippers,

Δp - pressure difference between gripper vacuum volume and environment.

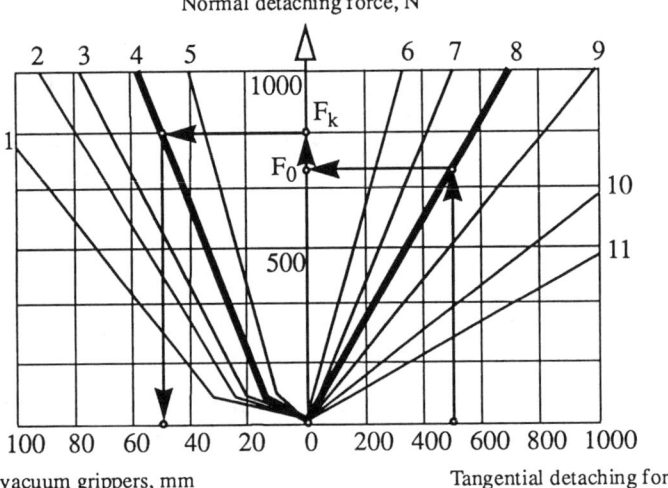

Normal detaching force, N

Diameter of vacuum grippers, mm   Tangential detaching force, N

Figure 5:   Calculation diagram of a vacuum gripper diameter for different numbers of the grippers and friction coefficients

Number of grippers: 1 - 2, 2 - 3, 3 - 4, 4 - 8, 5 - 15.
Friction coefficient: 6 - 0.3, 7 - 0.5, 8 - 0.7, 9 - 1.0, 10 - 1.5, 11 - 2.0.

The characteristics in Figure 5 correspond to the pressure difference between the gripper system and the environment equal to 0.5 MPa. It is assumed that the height of the WCR is relatively small, so the moment of the tangential detaching force is neglected.

A calculation example is given in Figure 5 for initial conditions with the maximum tangential detaching force F_t = 500 N and friction coefficient μ = 0.7. The corresponding theoretical normal detaching force F_0 is equal to 715 N.

Figure 6 shows the characteristics of the theoretical and experimental normal cohesion force as a function of the pressure difference. It follows from these characteristics that, in addition to the theoretical normal detaching force, it is necessary in practice to add an insurance coefficient K = 0.1 from F_0 to F_k because of the difference between theoretical and experimental cohesion curves.

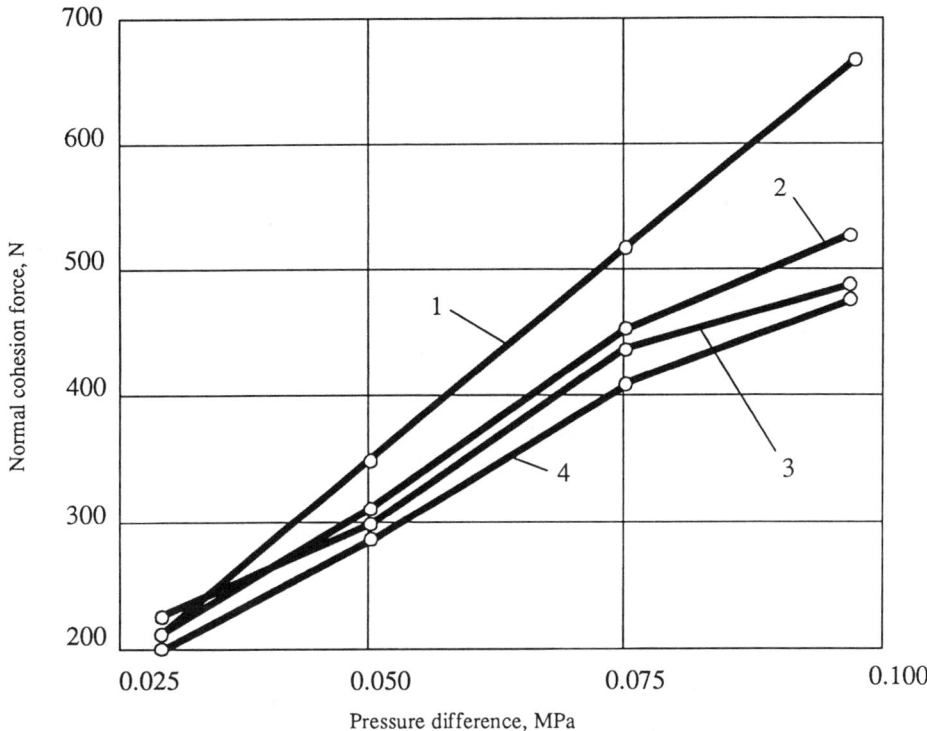

Figure 6: Theoretical and experimental normal gripper cohesion force for different gripper materials

1 - theoretical, 2 - Vulkollan (PU), 3 - Perbunan (NBR), 4 - Silopren (SI).

After taking into account the insurance coefficient K, in accordance with above rule, we have the normal force for calculation equal to about 787 N. Now it is possible to find the diameter of one gripper for the desired numbers of grippers. In the example of eight grippers we have the diameter of one gripper equal to 50 mm.

It is possible to calculate the number and diameter of vacuum gripper for other initial conditions.

Relation between vertical and normal cohesion forces as a function of the gripper pressure difference for different gripper materials is shown in Figure 7. The experimental normal cohesion force increases more with the pressure difference than the vertical cohesion force.

Figure 8 presents the vertical cohesion forces as function of the pressure difference for different levers of the force and gripper materials on the steel motion surface with roughness Rz = 67.64 μm. It follows from this diagram that the vertical cohesion force increases with the level of detaching force.

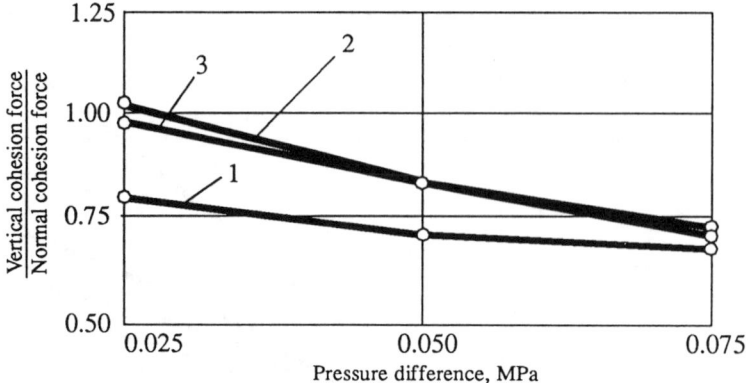

Figure 7: Relation between vertical and horizontal cohesion forces as a function of the gripper pressure difference for different gripper materials

1 - Perbunan (NBR), 2 - Vulcollan (PU), 3 - Silopren (SI).

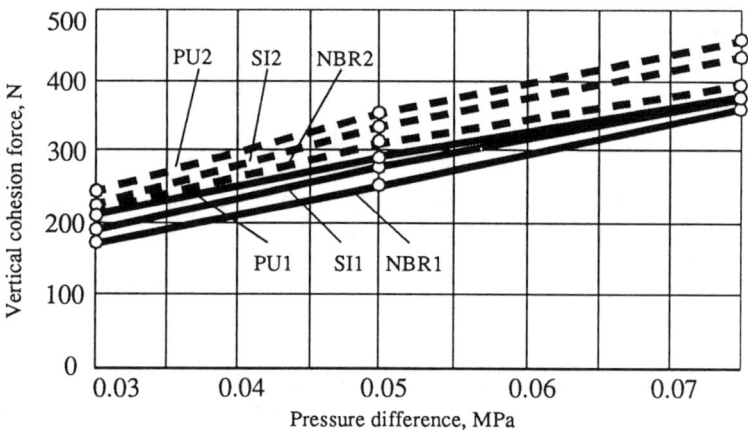

Figure 8: Vertical cohesion forces as function of the pressure difference for different levers of the force and gripper materials on the steel motion surface with roughness Rz = 67.64 μm

Gripper materials: NBR - Perbunan, PU - Vulkollan, Si - Silopren
Value of lever: 1 - 0 mm, 2 - 40 mm
Gripper design: PFYN 95 (SCHMALZ Company).

Roughness of motion surfaces has a significant influence on the friction coefficient of the grippers. Figure 9 shows an experimental diagram of gripper friction coefficient as function of the pressure difference for different steel surface roughness and gripper materials. The higher the motion surface roughness the higher is the friction coefficient for the same gripper material.

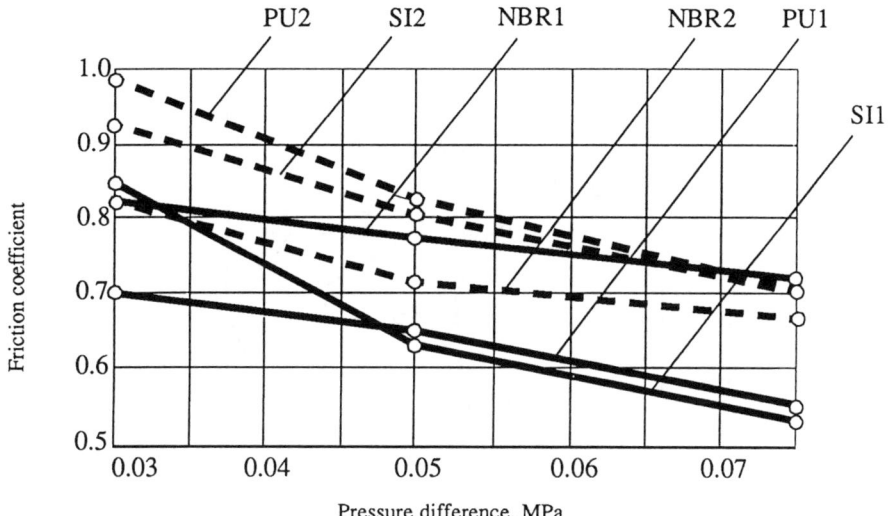

Figure 9: Friction coefficient of the grippers as function of the pressure difference for different motion steel surface roughness and gripper materials

Surface roughness: 1 - Rz = 21.64 µm, 2 - Rz = 63.98 µm,
Gripper materials: NBR - Perbunan, PU - Vulkollan, Si - Silopren.

It is necessary to take the obtained results into account to modify the design of the WCRs in accordance with different conditions of the robots application.

5   Technological possibilities of the robots

The technological possibilities of the designed robots are determined by their load capacity and mobility.

In Figure 10 possible technological motion trajectories of Hydra II are shown.

A technological device, for example maintenance equipment is installed on the robot's platform. Transport and technological motion can be performed at the same time for such maintenance operations as decontamination. In this case a cleaning tool is used as a technological equipment. During the technological motion a quality of decontamination can be proved by radiotracer as checking device. The video camera performs the orientation of the robot.

In Figure 11 the possible technological motion trajectories of Hydra III are presented.

A technological device, for example inspection equipment [9] has the possibility to move with the robot's platform along vertical, horizontal and diagonal trajectories. Desired angle of the diagonal trajectory can be chosen by setting a corresponding relation between velocities of the vertical and horizontal motion drives. Any desired common trajectory of the robot's motion can be composed by combination of the described trajectories.

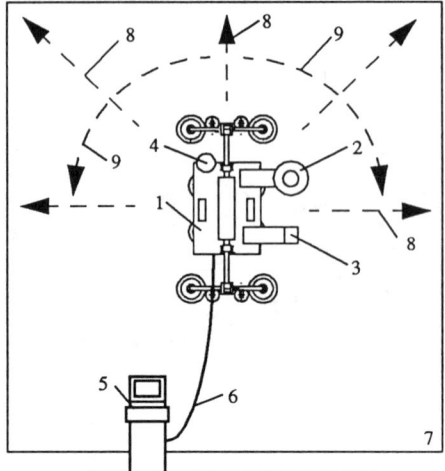

Figure 10: Possible technological motion trajectories of Hydra II

1 -WCR's platform, 2 - technological device, 3 - checking device, 4 - video camera, 5 - control unit , 6 - control-supply line, 7 - vertical motion surface,
8 - longitudinal motion, 9 - rotation motion.

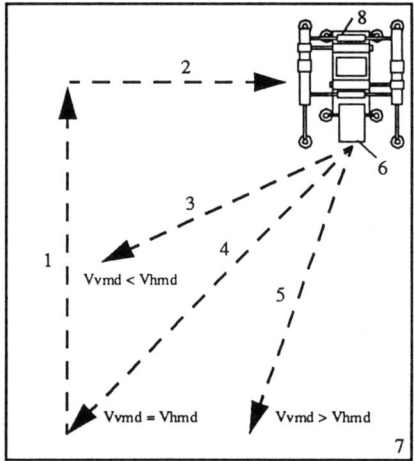

Figure 11: Possible technological motion trajectories of the Hydra III

1 - vertical motion, 2 - horizontal motion, 3,4,5 - diagonal motions,
6 - inspection device, 7 - vertical motion surface, 8 - WCR's platform,

Vvmd - velocity of the vertical motion drives,
Vhmd - velocity of the horizontal motion drives

6 Conclusions

The maintenance and inspection tasks are relatively expensive and sometimes hazardous processes. The new designs of the WCRs provide automating of these processes.

Robots designs of increased load capacity and light weight give a possibility of performing a wide range of the maintenance and inspection tasks on different vertical surfaces. A sensor-based computer control system provides flexibility and reliability of the robots transport and technological motion.

The considered methods of the vacuum gripper design calculations allow to modify the design of the gripper system in an optimal way according to different working conditions.

7 References

[1] Cusack, M., Thomas, J.: Robotics for the Inspection of Vertical Surfaces of Buildings and Structures, Proceedings of the 25th ISIR, Hanover, 1994, pp. 287-295.

[2] Haferkamp, H., Bach, Fr.-W., Ogawa, Y., and Rachkov, M.: Climbing Robot for Underwater Cutting, Proceedings of the International Conference on Oceans Engineering, OCEANS, Brest, France, 1994, Vol 1, pp 602-607

[3] Gradetsky, V., Rachkov, M.: Wall Climbing Robot and Its Application for Building Construction, Mechatronic Systems Engineering, Kluwer Academic Publishers, 1990, no. 1, pp. 225-231

[4] Rachkov, M.: Wall Climbing Robots: Development and Trends of Application in Russia, Proceedings of the International Conference on Mechatronics and Robotics, Aachen, 1994, pp. 157-174.

[5] Teleman - Robotics and remote systems in hazardous or disordered nuclear environments, European Commission, Second edition, EUR 15900 EN, 1995, 63 p.

[6] Kojima, H., Toyama, R., Kobayashi, K.: Development of Wall Climbing Robot, Technical Review of Ishikawajima Harima Heavy Industry, 1992, 32 (2), pp. 123-128

[7] Ikeda, K., Nozaki, T., Shimada, S., and Tajima, Y.: Basic Study on a Wall Climbing Robot, Journal of Mechanical Engineering Laboratory, Japan, 1989, 43, pp. 183-194

[8] Gradetsky, V., Rachkov, M., Nandi, G.: Vacuum Pedipulators for Climbing Robots, Proceedings of the 23rd ISIR, Barcelona, 1992, pp. 517-522.

[9] Schreck, G., Bach, Fr.-W., Haferkamp, H.: Remotely controlled inspection and handling systems for decommissioning tasks in nuclear facilities, Remote Techniques for Nuclear Plant, British Nuclear Energy Society, London, UK, 1993, pp. 336-343

Prof. Dr.-Ing. Fr.-W. Bach, J. Seevers, M. Hahn,
Dr. M. Rachkov (Humboldt Research Fellow)
Institut fuer Werkstoffkunde, University of Hanover
D - 30167 Hanover, Appelstrasse 11A
Tel: +49 511 / 762 - 4316
Fax: + 49 511 / 762 - 5245
email: bach@iw.uni-hannover.de

VII. Modelling and Simulation

O. Enge, G. Kielau, P. Maißer
Modelling and Simulation of Discrete Electromechanical Systems

M. Kaltenbacher, F. Lindinger
Software Environment for the Computer Modeling of Magnetomechanical Systems

H. Freudenberg, P. A. Tuan
Modeling and Simulation of a Flexible Shuttle-Robot

H. Hesse, J. Wallaschek
Optimization of the Dynamic Behavior of a Wire Bonder Using the Concept of Mechatronic Function Modules

Modelling and Simulation of Discrete Electromechanical Systems

Olaf Enge, Gerald Kielau, Peter Maißer
Institute of Mechatronics at the Technical University Chemnitz,
Chemnitz, Germany

Abstract: Electromechanical systems can be regarded as physical structures characterized by interaction of electromagnetic fields with inertial bodies. Constitutive equations describing the coupling of multibody dynamics with Kirchhoff's theory define discrete electromechanical systems. The motion of an electromechanical system will be understood as the motion of its representing point in its configuration space. Based on the principle of virtual work the motion equations are Lagrange's equations of second kind. The main goal is to show the automatic generation of these model equations based on a unique approach using a differential-geometric frame.

1   Introduction

Electromechanical systems (EMS) are physical structures characterized by interaction between electromagnetic fields and inertial bodies [5]. The interaction can be expressed by constitutive equations (force law) describing the coupling of Maxwell's theory and mechanics. Constitutive equations describing the coupling between the dynamics of multibody systems (MBS) with a finite degree of freedom and Kirchhoff's theory (as quasi stationary approximation of Maxwell's theory) define discrete EMS. A mathematical description oriented by classical analytical mechanics and completed by some basic concepts and methods of graph theory to characterize topological properties of electrical networks plays a fundamental role for a unique modelling and simulation of discrete EMS. In view of such a unique approach, concepts, definitions and notations of analytical multibody dynamics will be used [1]. That is justified because on the one hand analytical multibody dynamics can be regarded as a possible special case of EMS and on the other hand its concepts and definitions are developed very well to describe other physical structures in the same manner.

2 Electrical Systems

Using some basic concepts and methods of multibody dynamics [2], the electrical system dynamics can be described in the same manner very practically. The application of Lagrange's approach to electrical systems (ES) with lumped parameters is based on the concept of a multipole and its representation by abstract 2-poles as well as Kirchhoff's theory and the principle of virtual work. The *internal* dynamical behaviour of electrical construction elements (resistor, capacitor, inductor etc.) can be represented by external measurements at the terminals using voltmeters and ammeters. This leads to a model of a construction element by using a *black box* endowed with a finite number of terminals. A black box having two terminals to measure voltage and current according to figure 1 is called an *electrical 2-pole* (e.g. a coil, capacitor etc.). The terminals of the voltmeter/ammeter have "+" and "−" symbols. Positive (negative) indication means voltage $V(t) \underset{(<)}{\gtrless} 0$ and current $I(t) \underset{(<)}{\gtrless} 0$. The conditions how to connect volt-/ammeters are defined by a *reference arrow*, so that the initial point of which corresponds to "+" terminal and the final point corresponds to "−" terminal. This way, a "coordinate system" (in the sense of mechanics) is defined at each 2-pole. The 2-pole-element defines a branch of the ES. The measured quantities $V(t)$ and $I(t)$ are called *branch voltage* and *branch current*. Their mathematical relationship in general given by a nonlinear operator equation is called a *branch relation* (constitutive equation).

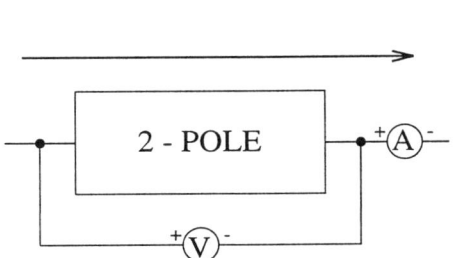

Figure 1:
Measurement instructions at 2-pole

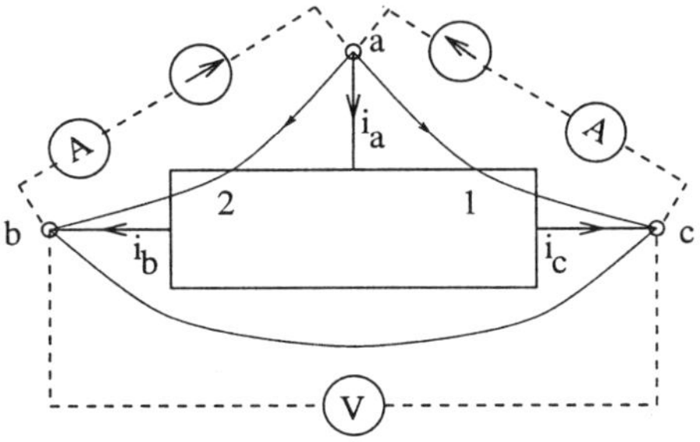

Figure 2:
External measurements at 3-pole
$$i_a = I^1 + I^2, \ i_b = I^2, \ i_c = I^1$$

Definition 1 *An electrical n-pole is defined by a black box with n, $2 \leq n < \infty$, terminals and the following properties:*

a) *Voltages and currents can be measured between any two terminals at the same time;*

b) *Kirchhoff's laws must be satisfied related to the measured voltages and currents.*

If the measurement results are independent on external influences then the black box is called *proper n-pole* else *pseudo n-pole*. The electrical properties of an n-pole can be characterized by the set of all relations (determined by external measurements) between $\binom{n}{2}$ voltages at any two terminals and the n currents through the terminals (figure 2). The $\binom{n}{2}$ external measurements of voltages at an n-pole define a so-called *complete measurement graph* Γ_M. Its nodes represent the terminals of the n-pole, and its branches represent the *fictitious 2-poles*, the voltages of which are measured. Prescribing the voltages in an arbitrary frame $G(\Gamma_M)$, the other voltages in Γ_M are uniquely determined due to Kirchhoff's voltage law. The currents measured in the branches of the frame $G(\Gamma_M)$ determine uniquely the currents through the terminals of the n-pole. Hence, the *electrical properties* of an n-pole can be characterized by (measured) relations between the $n-1$ voltages and the $n-1$ currents in $G(\Gamma_M)$.

Definition 2 *A complete measurement graph* Γ_M *with exactly two nodes and one relation is called an* abstract 2-pole *if it contains at least one of the both variables V, I of the measurement graph.*

Abstract 2-poles are called connected if their graphs have common nodes. Following, an n-pole is represented by $n-1$ abstract 2-poles which are connected tree-like. The abstract 2-poles are in general *pseudo 2-poles*, because its branch relations can also contain variables V, I of other 2-poles of the n-pole. Removing branches of *isolators* $(I \equiv 0)$ from the representing tree of an n-pole and identifying the nodes of *short-circuit elements* $(V \equiv 0)$, one can get a representation of an n-pole by a forest of not more then $n-1$ abstract 2-poles. This forest is called the *pole-graph* of the n-pole.

2.1 Kirchhoff's Theory of Electrical Systems

Definition 3 *An electrical system (ES) is a finite set of galvanically connected electrical multipoles (finite multipole network).*

After defining all pole-graphs, an ES is represented by a finite network of abstract 2-poles with the network graph Γ (containing B branches, N nodes, p components). Let G denote an arbitrary (but fixed) frame of Γ and $H(G)$ the coframe of G in Γ. The corresponding fundamental cut and the fundamental loop matrices are denoted by Q and A, respectively.

Then, *Kirchhoff's laws* become:

$$\sum_{i \in \Gamma} A^i{}_j V_i = 0, \quad j \in H \quad (\text{"voltage law"}), \tag{2.1}$$

$$\sum_{i \in \Gamma} Q_i{}^j I^i = 0, \quad j \in G \quad (\text{"current law"}). \tag{2.2}$$

(2.1) describes the vanishing of the sum of all (oriented) voltages across each fundamental loop $j \in H$. (2.2) describes the vanishing of the sum of all (oriented) currents across each fundamental cut $j \in G$. (Symbols of subgraphs are used simultaneously to denote corresponding index sets.) Both linear equation systems have full rank

$(rank(A) = B - N + p, rank(Q) = N - p)$ and are equivalent to Kirchhoff's voltage law or Kirchhoff's current law, respectively $(b^i{}_j V_i = 0, a^i{}_j I^j = 0; b^i{}_j$: mesh-branch-inzidence-matrix, $a^i{}_j$: node-branch-inzidence-matrix). Together with the B branch relations of the abstract 2-poles (constitutive equations), (2.1) and (2.2) constitute a system of $2B$ equations in order to determine the $2B$ functions $V_i(t)$, $I^i(t)$, $i \in \Gamma$.

A and Q are related to each other by

$$A^i{}_j = \begin{cases} \delta^i{}_j , & i, j \in H \\ -Q_j{}^i , & i \in G, \ j \in H. \end{cases} \tag{2.3}$$

Hence, $A^T Q = 0$.

2.2 Lagrange's Equations for ES in Charge Formulation

The Principle of Virtual Work

Resolving (2.1) and (2.2), *three different sets of generalized coordinates* can be introduced:

- $V_i = Q_i{}^j V_j$, $\quad i \in \Gamma$, $j \in G$, where V_j are the independent variables [3],
- $I^i = A^i{}_j I^j$, $\quad i \in \Gamma$, $j \in H$, where I^j are the independent variables [6],
- and a mixed one of that two forms [7].

In the following, the second form (the so-called "charge-approach") will be presented. (2.2) yields

$$I^j = -Q_i{}^j I^i, \quad i \in H, \ j \in G,$$

and due to (2.3)

$$I^j = A^j{}_\mu i^\mu, \quad \mu \in H, \ j \in \Gamma. \tag{2.4}$$

That means, each current I^j, $j \in \Gamma$, can be represented by a linear combination of coframe currents i^μ, $\mu \in H$. (2.4) is called *mesh-transformation* (MT), it defines the kinematics of the ES.

Generalized coordinates of ES are introduced as follows:
Defining functions

$$\bar{q}^j(t) := \int_0^t I^j(\tau) \, d\tau, \quad \bar{\lambda}_j(t) := \int_0^t V_j(\tau) \, d\tau \tag{2.5}$$

as *charge* and *flux* of the abstract 2-pole j, $j \in \Gamma$, respectively, its branch relation has the form

$$\dot{\bar{\lambda}}_j \equiv V_j = f_j(\ddot{\bar{q}}, \dot{\bar{q}}, \bar{q}, t) \quad \text{or} \quad \bar{\lambda}_j = f_j(\dot{\bar{q}}, \bar{q}, t) \quad \text{or} \quad \bar{q}^j = g_0^j(t), \tag{2.6}$$

where f_j and g_0^j are given, differentiable functions (notation: $\bar{q} := (\bar{q}^1, \ldots, \bar{q}^B)$, $\bar{\lambda} := (\bar{\lambda}_1, \ldots, \bar{\lambda}_B)$). $\bar{q}^j$ denotes the charge which has been moved in (0,t) caused by the current I^j; $\bar{\lambda}_j$ has not necessarily the meaning of "magnetic flux". An abstract 2-pole with the relation $\bar{q}^j = g_0^j(t)$ is called *current generator*, $g_0^j(t)$ denotes its charge source. Hence, the equations (2.6) are the constitutive equations of an ES (in the sense of mechanics).

Due to the mesh-transformation (2.4), a frame G must not contain current generators. Consequently, all current generators belong to the coframe $H(G)$ and decomposite the index set

$$H(G) \;=\; H^* \cup H_0 , \tag{2.7}$$

$$H^* : \text{coframe branches not containing current generators,}$$
$$H_0 : \text{coframe branches containing current generators,}$$

so that (2.4) with (2.5) yields

$$\bar{q}^j = A^j{}_\mu q^\mu + q_0^j(t), \quad q_0^j := A^j{}_\lambda \bar{q}^\lambda \equiv A^j{}_\lambda g_0^\lambda(t), \quad j \in \Gamma, \; \mu \in H^*, \; \lambda \in H_0. \tag{2.8}$$

The set of branch charges $\{\bar{q}^j, j \in \Gamma\}$ is called a *configuration* of the ES. All charges $\bar{q}$ fulfilling Kirchhoff's current law in integrated form $a^i{}_j \bar{q}^j = 0$ at time t determine the set

$$\mathfrak{L}_t := \{\bar{q}^j | \, \bar{q}^j = A^j{}_\mu q^\mu + q_0^j(t); \; q^\mu \in \mathbb{R}\} \tag{2.9}$$

of all *admissible configurations* of the ES at time t.

In the following, latin indices are related to Γ, greek indices to H^*, and the well-known summation convention will be used. Denoting the power of H^* by m, (2.9) defines a 1-1 map of $\mathfrak{L}_t$ to $\mathbb{R}^m$.

Definition 4 *The ES is said to be* holonomic *having the* quasi degree of freedom *m.* $q = (q^\mu, \; \mu \in H^*)$ *is called its* representing point, $q \in \mathbb{R}^m$; $\mathbb{R}^m$ *is called its* configuration space, *the q^μ, $\mu \in H^*$, are called (topologically generated)* generalized coordinates *of the ES.*

The motion of the ES is described by C^2-functions $q = q(t)$. The state of the ES is given by $(\dot{q}, q)$. The ES is called *scleronomic* if $\mathfrak{L}_t$ does not depend on t explicitly, otherwise it is called *rheonomic* ($H_0 \neq \emptyset$).

A *virtual displacement* of an ES is defined by a set of differential increments of charges $\bar{q}^j$ belonging to a variation δq^μ, $\mu \in H^*$, at fixed time t:

$$\{\delta\bar{q}^j | \, \delta\bar{q}^j = A^j{}_\mu \delta q^\mu, \quad j \in \Gamma, \; \delta q^\mu \text{ arbitrary}\}. \tag{2.10}$$

(Because $A^j{}_\mu = \delta^j_\mu$, $j, \mu \in H$, it is $\delta\bar{q}^j = 0 \quad \forall j \in H_0$.) Motion and state of an ES are defined analogously to mechanical systems [2].

Axiom 1 *(Principle of virtual work in view of the charge approach)*
Let the constitutive equations be given as

$$\dot{\lambda}_i \equiv V_i = f_i(\ddot{\bar{q}}, \dot{\bar{q}}, \bar{q}, t), \quad i \in \Gamma \setminus H_0.$$

Then

$$\delta' A := -V_i \delta \bar{q}^i = 0 \quad \forall \delta \bar{q}^i \; virtual \tag{2.11}$$

defines the actual motion of an ES.

Assuming that the constitutive equations satisfy certain necessary and sufficient conditions with respect to the existence of a kinetic potential of first order (Helmholtz-conditions), the axiom 1 yields the Lagrange equations of motion of the ES

$$(\partial_\mu \Lambda)\dot{} - \partial_\mu \Lambda + \partial_\mu D = Q_\mu^{(S)}, \tag{2.12}$$

where $\Lambda(\dot{q}, q, t) := W'_m - W_e - V^h$ is the Lagrangian (W'_m - magnetic coenergy, W_e - electric energy, V^h - generalized potential) and $D(\dot{q}, q, t) := D^{(0)} + D^{(1)}$ denotes the dissipation function of the ES. $Q_\mu^{(S)} := -A^i{}_\mu V_i^{(S)}|_{MT}$ are those voltages in the fundamental circuit μ that cannot or should not be represented by Λ or D.

A Lagrange Model for ES

Let the constitutive equations of an ES be given by

$$\begin{aligned}
V_i^{(L)} &= \dot{\Psi}_i, \quad \Psi_i(\dot{\bar{q}}, t) = L_{ij}(t)\dot{\bar{q}}^j + \Psi_{i0}(t), \\
V_i^{(R)} &= R_{ij}(t)\dot{\bar{q}}^j & \text{(rheolinear resistors)}, \\
V_i^{(C)} &= C_{ij}(t)\bar{q}^j + V_{i0}^{(C)}, \\
V_{i0} &= V_{i0}(t) & \text{(voltage generators)}
\end{aligned}$$

with

$$L_{ij} = L_{ji}, \quad C_{ij} = C_{ji}, \quad \partial_0 R_{[ij]} \equiv 0, \tag{2.13}$$

where $\Psi_{i0}(t)$, $V_{i0}(t)$, $V_{i0}^{(C)}$ (initial voltage of a capacitor) are arbitrary and $\bar{q}^i(t) := g_0^i(t)$, $i \in H_0$.

Hence, the magnetic copotential reads

$$\begin{aligned}
W'_m(\dot{q}, t) &:= \frac{1}{2} A^i{}_\mu A^j{}_\nu L_{ij} \dot{q}^\mu \dot{q}^\nu + A^i{}_\mu [L_{ij}\dot{q}_0^j(t) + \Psi_{i0}(t)]\dot{q}^\mu, \tag{2.14} \\
\dot{q}_0^j(t) &:= A^j{}_\lambda \dot{g}_0^\lambda(t),
\end{aligned}$$

and the electric potential reads

$$W_e(q, t) := \frac{1}{2} A^i{}_\mu A^j{}_\nu C_{ij} q^\mu q^\nu + A^i{}_\mu [C_{ij} q_0^j(t) + V_{i0}(t) + V_{i0}^{(C)}] q^\mu. \tag{2.15}$$

The dissipation function

$$D^{(1)}(\dot q, t) := \frac{1}{2} A^i{}_\mu A^j{}_\nu R_{(ij)} \dot q^\mu \dot q^\nu + A^i{}_\mu R_{ij} \dot q_0^j(t) \dot q^\mu \tag{2.16}$$

and the generalized (gyroscopic) potential

$$V(\dot q, q) := \frac{1}{2} A^i{}_\mu A^j{}_\nu R_{[ij]} q^\mu \dot q^\nu \equiv \frac{1}{2} r_{[\mu\nu]} (q^\mu \dot q^\nu - q^\nu \dot q^\mu)|_{\mu < \nu} \tag{2.17}$$

can be assigned to the rheolinear resistors. Hence, the Lagrange model $\{\Lambda, D\}$ of the ES is

$$
\begin{aligned}
\Lambda(\dot q, q, t) \ :=&\ W'_m - W_e - V \equiv \tfrac{1}{2} A^i{}_\mu A^j{}_\nu [L_{ij}(t) \dot q^\mu \dot q^\nu - C_{ij}(t) q^\mu q^\nu - R_{[ij]} q^\mu \dot q^\nu] + \\
&\ + A^i{}_\mu \{ [L_{ij}(t) \dot q_0^j(t) + \Psi_{i0}(t)] \dot q^\mu - [C_{ij}(t) q_0^j(t) + V_{i0}(t) + V_{i0}^{(C)}] q^\mu \} \\[4pt]
D(\dot q, t) \ :=&\ \tfrac{1}{2} A^i{}_\mu A^j{}_\nu R_{(ij)} \dot q^\mu \dot q^\nu + A^i{}_\mu R_{ij} \dot q_0^j(t) \dot q^\mu.
\end{aligned}
$$

Both functions are quadratic in $\dot q$:

$$\Lambda = \frac{1}{2} g_{\mu\nu} \dot q^\mu \dot q^\nu + g_{\mu 0} \dot q^\mu + \frac{1}{2} g_{00}, \quad D = \frac{1}{2} r_{(\mu\nu)} \dot q^\mu \dot q^\nu + r_\mu \dot q^\mu$$

with

$$
\begin{aligned}
g_{\mu\nu}(t) \ &:=\ l_{\mu\nu}(t) \equiv A^i{}_\mu A^j{}_\nu L_{ij}(t), \\
g_{\mu 0}(q, t) \ &:=\ A^i{}_\mu [L_{ij}(t) \dot q_0^j(t) + \Psi_{i0}(t) + \tfrac{1}{2} A^j{}_\nu R_{[ji]} q^\nu], \\
g_{00}(q, t) \ &:=\ -A^i{}_\mu \{ C_{ij}(t) [A^j{}_\nu q^\nu + 2 q_0^j(t)] + 2[V_{i0}(t) + V_{i0}^{(C)}] \} q^\mu, \\
r_{(\mu\nu)}(t) \ &:=\ A^i{}_\mu A^j{}_\nu R_{(ij)}(t), \qquad r_\mu(t) := A^i{}_\mu R_{ij} \dot q_0^j(t).
\end{aligned}
$$

Then, the Lagrange's equations can be written in their explicit form

$$g_{\mu\nu} \ddot q^\nu + \Gamma_{\mu\alpha\beta} \dot q^\alpha \dot q^\beta + r_{(\mu\nu)} \dot q^\nu + r_\mu = 0$$

$(\mu, \nu \in H^*; \ \alpha, \beta \in H^* \cup \{0\}; \ q^0 = t; \ \dot q^0 = 1)$ with

$$
\begin{aligned}
\Gamma_{\mu\nu\varrho} \ &\equiv\ 0 \text{ for } \nu, \varrho \in H^*, \\
\Gamma_{\mu\varrho 0} = \Gamma_{\mu 0 \varrho} \ &=\ \tfrac{1}{2} A^i{}_\mu A^j{}_\varrho \dot L_{ij} + \tfrac{1}{4} A^i{}_\mu A^j{}_\varrho R_{[ij]}, \\
\Gamma_{\mu 00} \ &:=\ \partial_0 g_{\mu 0} - \tfrac{1}{2} \partial_\mu g_{00}.
\end{aligned}
$$

3 Discrete Electromechanical Systems

The Lagrange formalism is described for electromechanical systems (EMS) [4]. This will be done based on both, analytical mechanics (i.e. its application in multibody dynamics) and the formalism described for electrical systems in section 2. EMS can be regarded as physical structures characterized by interactions of electromagnetic fields with inertial bodies. The interaction is expressed by constitutive equations describing the coupling between Maxwell's theory and mechanics. Constitutive equations, describing the coupling of multibody dynamics with Kirchhoff's theory (as quasi stationary approximation of Maxwell's theory), define discrete EMS. In the following, only such systems will be regarded.

Definition 5 *An EMS is a finite set of physical objects with mechanical and/or electrical multipole-properties interacting among themselves by electrodynamical and/or electro-magneto-mechanical coupling:*

$$\text{multibody dynamics} \cup \text{Kirchhoff's theory.}$$

3.1 Kinematics of Electromechanical Systems

The kinematics of an EMS is defined by the geometric constraints between the rigid bodies and the topology of the electrical network represented by abstract 2-poles. The set $\{\bar{q}^i, \bar{x}^s_k | \ i \in \Gamma, \ s = 1,\ldots,6; \ k = 1,\ldots,K\}$ is called a *configuration* of the EMS, where the $\bar{q}^i$ denote the branch charges (Γ = graph of the 2-pole-network) and the $\bar{x}^s_k$ denote the mechanical coordinates. All branch charges $\bar{q}^i$ and coordinates $\bar{x}^s_k$, which fulfil all constraints of the EMS at given time t (i.e. geometric constraints and Kirchhoff's current law), determine the set

$$\mathfrak{L}_t := \left\{ \bar{q}^i, \bar{x}^s_k \Big| \bar{q}^i = A^i{}_\mu q^\mu + q^i_0(t), \ \bar{x}^s_k = \bar{x}^s_k(q^\kappa, t); \ \begin{matrix} i \in \Gamma, s = 1, ..., 6; k = 1, ..., K; \\ (q^\mu, q^\kappa) \in \mathbb{R}^n, \mu \in H^*, \kappa \in J \end{matrix} \right\} \quad (3.1)$$

of all *admissible configurations* of an EMS at time t. H^* denotes the subset of the coframe $H(G)$ of Γ not containing current generators, J is an index set of mechanical coordinates q^κ; $J \cap H^* = \emptyset$. Convention: In the following, $\mu, \nu, \omega \in H^*$, $\kappa, \lambda, \rho, \in J$, $a, b, c \in H^* \cup J$, $\alpha, \beta \in H^* \cup J \cup \{0\}$, and $q^0 = t$, $\dot{q}^0 = 1$ have to be assumed. $\mathfrak{L}_t$ is a 1-1 map to a cylindrical domain $D \subset \mathbb{R}^n$ of the configuration space, $n := |H^*| + |J|$. n is called quasi degree of freedom of the EMS, and $q = (q^a) = (q^\mu, q^\kappa)$ denotes its representing point. The mesh charges q^μ and the mechanical coordinates q^κ are the generalized coordinates of the EMS. The motion of the EMS is given by C^2-functions $q = q(t)$, and the state of the EMS is given by $(\dot{q}, q)$. The EMS is called *holonomic* if there are no nonintegrable kinematic constraints, otherwise it is called *anholonomic*. The EMS is called *scleronomic* if $\mathfrak{L}_t$ does not depend on t explicitly, otherwise it is called *rheonomic*.

3.2 Constitutive Equations

Let $d\mathfrak{k}(\xi)$ be the applied forces acting on the bodies of the EMS, V_i denote the voltages of abstract 2-poles, and Q_κ and v_μ are the generalized forces and the mesh voltages, respectively. Hence,

$$Q_\kappa := S\partial_\kappa \mathfrak{x} d\mathfrak{k}, \quad v_\mu := A^i{}_\mu V_i|_{MT}. \tag{3.2}$$

Assumtion:
The constitutive equations of an EMS

$$Q_\kappa = Q_\kappa(\dot{q}^a, q^a, t), \quad v_\mu = v_\mu(\ddot{q}^a, \dot{q}^a, q^a, t) \tag{3.3}$$

are given by sufficient smooth functions Q_κ and v_μ.

The simultaneous presence of q^λ and q^ν and their derivatives in these equations indicates the electromechanical interaction. The description of Q_κ and v_μ by state functions $\Omega(\dot{q}, q, t)$ and $D(\dot{q}, q, t)$ is necessary for a unique representation of the motion equations in Lagrange's notation:

$$Q_a^* \equiv \delta_a \Omega - \dot{\partial}_a D \tag{3.4}$$

with

$$Q_a^* := \begin{cases} -v_a \,, & a \in H^* \\ Q_a \,, & a \in J. \end{cases} \tag{3.5}$$

Hence, the structure of the functions Q_a^* and Ω follows (not considering any physical background):

a) $v_\mu = \dot{\psi}_\mu + u_\mu$ with $\psi_\mu = \psi_\mu(\dot{q}^\nu, q^a, t)$, $\dot{\partial}_{[\nu}\psi_{\mu]} = 0$ and $u_\mu = u_\mu(\dot{q}, q, t)$,

b) $\Omega = -\Psi + V$ with $\Psi(\dot{q}^\nu, q^a, t) := \int \psi_\mu d\dot{q}^\mu$, $\quad V(\dot{q}, q, t) := \omega_a(q, t)\dot{q}^a + \omega_0 \equiv \omega_a \dot{q}^a$,

c) $Q_a := \partial_a \Psi + \delta_a V - \dot{\partial}_a D$ with $Q_a := \begin{cases} -u_\mu \,, & a = \mu \in H^* \\ Q_\kappa \,, & a = \kappa \in J \end{cases}$ and $\delta_a V \equiv 2\partial_{[\alpha}\omega_{a]}\dot{q}^\alpha$.

The class representation

$$\begin{aligned} Q_\kappa &= & Q_\kappa^{(0)} &+ Q_\kappa^{(1)} &+ Q_\kappa^{(2)}, \\ -v_\mu &= - \dot{\psi}_\mu &- u_\mu^{(0)} &- u_\mu^{(1)} &- u_\mu^{(2)} \end{aligned} \tag{3.6}$$

with

$$\begin{aligned} \psi_\mu &= \dot{\partial}_\mu \Psi, & u_\mu^{(0)} &= -\partial_\mu \Psi, & Q_\kappa^{(0)} &= \partial_\kappa \Psi, & (Q_a^{(0)} &= \partial_a \Psi), \\ & & u_\mu^{(1)} &= -\delta_\mu V, & Q_\kappa^{(1)} &= \delta_\kappa V, & (Q_a^{(1)} &= \delta_a V), \\ & & u_\mu^{(2)} &= \dot{\partial}_\mu D, & Q_\kappa^{(2)} &= -\dot{\partial}_\kappa D, & (Q_a^{(2)} &= -\dot{\partial}_a D) \end{aligned} \tag{3.7}$$

is a sufficient condition in order to represent Q_κ and v_μ by state functions Ω, D due to (3.4).

3.3 State Functions

The arbitrarly given state functions Ψ, V, D describe the classes K^0, K^1 and K^2 according to (3.7). On the other hand, if the functions Q_a are given by (3.3) and if there is a decomposition $Q_a = Q_a^{(0)} + Q_a^{(1)} + Q_a^{(2)}$ fulfilling the integrability conditions belonging to (3.7), the state functions can be calculated:

the magnetomechanical copotential

$$\Psi := \int\limits_{(0,0,0)}^{(\dot{q}^\mu, q^\mu, q^\kappa)} \psi_\mu d\dot{q}^\mu - u_\mu^{(0)} dq^\mu + Q_\kappa^{(0)} dq^\kappa; \tag{3.8}$$

the generalized electromechanical potential

$$V = \omega_\alpha(q,t)\dot{q}^\alpha \equiv \omega_a \dot{q}^a + \omega_0, \tag{3.9}$$

ω_a, ω_0 defined by a PDE-system [4];

the dissipation function

$$D := \int\limits_{(0,0)}^{(\dot{q}^\mu, \dot{q}^\kappa)} u_\mu^{(2)} d\dot{q}^\mu - Q_\kappa^{(2)} d\dot{q}^\kappa. \tag{3.10}$$

3.4 Kinetics of Electromechanical Systems

The kinetics of EMS is based on the principle of virtual work in Lagrange's notation.

Axiom 2 *The actual motion of an EMS is characterized by the vanishing of the virtual work*

$$\delta' A := -V_i \delta\bar{q}^i + S\delta\mathfrak{x}(d\mathfrak{k} - \ddot{\mathfrak{x}}dm) = 0 \quad \forall\, \delta\bar{q}^i, \delta\mathfrak{x} \; virtual \tag{3.11}$$

at any time t.

(S: Summation over all material points ξ of the EMS, $d\mathfrak{k}$: applied forces arbitrarly distributed, dm: mass element, $\dot{\mathfrak{x}} = \frac{d}{dt}\mathfrak{x}(\xi, q^\kappa, t)$: velocity, $\ddot{\mathfrak{x}} = \frac{d}{dt}\dot{\mathfrak{x}}$: acceleration of ξ related to an inertial frame, V_i: voltages of abstract 2-poles of the representing graph Γ of the electrical network, $\bar{q}^i$: branch charges of Γ.)

Using the kinetic energy of the EMS

$$T(\dot{q}^\lambda, q^\lambda, t) := \frac{1}{2} S\dot{\mathfrak{x}}^2 dm$$

and the generalized forces Q_κ and mesh voltages v_μ

$$S\delta\mathfrak{x}d\mathfrak{k} \equiv S\partial_\kappa\mathfrak{x}d\mathfrak{k}\delta q^\kappa \equiv Q_\kappa\delta q^\kappa, \quad V_i\delta\bar{q}^i \equiv V_iA^i{}_\mu\delta q^\mu \equiv v_\mu\delta q^\mu,$$

(3.11) yields

$$\delta'A = -v_\mu\delta q^\mu + [-(\dot{\partial}_\kappa T)\dot{} + \partial_\kappa T + Q_\kappa]\delta q^\kappa = 0 \quad \forall \, \delta q^\mu, \, \delta q^\kappa.$$

The motion equations of the EMS

$$v_\mu = 0, \quad (\dot{\partial}_\kappa T)\dot{} - \partial_\kappa T = Q_\kappa \tag{3.12}$$

are Kirchhoff's mesh equations and "mechanical" Lagrange's equations. With respect to (3.4) and $\partial_\mu T \equiv 0$, $\dot{\partial}_\mu T \equiv 0$, they read in Lagrange's form

$$(\dot{\partial}_a\Lambda)\dot{} - \partial_a\Lambda + \dot{\partial}_a D = 0, \quad a \in J \cup H^*, \tag{3.13}$$

$$\Lambda := T - \Omega = T(\dot{q}^\lambda, q^\lambda, t) + \Psi(\dot{q}^\nu, q^a, t) - V(\dot{q}, q, t). \tag{3.14}$$

If v_μ, Q_κ are given, Ψ, V und D can be calculated according to (3.8), (3.9) and (3.10). Parts of the generalized forces $v_\mu^{(S)}$, $Q_\kappa^{(S)}$, which cannot or should not be represented by these state functions, have to be taken into account on the right-hand side of (3.13):

$$(\dot{\partial}_a\Lambda)\dot{} - \partial_a\Lambda + \dot{\partial}_a D = Q_a^{(S)}, \tag{3.15}$$

$$Q_a^{(S)} := \begin{cases} -v_\mu^{(S)}, & a = \mu \in H^* \\ Q_\kappa^{(S)}, & a = \kappa \in J. \end{cases}$$

Λ and D together with $Q_a^{(S)}$ determine the motion equations of the EMS. The formalism to get these equations is independent of the EMS itself. $\{\Lambda, D, Q_a^{(S)}\}$ describes a mathematical model of the EMS. Due to (3.15) a trivial model $\{T, 0, Q\}$ always exists. $\{\Lambda, D\} := \{\Lambda, D, 0\}$ is called a *Lagrange model* of the EMS.

3.5 The Structure of Lagrange's Equations of Motion

Starting from the Lagrange model $\{\Lambda, D\}$

$$\Lambda = \tfrac{1}{2}g_{\alpha\beta}(q, t)\dot{q}^\alpha\dot{q}^\beta, \quad D = \tfrac{1}{2}s_{\alpha\beta}(q, t)\,\dot{q}^\alpha\dot{q}^\beta$$

with

$$\begin{aligned}
(\dot{\partial}_a\Lambda)\dot{} &= (g_{\alpha\beta}\dot{q}^\beta)\dot{} = g_{ab}\ddot{q}^b + \partial_\alpha g_{\alpha\beta}\dot{q}^\alpha\dot{q}^\beta, \\
\partial_a\Lambda &= \tfrac{1}{2}\partial_a g_{\alpha\beta}\dot{q}^\alpha\dot{q}^\beta, \\
\dot{\partial}_a D &= s_{\alpha\beta}\dot{q}^\beta
\end{aligned}$$

and

$$\Gamma_{a\alpha\beta} := \frac{1}{2}(\partial_\alpha g_{a\beta} + \partial_\beta g_{a\alpha} - \partial_a g_{\alpha\beta}),$$

the Lagrange's equations in explicit form read:

$$g_{ab}(q,t)\ddot{q}^b + \Gamma_{a\alpha\beta}(q,t)\dot{q}^\alpha\dot{q}^\beta + s_{a\beta}\dot{q}^\beta = 0. \tag{3.16}$$

Here is

$$g_{ab} := \dot{\partial}_a\dot{\partial}_b\Lambda = g_{ba},$$

and with $\Lambda = T + \Psi - V$ follows

$$(g_{ab}) = \begin{bmatrix} g_{\mu\nu} & 0 \\ 0 & g_{\kappa\lambda} \end{bmatrix}, \quad \begin{aligned} g_{\mu\nu} &= \dot{\partial}_\mu\dot{\partial}_\nu\Psi = l_{\mu\nu} = A^i{}_\mu A^j{}_\nu L_{ij}, \\ g_{\kappa\lambda} &= \dot{\partial}_\kappa\dot{\partial}_\lambda T = \sum_k [m\, u_{ik}\, u_{i\lambda} + (1-s_\kappa)(1-s_\lambda)\Theta^{ij}\Omega_{ik}\Omega_{j\lambda}]. \end{aligned}$$

Hence, g_{ab} is a direct sum of the two matrices $g_{\mu\nu}$ and $g_{\kappa\lambda}$. Because of the quasi stationary approximation of Maxwell's theory $(\partial_\mu\Psi \equiv 0)$, g_{ab} is independent of q^μ. If $g_{\mu\nu}$ is a regular matrix, g_{ab} is regular, too ($g_{\kappa\lambda}$ is regular because of the positive definiteness of the kinetic energy T). Assuming $g_{\mu\nu}$ is regular, $g_{ab}(q,t)$ defines (in general) a time-dependent Riemannian metric in $\mathbb{R}^n$. The $\Gamma_{abc}(q,t)$ are the time-dependent Christoffel symbols of first kind.

It is

$$\begin{aligned} \Gamma_{\mu\nu\lambda} &= \tfrac{1}{2}\partial_\lambda g_{\mu\nu} = \tfrac{1}{2}\partial_\lambda l_{\mu\nu}, & \Gamma_{\kappa\nu\omega} &= -\tfrac{1}{2}\partial_\kappa g_{\nu\omega} = -\tfrac{1}{2}\partial_\kappa l_{\nu\omega}, \\ \Gamma_{\mu\lambda\varrho} &\equiv 0, \quad \Gamma_{\mu\nu\omega} \equiv 0, \quad \Gamma_{\kappa\lambda\nu} \equiv 0, & \mu,\nu,\omega &\in H^*; \kappa,\lambda,\varrho \in J. \end{aligned}$$

Using this, (3.16) yields

$$\begin{aligned} g_{\mu\nu}\ddot{q}^\nu & & +\partial_\lambda g_{\mu\nu}\dot{q}^\lambda\dot{q}^\nu & & +(2\Gamma_{\mu b0} + s_{\mu b})\dot{q}^b & & +\Gamma_{\mu 00} & +s_{\mu 0} & = 0 \\ g_{\kappa\lambda}\ddot{q}^\lambda & +\Gamma_{\kappa\lambda\varrho}\dot{q}^\lambda\dot{q}^\varrho & -\tfrac{1}{2}\partial_\kappa g_{\nu\omega}\dot{q}^\nu\dot{q}^\omega & & +(2\Gamma_{\kappa b0} + s_{\kappa b})\dot{q}^b & & +\Gamma_{\kappa 00} & +s_{\kappa 0} & = 0. \end{aligned} \tag{3.17}$$

This structure of Lagrange's equations follows from (3.16) with respect to a partitioning of the generalized coordinates $q = (q^\mu, q^\kappa)$ and the special representation of the matrix g_{ab} as direct sum of $g_{\mu\nu}$ and $g_{\kappa\lambda}$ due to the quasi stationary approximation of Maxwell's theory.

The Christoffel symbols appearing in (3.17) can be calculated as:

$$\Gamma_{\kappa\lambda\varrho} = (1-s_\lambda)\sum_{k=\varrho}^{K}\varepsilon_i{}^{qr}[\underset{k}{m}\,\delta^{ij}\underset{k}{u}_{j\kappa}\underset{k}{\Omega}_{q\lambda}\underset{k}{u}_{r\varrho} + (1-s_\kappa)(1-s_\varrho)\underset{k}{\vartheta}^{ij}\underset{k}{\Omega}_{r\kappa}\underset{k}{\Omega}_{j\lambda}\underset{k}{\Omega}_{q\varrho}],$$

$$\Gamma_{\mu\nu0} = \frac{1}{2}\partial_0 l_{\mu\nu} - \partial_{[\nu}\omega_{\mu]},$$

$$\Gamma_{\mu\lambda0} = \frac{1}{2}\partial_\lambda(\psi_{\mu0} - \omega_\mu),$$

$$\Gamma_{\mu00} = \partial_0\psi_{\mu0} + \partial_\mu\omega_0,$$

$$\Gamma_{\kappa\nu0} = -\Gamma_{\nu\kappa0},$$

$$\Gamma_{\kappa\lambda0} = \frac{1}{2}\partial_0\partial_\kappa\partial_\lambda T_2 - \partial_{[\lambda}\partial_{\kappa]}T_1,$$

$$\Gamma_{\kappa00} = \partial_0\partial_\kappa T_1 - \partial_\kappa(T_0 + \Psi_0 - V_0).$$

Some terms in (3.17) have a simple physical meaning. $g_{\mu\nu}\ddot{q}^\nu$ describes induced voltages as a result of changing currents and $\partial_\lambda g_{\mu\nu}\dot{q}^\lambda\dot{q}^\nu$ is the Coriolis voltage. The term $\frac{1}{2}\partial_\kappa g_{\nu\omega}\dot{q}^\nu\dot{q}^\omega$ describes generalized forces with electrical origin (Lorentz forces).

A special EMS (with inductivities, permanent magnets, resistors, current and voltage generators) with the constitutive equations

$$\Psi_i = L_{ij}\dot{q}^j + \Psi_{i0}, \qquad i,j \in H,$$

$$L_{ij} = L_{ij}(q^\kappa), \qquad \Psi_{i0} = \Psi_{i0}(q^\kappa),$$

$$V^{(R)} = R_{ij}\dot{q}^j,$$

$$V_{i0} = V_{i0}(t)$$

and the force laws

$$\underset{r}{\mathfrak{k}} = \underset{k}{K}^{(i)}e_{(i)} = \underset{r}{K}^i\mathfrak{E}_i, \qquad \text{"mechanical" forces,}$$

$$Q_\kappa = -\delta_\kappa(\Psi^* - V^*), \qquad \text{"electromagnetic" forces,}$$

has Lagrange's equations of the following form:

$$A^i{}_\mu A^j{}_\nu[L_{ij}\ddot{q}^\nu + \partial_\lambda L_{ij}\dot{q}^\lambda\dot{q}^\nu] + A^i{}_\mu\partial_\kappa[L_{ij}\dot{q}_0^j(t) + \Psi_{i0}]\dot{q}^\kappa + A^i{}_\mu V_{i0}(t)$$

$$= -A^i{}_\mu A^j{}_\nu R_{ij}\dot{q}^\nu - A^i{}_\mu R_{ij}\dot{q}_0^j(t)$$

$$g_{\kappa\lambda}\ddot{q}^\lambda + \Gamma_{\kappa\lambda\varrho}\dot{q}^\lambda\dot{q}^\varrho - \frac{1}{2}A^i{}_\mu A^j{}_\nu\partial_\kappa L_{ij}\dot{q}^\mu\dot{q}^\nu - A^i{}_\mu\partial_\kappa[L_{ij}\dot{q}_0^j(t) + \Psi_{i0}]\dot{q}^\mu$$

$$= \sum_{k=\kappa}^{K}[\underset{k}{K}^i\underset{k}{u}_{i\kappa} + \underset{k}{M}^i\underset{k}{\Omega}_{i\kappa}] + \frac{1}{2}\partial_\kappa L_{ij}\dot{q}_0^i(t)\dot{q}_0^j(t) + \partial_\kappa\Psi_{i0}\dot{q}_0^i(t).$$

4 Examples

Generalized Electric Machine

The unique Lagrange approach to dynamic simulation of discrete EMS will be shown shortly by the example of the generalized electric machine.

An idealized 2-pole 2-phase-machine is called generalized electric machine (figure 3). The mechanical submodel consists of two bodies connected by a rotational joint.

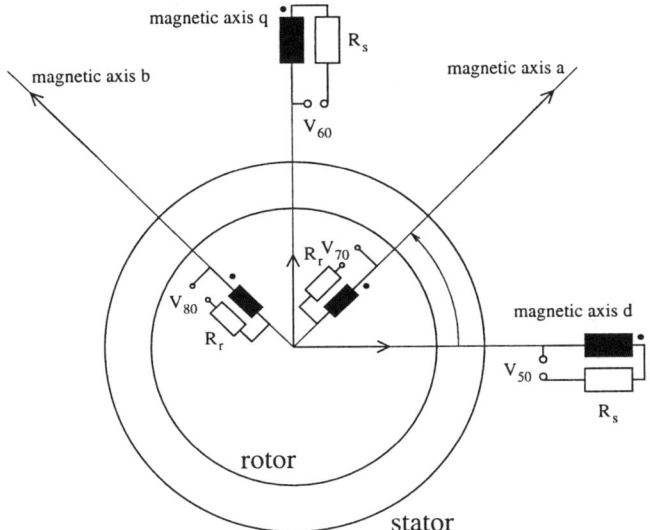

Figure 3: Generalized electric machine

Both, rotor and stator should have two lumped inductances, displaced by $\pi/2$, which summarize all inductances of rotor and stator respectively. The quasi stationary approximation of Maxwell's theory is used. The magnetomechanical interaction is defined only by field distribution in the air gap. The magnetic material of rotor and stator should have a linear $\mathfrak{B}$-$\mathfrak{H}$-curve and no saturation. The structure graph consists of four fundamental loops, each of them containing an inductance, a voltage generator and a resistor (figure 4). The relative angle between rotor and stator is the only mechanical coordinate. The four charges in the fundamental loops of the structure graph are the generalized electrical coordinates.

The physical model yields the topology with the fundamental loop matrix

$$A^i{}_\mu = \delta^{i-4}_\mu, \quad i \in G, \mu \in H, H = H^*, H_0 = \emptyset, |H^*| = 4.$$

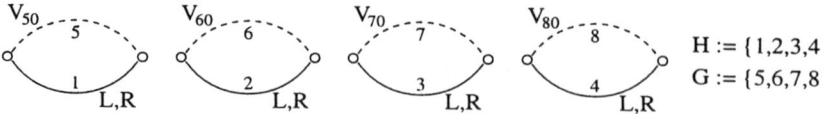

Figure 4: Graph of the generalized electric machine

The generalized coordinates are denoted by

$$q \equiv (q^a) \equiv (q^\mu; q^\kappa) \equiv (q^{ds}, q^{qs}, q^{ar}, q^{br}; \varphi) \equiv (q^1, q^2, q^3, q^4; q^5)$$

(q^μ: charge in fundamental loop μ), and $\dot{q}^a$ are the generalized velocities of the electric machine. The pseudo degree of freedom is five.

The constitutive equations for a model of a smooth-air-gap machine read

$$
\begin{array}{llll}
\Psi_i & = & L_{ij}I^j, & L_{ij} = L_{ij}(\varphi) = L_{ji}; \\
M_{el} & = & \frac{1}{2}L'_{ij}(\varphi)I^iI^j, & L'(\varphi) = \frac{\partial L}{\partial \varphi}; \\
V_i^{(R)} & = & R_sI^i, & i = 1, 2, \quad R_s = const.; \\
V_i^{(R)} & = & R_rI^i, & i = 3, 4, \quad R_r = const.; \\
V_{i0} & = & V_{i0}(t), & i = 5, 6, 7, 8; \\
M_L & = & M_L(\dot{\varphi}, \varphi, t) & = \text{ loading torque}, \quad M_r = -k\dot{\varphi} \quad (k > 0);
\end{array}
$$

$$
(L_{ij}) = \begin{pmatrix}
\bar{a} & 0 & \bar{c}\cos\varphi & -\bar{c}\sin\varphi \\
0 & \bar{b} & \bar{d}\sin\varphi & \bar{d}\cos\varphi \\
\bar{c}\cos\varphi & \bar{d}\sin\varphi & e + f\cos 2\varphi & -f\sin 2\varphi \\
-\bar{c}\sin\varphi & \bar{d}\cos\varphi & -f\sin 2\varphi & e - f\cos 2\varphi
\end{pmatrix}
$$

with the simplifying assumtions

$$\bar{a} = \bar{b} = L_s, \quad \bar{c} = \bar{d} = M, \quad e = L_r, \quad f = 0$$

(L_r, L_s - self-inductances of the rotor and the stator; M - mutual inductance).

Using this approach, the simulation tool **alaska** is able to generate automatically the complete system of motion equations of the generalized electric machine. In this case, these are five differential equations of second order. Hence, various investigations of the dynamics of different rotational electric machines, like induction or synchronous drives, even DC-machines, can be done. The unique Lagrange approach always guarantees the correct results with respect to generation of motion equations.

Other Applications

Some other applications, which have been investigated using the simulation tool **alaska**, should be shortly mentioned. In the case of the MAGLEV Transrapid, the main goal was to study some aspects of controlling electromagnets which are unstable in the opened loop. The carrying and tracking system is based on the attraction between the electromagnets in the vehicle and the ferromagnetic material in the track. It works contactlessly. This MAGLEV-model consists of five rigid bodies, the cabin and four suspension bodies. The cabin and the suspension bodies are coupled by springs. One magnet to carry the vehicle and one to track it are fixed at each suspension body. The drive is realized by stator windings in the track producing a travelling field (principle of linear synchronous motor). The parameters of the model (masses, stiffnesses etc.) are comparable to the technical data of the Transrapid TR06/TR07. The electrical network of this example consists of 28 branches (two branches for each magnet, six branches for drive windings on each side of the track). The degree of freedom of the mechanical subsystem is 30, the quasi degree of freedom of the complete system is 44. The control of the distance between the magnets and the track is implemented by control of the current of each magnet separately. PD-, PID- and state controls have been used in the entire model of the MAGLEV.

Another example is that of a planar motor driven directly. Some types of such drives with different constructions and working principles are investigated at our institute. One of the planar DC-motors, e.g., consists of a stator with permanent magnets and a slide with inductivities. The driving force is produced due to Lorentz's law. Another one is a hybrid stepper motor. Its stator is a comb-like structured ferromagnetic plate being magnetically passive. The permanent magnets and inductivities of the slide are arranged in such a way that the magnetic flux can be increased and decreased properly. The reluctance principle is used to produce the force in every motion direction.

In some special cases simplifications usually done in the theory of electrical machines can cause a mathematical problem. For instance, modelling of a 3-phase-machine with the same assumptions like in section 4 yields a singular matrix of inductivities. That's why the metric of the whole EMS is not regular and, if the simulation tool uses explicit integration codes only (like **alaska**), a new model of the 3-phase-machine has to be designed. This can bo done using a linear transformation with respect to generalized electrical coordinates.

The Lagrange approach allows to describe the "mechanical" and "electrical" subsystems of every discrete EMS and the electro-magneto-mechanical interactions between them in a unique way. Then, a coupling of different simulation tools can be avoided.

References

[1] Enge, O., Kielau, G., Maißer, P.: *Dynamiksimulation elektromechanischer Systeme*. VDI-Fortschritt-Berichte, Reihe 20: Rechnerunterstützte Verfahren, Heft 166. VDI-Verlag, Düsseldorf, to appear.

[2] Maißer, P.: Analytische Dynamik von Mehrkörpersystemen. *ZAMM*, 68(10):463–481, 1988.

[3] Maißer, P., Steigenberger, J.: Zugang zur Theorie elektromechanischer Systeme mittels klassischer Mechanik, Teil 1: Elektrische Systeme in Ladungsformulierung. *Wissenschaftliche Zeitschrift TH Ilmenau*, 20(6):105–123, 1974.

[4] Maißer, P., Steigenberger, J.: Lagrange-Formalismus für diskrete elektromechanische Systeme. *ZAMM*, 59:717–730, 1979.

[5] Miu, D.K.: *Mechatronics - Electromechanics and Contromechanics*. Springer-Verlag, New York-Berlin, 1993.

[6] Steigenberger, J., Maißer, P.: Zugang zur Theorie elektromechanischer Systeme mittels klassischer Mechanik, Teil 2: Elektrische Systeme in Flußformulierung. *Wissenschaftliche Zeitschrift TH Ilmenau*, 22(3):157–163, 1976.

[7] Steigenberger, J., Maißer, P.: Zugang zur Theorie elektromechanischer Systeme mittels klassischer Mechanik, Teil 3: Elektrische Systeme in gemischter Formulierung. *Wissenschaftliche Zeitschrift TH Ilmenau*, 22(4):123–139, 1976.

Adresses of the authors

Dipl.-Ing. Olaf Enge
Dr. rer. nat. Gerald Kielau
Prof. Dr. sc. nat. Peter Maißer
Institute of Mechatronics at the Technical University Chemnitz
Reichenhainerstraße 88
D-09126 Chemnitz, Germany

Tel.:   +49 371 4671
Fax: +49 371 4669
e-mail: O.Enge@IfM.TU-Chemnitz.de
        G.Kielau@IfM.TU-Chemnitz.de
        P.Maisser@IfM.TU-Chemnitz.de

Software Environment for the Computer Modeling of Magnetomechanical Systems

Manfred Kaltenbacher, Franz Lindinger
Institut für Elektrische Meßtechnik, Universität Linz,
A-4040 Linz, Austria

Abstract: This paper presents a new numerical technique for the computer simulation of magnetomechanical systems taking fully into account the reaction of movement of structural parts on the electromagnetic field. The wellknown problem of mesh distortion due to moving parts which occurs in finite element techniques has been overcome by utilizing a coupled Finite-Element-Boundary-Element formulation, which is introduced here. The use of an explicit time step algorithm allows the direct coupling of magnetical and mechanical quantities in one model. To efficiently handle this complex calculation scheme, a special preprocessing software has been developed. Therewith, the specifications of different materials, magnetical and mechanical boundary conditions, electrical current and voltage sources as well as mechanical loads are already performed at the geometry level and not on the basis of the finite element mesh, as it is usual.

1 Introduction

Nowadays process automation often asks for high performance electromechanical sensors and actuators. The electromechanical sensors and actuators considered in this paper are restricted to magnetomechanical transducers. The transducing mechanism of these components is based on the coupling of mechanical and magnetical fields. The computer modeling of the dynamic behaviour of such magnetomechanical components is of great interest for modern transducer development since it allows an optimization of the devices under consideration and, furthermore, reveals their design limits without fabricating numerous prototypes.

Since the complexity of these magnetomechanical systems with respect to their geometric shape, technical features and underlying physics is high, the precise computer modeling of such devices requires the application of numerical calculation schemes such as the Finite Element Method (FEM) and the Boundary Element Method (BEM). In the case of magnetomechanical transducers, especially in the case of magnetomechanical actuators, a modeling scheme must allow to handle the direct and full coupling of the mechanical and electromagnetical field quantities. The analysis schemes known up to now are not able to handle this coupling in a direct manner but only by iterative switching between the solutions of the two separated systems, i.e. the solution for the mechanical field and electromagnetical field. The calculation scheme introduced in this paper, however, allows the simultaneous solution of the fully coupled problem. The theory of magnetomechanical systems and the related numerical calculation scheme are described in chapters 2 and 3.

To apply the above mentioned numerical calulation schemes to magnetomechanical systems, it is furthermore necessary to perform the difficult task of setting up an appropriate computer model of the device to be analysed. This preprocessing includes geometric modeling, meshing of the calculation area, definition of boundary conditions and loads. The dedicated preprocessing asks for the following special features:

- simultaneous handling of magnetical _and_ mechanical quantities
- specification of material exhibiting magnetical _and_ mechanical features
- simultaneous definition of magnetical _and_ mechanical boundary conditions
- definition of electrical current and voltage sources as well as mechanical loads

Since available Computer Aided Engineering (CAE) software does not provide the above features, an object oriented software tool has been developed which allows the geometric modeling and all further preprocessing for the computer simulation of magnetomechanical systems in an unified environment (chapter 4).

2 Theory of Magnetomechanical Systems

The main equations describing the magnetical part of magnetomechanical systems can be derived from Maxwell equations [1] by neglecting the displacement current $\partial \vec{D}/\partial t$, which has just to be considered in electromagnetic wave propagation.

$$\nabla \times \vec{H} = \vec{J} \tag{2.1}$$

$$\nabla \times \vec{E} = -\frac{\partial \vec{B}}{\partial t} \tag{2.2}$$

$$\nabla \cdot \vec{B} = 0 \tag{2.3}$$

$$\vec{B} = \mu \vec{H} \tag{2.4}$$

$\vec{H}$ intensity of magnetic field
$\vec{J}$ current density
$\vec{E}$ intensity of electric field
$\vec{B}$ magnetic induction
μ permeability.

According to equation 2.3, the magnetic field is always solenoidal and consequently the magnetic induction $\vec{B}$ can be represented as the curl of a vector potential $\vec{A}$.

$$\vec{B} = \nabla \times \vec{A} \tag{2.5}$$

To guarantee the uniqness of the vector potential $\vec{A}$, the divergence of $\vec{A}$ is set to zero, which is known as _Columb's gauge_ [2]. By introducing the magnetic vector potential $\vec{A}$, equation 2.2 can be expressed as

$$\nabla \times \left(\vec{E} + \frac{\partial \vec{A}}{\partial t}\right) = 0. \tag{2.6}$$

Thus, the field of the vector $\vec{E} + \partial\vec{A}/\partial t$ is irrotational and can be set equal to the gradient of a scalar function V

$$\vec{E} = -\nabla V - \frac{\partial \vec{A}}{\partial t}. \tag{2.7}$$

Equations 2.1 - 2.7 and the generalized Ohm's law [3],

$$\vec{J} = \gamma(\vec{E} + \vec{v} \times \vec{B}) \tag{2.8}$$

γ conductivity
$\vec{v}$ velocity of moving parts

constitute the partial differential equation for the vector potential $\vec{A}$, capable to fully describe magnetical systems with stationary and moving parts.

$$\nabla \times \left(\frac{1}{\mu}\nabla \times \vec{A}\right) = -\gamma\frac{\partial \vec{A}}{\partial t} - \gamma\nabla V + \gamma\vec{v} \times (\nabla \times A) \tag{2.9}$$

The velocity $\vec{v}$ can be expressed by the partial time derivation of the mechanical displacement vector $\vec{u}$ with respect to the time variable t. Thus, the term $\gamma\vec{v} \times (\nabla \times \vec{A})$ represents a coupling between the magnetical and mechanical part of magnetomechanical systems. This coupling is due to the mechanical movement of conductors in a magnetic field.

On the other hand, the general equations for linear elastodynamic systems [4] can be expressed in a compact form by using the Nabla operator

$$\frac{E}{2(1+\nu)}\left((\nabla \cdot \nabla)\vec{u} + \frac{1}{1-2\nu}\nabla(\nabla \cdot \vec{u})\right) + \vec{f}_V = \rho\frac{\partial^2 \vec{u}}{\partial t^2} \tag{2.10}$$

$\vec{u}$ vector of mechanical displacement
$\vec{f}_V$ volume force
E modulus of elasticity
ν Poisson's ratio
ρ density.

The volume force f_V due to a magnetic field is given [3]

$$\vec{f}_V = \vec{J} \times \vec{B} - \frac{1}{2}(\vec{H} \cdot \vec{H})\nabla\mu \tag{2.11}$$

By using the vector potential $\vec{A}$ we obtain

$$\vec{f_V} = \left(-\gamma\frac{\partial\vec{A}}{\partial t} - \gamma\nabla V + \gamma\vec{v} \times (\nabla \times A)\right) \times (\nabla \times \vec{A})$$

$$- \frac{1}{2}\left((\frac{1}{\mu}\nabla \times \vec{A}) \cdot (\frac{1}{\mu}\nabla \times \vec{A})\right)\nabla\mu. \tag{2.12}$$

Equation 2.12 represents a further coupling between the mechanical and magnetical part of magnetomechanical systems. This coupling is due to mechanical forces produced by currents in a magnetic field as well as mechanical forces between materials with different permeability.

Due to the physical complexity, a numerical calculation scheme such as the Finite Element Method (FEM) or the Boundary Element Method (BEM) has to be applied for computer modeling of these devices. When adapting such numerical calculation methods to magnetomechanical systems (equations 2.9, 2.10 and 2.12), the following problems arise:

- By using FEM, the moving parts cause mesh distortion and, therefore, introduce a geometric nonlinearity. In addition, if the displacement exceeds the element size, a new meshing of the whole simulation area is required.

- Due to the velocity term in equation 2.9, which has to be expressed by the partial derivation of the displacement vector $\vec{u}$ with respect to the time variable t, a multiplication of the two required variables $\vec{A}$ and $\partial\vec{u}/\partial t$ arises, so that equation 2.9 gets bilinear.

- The volume force f_V in equation 2.10 is a nonlinear term with respect to the magnetic vector potential $\vec{A}$.

3 FEM-BEM Formulation of Magnetomechanical Systems

Applying the FEM-formulation to equation 2.9, leads to the FE-matrix equation [5]

$$\mathbf{L}\{\dot{A}\} + \mathbf{P}\{A\} - \mathbf{C}\{q\} = \{Q\} \tag{3.1}$$

$\mathbf{L}$	conductivity matrix
$\mathbf{P}$	permeability matrix
$\mathbf{C}$	boundary integral matrix of the FEM regions
$\{A\}$	unknown nodal magnetic vector potentials
$\{\dot{A}\}$	unknown nodal time derivatives of magnetic vector potential
$\{q\}$	unknown nodal normal derivatives of magnetic vector potential
$\{Q\}$	vector due to current sources.

Applying the BE-method to a region, which does not contain electrical currents, yields the following BE-matrix equation [5]

$$\mathbf{H}\{A\} - \mathbf{G}\{q\} = \{0\} \tag{3.2}$$

$\mathbf{H}, \mathbf{G}$ boundary element matrices
$\{A\}$ unknown nodal magnetic vector potentials
$\{q\}$ unknown nodal normal derivatives of magnetic vector potential.

The above described problem of distortion of the finite element mesh due to moving parts can be avoided by using a coupled FEM-BEM formulation. Therewith, a total separation of stationary and moving parts can be achieved, as shown in Fig. 3.1. If the nodal vector $\{A\}$ is splitted analogous to [6] into inner degrees of freedom $\{A_{11}\}$, $\{A_{22}\}$ and degrees of freedom along the boundaries $\{A_{1\Gamma_1}\}$, $\{A_{2\Gamma_2}\}$, we obtain the modified FEM- and BEM-matrix equations, according to the different calculation areas.

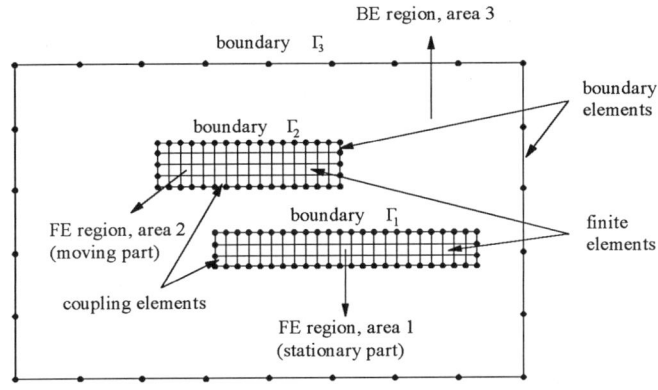

Figure 3.1: Principle of FEM-BEM Discretization

FEM formulation for stationary parts (area 1)

$$
\begin{pmatrix} \mathbf{L}_{11} & \mathbf{L}_{1\Gamma_1} \\ \mathbf{L}_{\Gamma_1 1} & \mathbf{L}_{\Gamma_1\Gamma_1} \end{pmatrix} \begin{pmatrix} \{\dot{A}_{11}\} \\ \{\dot{A}_{1\Gamma_1}\} \end{pmatrix} + \begin{pmatrix} \mathbf{P}_{11} & \mathbf{P}_{1\Gamma_1} \\ \mathbf{P}_{\Gamma_1 1} & \mathbf{P}_{\Gamma_1\Gamma_1} \end{pmatrix} \begin{pmatrix} \{A_{11}\} \\ \{A_{1\Gamma_1}\} \end{pmatrix}
$$
$$
- \begin{pmatrix} 0 & 0 \\ 0 & \mathbf{C}_1 \end{pmatrix} \begin{pmatrix} \{0\} \\ \{q_{1\Gamma_1}\} \end{pmatrix} = \begin{pmatrix} \{Q_{11}\} \\ \{Q_{1\Gamma_1}\} \end{pmatrix} \qquad (3.3)
$$

FEM formulation for moving parts (area 2)

$$
\begin{pmatrix} \mathbf{L}_{22} & \mathbf{L}_{2\Gamma_2} \\ \mathbf{L}_{\Gamma_2 2} & \mathbf{L}_{\Gamma_2\Gamma_2} \end{pmatrix} \begin{pmatrix} \{\dot{A}_{22}\} \\ \{\dot{A}_{2\Gamma_2}\} \end{pmatrix} + \begin{pmatrix} \mathbf{P}_{22} & \mathbf{P}_{2\Gamma_2} \\ \mathbf{P}_{\Gamma_2 2} & \mathbf{P}_{\Gamma_2\Gamma_2} \end{pmatrix} \begin{pmatrix} \{A_{22}\} \\ \{A_{2\Gamma_2}\} \end{pmatrix}
$$
$$
- \begin{pmatrix} 0 & 0 \\ 0 & \mathbf{C}_2 \end{pmatrix} \begin{pmatrix} \{0\} \\ \{q_{2\Gamma_2}\} \end{pmatrix} = \begin{pmatrix} \{Q_{22}\} \\ \{Q_{2\Gamma_2}\} \end{pmatrix} \qquad (3.4)
$$

BEM formulation (area 3)

$$
\begin{pmatrix} \{0\} \\ \{0\} \\ \{0\} \end{pmatrix} = \begin{pmatrix} \mathbf{H}_{\Gamma_1\Gamma_1} & \mathbf{H}_{\Gamma_1\Gamma_2} & \mathbf{H}_{\Gamma_1\Gamma_3} \\ \mathbf{H}_{\Gamma_2\Gamma_1} & \mathbf{H}_{\Gamma_2\Gamma_2} & \mathbf{H}_{\Gamma_2\Gamma_3} \\ \mathbf{H}_{\Gamma_3\Gamma_1} & \mathbf{H}_{\Gamma_3\Gamma_2} & \mathbf{H}_{\Gamma_3\Gamma_3} \end{pmatrix} \begin{pmatrix} \{A_{3\Gamma_1}\} \\ \{A_{3\Gamma_2}\} \\ \{A_{3\Gamma_3}\} \end{pmatrix}
$$

$$- \begin{pmatrix} \mathbf{G}_{\Gamma_1\Gamma_1} & \mathbf{G}_{\Gamma_1\Gamma_2} & \mathbf{G}_{\Gamma_1\Gamma_3} \\ \mathbf{G}_{\Gamma_2\Gamma_1} & \mathbf{G}_{\Gamma_2\Gamma_2} & \mathbf{G}_{\Gamma_2\Gamma_3} \\ \mathbf{G}_{\Gamma_3\Gamma_1} & \mathbf{G}_{\Gamma_3\Gamma_2} & \mathbf{G}_{\Gamma_3\Gamma_3} \end{pmatrix} \begin{pmatrix} \{q_{3\Gamma_1}\} \\ \{q_{3\Gamma_2}\} \\ \{q_{3\Gamma_3}\} \end{pmatrix} \tag{3.5}$$

The FEM-BEM coupling is performed by introducing the continuity relations which are valid at the boundaries Γ_1 and Γ_2. These boundaries are common to FEM and BEM regions.

$$A_{3\Gamma_1} = A_{1\Gamma_1} \tag{3.6}$$
$$A_{3\Gamma_2} = A_{2\Gamma_2} \tag{3.7}$$
$$q_{3\Gamma_1} = -q_{1\Gamma_1} \tag{3.8}$$
$$q_{3\Gamma_2} = -q_{2\Gamma_2} \tag{3.9}$$

This coupling also guarantees the uniqness of the solution of equations 3.3 - 3.5. Applying the FE-formulation to equation 2.10 leads to the wellknown matrix equation for the mechanical quantities [7].

$$\mathbf{M}\{a\} + \mathbf{C}\{v\} + \mathbf{K}\{u\} = \{F\} \tag{3.10}$$

$\mathbf{M}$ mass matrix
$\mathbf{C}$ damping matrix (damping of mechanical systems can be introduced by the method of Rayleigh [7]
$\mathbf{K}$ stiffness matrix
$\{F\}$ force vector
$\{a\}$ unknown nodal accelerations
$\{v\}$ unknown nodal velocities
$\{u\}$ unknown nodal displacements

The remaining two problems, (i) volume force $\vec{f}_V$ being a nonlinear term in respect to $\vec{A}$ and, (ii) the multiplication of variables $\vec{A}$ and $\partial \vec{u}/\partial t$, can be avoided using an explicit time step algorithm [7] for the time discretization of the matrix equations. Therefore, the force vector $\{F\}$ in equation 3.10 as well as the permeability matrices $\mathbf{P}_{22}, \mathbf{P}_{2\Gamma_2}, \mathbf{P}_{\Gamma_2 2}$ and $\mathbf{P}_{\Gamma_2\Gamma_2}$ in equation 3.4 can be evaluated for time-step $n+1$ by using the solutions for the magnetic vector potential and the velocity calculated at time-step n. According to Dalquist's Theorem [8] no absolute stable explicit time step algorithm exists. This fact has to be considered by choosing an appropriate time step value Δt.

Now, discretization in respect to time can be applied to the whole magnetomechanical system, defined by the matrix equations 3.3-3.5 and 3.10, and written in one single matrix equation

$$\mathbf{A}\{x\} = \{b\}. \tag{3.11}$$

It should be mentioned, that the two matrices $\mathbf{H}$ and $\mathbf{G}$ of the BEM-scheme, both a part of the system matrix $\mathbf{A}$, are non-symmetric. Furthermore, these matrices have to be set up for each time step because their elements vary from time step to time step due to moving parts. The elements of the matrices $\mathbf{P}_{22}, \mathbf{P}_{2\Gamma_2}, \mathbf{P}_{\Gamma_2 2}, \mathbf{P}_{\Gamma_2 \Gamma_2}$ and $\mathbf{C}_2$ (simulation region 2) remain constant when the mechanical part of the system can be described by the theory of linear elasticity. Therefore the matrix equation of simulation area 3 should be calculated separately.

4 Geometric Modeling and Preprocessing

CAD systems can be divided in systems which only allow the interactive creation of engineering drawings and in systems which provide solid modeling, too. Solid modeling is a branch of geometric modeling that emphasizes the general applicability of models, and insists on creating complete representations of physical solid objects. These complete representations allow the algorithmical determination of arbitrary geometrical properties of the modeled part. A numerical calculation scheme such as FEM needs a complete description of the geometry of the simulated device as a part of its input data. In order to provide this input data for devices with complex shapes, a CAD system based on solid modeling should be used.

Commercially available CAD software typically supports the geometry definition very well while all the other information needed for numerical simulation (boundary conditions, constraints, excitations etc.) must often be specified on the basis of the finite element mesh, meaning that the user has to define these quantities at the level of the nodes of the mesh. Therewith, the specification of the input data is elaborating and, the probability of user errors during the input phase is high. Further problems may arise, if a coupled numerical calculation scheme is used such as the above FEM-BEM-scheme. Then, different kinds of elements (in respect to their geometric shape and associated physical quantities) must be created, according to the different numerical schemes which are involved.

Since an efficient computer modeling of magnetomechanical devices asks for the features mentioned in chapter 1, our CAD system was equipped with these features in order to obtain an efficient modeling of such devices. This modeling includes the simultaneous definition of the geometrical shape and the underlying physics with all technical features. Therewith, the specification of all required attributes is already performed at the geometry level and not at the node level which is much more convenient for the user and much more efficient, too.

A standard technique for representing solid objects is to represent the boundary of the object. A wellknown boundary representation form is based on Euler's formula [9], [10] defining a so called manifold representation form. This manifold boundary representation form is sufficient for representing single mechanical parts such as gears, crane hooks, etc. The major drawback of this representation form, however, is that representable models must fulfill the following manifold conditions:

- all portions of the model must be three dimensional (wireframes and dangling faces are not allowed)
- the boundary of the model must not touch itself
- a model must not contain regions of different materials

A model, which violates only one of the above mentioned conditions is called a non-manifold model. Since magnetomechanical devices are inherent non-manifold - e.g. consist of different adjacent parts made of different materials - it is advantageous to use a non-manifold representation form for the computer modeling of such devices. A non-manifold representation form is an extension to the manifold representation form which allows the simultaneous representation of the wireframe, the surface and the solid portions of the model. Recently developed data structures and modeling operators [11], [12], [13] which support the creation and representation of non-manifold models are very complex compared to their manifold counterparts (Winged-Edge data structure [14] and Euler-Operators [9]). Therefore, it is desireable to hide the complexity of the underlying data structures and modeling procedures by utilizing an object oriented concept. Furthermore, this object oriented concept permits the easy specification of all relevant physical and technical features of the device by directly attaching attributes such as magnetic and mechanic boundary conditions to the associated geometric part (object).

Since all relevant information is specified at the geometry level, it is possible to establish links between created finite elements (finite element nodes) and the associated geometric parts during mesh generation. An example for such a link is an unique reference to a face of the model, which is associated with a node and determines the face where the node is lying on. These links in conjunction with the fact, that non-manifold representation forms can represent a wireframe (mesh) provide the opportunity to house the geometric model <u>and</u> the simulation model (boundary conditions, loads, etc.) in an unified data structure. The major advantages of this approach are:

- all specified information is accessible during the numerical calculation and therefore can be inquired on demand, as it is necessary for the generation of adaptive meshes, for example
- all functionality which is provided for geometric modeling can be used for mesh generation, too.

5 Applications

5.1 Eddy Current Sensor

Eddy current sensors are often used in the field of non destructive testing of materials. Figure 5.1 shows a typically arrangement of an eddy current sensor for material

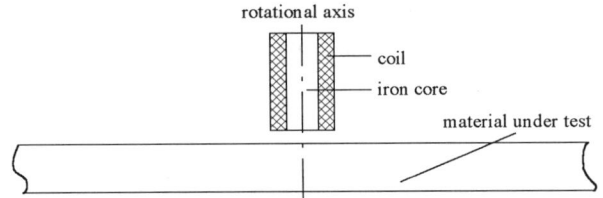

Figure 5.1: Eddy current sensor and material under test

testing. The sensor consists of a coil and an iron core which is used for guiding the magnetic field. The principle of an eddy current sensor, as its name implies, is based on the induction of eddy currents in the material under test. These eddy

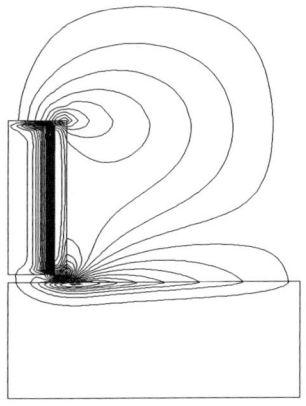 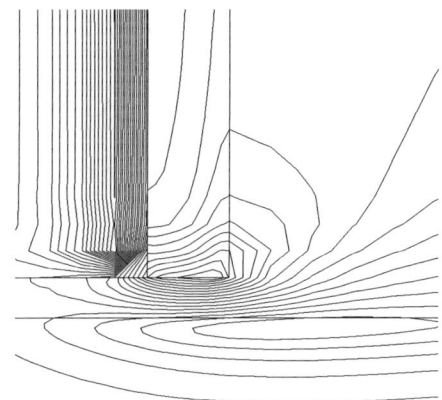

Figure 5.2: Distribution of the magnetic field $\vec{B}$

Figure 5.3: Detail of Figure 5.2

currents produce a magnetic field which is superimposed to the original magnetic field and, therewith, influences the electrical input impedance of the sensor. Since the strength of the magnetic field entering the material under test is determined by the conductivity of the material, the impedance of the sensor is a direct measure of

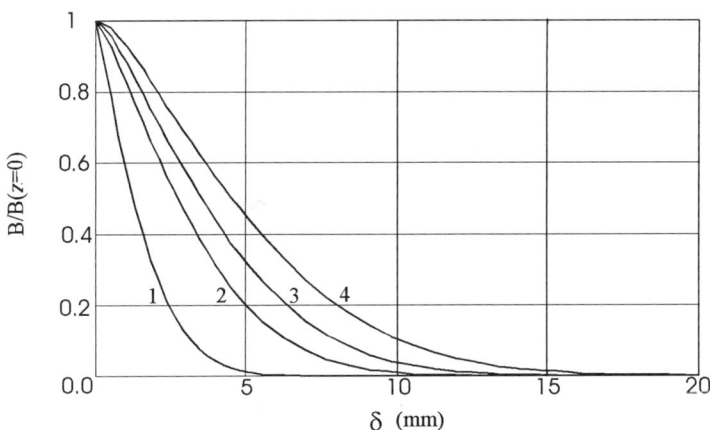

Figure 5.4: Intensity of the magnetic induction $\vec{B}$ penetrating into the material under test for four different conductivities γ ($\gamma_1 = 3.8 \cdot 10^7$ S/m, $\gamma_2 = 1.7 \cdot 10^7$ S/m, $\gamma_3 = 7.8 \cdot 10^6$ S/m, $\gamma_4 = 5.8 \cdot 10^6$ S/m) and constant permeability ($\mu = \mu_0$).

the conductivity and therefore can be used for the determination of material properties. Figure 5.2 and Figure 5.3 show the computed distribution of the magnetic induction for an eddy current sensor with 8 mm radius, 20 mm length and a coil with 500 turns. The intensity of the magnetic induction $\vec{B}$ penetrating into the material under test is shown in Figure 5.4 as a function of the penetration depth and the conductivity of material.

5.2 Electromagnetic Acoustic Source

The non-invasive destruction of concretements in human kidneys and urinaray tracts by shock waves has assumed world wide importance.

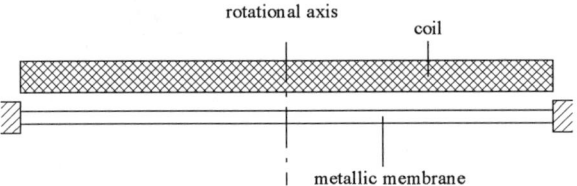

Figure 5.5: Schematic of an electromagnetic actuator used as a source for the emission of pressure pulses into a fluid medium

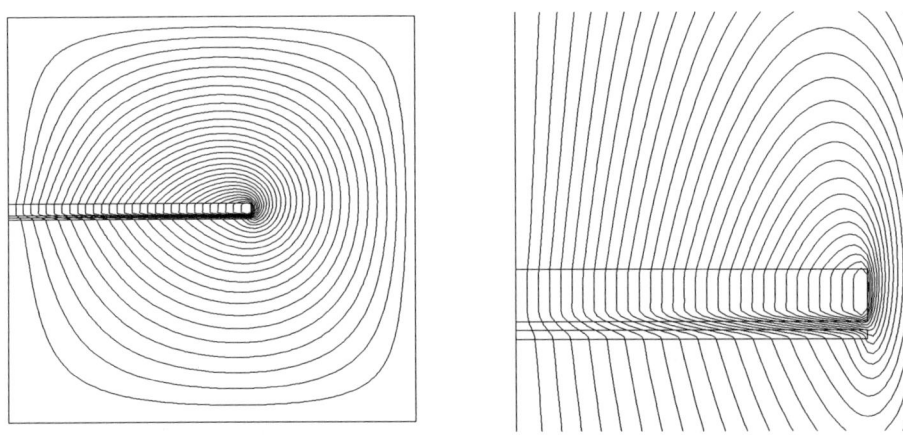

Figure 5.6: Magnetic field $\vec{B}$ of the actuator and in surrounding air

Figure 5.7: Detail of Figure 5.6

One advanced method to generate the high-intensive mechanical pressure pulse is based on an electromagnetic source [15], schematically shown in Fig. 5.5. When the slab coil is loaded by an electrical current pulse, eddy currents are induced in the membrane. The interaction between the eddy currents in the membrane and the

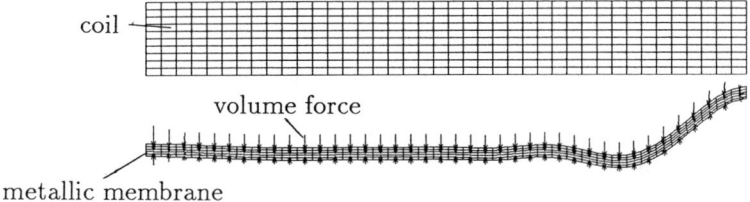

Figure 5.8: Mechanical deformations of the metallic membrane. The vector arrows represent the generated volume forces

magnetical field results in an mechanical volume force (see equation 2.12). Therewith, the metallic membrane is deformed and, an acoustic pulse is radiated into the surrounding medium. Figure 5.6 and 5.7 show the resultant magnetic field $\vec{B}$ after $0.5\,\mu s$ due to a current pulse with a duration of $10\,\mu s$ and peak value of 3.7 kA. The acting volume forces in the metallic membrane (60 mm radius and 1 mm thickness) cause the mechanical deformation shown in Figure 5.8.

6 Conclusions

A new coupled FEM-BEM calculation scheme for the simulation of magnetomechanical transducers has been introduced. By utilizing this method, the dynamic behaviour of such devices can be precisely computed. Therewith, the complex interaction of magnetical and mechanical fields can be studied and, furthermore, an optimization of the devices under consideration can be performed.

In the case of the eddy current sensor, the penetration of magnetic field into the material under test can be determined and used for optimizing the sensor with respect to its geometric design and its core material.

The precise simulation of the electromagnetic acoustic source allows a deeper understanding of the physical mechanisms generating the acoustic pulse. The detailed study of the membrane deformations can be used for an optimization in respect to the maximization of the radiated acoustic pressure pulse.

7 Acknowledgement

The authors are grateful to Prof. Dr.–Ing. R. Lerch and Dipl. Math. H. Landes for performing valuable support and providing stimulating discussions.

8 References

[1]   J.A. Stratton: *Electromagnetic Theory*, New York: McGraw-Hill, 1941

[2]   O. Bíró / K.Preis: *On the Use of Magnetic Vector Potential in the Finite Element Analysis of Three-Dimensional Eddy Currents*, IEEE Transactions on Magnetics, 1989

[3]   K. Simonyi: *Theoretische Elektrotechnik*, Leipzig: J.A. Barth, 1993

[4]   H. Parkus: *Mechanik der festen Körper*, New York: Springer, 1988

[5]   K.J.Binns / P.J.Lawrenson / C.W.Trowbridge: *The Analytical and Numerical Solution of Electric and Magnetic Fiels*, New York: John Wiley & Sons, 1992

[6]   R.Lerch / H.Landes / W.Friedrich / R.Hebel / A.Höß / H.Kaarmann: *Modelling of Acoustic Antennas with a combined Finite-Element-Boundary-Element-Method*, Proc. IEEE Ultrasonics Symbosium, 1992, p. 581-584

[7]   T.J.R. Hughes: *The Finite Element methode*, New Jersey: Prentice-Hall, 1987

[8]   G.Dalquist: *A special Stability Problem for Linear Multistep Methods*, BIT, 3, 1963

[9]   M. Mäntylä: *An Introduction to Solid Modeling*, Computer Science Press, 1988.

[10] H. Toriya / H. Chiyokura: *3D CAD Principles and Applications*, Springer, 1993.

[11] K. Weiler: *Topological Structures for Geometric Modeling*, PhD thesis, Rensselar Polytechnic Institute, 1986.

[12] E.L. Gursoz / Y. Choi / F.B. Prinz: *Vertex-Based Representation of Non-manifold Boundaries*, Geometric Modeling for Product Engineering, North-Holland, Amsterdam, 1990, pp. 107-130.

[13] Y. Yamaguchi / K. Kobayashi / F. Kimura: *Geometric Modeling with Generalized Topology and Geometry for Product Engineering*, Product Modeling for Computer Aided Design, North-Holland, Amsterdam, 1991, pp. 97-115.

[14] G. Baumgart: *Geometric modelling for computer vision*, technical report, Report STAN-CS-74-463, Stanford University: Stanford Artificial Intelligence Laboratory, 1974.

[15] H.Reichenberger / G. Naser: *Electromagnetical Acoustic Source for the Extracorporeal Generation of Shock Waves in Lithotripsy*, Siemens Forsch,- u. Entwickl.-Ber. Bd. 15 (1986), Springer

Modeling and Simulation
of a Flexible Shuttle-Robot

Heiko Freudenberg and Pham Anh Tuan

Institute of Mechatronics, Chemnitz - Germany

Abstract: The dynamic behaviour of a flexible Shuttle-Robot which is used for the docking of an APM (Attached Pressurized Module) to an ISS (International Space Station Freedom) was analysed based on superelement technique which is the most simple approach for modeling of elastic deformable bodies in Hybrid Multibody Systems (HMBS).
The docking of the modul with the robot was carried out in various scenarios which differ in robot movement and the attached point of the end effector on the modul. Alternativ profiles of velocity for the docking process are discussed and used for simulation. From the point of view of the oscillation and vibration theory the results from the geometric nonlinear coupling of torsional and bending vibration of the flexible arm appearing when the robot undergoes a "large" motion are discussed. Additional the natural vibrations behaviour of the system robot-modul in various positions of the scenarios are investigated.

1 Introduction

Requirements with respect to the dynamic behaviour of a flexible Shuttle-Robot relating to accuracy, speed and the balance between the permitted load and its own mass require a model describing properties of the real system sufficiently correct. In particular difficulties appear when the Shuttle-Robot with flexible upper and lover parts of the arm undergoes large displacements. In this cases the modeling based on rigid body systems is not allowed and the finite element method can not be used because of the extrem nonlinear dynamics.

The objective of this investigation is to model and simulate the flexible Shuttle-Robot using the program system **alaska** (acronym **a**dvanced **l**agrangian **s**olver in **k**inetic **a**nalysis) [1], [2] which takes into account most simple elastic deformable bodies - homogeneous slender beams with uniform cross section. In consideration of real dates a model of robot with rigid bodies is designed and used to determine the joint control forces that produce the desired motion. A model of an elastic robot is used to study the dynamical behaviour of the Shuttle-Robot whose flexibility of its parts is described by means of superelements. Determining time histories of joints and elastic vibrations of the robot during the docking process in various scenarios could be the base of a control by vibration damper or to keep a designed motion.

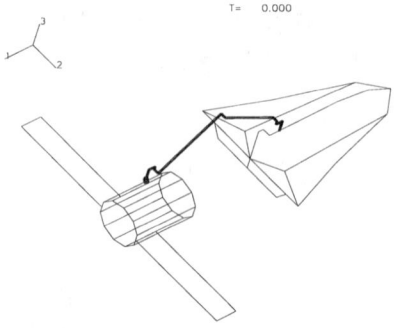

Figure 1: Shuttle-Robot with modul

2 Theoretical Background - Superelement Technique

An elastic beam is modeled due to RAUH [3] using a finite number of rigid bodies and inertialess coupling elements (force coupling elements, ideal joints) connecting them. The elastic deformations of a superelement model agree with well-known results from elastostatics for beams under static load. Moreover, the superelement in undeformed state should be possess the same inertia properties as the undeformed beam with identical dimensions.

2.1 Kinematics of a Superelement

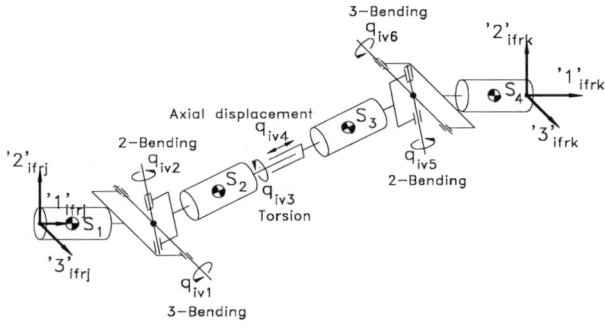

Figure 2: Kinematics of a superlement

A superelement is a spatial joint-beam-element consisting in general of four rigid bodies, where two of which neighbouring are geometrical connected by a universal joint or cylindrical joint.

The degree of freedom of a superelement is six. Three translations and three rotations of the boundary cross section are associated to that degree of freedom. Hence the elastic deformations of flexible beam (axial displacement, torsion and bending) can be described sufficient exactly.

2.2 Inertial and Elastic Properties of a Superelement

Let k $(0.0 < k < 0.5)$ be the coefficient of dividing the beam into four parts, then the lengths of partial bodies can be obtained by $L_1 = L_4 = kL$ and $L_2 = L_3 = (0.5-k)L$, where L is the length of the beam.

Mass distribution :

Mass of the beam	: M
Mass of superelement partial bodies	: $M_1 = M_4 = kM$ and $M_2 = M_3 = (0.5-k)M$

Inertial tensor :
Inertial moment of beam with respect to longitudinal axis   : Θ^{11}
Inertial moments of beam with respect to cross axis   : Θ^{22} ; Θ^{33}
Inertial moments with respect to the body-fixed principal axis of superelement parts:

$$\Theta_1^{11} = \Theta_4^{11} = k\Theta^{11} \qquad ; \Theta_1^{22} = \Theta_4^{22} = k^3\Theta^{22} \qquad ; \Theta_1^{33} = \Theta_4^{33} = k^3\Theta^{33}$$

$$\Theta_2^{11} = \Theta_3^{11} = (0.5-k)\Theta^{11} \qquad ; \Theta_2^{22} = \Theta_3^{22} = (0.5-k)^3\Theta^{22} \qquad ; \Theta_2^{33} = \Theta_3^{33} = (0.5-k)^3\Theta^{33}$$

Spring constants :
The elastic properties of the beam can be represented by spring constants in joints and spring constants between the two boundary bodies of the superelement.

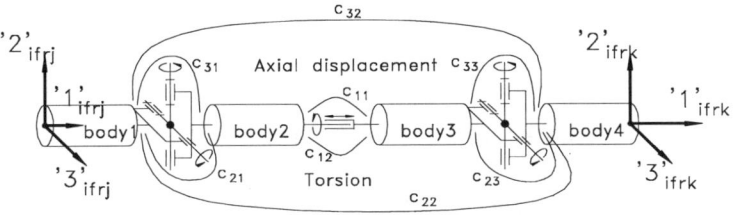

Figure 3: Elastic properties of a superelement

Spring constants can be obtained from elementary structure mechanics as

$$C_{11} = EA/L \qquad ; \qquad C_{12} = GI_t/L$$
$$C_{21} = C_{23} = 6EI_2(1-2k)^2/L \qquad ; \qquad C_{22} = 2EI_2(-1+6k-6k^2)/L$$
$$C_{31} = C_{33} = 6EI_3(1-2k)^2/L \qquad ; \qquad C_{32} = 2EI_3(-1+6k-6k^2)/L$$

where E and G are, respectively, the elastic and shear modulus of the beam, A is the area of beam cross section, I_2 and I_3 are the centroidal second moments of area of the cross section relativ to cross axis and I_t is the torsional constant.

3 Modeling and Simulation

3.1 Modeling

Informations about the mass distribution and the geometrical dimensions of the robot and the specific material for the parts of the robot arm are available [4]. This and the knowledge of the elastic behaviour in the joints are the base for a kinematic flexible model nearby the reality. The drawing in Figure 4 illustrates the dimensions of the robot in a construction configuration.

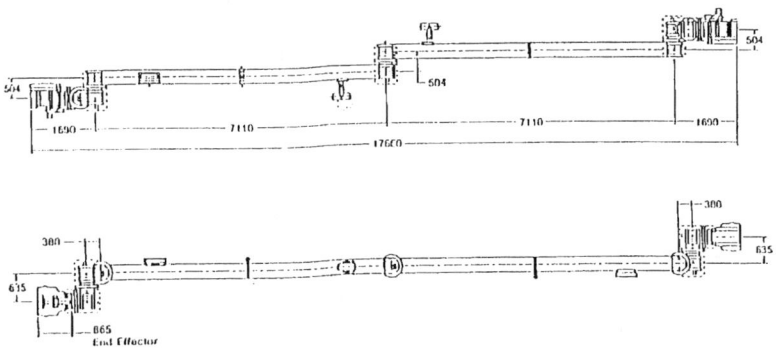

Figure 4: Robot in a construction configuration

The use of superelement technique makes possible the modeling of the superposition of large rigid body motion and elastical vibrations in parts of the robot arm. Upper arm and forearm of the robot are described with each one superelement. This allows the research of the oscillations due to bending and torsion. Axial displacement deformations of the robot arm is not taking into account. Upper arm and forearm consist of a tube with a diameter of 400 mm and a wall thickness about 2 mm.

Today's robot technology allows the production of joints without clearance and friction. Clearance and friction in joints was disregarded during the modeling process. The robot consists of an upper arm and a forearm which are connected with a elbow joint. This is a simple revolute joint. The shoulder consists of three revolute joints. The axis of rotation of this joints are orthogonal to each other (pitch, yaw and roll axis). The shoulder is the connection between the Docking Port (the ground) and the robot. The end effector is connected with the forearm over a wrist.

For the realisation of the berthing motion only one revolute joint in shoulder and wrist are sufficient. However, for a complete description of the elastical behaviour in shoulder and wrist also these joints are modeled which for the berthing process are not necessary.

To describe an elastical joint with additional torques in shoulder, elbow and wrist two revolute joints are modeled one after the other. The location of both joint axis are the same. Because of the additional elastical coordinates, the degree of freedom of the model is increased.

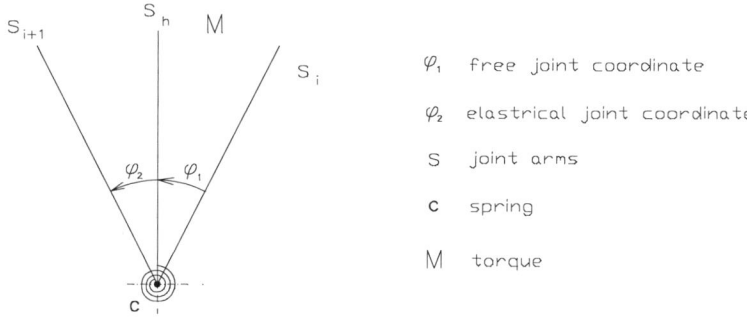

φ_1	free joint coordinate
φ_2	elastrical joint coordinate
S	joint arms
c	spring
M	torque

Figure 5: Sketch of an elastical joint

3.2 Simulation

Three various berthing processes of the modul to the docking port of the space station by means of a flexible Shuttle-Robot are simulated. The differences between this processes are the manipulator's movements and the fixing point of the end effector on the modul. The distances between modul and docking port, that must be overcome during the berthing processes are 5 and 7,5 metres, respectively. Both upper arm and forearm or only the forearm are used for the simulations. The berthing process take place only in a working plane, so not all joints of robot's shoulder and wrist are required for the docking motion. The modul is modeled only by a rigid inertial body. The connection point between end effector and modul is not in the moduls plane of symmetry. That plays an important role for the dynamic behaviour of the robot.

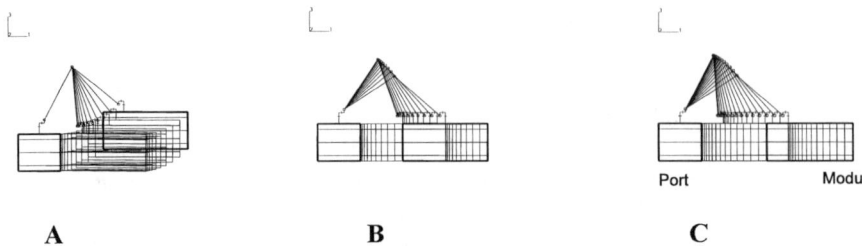

A   B   C

Figure 6: Schematical illustration of the docking processes A, B and C

The time histories of the various processes are described with a velocity profile of a point that is defined in the centre of moduls docking face. The processes are going over a period of 200 and 300 sec., respectively. This linear profile is shown in Figure 7a. Beside this profile an alternative velocity profile for the modul motion is discussed and used. It is defined as:

$$v(t) = \frac{v_0}{2} \left(1 - \cos\left(2\pi \cdot \frac{t-t_0}{t_{end}-t_0}\right)\right)$$

$v_0 = 0.05$ m/sec.,
$t_0 = 0.0$,
$t_{end} = 200$ sec.,

and has the following advantage: at the beginning and at the end of the process, the accelaration of the modul is equal zero. Figure 7b illustrates this fact.

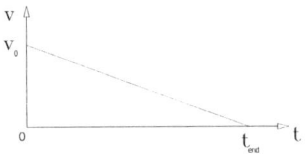

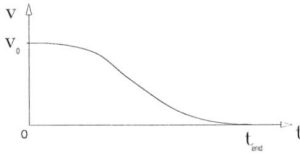

Figure 7a: Linear velocity profile

Figure 7b: Nonlinear velocity profile

These selected velocity profiles can be realized by functions which are available in **alaska**. The basis for the simulation of the docking processes is the determination of this driving torques which are required for a designed motion of the modul with corresponding velocity. Following, the problem of inverse kinematics has to be solved at first. For this purpose a rigid model of the robot is used. The results of the inverse kinematics problem are the time histories of the joint coordinates in shoulder, elbow and wrist. Using this time histories the needed driving torques can be calculated. For the problem of inverse dynamics a rigid model is used, too. Both problems, inverse kinematics and inverse dynamics can be solved with the tool **alaska**. The torques are determined in such a way, that the system robot-modul at the end of the docking manoever is without motion.

3.3 Results of Simulation

The nonlinear simulation of this docking processes with the Multibody System-tool **alaska** allows to obtain a lot of interesting informations about the dynamical behaviour of the system robot-modul during the process. Using the linear velocity profiles (Figure 7a) the accelarations of the modul at the beginning and at the end of process do not vanish. That is one of the reasons for an excitation of oscillations in the system. With the help of the time histories of joint coordinates of superelements the elastic deformations of the parts of the robot arm which are modeled with superelements can be judged. The amplitudes of this deformations depend on the corresponding working position of the robot. The bending and torsion stress are caused by the lateral displacement of the robot. In the following figures some time histories of superelement coordinates which represent the bending and torsion of the upper arm and forearm are shown. Figure 8 presents time histories of coordinates which describe the transversal bending and the torsion of the forearm during simulation of process C with linear velocity profile. This curves illustrate the increase of dynamic load in the working positions at the end of the process, then the forearm is nearly vertical to the modul. The torsional behaviour of the forearm is reflected in the yawing of the modul.

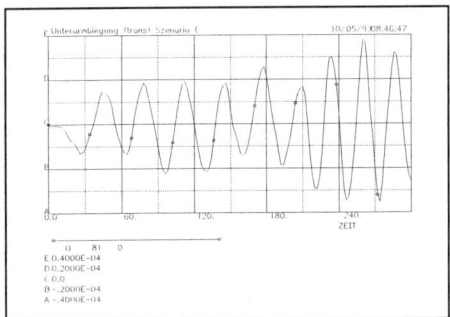 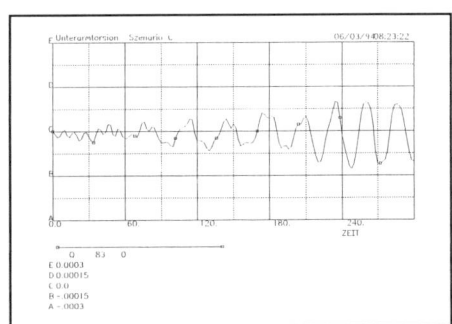

Figure 8: Forearm bending transversal to the working plane
and torsion of forearm, process C, linear profile

Figure 9 shows time histories of coordinates which describe the bending stress of upper arm and forearm in the working plane during process A with linear velocity profile. The deformation of the forearm (right figure) is embossed from the torques in the elbow and the wrist joint.

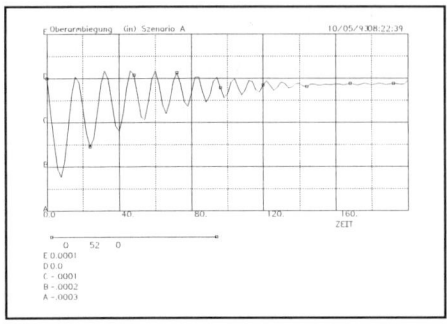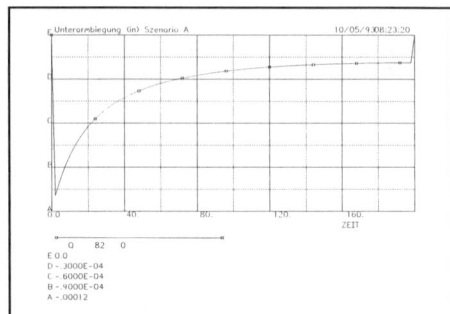

Figure 9: Bending coordinates of upper and forearm in working plane,
process A, linear velocity profile

3.4 Comparison and Valuation of the Simulation Results

The used alternativ velocity profiles cause dynamic load in the system robot-modul which is very small in comparison to the linear profiles. The following figures show confrontations of corresponding curves from simulations with different velocity profiles. (The curve with the small amplitudes belong to the alternativ profiles.) This figures illustrate the large difference in the dynamic load.

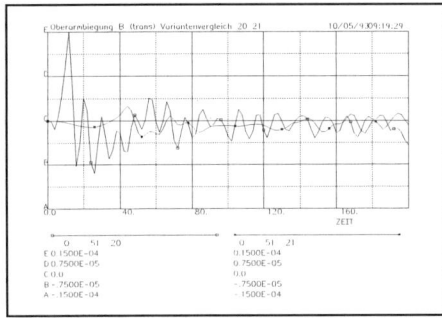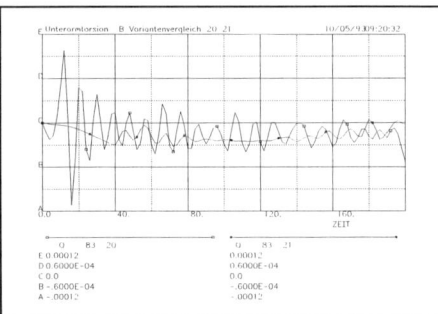

Figure 10: Dynamic stress comparison between linear and alternative
velocity profile - bending and torsion

In the joints act constant torques when a linear profile is used (with the exception of process A). The bending load of upper arm and forearm in the working plane are caused by torques which acts at the ends of the arm parts. This quasistatical bending load is smaller when a linear velocity profile is used because of the smaller maximal acceleration and therefore smaller maximum torques.

Another result of the decreasing of the dynamic strain due to alternatively profiles is the decreasing of the deviation from the theoretical way.
As a comparative value for the deviation from the theoretical way the distance of a point in moduls docking face to the port at the end of docking process can be used. In the case of a rigid model, this distance is equal to zero. By using a linear velocity profile the distance is about 0,1 metres (in process B). By using an alternativ profile, this distance decreases to 0,04 metres. (In process C this distance is already with linear profile only 0,001 metres.)

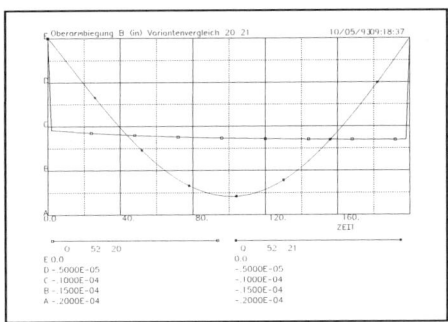

Figure 11: Quasistatical bending deformation of the upper arm in the working plane as a comparison between linear and alternative velocity profile

In spite of the large mass of the modul (21 t) and a very flexible material (under force of gravity) for the arm parts, the elastic deformations of the arm during the docking process without gravitational force are relative small. The maximal occured value for torsion for example is $3*10^{-4}$ rad. After the docking processes the flexible system robot-modul vibrates further.

4 Acknowledgements

This research was supported by the German Space Agency (DARA) and in corporation with MBB/ERNO Bremen in the project "Modeling and Simulation of Complex Hybrid Multibody Systems based on Superelement Technique" [5].

5 References

[1] **alaska** - User Manual : Institute of Mechatronics, Chemnitz, June 1994.

[2] Maißer, P. : Analytische Dynamik von Mehrkörpersytemen, ZAMM 68(1988)10, pp. 463-481.

[3] Rauh, J. : Ein Beitrag zur Modellierung elastischer Balkensysteme, VDI-Fortschritt-Berichte, Reihe 18: Mechanik/Bruchmechnik, Düsseldorf, 1987.

[4] MBB/ERNO internal Project Informations

[5] Maißer, P. and others: Modellierung und Simulation komplexer hybrider Mehrkörpersysteme mittels Superelementtechnik,Technischer Bericht, Institut für Mechatronik, Chemnitz, März 1993.

Dipl.-Ing. Heiko Freudenberg
Institute of Mechatronics
Reichenhainer Str. 88
D-09126 Chemnitz - Germany
Telephone: +49 - (0)371 - 5314672
Fax: +49 - (0)371 - 5314669
e-mail: H.Freudenberg@IfM.TU-Chemnitz.de

Dr.-Ing. Pham Anh Tuan
Institute of Mechatronics
Reichenhainer Str. 88
D-09126 Chemnitz - Germany
Telephone: +49 - (0)371 - 5314680
Fax: +49 - (0)371 - 5314669
e-mail: Tuan@IfM.TU-Chemnitz.de

Optimization of the dynamic behavior of a wire bonder using the concept of mechatronic function modules

Hans Hesse
Hesse & Knipps GmbH
Vattmannstr. 6
33100 Paderborn
Germany

Jörg Wallaschek
Heinz Nixdorf Institut
Universität-GH Paderborn
33095 Paderborn
Germany

Abstract: During the past years, ultrasonic wedgebonding has attracted much attention because it is one of the keystones of the Chip-On-Board technology, which allows to considerably increase the packaging density of complex electronic devices. In order to achieve high quality and productivity, fully automated ultrasonic wedgebonding machines with small cycle time are required. In the process of product development, special attention must be paid to meet the demands of very accurate high speed positioning in combination with the delicate handling of the bonding wire. In this paper a mechatronic design approach is described which is based on functional modelling of the main components of a bonding machine in terms of so called mechatronic function modules. In particular the modelling of a piezoelectric bimorph actuator is studied in detail.

1 Introduction

In modern electronic devices many functions are integrated into a small volume. Applications like mobile telecommunication, integrated sensor modules or chip-cards require extremely high packaging density [1]. One possibility to achieve this density is to use „naked" chips which are directly mounted on the printed circuit board (PCB). These chips are then interconnected via the PCB by aluminum or gold wires which are bonded to the contact pads of chip and PCB, see Fig. 1.

The wire used in COB usually is very thin (25 μm - 100 μm). It is bonded to the pads of chip and PCB either by the ball-wedge or by the wedge-wedge process, see Fig. 2. Wedge bonding has the advantage that thermal heating is much lower than in ball bonding and that higher bonding quality can be achieved. Usually aluminum wire is used in wedge bonding which is a solid state process joining two pieces of metal using a process of diffusion. The bond is formed by rubbing the surfaces of wire and metalization together. To this end the wire is pressed to the pad by the wedge which vibrates at ultrasonic frequency driven by a piezoelectrically excited sonotrode. The rubbing disturbs surface oxide films and exposes clean metallic surfaces so that a metallurgical weld is formed by diffusion [2].

Fig.1: **Chip On Board** [3].

a)

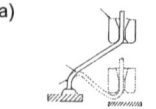

b)

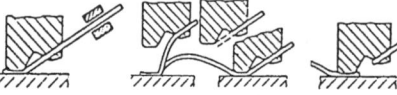

Fig.2: a) Ball-wedge bonding.
b) Wedge-wedge bonding [4].

2 Kinematic requirements

Fully automated ultrasonic wire bonders usually are integrated in manufacturing lines. Image processing and pattern recognition are used to determine the relative position and orientation of chip, substrate and bonding machine at the start of a bonding sequence. After initialization, bonding wedge and wire are positioned at the first bond pad (source) and the wire is bonded to the pad, see Fig. 2. Then wedge and wire are positioned at the second bond pad (destination). They must move on a well defined trajectory in order to obtain the correct loop form of the wire. After the second bond has been made, the wire clamp is used to tear the wire from the bond. Then, wedge and wire are positioned at the source bond of the second connection to be made, while the wire clamp feeds wire under the wedge and the next bond cycle starts. Due to the fact that the wire must always be pulled in tangential direction relative to the wedge, a minimum of 4 axes (3 translations, 1 rotation) is required in an ultrasonic wedgebonding machine. State of the art bonding machines achieve cycle times as short as 500 ms for a typical contact distance of 2 mm, the positioning accuracy being about 5 μm [5].

In most ultrasonic wedgebonding machines 4 degrees of freedom positioning is realized by two degrees of freedom of a horizontally moving table in combination with two degrees of freedom of the vertically moving rotating bonding head which carries transducer and wedge, see Fig. 3a. This design however is difficult to integrate into fully automated manufacturing lines, so that the integration of all axes into the bonding head is advantageous, Fig. 3b. During a bond cycle all four axes must perform a controlled coordinated motion. Additionally the wire clamp action and the power flow of the ultrasonic transducer must be synchronized.

a)　　　　　　　　　　　　　b)

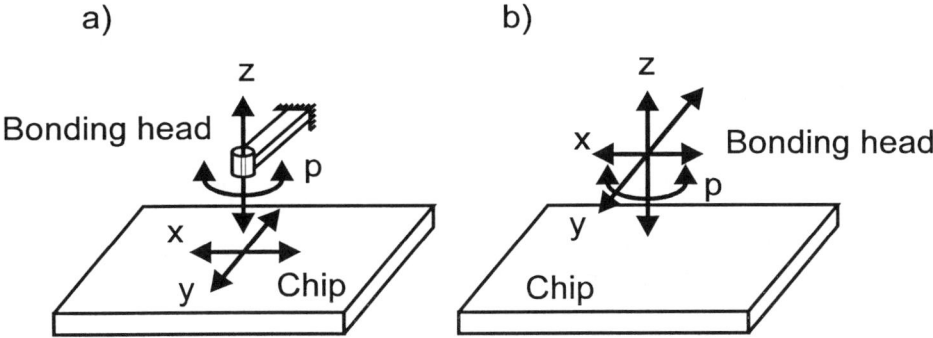

Fig.3: Four degrees of freedom positioning mechanisms used in bonding machines:
　　　a) Horizontally moving table, vertically moving rotating bonding head.
　　　b) Fixed table, all axes integrated into the bonding head.

3 Dynamic requirements

Most important for good bonding quality is the vibration of the bonding tool and the normal force between wedge and wire during the bond process. The normal force has to be controlled at a constant value which depends on the wire thickness and other process parameters. For 25 μm aluminum wire 50 cN would be a typical value of the bond force. In order to meet the requirement of controlled normal force, transducer and wedge are mounted elastically to the bond head, see Fig. 4. When the bond head is lowered and the wedge approaches the bond pad, a so-called „touch-down sensor" is used to detect the exact moment when wedge and wire come in touch with the bond pad. Then the motion of the z-axis continues until a well-defined distance has been traveled which is chosen such as to elastically preload the wedge with the correct contact force. During the bond process the contact force can be controlled by an additional bond force actuator and the ultrasonic vibration of the wedge is controlled by a PLL circuit and power control. There have been many attempts to improve bonding quality by means of on-line process control [6].

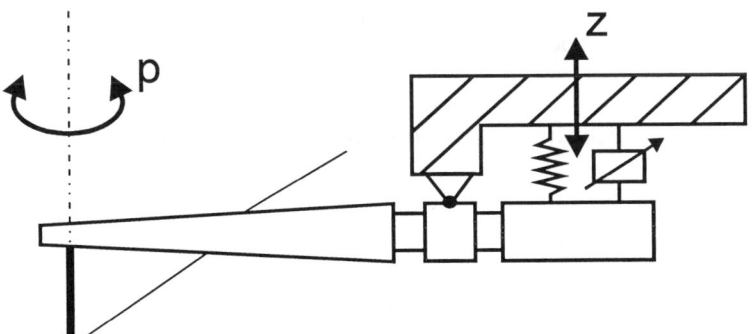

Fig.4: Elastically mounted transducer with wedge and bond force actuator.

4 Design considerations

Ultrasonic wedgebonding machines are complex mechatronic systems. With respect to the strong demands for bond cycle time, bond force and other items as well as with respect to the importance of the software used in motion control and the marked influence of the actuator dynamics (like e. g. magnetic rise time in the bond force actuator or wire clamp) on the overall system performance a functional design approach must be chosen. At the same time geometric bounds must also be taken into account so that the whole design process requires an approach based on geometric and functional modelling.

One method which seems to be suitable for simultaneous engineering integrating geometric and functional modelling is the concept of mechatronic function modules [7]. It is based on the idea of hierarchically structuring of the overall system in modules. Each module is then described by a block-oriented input-output relationship in the form of a standardized mathematical model complementing the geometry information of a CAD-model. As each module contains a mechanical base system, sensors, actuators as well as local information processing, the whole system can be described as a combination of mechatronic function modules with well defined input-output relationships. This approach facilitates the development of mechatronic products, because the modularity

1. provides a methodological base for simultaneous engineering,
2. supports a „combine and compose" type system design method [8],
3. saves modelling cost when applied during the whole system design process[9].

Although many efforts have been made during the past years, no universal software-tools for the integrated functional and geometric design are available today. Present state-of-the-art is to combine different software-tools into a common workbench environment, object oriented programming techniques being increasingly used [9]. In this paper we will, however, not discuss questions related to software-tools for the design process. Instead we will concentrate on the process of structuring the ultrasonic wegebonding machine into modules and on the functional modelling of one of these, in order to illustrate the design method.

5 Structuring of the bonding machine into mechatronic function modules

In principle there are two basic methods which can be used in the process of structuring. In the classical top-down approach the process starts with the overall system and breaks it apart into subsystems. Each subsystem is then divided into smaller subsystems until the smallest units in terms of the mechatronic function modules are obtained. In the bottom-up approach, the process starts with the smallest units and combines them to larger systems until the overall system is completely described. Both methods have their relative merits and often they are used simultaneously [10]. In the present case it turned out that a combination of top-down structuring in order to find the basic system structure and bottom-up synthesis in order to generate alternative system structures had most benefits.

Fig. 5 shows a possible hierarchical structure of the ultrasonic wedgebonding machine in a simplified form. Only some of the most important subsystems and components are shown. As can be easily seen, the bonding machine consists of many different modules and each module offers a certain degree of freedom in the design process, so that the optimal design of the overall system is a challenge.

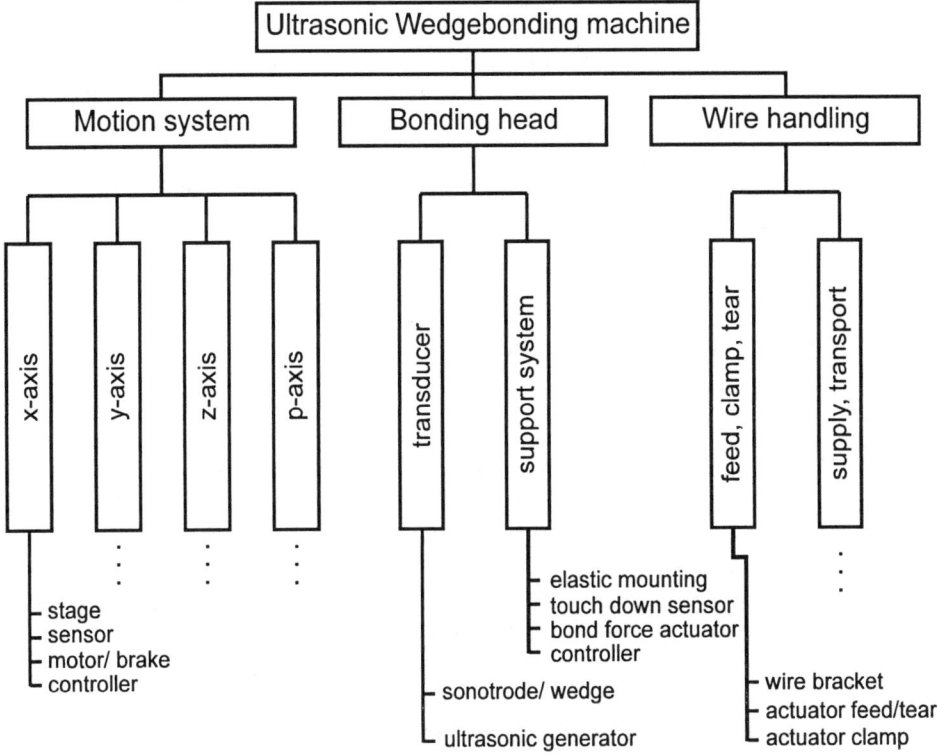

Fig.5: Hierarchical structure of the ultrasonic wedgebonding machine.

Although the structure as a whole looks very complicated, each module is a simple system and its mathematical modelling can be performed in a straightforward manner. Also the input-output coupling of the modules can be described without difficulties if certain rules are followed. It has turned out to be very convenient to describe the input-output relations of mechatronic function modules explicitly by means of the state-space equations

$$\dot{\vec{x}} = \vec{f}(\vec{x}, t, \vec{u}),$$ (1)

$$\vec{y} = \vec{g}(\vec{x}, t, \vec{u}).$$ (2)

Here $\vec{u}(t)$ is the input, $\vec{x}(t)$ is the state vector and $\vec{y}(t)$ is the output of the mechatronic function module. Problems may arise if the coupling between mechatronic function modules is given by ideal mechanical constraints. In this case the state variables of the coupled subsystems are not independend from each other and the interaction forces can not be calculated from constitutive equations. Thus the concept of modularity which is based on the assumption of locality is violated. There are two ways of handling this difficulty. The engineering approach is to approximate the ideal mechanical constraint by a sufficiently strong physical constraint at the price of obtaining a stiff system of state equations for the overall system. The mathematical approach is to explicitly formulate the constraint equation and make it part of the overall system modelling. In this way a set of differential-algebraic equations is obtained which can either be solved directly by appropriate numerical methods or must be reduced to minimal coordinates by a suitable elimination or projection algorithm.

6 Modelling of the bonding head

6.1 Equations of motion

The bonding head consists of a carrier which is mounted to the z- and p-axis respectively and the transducer which is hinged to the carrier and pivots around the connection point, see Fig. 4. In present state-of-the-art ultrasonic wedgebonding machines, transducer and carrier are connected by a spring. An electromagnetic actuator, like e. g. a voice coil motor is used to generate the bond force as well as to press the transducer against a restraint during fast motions of the bonding head in order to prevent large uncontrolled motions of the transducer.

The first step in obtaining the explicit state-space equations, which are needed for the mechatronic function modelling, is the formulation of the equations of motion. Here they were obtained using KANE'S method [11] with the help of the symbolic formula manipulator AUTOLEV [12]. The equations of motion are of the form

$$\dot{q}_i = u_i \quad , \quad i = 1, \ldots, 5,$$ (3)

$$\sum_{j=1}^{5} M_{ij}(q_k, u_k, t)\dot{u}_j = f_i(q_k, u_k, t) \quad , \quad i = 1, \ldots, 5.$$ (4)

with $q_i(t)$ being the geralized coordinates and $u_i(t)$ the generalized speeds of the system. After solving for the time derivatives of the generalized speeds, the explicit state-space equations of the mechanical part of the system are readily available. In these equations, however, the dynamics of the electric motors as well as the dynamics of the drive train has not yet been included and also no modelling of the sensors used in motion control has been performed at this point. It is however not difficult to take these effects into account step by step. In the present case, detailed models for the drive system were developed and have been used in the optimization of the drives, where a balanced compromise between high speed dynamics and stable operation needed to be guaranteed. In the resulting model also the dynamics of the electrical system and controllers was formulated. It should be noted here that mechanical and electrical state-space equations are also coupled through the displacement of the actuator-coil, and by the actuator force which are functions of the pivot angle and the magnetic field.

6.2 Modelling of a piezoelectric bimorph actuator

One of the most important functions of the bonding head is to provide a controlled contact force between wedge, bonding-wire and substrate. In present state-of-the-art bonding machines this is achieved by using a spring and an electro-magnet. The spring is preloaded when wedge and wire touch the substrate, while the z-axis continues its displacement until the desired contact force is reached. The electro-magnet provides an additional bonding force which might be used in an active bond process control. It is obvious that spring, electro-magnet, touch down sensor and bond force controller can be considered as a mechatronic function module and that they can be substituted by any other system of the same functionality.

In the design process for a new improved ultrasonic wedgebonding machine it was suggested to replace the spring and the electromagnetic actuator by a piezoelectric bimorph actuator, thus combining two functions into one single component. With respect to the functional modelling of the bonding head, the general structure of the mechatronic function module remains unchanged, the only difference being in the state space-equations of the system. Instead of the spring force and the electromagnetic actuator force, now the tip force of the piezoelectric bimorph enters the equation of motion.

The piezoelectric bimorph combines the elasticity of a passive spring with the capability of active force generation. Therefore it is a promising alternative to the present design. In the following the piezoelectric bimorph will be investigated with the aim of deriving the mathematical model required in the functional modelling.

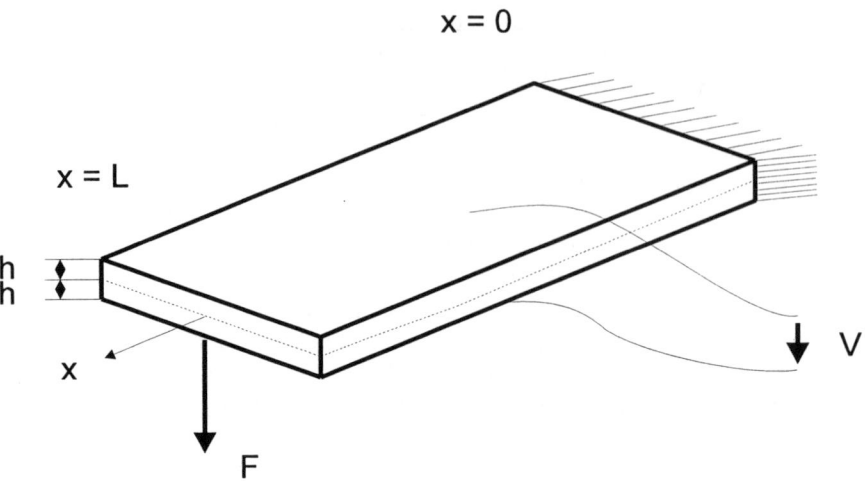

Fig.6: Piezoelectric bimorph.

The piezoelectric bimorph consists of two flat strips of piezoceramics which are bonded over their long surfaces, see Fig.6. The piezo-strips are polarized in such a way that one strip elongates and the other strip contracts when an electric field is applied via the electrodes. This results in a bending motion of the overall system when the electrodes are excited electrically. According to the EULER-BERNOULLI hypothesis the longitudinal strain in the bimorph is given by

$$S_{11} = -z\,w'' \quad ,$$ (5)

where $w(x,t)$ denotes the transverse displacement of the neutral plane of the beam. The corresponding stress is then given by

$$T_{11} = c_{11}^E S_{11} - e_{13} E_3$$ (6)

where E_3 is the electric field, c_{11}^E is the stiffness and e_{13} describes the piezoelectric coupling. In (5) and (6) the notation according to [13] was used.

Taking into account that upper and lower piezo are polarized differently, the resulting bending moment can be calculated from

$$M = -\int_{-h}^{h} c_{11}^E\, z^2\, w'' b\, dz \;\; - \int_{-h}^{0} e_{13} \frac{V}{h} z\, b\, dz \;\; + \int_{0}^{h} e_{13} \frac{V}{h} z\, b\, dz,$$ (7)

where h is the thickness of one piezoelectric layer and b is the width. Evaluating (7) the bending moment is given by

$$M = -\frac{2}{3}c_{11}^E b h^3\, w'' + e_{13}\, V\, b h. \tag{8}$$

The piezoelectrically induced bending moment is constant along the bimorph and does not enter into the equation of motion

$$\rho\, A\ddot{w}(x,t) + \frac{2}{3}b h^3\, c_{11}^E\, w^{IV}(x,t) = 0 \tag{9}$$

of the bimorph. The piezoelectrically induced bending moment does however show up in the time-dependent boundary conditions

$$w(0,t) = 0$$
$$w'(0,t) = 0$$
$$\frac{2}{3}c_{11}^E b h^3\, w''(L,t) = e_{13}b h V(t) \tag{10}$$
$$\frac{2}{3}c_{11}^E b h^3\, w'''(L,t) = -F(t)$$

where the third equation describes the vanishing bending moment at the tip and the last equation relates the tip force to the transverse shear force in the beam. The equation of motion and the corresponding boundary conditions form a boundary value problem with inhomogeneous boundary conditions.

The deformation of the piezoelectric bimorph can be considered as a superposition of a quasistatic deformation $w_0(x,t)$ and an elastodynamic deformation $w_d(x,t)$, where $w_0(x,t)$ is calculated from (9) without the inertia term. In the present case we find

$$w_0(x,t) = \frac{3L^3}{2b h^3 c_{11}^E}\left\{\frac{e_{13}b h}{L}\frac{x^2}{2L^2}V(t) - \left[\frac{x^3}{6L^3} - \frac{x^2}{2L^2}\right]F(t)\right\}. \tag{11}$$

Substituting the transformation

$$w(x,t) = w_o(x,t) + w_d(x,t) \tag{12}$$

into (9), the elastodynamic deformation $w_d(x,t)$ is obtained as the solution of the inhomogeneous equation of motion

$$\rho\, A\ddot{w}_d(x,t) + \frac{2}{3}b h^3 c_{11}^E\, w_d^{IV}(x,t) = \frac{-3L^3}{2b h^3\, c_{11}^E}\left\{\frac{e_{13}b h}{L}\frac{x^2}{2L^2}\ddot{V}(t) - \left[\frac{x^3}{6L^3} - \frac{x^2}{2L^2}\right]\ddot{F}(t)\right\} \tag{13}$$

with homogeneous boundary conditions

$$w_d(0,t) = 0,$$
$$w'_d(0,t) = 0,$$
$$w''_d(L,t) = 0,$$
$$w'''_d(L,t) = 0.$$

(14)

This boundary value problem is of standard type. It can be solved analytically for the special case of harmonic motion. In this way the dynamic stiffness matrix of the piezoelectric bimorph can be calculated [14].

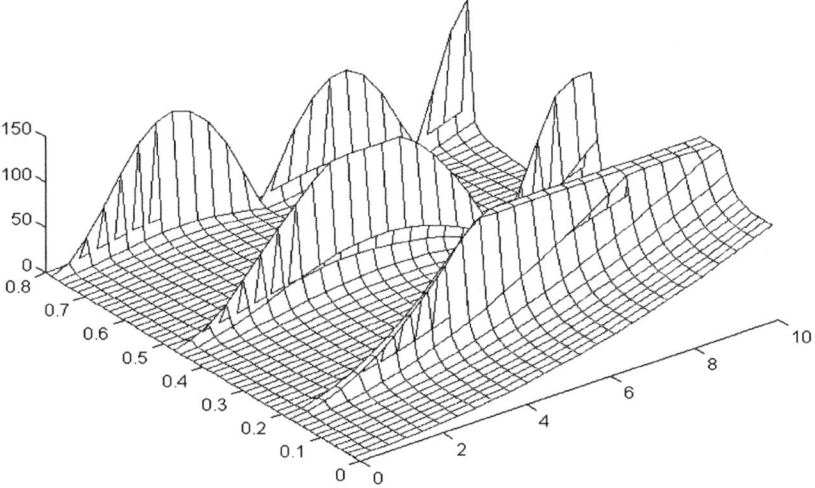

Fig.7: Calculated amplitude of the piezoelectric bimorph displacement under harmonic excitation as a function of the frequency of excitation (GREEN's function)

In Fig.7 the first three eigenfrequencies and the corresponding eigenmodes can easily be recognized. For the purpose of mechatronic function modelling, however, an input-output description in state-space form is required. This description can be obtained from the dynamic stiffness matrix by inverse Fourier-transformation or by direct parameter estimation. It is however more convenient in the present case to work directly with the partial differential equation and to use discretization techniques like e. g. the Rayleigh-Ritz method. For frequencies well below the first natural frequency of the bimorph, the quasistatic solution can be used. If however also higher frequencies have to be taken into account, an expansion of $w_d(x,t)$ in the form

$$w_d(x,t) = \sum_{i=1}^{N} q_i(t) W_i(x) \tag{15}$$

with given functions $W_i(x)$ and unknown amplitudes $q_i(t)$ must be made. It results in discretized equations

$$M\ddot{\vec{q}} + C\dot{\vec{q}} = \vec{F} \tag{16}$$

for the amplitude functions $q_i(t)$. In general, the electric field also must be approximated by a comparable expansion. In the present case however this is not necessary since due to the geometry of the problem, the quasistatic approximation of the electric field is valid in the whole frequency range of interest. The functions $W_i(x)$ must satisfy the geometric boundary conditions [15]. If the functions $W_i(x)$ are chosen as the eigenfunctions of the underlying homogeneous boundary value problem, the matrices M and C become diagonal.

The transformation of (16) into state-space form is straightforward and completes the description of the mechanical system behavior. In addition, however, also the electrical behavior must be described. To this end we consider the constitutive equation

$$D_3 = e_{13} S_{11} + \varepsilon_{33}^S E_3 \tag{17}$$

for the dielectric displacement. The electrode charge is then given by

$$Q(t) = -\int_0^L D_3(x,h,t) b \, dx \tag{18}$$

and can be calculated directly in analytical form, since the mechanical displacement field has already been determined.

If we are only interested in the quasistatic behavior of the bimorph, the resulting input-output description is

$$y_1 = \frac{3 e_{13} L^2}{4 c_{11}^E h^2} u_1 + \frac{L^3}{2 b h^3 c_{11}^E} u_2, \tag{19}$$

$$y_2 = \left[\frac{3 e_{13} bL}{2 h c_{11}^E} - \frac{\varepsilon_{33}^S bL}{h} \right] u_1 + \frac{3 e_{13} L^2}{4 c_{11}^E h^2} u_2, \tag{20}$$

where the electric voltage $u_1 = V(t)$ and the tip force $u_2 = F(t)$ are considered as input variables and the tip displacement $y_1 = w(L,t)$ and the electrode charge $y_2 = Q(t)$ are considered as output variables. This concludes the modelling of the piezoelectric bimorph. The mathematical model for the bimorph now has to be integrated into the model of the bonding head and of the overall system.

6.3 Model validation

Of course the mathematical models used in the mechatronic function modelling must be validated against experimental results. Fig. 8 shows the measured amplitude of the piezoelectric bimorph at the first and second eigenfrequency as well as the transfer function between electrical excitation and tip displacement for an unloaded actuator. Good agreement with the mathematical model can be observed, compare Fig. 7.

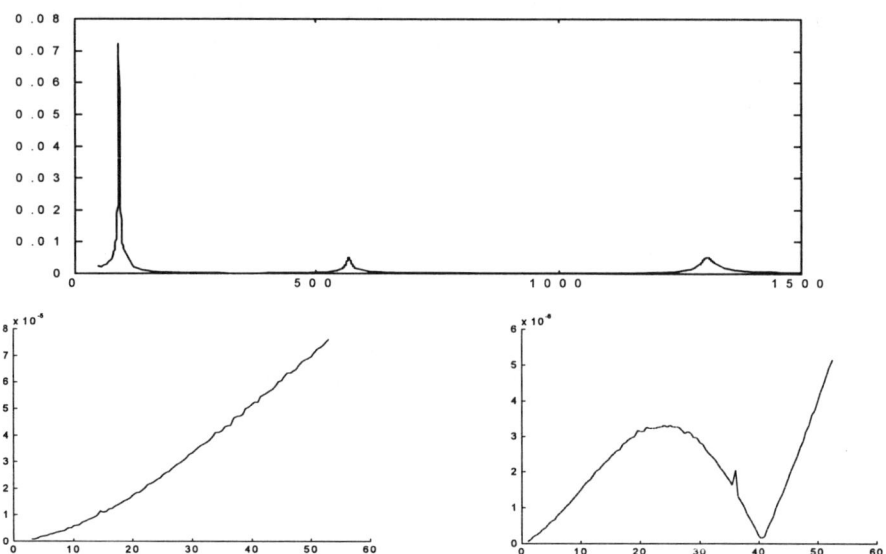

Fig. 8 Experimentally obtained transfer function and amplitude distributions of the piezoelectric bimorph.

7 Summary and outlook

The concept of mechatronic function modules was applied to the modelling of an ultrasonic wedgebonding machine. After an introductory explanation of the Chip On Board technology the general demands on wedgebonding machines were pointed out. A hierarchically structured model of the bonding machine was developed and the principles of mechatronic function modelling were explained briefly. Special emphasis was put on the modelling of the bonding head and a piezoelectric bimorph which might replace the electromagnetic bond force actuator in the future was studied in detail.

8 References

[1]   **Tiederle, V.**: Markt, Potential und Wirtschaftlichkeit der Chip-on-Board-Technik. VTE (Verbindungstechnik in der Elektronik) 2/94, pp. 68 - 73.

[2]   **Falk, J.; Hauke, J.; Kyska, G.**: Thermodynamik des Drahbondprozesses. VTE (Verbindungstechnik in der Elektronik), 3/93, pp. 110 - 118.

[3]   **GAISER Tool Company**: Catalogue of tools for wire bonding.

[4]   **Deutscher Verband für Schweißtechnik e. V.**: Merkblatt DVS 2810 (September 1992).

[5]   **Rüdiger, T.**: Stand der Technik bei Ultraschallbonden. Studienarbeit am Heinz Nixdorf Institut der Universität-GH Paderborn, 1995.

[6]   **Draugelates, U.; König, K.H.**: Untersuchungen zur prozeßintegrierten Kontrolle des Bindungsvorganges beim Ultraschalldrahtbonden. 6. Intern. Kolloquium Verbindungstechnik in der Elektronik. Fellbach, 1992, DVS-Berichte 141, pp. 56 -61.

[7]   **Lückel, J.**: The concept of mechatronic function modules applied to compound active suspension systems. „Research Issues in Automotive Integrated Chassis Control Systems". International Symposium for Vehicle System Dynamics, Herbertov, CSFR, 1992.

[8]   **Gausemeier, J.; Frank, T.; Sabin, A.**: Lösungselemente als Grundlage des zukünftigen CAE-Prozesses. CAD '94, Carl Hanser Verlag, München 1994.

[9]   **Wallaschek, J.**: Modellierung und Simulation als Beitrag zur Verkürzung der Entwicklungszeiten mechatronischer Produkte. VDI-Tagung „Simulation in der Praxis - Produkte effizienter entwickeln", Fulda, 1995.

[10]   **Pahl, G.; Beitz, W.**: Konstruktionslehre. Springer Verlag, 3. Auflage, 1993.

[11]   **Kane, T. R.; Levinson, D.**: Dynamics - Theory and Applications. McGraw Hill, 1985.

[12]   **AUTOLEV**: A symbolic formula manipulator for dynamics. OnLine Dynamics Inc., USA.

[13]   **IEEE Standard on Piezoelectricity**, 1988. ANSI/IEEE Std. 176 - 1987.

[14]   **Smits, J.G.,; Ballato, A**.: Dynamic behavior of piezoelectric bimorphs. Proc. of 1993 IEEE-Ultrasonics Symposium, pp. 463 - 465.

[15]   **Hagood, N.W.; Chung, W.H.; von Flotow, A**.: Modelling of piezoelectric actuator dynamics for active structural control. Journal of Intelligent Material Systems and Structures, July 1990, pp. 327 - 354.

Dr.-Ing. Hans Hesse
Hesse & Knipps GmbH
Vattmannstraße 6
33100 Paderborn
Germany
Tel.: ++49-5251-156010
Fax: ++49-5251-156099

Prof. Dr.-Ing. Jörg Wallaschek
Heinz Nixdorf Institut
Universität-GH Paderborn
33095 Paderborn
Germany
Tel.: ++49-5251-603257
Fax: ++49-5251-603430

VIII. Vehicles

R. Busch
Development of the Control of a Fully Automized Hybrid Drive

A. Daberkow, M. Koch, N. Ott
*Mechatronic System Elements for Traction Control of Light
Railway Vehicles*

M. Hahn, J. Richert, J. Seuss
*Mechatronic Object-Oriented Modelling and Control Strategies
for Vehicle Convoy Driving*

A. Rükgauer, U. Petersen, W. Schiehlen
Mechatronic Steering of a Convoy Vehicle

Development of the control of a fully automized hybrid drive

Rainer Busch

Institut für Kraftfahrwesen Aachen, RWTH Aachen, Germany

Abstract: This paper presents the control system of a fully automatized hybrid vehicle which has been realized as a prototype. The hybrid vehicle has been developed at the ika for the Ford Motor Company. It is well-known under the term Ford-EHV. The "electric hybrid vehicle" contains a parallel hybrid driveline which is either powered by a combustion engine or by an electric motor. The power can even be combined in a torque addition gear box. The output is transferred via an automatic transmission to the wheels. This mechanical system has many degrees of freedom concerning the operating strategy. In intensive simulations using longitudinal simulation tools like "STAR", a so-called 2-dimensional strategy was developed to achieve best efficiency and drivebility.

The simulation forms the basis for the final "run-time" operating strategy. Because of the complexity of this mechatronical system, the implementation requires special standards for the hardware and software basis. To achieve great modularity and best efficiency for the implementation and maintenance process, a special high-end microcontroller system, based on the microcontroller SAB80C166 was developed. An optimized multitasking operating system is the basis of the control system, which realizes a task-depending observation and control of the various units in the powertrain. An additional element in this control forms the finite state description.

This great effort leads to the realized Ford EHV which has already shown its qualities on various test bench drives and in real life situations.

1 Introduction

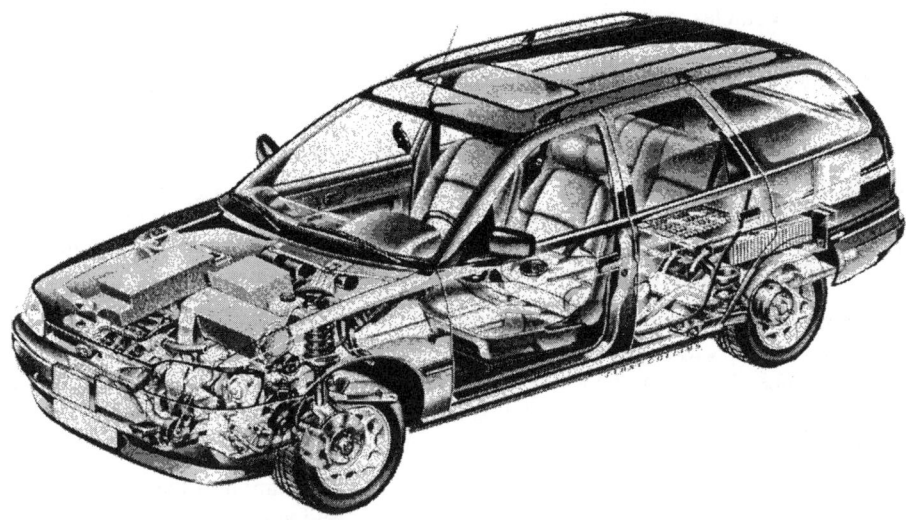

Figure 1: The Ford electric hybrid vehicle

The Ford electric hybrid vehicle was developed in close cooperation with the Institute of Automotive engineering in Aachen (ika) and the Ford Motor Company in Cologne. The development and the final realization in form of a prototype, required several years of work. The prototype was presented on the international automotive show (IAA) in Frankfurt ,1993.

Since this presentation additional tests have been performed. This includes tests based on standard driving cycles like the ECE75 or the FUDS-cycle, as well as the investigation of the behaviour in situations of daily usage.

The development was divided into four phases.

- Specification
- Simulation
- Prototyping
- Testing

Figure 2: Development steps

Forced by the laws of California, which commit the automotive companies to produce 2% of zero emission vehicles from 1998 on, new vehicle technologies are beeing investigated. These investigations are done especially for the US-market, but also for the European market. In this project we focussed our work to meet the European requirements.

It is obvious, that up to date zero emission vehicles can only be realized as electric driven vehicles. The main problem with the electric vehicles is the "tank". The storage of the electrical energy could not yet be resolved sufficiently. Therefore, the main problem with pure electrical vehicles is their range. Depending on the type and size, the range can vary up to 200 km or more. This is certainly dependant on various parameters like the weight of the vehicle, the driving topology and -style. The influence can be so extreme, that the possible driving range can be reduced to half or even less.

If we take into account that in European countries the share of second cars is much smaller than in the US, the range is one of the main factors which determines the overall customer value in use. Facing the fact that in European countries vehicles must allow daily short distance driving e.g to work, as well as weekend or holiday long distance journeys, a conventional powertrain is mandatory.

These facts lead to the specifications of the vehicle. After they had been developed, intensive simulations were performed to find out a drive line structure for best efficiency and performance. These simulations were done with the longitudinal simulation program STAR, which had been developed at the ika. The result of the simulation was the drive-line structure and a basic operating strategy.

The next step was to build up a prototype. For the implementation of the operating strategy a special controller hardware and software was developed.

Afterwards, final extensive tests and usability checks were performed to validate the simulation results.

2 Specifications

The specifications for this hybrid development are divided into several parts. Two of them shall be presented. These are:

- General vehicle specifications
- Basic control specifications

Figure 3: Specifications

The general vehicle specifications state the basic overall vehicle requirements. These are:

- Long distance journeys must be possible.
- Zero emission driving must be possible up to 60 kph.
- The vehicle must be user friendly.
- All systems needed to maintain the general functions must be in the car.
- The performance characteristics must be predictable.
- The vehicle should have the same performance data as conventional cars:
  - ➢ max. speed
    - ≥ 130 kph with the IC engine
    - ≥ 60 kph with the electric motor
  - ➢ acceleration

0 - 130 kph in 14 seconds (both aggregates)
0 - 100 kph in 16 seconds (only IC engine)
0 - 60 kph in 9 seconds (only electric motor)
➢ range
30 km in the ECE-cycle and additional 20 km with constant
velocity (50 kph)

Figure 4: General vehicle specifications

These requirements lead to those for the basic control algorithms, which have to be implemented into the central controller.

- The vehicle must be equipped with an 'intelligent' operating strategy, which selects the optimal driving source.
- The operating system should control the vehicle totally autonomously.
- The operating strategy must be designed in a way, that the driver can predict and influence the control scheme.
- The driver must be informed about the selected strategy.
- The vehicle must always be transferred into a safe state in case of a malfunction.

Figure 5: Basic control specifications

To fulfill the general vehicle specifications an intelligent operating strategy is demanded. This allows to decouple the driver from the complex drive line. Taking the demand of conventional driving into account, this leads to the requirement of an autonomous control system. However the driver must always be able to predict the vehicle's behaviour and have the possibility to influence the operating strategy. To give the driver a feedback from the control functions, the driver must get information about the actual driving strategy. Especially due to the high voltage of the battery, the vehicle must fulfill a high safety standard. This affects normal driving situations as well as the crash behaviour.

3 The hybrid drive train

In intensive simulations the presented hybrid driveline has been evaluated as the best compromise meeting the general vehicle requirements.

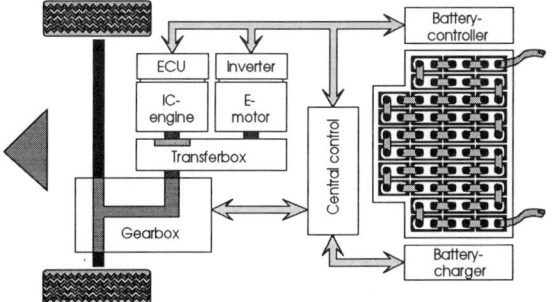

Characteristics	Data
type	Ford Escort
max. power	90 kW
max. speed	160 kph
max. range	unlimited
max. el. range	20-25 km
weight	1600 kg

Figure 7: Data of the EHV

Figure 6: General drive train structure

The drive train consists of six main components:

- IC engine (ICE)
- Electric motor (EM)
- Traction battery
- Transferbox
- Gearbox
- Central control

Figure 8: Main components of the drive train

Depending on the driver's power demand, the central control unit controls the IC-engine (ICE) and electric motor (EM). The power of the motors is added in the transferbox. This includes a one way clutch to decouple the ICE from the powertrain, when electrical driving is demanded. The output of the transferbox is led to the automatic gearbox, which feeds the wheels. Consequently, the electric motor can be used as a braking device and the battery can be charged by the EM. This is always done during braking or on a special driver demand via a separate button in the dashboard.

All components mentioned previously are located in the engine's compartment. They form one powertrain system which can be assembled outside the vehicle. The electric motor lies on top of the gearbox and is mounted to the transferbox which is located between the ICE and the gearbox. On top of the ICE the inverter of the EM is placed.

The electric motor is supplied with power from a NiCd battery, which is located below the trunk of the vehicle. It has a total energy of 7 kWh by a nominal voltage of 196.8 Volts. It consists of 164 sealed cells connected in one line.

This driveline structure allows three different driving modes:

- IC-Mode (only ICE is enabled)
- E-Mode (only EM is enabled)
- C-Mode (EM and ICE are enabled)

Figure 9: Driving modes

In IC-Mode or E-Mode, only one power source can be used to drive the vehicle. This is the ICE in the IC-Mode and similarly the EM in the E-Mode. Whereas the C-Mode allows the activation of both engines, they need not to be activated simultaneously. The use is defined in the main operating strategy.

4 The implementation of the operating strategy

The main control structure is given in the following figure:

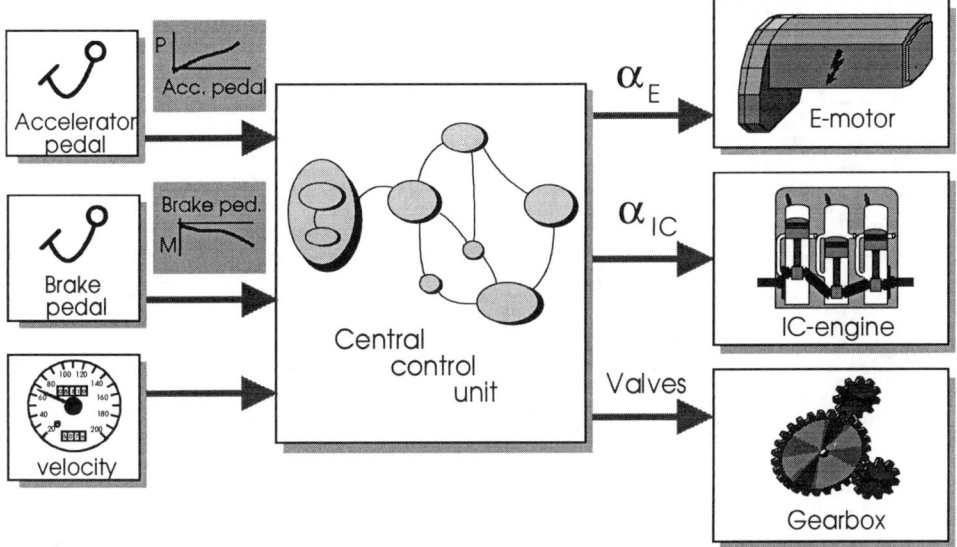

Figure 10: The main control structure

If the control task is limited to the basics, the above figure can be derived. By applying the accelerator or brake pedal the driver demands driving power or braking torque. These are splitted between those on the EM and ICE. The vehicle speed presents an additional parameter in the control, because decisions for the activation of the units are made as a function of the vehicle's speed. Furthermore the gearbox has to be controlled by the main controller due to the actual output torque and speed, as well as the powertrain's temporary state.

Beneath a high performance controller hardware, a special software basis has been developed to realize real time control systems.

The basis of the software implementation forms a multitasking operating system called MOS, which was developed at the ika. This allows multiple tasks in the system executing quasi concurrently the control of the whole vehicle. The tasks can have different priorities, so that intensive and important control loops get enough CPU time. The multitasking system has a time slice of 1 ms. It possesses fast dispatching algorithms and task synchronisation methods. A flexible message system allows task intercommunication. The scalebility of the program leads to small overhead due to the system itself.

This multitasking approach is used to divide the control of the whole vehicle into smaller parts which are defined analog to the existing main aggregates in the vehicle.

The control is separated into three layers. They are shown in the next figure.

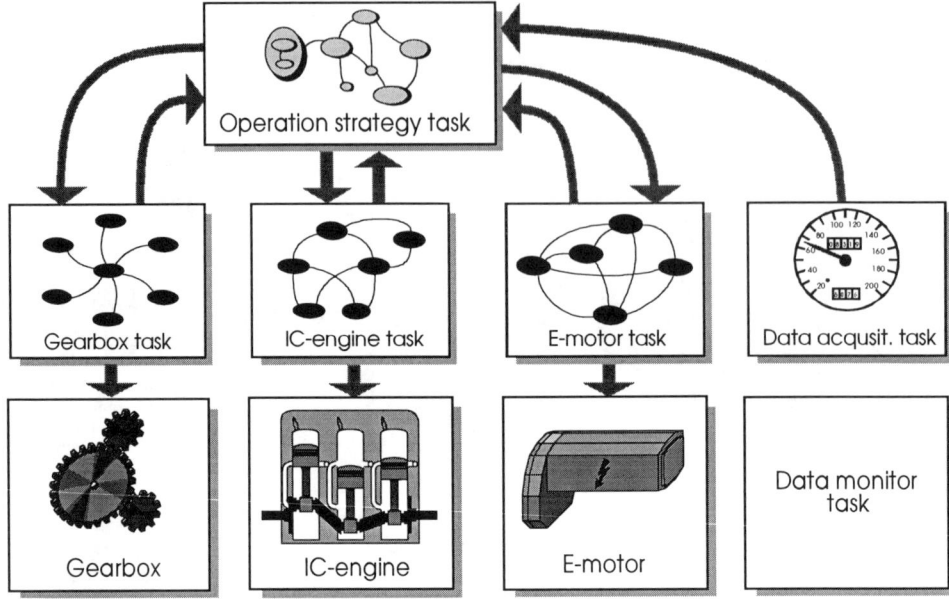

Figure 11: The software implementation

The first layer is determined by the aggregates themselves. Both the EM and ICE possess their own local controller, which realizes the inner fast control loop.

Above this layer, we find a second control layer, which is realized inside the central controller. For each of the main aggregates a separate control task is defined. These tasks are called:

- Gearbox task
- IC engine task
- E motor task

Figure 12: The aggregate tasks of the second layer

Additionally to the control of the aggregates, an observation of malfunctions is implemented in these tasks. This has to be done because the interface between central controller and aggregate is a low level one and not optimized for such a control concept. By observing additional signals in combination with high cycle frequencies, unexpected malfunctions are detected very fast and corresponding reactions can be performed.

Two more tasks of the second layer are shown in the figure. These are:

- Data acquisition task
- Data monitoring task

Figure 13: Additional data tasks of the second layer

The data acquisition task acquires global signals, which are relevant for all tasks including the main operating task. These are signals like:

- main ignition status
- speed of ICE and EM
- velocity
- battery s.o.c.
- battery voltage
- malfunction signals

Figure 14: Input signals of the data acquisition task

These values are transferred to the top layer (layer 3), which is presented by the operating strategy task. This task realizes the operating strategy. A well defined interface to the aggregates task on layer 2 let the information flow to and from this tasks. For example the main interface signals to or from the EM or ICE task are as follows:

- to the ICE or EM task:
  - ➢ demanded value
  - ➢ torque, power or speed demand (boolean flags)
- from the ICE or EM task:
  - ➢ actual torque, power or speed
  - ➢ max. available torque or power
  - ➢ error flags

Figure 15: Interface to the EM and ICE task

As one can imagine, the top layer task must always be informed about the control of the components. It can take the values fed back from the tasks of level 2, to decide which aggregates should produce positive or negative power.

The data monitoring task shown in the lower part, is located on the second layer. It allows the monitoring of the function of the central controller. In the next subsection some more information will be given.

As indicated in the previous picture, a further implementation concept is used. This concept is given by the use of a finite state description of the control laws.

This description is used for controlling the subsystemsas well as for the implementation of the main operating strategy.

If the ICE is analysed, the following six states are suited to describe the ICE's function:

- OFF
- START
- SYNC
- ACTIVE
- IDLE
- ERROR

Figure 16: States of the ICE task

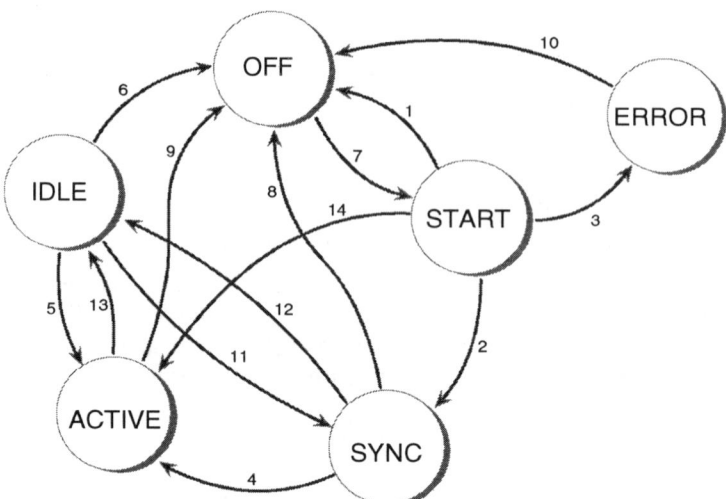

Figure 17: Finite state diagram of the IC engine control task

Like in the sequential logic, the states describe possible conditions in which the ICE can be. If the ICE is shut off, the ICE changes to the state OFF. The state START is entered, when the main operating task demands power from the ICE. After starting, the engine must be synchronized with the speed of the EM. This is done in the SYNC state. Then the state ACTIVE is entered and the ICE enabled to produce power. If the power demand is zero, the IDLE state is entered. After a determined time it will be left into the state OFF, if the operating strategy allows electric driving (C-Mode). In IC-Mode the OFF state will only be entered, when the ignition key is turned off.

In case of driving in IC-Mode, the electric motor always has the same speed as the ICE. Therefore no synchronization has to be performed and the transition from START to ACTIVE is taken.

If the cranking fails, the state ERROR is entered and not left until the ignition key is turned off.

To explain the main operating strategy's implementation, we first should have a look at one implemented driving strategy. Among the possible driving modes the combined Mode (C-Mode) is of main interest, because it has the greatest degree of freedom. This is caused by the free usability of the two motors by the main controller.

The next figure shows the 2-dimensional description of one strategy as an example.

It is called "2-dimensional strategy", because only the two parameters vehicle speed and power demand determine the operating of the subsystems in this simplified view.

If the driver demands a power, less than the maximum power of the EM and the vehicle's speed is below 65 kph, the EM works as the master engine. Above 65 kph and for a power demand less than 50 kW the vehicle is powered by the ICE engine only. If the driver increases his demand above the limit of the master engine, the slave engine will come in. However, above 105 kph to 121 kph the EM will be faded out, not to drain the battery too much. In braking situations the EM works as a generator and charges the battery.

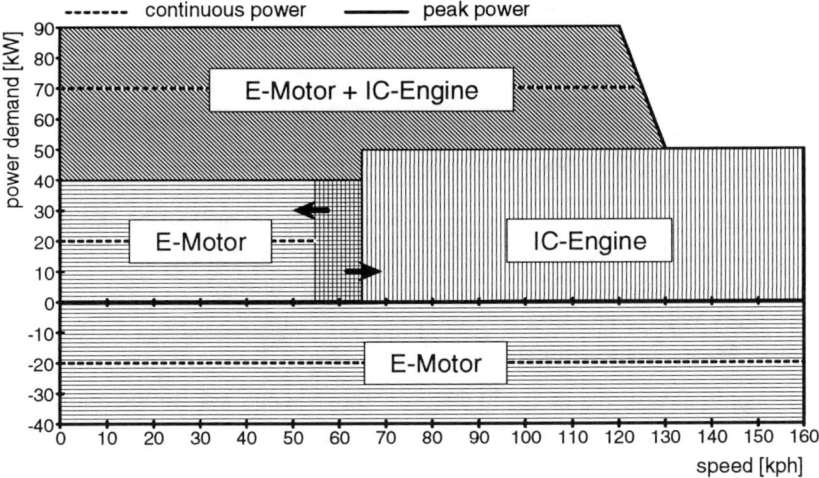

Figure 18: Basic driving strategy (C-Mode)

The implementation of this strategy is shown next, if we assume the driving is in C-Mode.

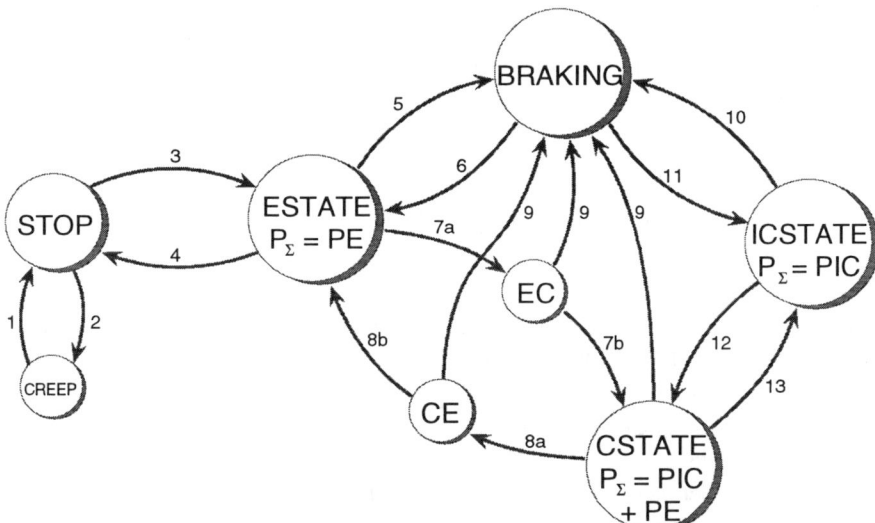

Figure 19: Finite state diagram of the main operating strategy task (C-Mode)

For each of the driving modes a separate state description is provided. As one can imagine, the transfer from one state to another needs special care, because smooth transition is demanded.

Eight states are necessary to implement the strategy of the C-Mode. These are:

- STOP/CREEP
- ESTATE/CSTATE/ICSTATE
- BRAKING
- EC
- CE

Figure 20: States of the main operating task

The state change is initialized depending on the following parameters:

- demanded driving power
- demanded braking torque
- cruising speed
- battery voltage and s.o.c.
- max. power of the EM and ICE
- shift lever position
- timing constants

Figure 21: Parameters for state transitions

The states should now be explained in the order they are entered, during normal driving.

The driver can only start the vehicle, if the shift lever is in position park. Afterwards, the operating system sets the C-Mode as the default mode. Therefore, the electric motor system is started and initialized. The main operating system sets the state STOP. Now the driver can shift the lever to a driving position. Usually he activates the brake in parallel. If he releases the brake pedal without applying the accelerator pedal, the state CREEP is entered. In this state the EM is idling with approx. 400 rpm. This is done for two reasons. Firstly, a sort of creeping is achieved and secondly the oil pressure of the torque converter is built up. If the driver now demands power, the state CREEP is left to the STOP state and the ESTATE is entered afterwards. Up to 65 kph, the vehicle will drive pure electrically, before the ICE takes over the power. For this the state CE is entered. The main operating task demands virtual power from the ICE. This leads the ICE task to start the engine. When the ICE has reached the ACTIVE state, the power is transferred from the EM to the ICE. If this is done, the main task changes to the CSTATE. In case the total demanded power is less than the ICE's maximum power, this state will be left in the next cycle to the ICSTATE.

If the driver brakes, the BRAKING state is set. It is reachable from every other state, if the velocity is over 5 kph. Otherwise electric braking is disabled.

If the driver accelerates after a braking situation, either the ESTATE or the ICSTATE will be entered. The next states will be the same as in conventional driving. This means, that depending on the demand, the ICE will be synchronized (and started, if it is OFF) and the EM power demand will be shifted to an ICE power demand.

5 The data debugging and logging system

For the development of this system a powerful data debugging and logging system is needed. It is called data monitor, although it allows tuning work as well.

This tool, consisting of two parts, was developed especially for this control. One part is implemented on a normal PC and works as a master. The slave part, called data monitor task, is implemented in the central controller as a separate task, which has access to all important variables, arrays etc.

It allows the following functions:

- On line data monitoring
- On line data logging
- Offline graphical or tabulated data presentation
- Parameter optimization
- Read task's error buffers
- Version number read functions

Figure 22: Functions of the data monitor

The main function is monitoring the important values during run time. Therefore the PC-part is requesting a transmission from the central controller. Since the data monitor task is implemented as the task with the lowest priority in the whole control system, the response will be not be done until enough CPU time is available for this function. Advantageously, the control will not be influenced by the logging work. Whereas it has the disadvantage, that its sampling frequency is limited. This limitation depends on various parameters. If the monitored values are shown on the PC screen the achievable rate is lower than by logging to disc. Generally, sampling rates from 10 to 20 Hz can be realized.

The online monitored data can be shown on four screens. These are:

- The main screen
- The ICE-task screen
- The EM-task screen
- The gearbox- task screen

Figure 23: Monitor screens

The main screen gives an overview of the whole control system. It's screen is shown as an example in the following:

In the upper part information is shown about the total demanded power and the internal demandings which are calculated from the main operating task divided for acceleration and braking.

Task information is given in the middle part. The state, speed and type of demand (power or torque) are presented.

Important data from the gearbox control is given in the lower left part. The selector position as well as the selected gear and lockup position are presented.

Signals, acquired by the data acquisition task, are illustrated in the lower right part.

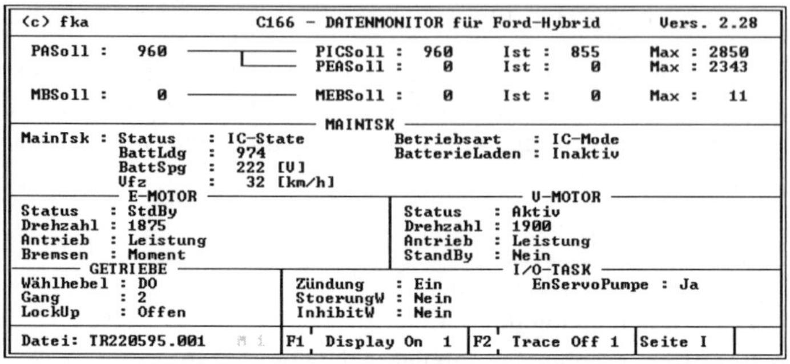

```
<c> fka                   C166 - DATENMONITOR für Ford-Hybrid      Uers. 2.28

PASoll :     960     ┌─────── PICSoll :   960    Ist :  855    Max : 2850
                     └────┐   PEASoll :     0    Ist :    0    Max : 2343

MBSoll :       0     ─────┘   MEBSoll :     0    Ist :    0    Max :   11
                        ─────── MAINTSK ───────
MainTsk : Status  : IC-State      Betriebsart  : IC-Mode
          BattLdg :  974          BatterieLaden : Inaktiv
          BattSpg :  222 [U]
          Ufz     :   32 [km/h]
       ── E-MOTOR ──                        ── U-MOTOR ──
Status    : StdBy                   Status    : Aktiv
Drehzahl  : 1875                    Drehzahl  : 1900
Antrieb   : Leistung                Antrieb   : Leistung
Bremsen   : Moment                  StandBy   : Nein
       ── GETRIEBE ──                        ── I/O-TASK ──
Wählhebel : DO          Zündung     : Ein        EnServoPumpe : Ja
Gang      : 2           StoerungW   : Nein
LockUp    : Offen       InhibitW    : Nein

Datei: TR220595.001      P1  Display On  1  P2  Trace Off 1  Seite I
```

Figure 24: The main screen of the data monitor

The status line of the screen gives information about the trace file, the data on which screen will be logged if logging is enabled and last but not least, the presented screen.

6 Results

For illustration of the control function, two driving situations shall be examined. These are:

- Power shift from EM to ICE
- Full power acceleration

Figure 25: Presented driving situations

The first figure shows the power shift, which is performed due to crossing the speed limit of 65kph.

The system handles this situation as follows. The main operating task switches from ESTATE to the EC state. In this state, a virtual power of '1' is demanded from the ICE-task. It is then forced to change to state START and the speed of the ICE increases. When the idle speed is crossed, the ICE task switches to the SYNC state. The speed of the ICE is increased by pure throttle control. If both aggregates are synchronized, the ICE task switches into the ACTIVE state. This is the start signal for the main operating task to shift the power from EM to ICE. After this has been performed, the main operating task enters the CSTATE and immediately the ICSTATE. This is done, because the ICE can fulfill the demanded power on it's own. At the end of the shifting procedure, the EM demand is zero, so that the EM task changes to STDBY.

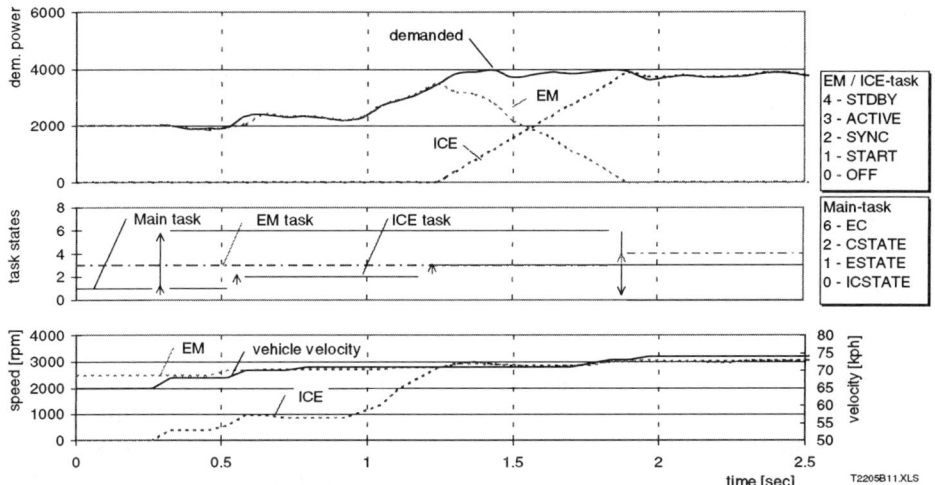

Figure 26: Power shift from EM to ICE

As one can see, this procedure needs about 1.7 seconds. This can be optimized by reducing the starting time of the ICE and by optimizing the synchronization. In parallel to the power shift, the gear shift strategy is changed to an optimized one for the ICE operation.

The next example shows the full power acceleration starting from zero speed.

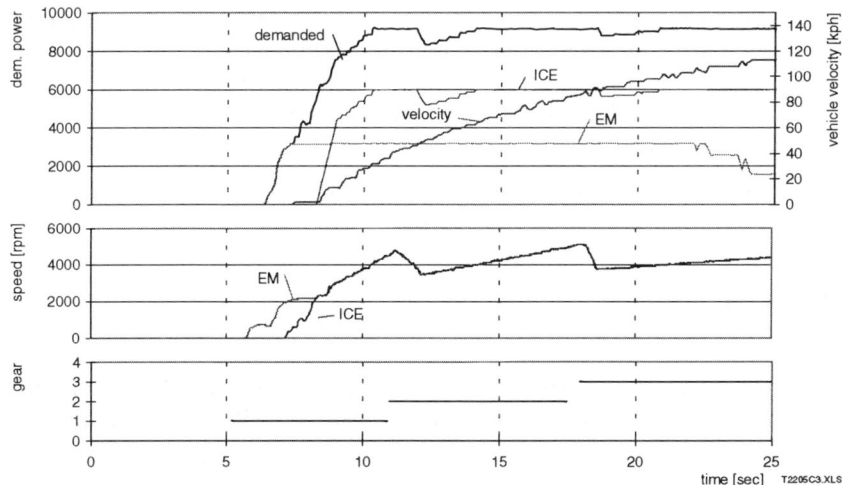

Figure 27: Full power acceleration

The demanded power is increased to 100 percent of available power which is calculated as the sum of maximum available power from the ICE and EM. The ordinate shows an internal representation of the power, which is not correlated with the real physical power. The reaction of the system is in that way, that the EM is run under full load. The ICE is started and synchronized. Its output is then added to the power of the EM. Due to the gear shifts from the 1st to the 2nd and 2nd to 3rd gear, the demanded power from the IC engine shows slight reductions. This is caused by the fact, that the ICE's maximum power is speed depending. Due to the fact, that the EM is faded out between 105 kph and 121 kph, the curve of the EM power demand decreases at the end of the cycle.

As one can see, the acceleration from 0 kph to 100 kph needs about 14 seconds because of a reduced output current of the battery. This reduction affects the available power from the EM as well.

7 Conclusions

In this paper a complex control system was presented, which was designed especially for real time control. Based on a powerful modular control hardware, the operating strategy was implemented. A specially developed and optimized multitasking operating system builds the basis of the control software. This allows the controlling of the aggregates in the powertrain quasi concurrently. The state space approach of the task structure helps to keep the control laws transparent. Easy optimizing work and maintenance are achieved. A data monitoring and logging system was developed to validate and optimize the behaviour of the control system. Additionally an error logging mechanism was implemented in every task, which stores error information in memory. This can be read by the data monitor to observe the system.

For future applications the control system hardware is redesigned. The Siemens micro-controller C167 will be used. This allows the interface of the CAN network to improve the hardware and software interface to the aggregates. Additional multiprocessor applications are supported, in which a fast (5Mbaud) serial synchronous link can be used for communication. The multitasking operating system will be extended to support the new features of the hardware.

Further improvements are made in the data monitoring and logging system to allow advanced observation and tuning work of the implemented control system.

8 References

[1] Buschhaus W   / Eggert, U.: Das Konzept des leistungsorientierten FORD-Hybri-dantriebs, Proceedings of the 4.th Aachener Kolloquium Fahrzeug und Motorentechnik, 1993

[2] Buschaus W.   Dissertation, "Entwicklung eines leistungsorientierten Hybridan-triebs mit vollautomatischer Betriebsstrategie", Institut für Kraftfahrwesen (ika), Aachen ISBN 3-925 194-29-0

[3] Wallentowitz H. /Busch R.: Proceedings of the Symposium "Connector 2000",
Mettingen, organizer: Stolberger Metallwerke
4.-5. May 1995,

[4] Ludes, R.   /Wallentowitz, H.: System Control Application For Hybrid
Vehicles, 3rd-IEEE Conference on Control applications (CCA)
Glasgow/Schottland

[5] Busch R.   /Renner C.: Fahrzeugtechnische Aspekte von Elektro- und Hybrid-
fahrzeugen, Haus der Technik, Essen, Tagung: "III. Essener
Elektroauto-Tagung", 26.10-27.10.94,

Autor: Dipl.-Ing. Rainer Busch
Institut für Kraftfahrwesen (ika),   Prof. Dr.-Ing. Henning Wallentowitz
RWTH Aachen
Steinbachstr. 10
52074 Aachen
Tel.: 0241 / 8056-18
Fax.: 0241 / 8888-147
e-mail: r.busch @ ika.rwth-aachen.de

Mechatronic system elements for traction control of light railway vehicles

Andreas Daberkow, Markus Koch und Norbert Ott

Forschungsinstitut AEG Automatisierungstechnik, Daimler-Benz AG,

F2A/M, Frankfurt; F2A/L, Berlin, Germany

Abstract: Due to costs and energy consumption, a weight reduction and new design of railway vehicles used in public transportation is necessary. Instead of using expensive track experiments, a simulation model serves to specify new, optimized vehicle concepts. Since light railway vehicles consist of 50% mechanical and 50% electronical equipment, a mechatronic simulation model is developed to investigate the vehicle and traction control dynamics. Thus, a multibody model is chosen for a dynamic analysis and combined with submodels for the asynchronous motors and traction slip and lock control. Furthermore, the control of independently driven wheels and its influence on the lateral dynamics is discussed by simulation examples.

1   Introduction

The need of an efficient public transportation in urban traffic systems results in high demands on light railway vehicles. Consequently, a low floor height, a noise and wear reduction down to a tolerable level is setting new standards to innovative railway vehicle and vehicle suspension concepts. Such new concepts with independently rotating wheels and a weight reduction by omitting a complete bogie lead to a new dynamic behaviour.

A traction system which consists of individual drive and brake units linked to each driven wheel demands an efficient traction control system. The traction control system of a light railway vehicle is a link between railway vehicle dynamics, sensors, data bus system, and drives, and therefore represents characteristic features of a mechatronical system. Its characteristics will be described by the interaction between lateral vehicle dynamics, drive system, and traction slip and lock control system.

To simulate the vehicle and traction dynamics, a mechanical model is derived for a low floor railway vehicle with three passenger wagons and three bogies. The equations of motion are calculated in symbolical form by commercial computer codes and are numerically solved within the interactive simulation tool KEASIM of the Daimler-Benz Research, Frankfurt. The mechanical model is completed by a time discrete traction slip and lock control system model developed by the Daimler Benz Research Berlin. To control the individual drive and brake units with respect to the railway track and traction condition, each drive unit gets primarily an equal desired value of torque. The interaction between railway vehicle dynamics, traction measurement, control, and coordination is finally described by a traction-enforced sinusoidal torque control for wear reduction on straight tracks.

2 Mechanical, electrical and electronical components of a light rail car vehicle

Light rail car vehicles in general consist of approximately 50 % mechanical and 50 % electrical and electronical components. The mechanical components include the car bodies, the joints between the car bodies, the bogies, the wheelsets, and spring and damper components of the suspension. The interface to the electrical equipment is given by the driving torques resulting from asynchronous motors and acting on the wheelset of each body. Speed sensors provide information about the actual wheel and motor speed for control purposes. A microprocessor compares the driver´s traction command and calculates the desired values of torques for each driven wheelset. Thus, together with the comunication system between sensor, microprocessor, actuator, and mechanical system all typical features of a *mechatronic system* are included [1].

2.1 Mechanical components

Figure 1 shows the light rail car of AEG Schienenfahrzeuge (ASF) Nürnberg with two joints and three car bodies. In opposite to traditional rail car design concepts with a bogie assigned under each joint and the front and rear car body, the ASF light rail car design concept saves up to 17 % weight by providing a bogie only under each car body.

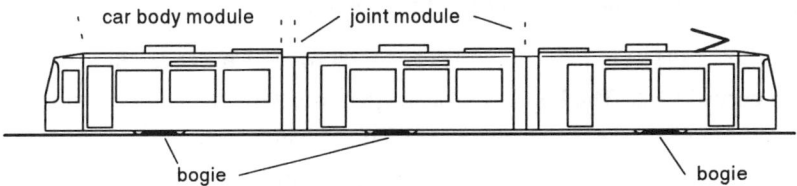

Figure 1: Light rail car of AEG Schienenfahrzeuge Nürnberg

The car body and joint modules can be assembled to larger units, e.g. a light rail car with four car bodies and four bogies. A low-floor design at about 300 mm entry floor height is achieved by omitting through-axles between the wheel centers of each wheelset. Figure 2 shows, that each bogie has two not-driven independently rotating wheels and two driven wheels, whose through-axle is located under the bogie frame. The connection of the through-axle to the wheels is realized by a spur gear. A cardanic shaft connects the drive and brake unit to the through-axle.

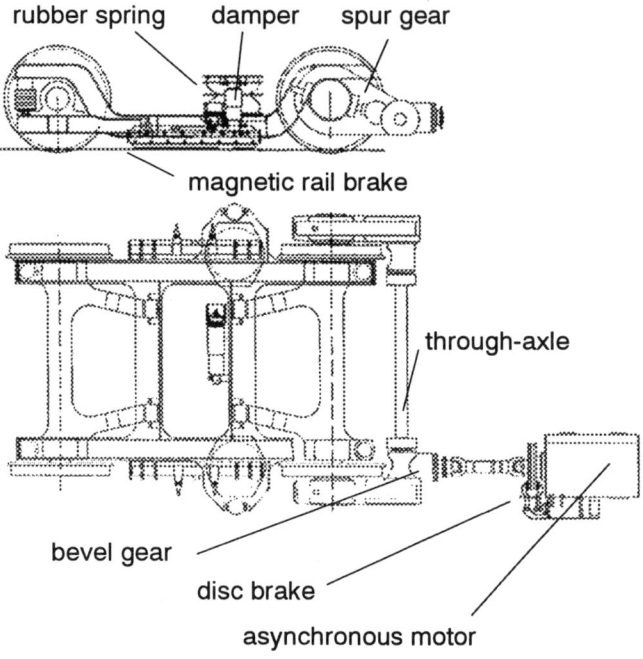

Figure 2: Bogie, drive and brake unit of ASF light rail car [2]

The secondary suspension is represented by rubber spring and hydraulic damper elements which connect car body and bogie. Besides comfort purposes, the two rubber spring elements serve to generate a torque along the vertical axis between the car bodies when angular deflections occur due to curved track conditions.

Figure 3 shows the top view of a typical car body and bogie position when running through a curved track. The bogie is located approximately in the middle of the track, the car bodiy is guided by the connecting joint modules and adjusted by the secondary spring torques. The traction torques at each driven wheel influence the longitudinal and lateral vehicle dynamics.

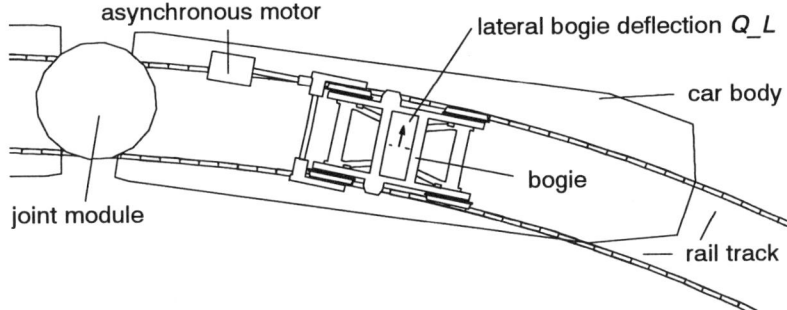

Figure 3: Top view of ASF light rail car running through a curved track

Different traction torques for each driven bogie results in longitudinal jerks on each car body. Different traction torques caused by wheel diameter differences or different slip conditions lead to lateral displacements of the bogie and the wheels.

2.2 Electrical and electronical components

Figure 4 shows the structure of the drive system with its electrical and electronical components. The interface between the mechanical and electrical system is given by the axle drive shaft that transmits torque and speed. The drive is realized with an induction motor supplied by a voltage source inverter. The inverter is linked to the d. c. voltage by a LC-filter that decouples inverter and line. Because of the comparatively low line voltage of 750 V the inverter could be realized with Insulated Gate Bipolar Transistors (IGBT) that allow high pulse frequency and therefore small torque ripples and a minimum of stress of the mechanical components.

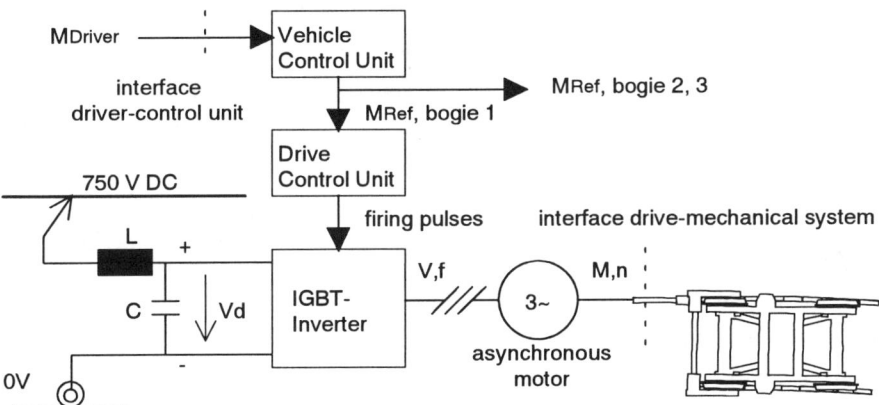

Figure 4: Electrical and electronical components of ASF light rail car

The control is composed of a vehicle control unit and a separate drive control unit for each drive. The traction torque demanded by the driver M_{Driver} is splitted and distributed to the drives by the vehicle control unit. In this connection the drive control of the front bogie gets a reference input M_{Ref} that is slightly higher than those of the following bogies in order to keep the vehicle always stretched. In the drive control unit the torque control and drive protection are realized with two microprocessors.

Figure 5 shows the main function blocks of the drive control unit. The reference input M_{Ref} which is given by the vehicle control first is corrected by the slip and lock control. The slip and lock control recognizes slip or lock by the speed difference between the driven and not driven wheel of the bogie. The controller will interact by reducing the torque reference input if the speed difference is higher than a given value. The motor torque can be reduced to zero in less than *200 ms* so that an effective slip and lock control is guaranteed.

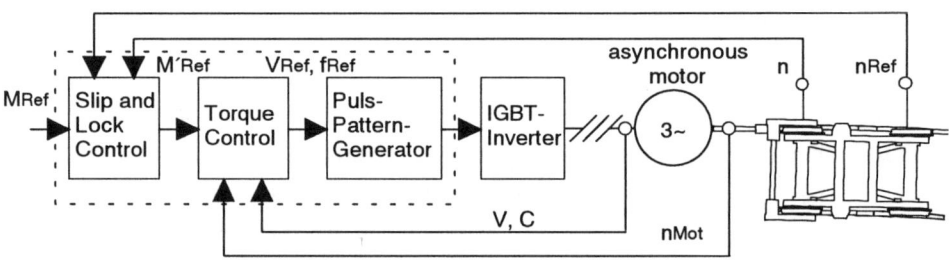

Figure 5: Function blocks of the drive control unit

The output M'_{Ref} is the reference input of the torque control. Since motor torque can not be measured is has to be calculated by measurable values of the motors voltage, current and speed. The torque controller calculates the necessary inverter output voltage V_{Ref} and output frequency f_{Ref} to realize the demanded motor torque. A pulse pattern generator determines the equivalent firing pulses. The step response time of the total torque control loop is 5 - 25 ms depending on the vehicle velocity.

To examine the lateral vehicle movements caused by dynamic drive and vehicle interactions, an overall analysis of vehicle and drive dynamics is necessary. This analysis will be performed by simulation and requires an appropriate *mechatronic simulation model*, which moreover serves to investigate single wheel drive concepts instead of the mechanical coupling of the driven wheels.

3   Simulation model for vehicle dynamics and traction control analysis

A dynamic and traction control analysis demands a proper simulation model. According to the low-frequent yaw motions of the car bodies when entering or leaving a curved track and the low-frequent lateral deflections of the bogies, a multibody model has been chosen for the dynamic investigation of the driven light rail car.

3.1   Multibody model

The car bodies, the bogie frame, the wheel suspensions and the wheels are rigid bodies which are connected by rigid and massless joint modules and coupled by massless spring and damper elements. Due to multibody modeling theory [4], the drive torques are modelled as applied torques acting at each driven wheel. Since the lateral and yaw motions are chosen to evaluate the dynamic behavior, a multibody system with planar degrees of freedom is sufficient for a dynamic analysis. Figure 6 illustrates the rigid bodies and force elements of the multibody system.

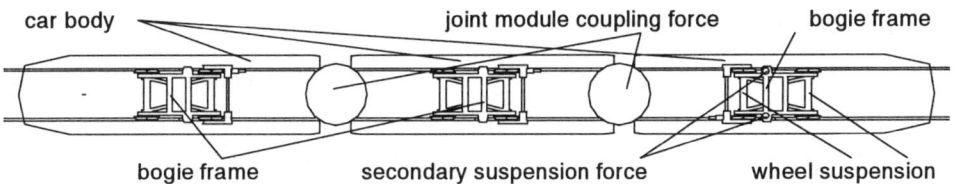

Figure 6: Multibody model of ASF light rail car

The detailed multibody model of the bogie is shown in figure 7. Each car body has a longitudinal, a lateral, and a rotational degree of freedom. The leading bogie frame has a longitudinal degree of freedom $Q1$, a lateral degree of freedom $Q2$, and a rotational degree of freedom $Q3$. Each wheel suspension has a lateral degree of freedom $Q4$ and $Q5$ with respect to the bogie frame and carries two wheels. The wheels have rotational degrees of freedoms $Q6$, $Q7$, $Q8$ and $Q9$ with respect to their axis of revolution. Creepage and wheel-rail contact forces guide the bogie in curved or straight tracks. The wheel-rail forces, the actual contact point location and force angle of each wheel are determined from a single-point contact computer code [5], which includes the numerical data of the digitized nonlinear wheel-rail geometry.

For each wheel, the creep forces are calculated from the relative velocities of the wheel-rail contact point with respect to the longitudinal and lateral bogie velocity [6].

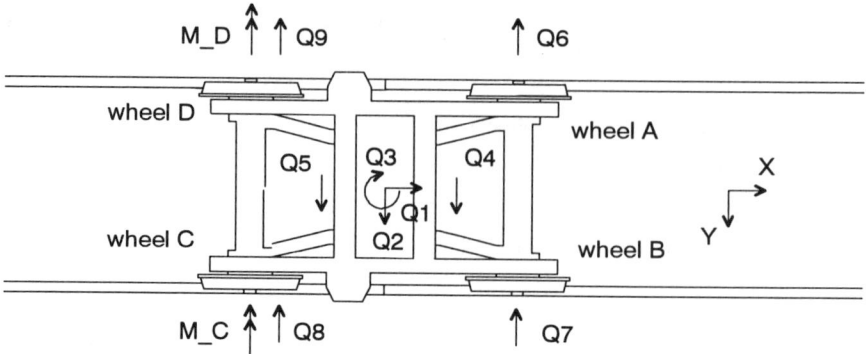

Figure 7: Multibody model of ASF light rail car bogie

The drive torques M_C and M_D act at wheel C and D, respectively, and result in longitudinal drive forces applied to the wheel suspensions. The equations of motion for the multibody system with $f=17$ degrees of freedom are derived from Jourdain´s principle, see e.g. [4], and can be written as

$$\dot{\mathbf{y}} = \mathbf{v}(t, \mathbf{y}, \mathbf{z}),\tag{1}$$

$$\mathbf{M}(\mathbf{y}, t)\dot{\mathbf{z}} + \mathbf{k}(\mathbf{y}, \mathbf{z}, t) = \mathbf{q}(\mathbf{y}, \mathbf{z}, t),\tag{2}$$

where M denotes the $f{\times}f$ mass matrix, y the $f{\times}1$ generalized coordinate position vector, and z the $g{\times}1$ generalized coordinate velocity vector. In case of holonomic constraints one obtains $f=g$ and v is a $f{\times}1$ kinematic relation vector between z and $\dot{y}$. The $f{\times}1$ vector k includes the centrifugal and coriolis acceleration. The $f{\times}1$ vector q of the generalized external forces includes the wheel drive torques M_C and M_D, which have to be provided by the drive and control system model.

3.2 Drive and control system model

The wheel drive torques that are used as an input of the mechanical multibody model are simulated with a simplified drive and control system model shown in figure 8. The simplification is useful with respect to simulation time. It could be done because of the large time constants of the mechanical system in comparison to the small time constants of the electrical system.

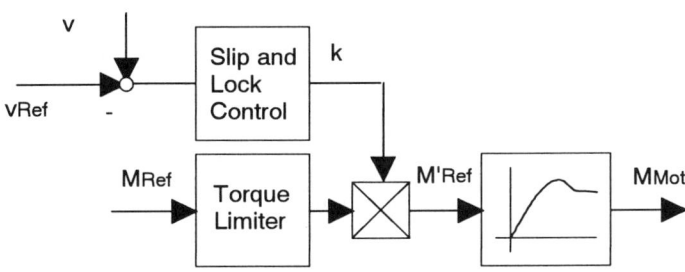

Figure 8: Simplified drive and control model

The torque transfer behavior of the controlled drive is modelled with a continious block of second order with time constants T_1 and T_2. Corresponding to the equations (1) and (2) the equation of the controlled drive can be written as follows:

$$T_2^2 \cdot \ddot{M}_{Mot} + T_1 \cdot \dot{M}_{Mot} + 1 = M'_{Ref} . \tag{3}$$

M_{Mot} is the sum of M_C and M_D (see above) in the case of axle coupled wheels. M'_{Ref} is the limited value od the demanded torque M_{Ref}.

The limitation is given by a torque limiter and the slip and lock control. The toque limiter simulates the maximum torque of the drive depending on the speed, the slip and lock control influences the torque reference multiplicatively. In regard to detailed simulation of the interaction of slip and lock control and lateral vehicle dynamics (see below), the the slip and lock control is modelled as a time discrete system with a sample rate of 10 ms.

3.3   Integrated multibody, drive and control simulation model

The symbolical equations of motion (1) and (2) including the drive control module (3) of the multibody system are solved numerically within the railway simulation system KEASIM [7],[8] of Daimler Benz Research, Frankfurt. KEASIM software modules include the numerical evalution of the wheel-rail contact and the track, a graphical result processing and allow a semi-automatically processing of a task-specific simulation module. The numerical integration is performed by a Runge-Kutta algorithm with variable stepsize, the simulation results are in compliance with measurements. To include the asynchronous motor dynamics, additional differential equations of the drive subsystem have to be solved and yield the drive torques from the input of the actual wheel velocity and of the desired value of torque for each time step. Figure 9 shows the software structure of KEASIM with the additional dynamic drive torque submodel.

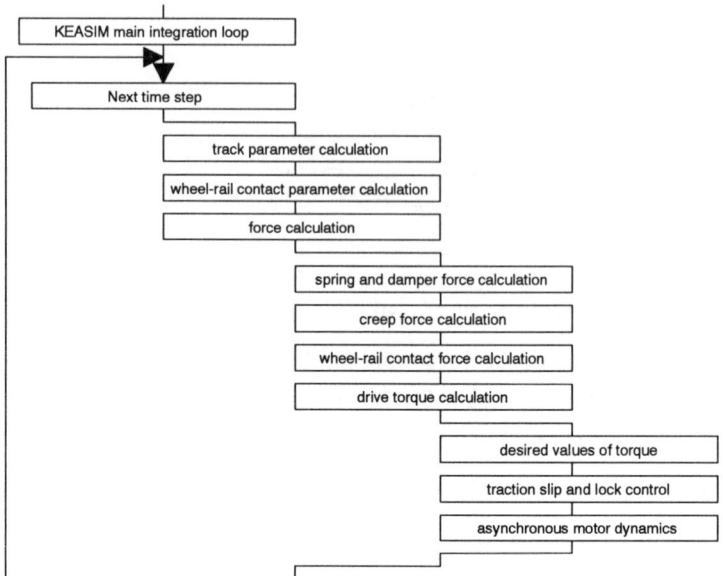

Figure 9: Software structure of KEASIM and drive submodel

From fig. 9 it becomes obvious that the sample rate of the time-discrete lock and slip control algorithm and the continuous dynamics of the asynchronous motor model reduces the step size of the numerical integration. A numerical integration of the asynchronous motor differential equations within the main KEASIM integration loop circumvents a very small integration stepsize, but increases preprocessing time because an automatic simulation module processing becomes more difficult.

Similar to the integration of the drive control subsystem within KEASIM, an integration of the symbolical equations of motion as a subsystem within a commercial interactive simulation tool like SIMULINK [9] was considered. From the view of runtime performance, software interface problems and software migration efford, see e.g. [10], it was decided to treat the drive subsystem as a submodel for the large railway vehicle model.

4   Applications and case studies

New railway vehicle concepts especially with independently rotating wheels lead to a new dynamic behaviour. On the one hand, individually driven wheels lack the lateral bogie oscillation enforced by a rigid wheelset in a straight track, which distributes wear over the wheel tread and recenters the bogie into the track. Consequently, one may consider a traction-enforced sinusoidal torque control for independently rotating wheels. On the other one hand, a loss of traction for one wheel may result in lateral jerks between railway vehicle and rail. In the following, these two case studies serve to discuss the dynamics of the mechatronic railway vehicle simulation model.

4.1 Traction-enforced sinusoidal torque control

To enforce a sinusoidal lateral bogie deflection, a preceeding analysis of the lateral dynamics of the vehicle with driven wheelset is necessary [11]. In fig. 10, the lateral oscillation of the railway vehicle from an initial lateral bogie deflection of 2 mm and an initial velocity of 16 m/s is illustrated.

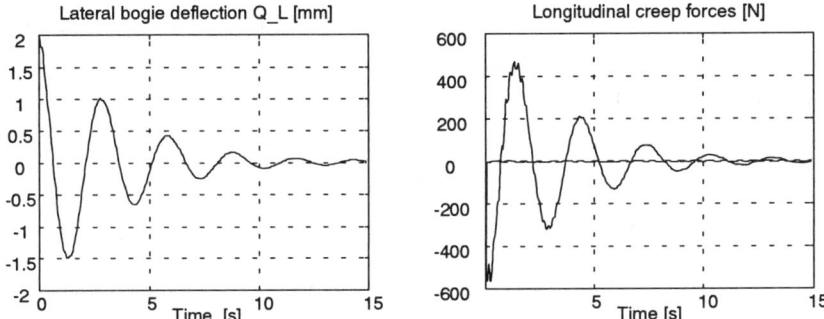

Figure 10: Initial lateral bogie deflection, creepage forces and lateral oscillation

It is obvious that the bogie is recentered by a yaw movement enforced by the different longitudinal creepage forces, see e.g. [6]. The velocity-dependent frequency and amplitude of an equivalent sinusoidal wheel torque can be estimated from fig. 10 and has to be superimposed to the driver´s desired value of torque. From [12], a square-wave torque with wavelenght l_{Sin} is chosen to superimpose the desired values of torque when the longitudinal vehicle velocity exceeds a predifined limit. Defining

$$M\_C = -M\_D = M_{Sin} \cdot \sin\left(\frac{2\pi \cdot v_{vehicle}}{l_{Sin}}\right) \qquad (3)$$

as the additional sinusoidal applied torque with amplitude M_{Sin}, one obtains the lateral bogie dynamics shown in fig. 11. The lateral deflection is sinusoidal but increases from zero to a mean value of approximately *4.2 mm*, indicating that the bogie is not recentered into the track and may lead to flanging within the available track clearance. This dynamic behaviour results from the small wheel-rail contact angles in the middle of the track.

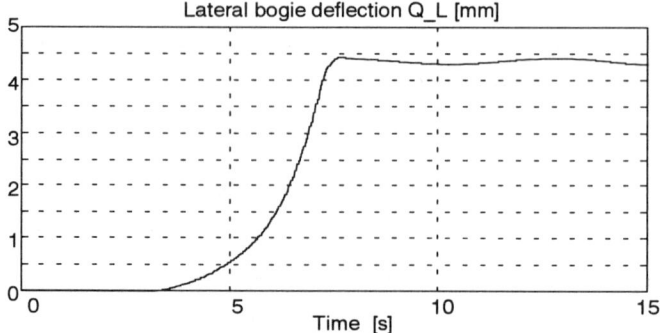

Figure 11: Sinusoidal applied torque and lateral bogie deflection

The wheels have a cone shaped contact geometry that leads to very small lateral contact forces and to a lateral bogie shift until the concave contact geometry of the flanging wheel increases the lateral contact force. Consequently, a sinusoidal torque control must take into account the actual lateral position of the bogie within the track. Good results which even shows a position-dependent amplitude of the torque are obtained by the asynchronous motor speed control algorithm suggested by [13].

4.2   Interaction of slip control and lateral vehicle dynamics

Slip and lock control is essential for light railway vehicles. A lock of a wheel results in local flattenings on the wheel surface and causes noisy and uncomfortable unbalances. A slip reduces the vehicle acceleration and increases wear of wheel and rail. Figure 12 shows the drive torques of wheel C and wheel D for an adhesion descent at $t= 4\ s$ and an adhesion of 100 % at $t=5\ s$.

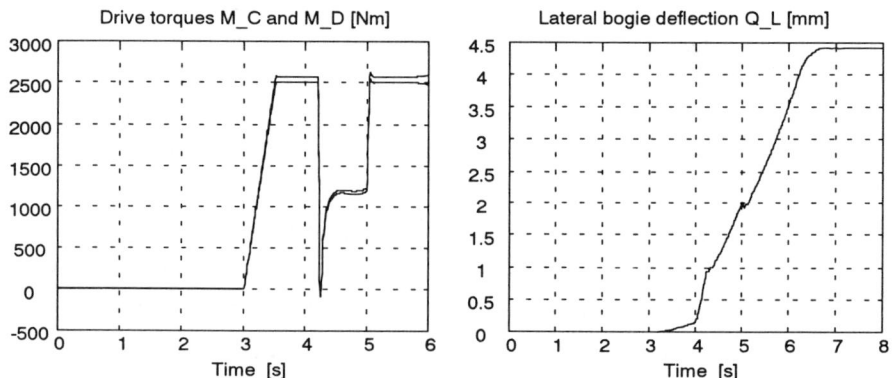

Figure 12: Adhesion decent, drive torques of wheel C and D and lateral bogie displacement

When the difference between wheel speed and vehicle velocity exceeds its upper limit, slip and lock control reduce the desired values of torque to the value calculated by the comparison of asynchronous motor speed and vehicle speed for both driven wheels and resets the desired values of torque at $t=5\ s$ when the adhesion has increased. The lateral deflection of the bogie shows a slight lateral movement. Further simulations yield to uncritical vehicle dynamics even in tracks with small curve radiuses and only one wheel driven.

5 Conclusions

A railway vehicle with approximately 50 % mechanical and 50 % electrical equipment is a typical application for a mechatronic system approach. Since track tests and measurements are very time consuming and expensive, enhanced capabilities in comfort, economy and price have to be investigated by simulation.

A multibody model is developed to simulate the longitudinal and lateral dynamics of a light railway vehicle with three passenger cars. The mechanical model is completed by a drive model and a time discrete traction slip and lock control system model. The drive torques applied to the wheels and the vehicle and wheel velocity yield the interface between mechanical and electrical model. Additional differential and difference equations are solved numerically to include the drive and traction control system dynamics.

The simulation example of a sinusoidal traction control shows that the lateral dynamics of the railway vehicle can be controlled by individually driven wheels. To avoid a lateral shift with nonzero mean value of lateral deflection, additional information of the bogie's track position has to be measured. Further simulations with different track adhesion conditions show that slip and lock control of individually driven wheels prevent critical vehicle dynamics and wheel wear.

6 References

[1] Bradley, D.A. / Dawson, D. / Burd, N.C. / Loader, A. J.: Mechatronics, London, Chapmann & Hall, 1991.

[2] Uebel, L.: Die Niederflurwagen München und Augsburg. In: Seminarband Nahverkehrsbahnen 2000. Hamburg: SNV, 1992.

[3]   Koch, M./ Liu, S.: Analyse der Antriebstechnik der Niederflur-Straßenbahn Zwickeau im Hinblick auf Einzelrad-Antriebskonzeptionen. Technische Notiz A86 IAL Nr. 7. Berlin: AEG, 1994.

[4]   Schiehlen, W.: Technische Dynamik. Stuttgart: Teubner, 1986.

[5]   Wallaschek, J. / Mahrt, R.: Die Berührgeometrie des Einzelrades aud idealem geradem Gleis. Technischer Bericht FAF 90.011. Frankfurt: Daimler Benz, 1991.

[6]   Frederich, F.: Möglichkeiten zur Hochausnutzung des Rad/Schiene-Kraftschlusses - Zusammenhänge, Einflüsse, Maßnahmen. In: Archiv für Eisenbahntechnik AET (38), 1983.

[7]   Himmelstein, G. / Herrmann, H.: Software-Konzept für das Simulationssystem KEASIM. Technischer Bericht F2A-92-017. Frankfurt: Daimler Benz, 1992.

[8]   Daberkow, A. / Himmelstein, G. / Krause, R.: Simulation und Visualisierung des dynamischen Hüllraumbedarfs von Straßenbahnen, 9. ASIM-Symposium Simulationstechnik 94, Stuttgart, Oct. 10-13, 1994, pp. 481-486.

[9]   N.N.: SIMULINK User´s Guide. Natick, USA: The Math Works Inc., 1993.

[10]   Rauh, J.: Fahrdynamik-Simulation in CASCaDE. In: VDI_Bericht Nr 816. Düsseldorf: VDI, 1991.

[11]   Daberkow, A.: Konzeptionelle Untersuchungen von Einzelrad- und Bremsmodulen für den Nahverkehr unter besonderer Berücksichtigung der Laufdynamik. Technische Notiz F2A-94-017. Frankfurt: Daimler Benz, 1994.

[12]   Lehotzky, P.: Niederflur-Straßenbahn-System für Wien. In: Der Nahverkehr 5/93. Freiburg: EK, 1993.

[13]   Koch, M. / Kipke, M.: Ein einfaches Einzelrad-Antriebs-Konzept (Basiskonzept) für Straßenbahnen. Berlin: Technischer Bericht F2-IAL-94.003, 1994.

Mechatronic Object-Oriented Modelling and Control Strategies for Vehicle Convoy Driving

Martin Hahn, Jobst Richert, Jürgen Seuss,
Mechatronics Laboratory Paderborn (MLaP), Paderborn, Germany

Abstract This paper reports on the contribution of the Paderborn group to a co-project[1] of five German research groups. By means of an example from vehicle dynamics, the vehicle convoy driving, modelling, analysis and synthesis approaches of a mechatronics integration project are presented. The new object-oriented modelling method for mechatronic systems, developed by the MLaP, is introduced in an exemplary way. The particular importance of analytical and therefore symbolic modelling is explained. Based on this approach control strategies for lateral and longitudinal dynamics and obtained control results are outlined. Applications in laboratory and field works are shown. Finally will be presented the software environment CAMeL, developed at the MLaP and used for the integrated design of mechatronic systems.

1 Introduction

Mechatronics claims to be more than just a conventional treatment of electro-mechanical systems. It does not imply additional and supplementary completion of passive mechanical structures by controlled active components but an integrated development of these systems with the aim of refining the mechanics using modern information processing in combination with intelligent sensors and actuators.

The great number of subsystems from heterogeneous disciplines (e. g. from mechanics, hydraulics, electronics and information processing) and the increased use of active components with suitable sensors explain the ever increasing complexity of these systems.

An indispensable first step in a systematic design of mechatronic systems is a unified and thus integrated description of all system components involved.

In this contribution the class based approach of the Mechatronics Laboratory Paderborn (MLaP) is introduced, which leads via different formalisms and categorization concepts to the mathematical model, used for analysis and the controller design. This methodology will be exemplified by the benchmark problem of *"Automatic Towing"* or, in more general terms, *"Vehicle Convoy Driving"* which is the focus of a compound project that five university groups are currently working upon [1].

"Automatic Towing" means the cooperative follow-up of two or more vehicles, of which only the leading one is guided by a human driver. The following car, in dependency on the development stage, is coupled to the leading one in one of various ways. In a first step, the two cars were connected by a mechanical connection with integrated sensors (*Automatic Towing*). Later, the project work concentrated on autonomous cars without

---
[1] The project has been sponsored by the VOLKSWAGEN foundation (AZ I/67 975-9 and AZ I/70 092-6).

a mechanical connection (*Vehicle Convoy Driving - VCD*). The distance between the two cars is measured by an image-processing system based on a CCD camera and a PC.

2 Mechatronic Object-Oriented Modelling

For the last three years, a new modelling technique with its corresponding software has been developed at the MLaP [2],[3]; it is called Mechatronic Object-Oriented Modelling (MOOMo). Three key concepts have been introduced into MOOMo:

- the description of mechatronic systems in a uniform manner,
- the unique integration of derivation formalisms of the mathematical model, and
- a unique internal description of systems topology.

In object-orientation, as used in this approach, all elements are objects and the common behaviour of objects is represented by their class. These classes are related by inheritance, a concept well known in object-orientation to support reuse and structuring of complex systems. The concepts are used for representing models in MOOMo, along with their accompanying modelling language, called Objective-DSS. At the center of interest in Objective-DSS models are the three modelling elements **part**, **connection**, and **aggregate**. They are represented by the modelling elements **basic element**, **couple element**, and **hierarchically composed structure**. These modelling elements represented by the classes

- *DssBasicElement*[2],
- *DssCoupleElement*, and
- *DssHierarchicalElement*

are introduced as basic classes of Objective-DSS. Subclasses of these basic classes are derived to represent parts, couple elements, and aggregates of different subject-oriented descriptions (multibody systems, control engineering systems, etc.) and to combine them in the same aggregate.

This chapter will show the advantages of the Objective-DSS modelling facilities and the unified approach for deriving the mathematical model with the help of different formalisms, exemplified by articulated vehicles. The underlying assumptions employed in the modelling of individual cars will be discussed in the next subchapter.

2.1 Modelling Assumptions

The single-track model approach of Riekert and Schunck [4] is very suitable for control strategy synthesis. The equations of motion at a constant longitudinal velocity are described by the lateral acceleration $\ddot{y}$ and the yaw acceleration $\ddot{\psi}$:

$$m\ddot{y} = \sum F_y = -F_{WR} - F_{WF} \cos\delta \tag{1}$$

$$I_{ZZ}\ddot{\psi} = \sum M_{ZZ} = -x_{WF} F_{WF} \cos\delta + x_{WR} F_{WR} \tag{2}$$

---

[2] In the following text class names and keywords of the description language are in italics.

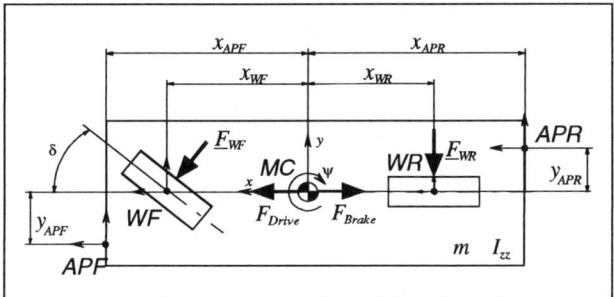

Fig. 1: Single track model of lateral dynamics

This very simple dynamics model is based on different assumptions of which only the essential ones will be outlined here. The center of gravity of the system is at road-level. Thus the dynamic load of the tyres is not influenced by the centrifugal forces. Vertical, pitch and roll motions are neglected. The tyres are considered to be without mass.

To obtain the tyre forces F_{WF}, F_{WR} different tyre models can be employed [5]. Small motions and therefore linear tyre forces are taken into account as follows:

$$F_{Wi} = C_i \, \alpha_i \tag{3}$$

At first glance the steering will be handled as an ideal gear without any elastic or damping characteristics, with an ideal efficiency of $\eta = 1$. This means that the transfer function from steering wheel to steering angle at the front wheels can be calculated simply by the transmission rate:

$$\delta = i_{Steering} \delta_{Steeringwheel} \tag{4}$$

In correspondence with the modelling depth of the other submodels, the steering actuator is treated as a first-order system with a high cut frequency or with a simple proportional system, representing an ideal actuator. This actuator can be seen as a subordinate *Mechatronic Function Module (MFM)* [6],[7] obtaining its steering angle input by digital information connection. By using the principles of MFM it will be possible to exchange this simple model for a model describing the dynamic behaviour in more detail. For a more refined actuator and a detailed steering model approach, see the contribution by a project partner [8].

2.2 Lateral Dynamics Model - A Combination of Multibody Systems and Control Engineering Systems

There are two different groups of parts used for the modelling of the Lateral Dynamics Model: first, parts and connections as used in multibody system dynamics and second, hierarchical block diagrams as used in control engineering. The lateral dynamics model of the car can be divided into a rigid body and a mathematical block containing the calculation of the wheel forces in dependence of the velocities of the rigid body and their connections.

Modelling elements of Objective-DSS: Hierarchical systems in Objective-DSS are a combination of hierarchical systems themselves (recursive formulation) and basic elements (the atomic models of the description), both connected via couple elements. The basic elements used for the implementation of the lateral dynamics model are the following:

- rigid body (class: *DssRigidBody*)
- multibody systems environment (class: *DssMultibodyEnvironment*)
- nonlinear state-space block (class: *DssMathBe*)

There are several features all modelling elements have in common. All elements are divided into an interface and an implementation. The interface of a model consists of the name of the element and the ports. Before ports are used to designate connections between elements, it has to be made sure if those ports which should be connected are even valid for the operation. The objects „ports" are divided into different groups: topological ports, input ports, output ports, and parameter ports.

Topological ports: Topological ports are used for describing physical/subject-oriented connection points. These connection points do not act as inputs or outputs. A connection of two topological ports means a physical/subject-oriented coupling, e.g. the topological port class introduced for multibody systems is *DssMbsAtp*. These ports of two bodies are connected via joints (rotational/translational/mixed joints) to tree- or closed-loop structures and finally to multibody systems.

```
DssPort
DssMbsPort
      DssMbsAtp
DssMbsInput
      DssMbsForce
            DssMbsForceICS
            DssMbsForceBCS
            DssMbsForceJCS
      DssMbsMoment...
DssMbsOutput
      DssMbsAngularVelocity
            DssMbsAngularVelocityAbs
                  DssMbsAngularVelocityAbsICS
                  DssMbsAngularVelocityAbsBCS
                  DssMbsAngularVelocityAbsJCS
            DssMbsAngularVelocityRel...
      DssMbsPosition
            DssMbsPositionAbs
                  DssMbsPositionAbsICS
                  DssMbsPositionAbsBCS
                  DssMbsPositionAbsJCS
            DssMbsAngularPositionRel...
      DssMbsOrientation
            DssMbsTransformationMatrix
            DssMbsCardan...
      DssMbsVelocity...
```

Fig. 2: Part of the port class hierarchy in O-DSS for multibody systems

Input ports: Input ports are used for the description of input signals, well-known in tools used for modelling hierarchical block diagrams. For multibody systems, different port classes are declared as input classes acting on a rigid body. The classes shown in Fig. 2 represent forces and moments in different coordinate systems of the body *(JCS* = attachment-point coordinate system; *BCS* = body coordinate system; *ICS* = inertial coordinate system). For the force and moment inputs it is necessary to be located on a point of the rigid body.

Output ports: Output ports are used for the description of output signals. Output ports acting on a rigid body are located via a frame on the element, in analogy to the inputs. There is a rich set of output port classes, shown in Fig. 2, to yield information on absolute and relative positions, velocities, orientations, and angular velocities. Different interpretations of the output ports denoting the orientations can be obtained as transformation matrix-, Euler angle-, cardan angle- or quaternion vector interpretation.

Parameter ports: Objective-DSS allows parametrization of the mathematical objects; parametrization of the physical/subject-oriented objects is under way. Parameter ports are objects containing mathematical objects, such as scalars, vectors, matrices or record-like components. The implementation varies in dependence of the modelling element. The modelling elements employed for the lateral dynamics model will be explained in the next sections.

Elements of the class *DssRigidBody:* As an example, Fig. 3 shows the Objective-DSS description of the rigid body used in the lateral dynamics model. The essential parts of the Objective-DSS description of the model are marked by circled numbers. ① The first keyword in every Objective-DSS description element is the class the object with the given name *carbodyType* is instanced of. ② The first element introduced into the interface is the definition of the ports used for topological couplings. Three ports are introduced here: MC, APF, and APR, connected to the class representing their behaviour, *DssMbsAtp*. ③ Then inputs have to be defined (optional). Inputs of this rigid body are forces in the body coordinate system attached to the designated frame. ④ The following outputs are declared for the model: the absolute velocities measured in the attachment-point coordinate

```
DssRigidBody named: carbodyType. ①

port:      #(MC APF APR)       on: DssMbsAtp; ②

input:     #(F_WF F_WR)        on: DssMbsForceBcs; ③

output:  #(v_WF v_WR)          on: DssMbsVelocityAbsJcs; ④
           #(r_MC)             on: DssMbsPositionAbsIcs;
           #(T_APF T_APR)      on: DssMbsOrientation;

auxiliar:  #(MC_Frame APF_Frame APR_Frame
             WF_Frame WR_Frame) on: DssVector size: 6; ⑤

auxiliarEquation:
   MC_Frame   := #( 0.000  0.0  0.0  0.0  0.0  0.0);
   APF_Frame  := #( 2.230  0.0  0.0  0.0  0.0  0.0);
   APR_Frame  := #(-2.185  0.0  0.0  0.0  0.0  0.0);
   WF_Frame   := #( 1.380  0.0  0.0  0.0  0.0  0.0);
   WR_Frame   := #(-1.380  0.0  0.0  0.0  0.0  0.0);

portsNamed:
   #(APF T_APF)  frame: APF_Frame; ⑥
   #(APR T_APR)  frame: APR_Frame;
   #(F_WF v_WF)  frame: WF_Frame;
   #(F_WR v_WR)  frame: WR_Frame;
   #(MC    r_MC) frame: MC_Frame;

mass:              1550.0; ⑦

inertiaTensor:   #(1.0  1.0  2725.0  0.0  0.0  0.0); ⑧
end. ⑨
```

Fig. 3: The rigid body model description

system *(JCS)*, the position vector measured in the inertial coordinate system *(ICS)*, and the orientation of the designated frame relative to the inertial coordinate system.
⑤ For the calculation of auxiliar numerical values, such as frames or angles, reduced masses, inertias, etc., mathematical variables called auxiliars can be used in the description elements. These variables can be calculated with in the section introduced by the keyword *auxiliarEquation:* . ⑥ Every port, input and output in a rigid body has to be assigned to a point on the body´s structure. This is done in the section introduced by the keyword *portsNamed:*. The port APF, for instance, is assigned to the frame APF_Frame, a vector designating the translational and rotational displacement relative to the mass center of the body measured in the body coordinate system. ⑦ and ⑧ contain the mass and inertia values of the body. ⑨ At the end of every description element, the keyword *end* is introduced.

Elements of the class *DssMathBe:* To calculate the wheel forces in dependence of the velocity of the wheel coordinate systems a mathematical block of the class *DssMathBe* is used, shown in Fig. 4. The elements of the class *DssMathBe* represent parametrizable nonlinear state-space models.

For the declaration of the parameters, inputs and outputs as well as for the calculation of the auxiliars, states and outputs vectors, matrices, scalars, booleans and record-like objects can be used. Here only some special properties will be explained, marked by circled numbers. ① The description is fully parametrizable. In this example the stiffness of the tyres (a simple linear tyre model is used here) is taken for a parameter. ② As inputs of the model, the velocities of the front and the rear wheel are declared as well as the steering angle of the front wheel. As output the wheel forces are declared as vectors of

```
DssMathBe named: wheelForcesType.
  parameter: #(c_WFL c_WRL)        on: DssScalar; ①
  input:       #(v_WF  v_WR)        on: DssVector size: 3; ②
               #(delta)             on: DssScalar;
  output:      #(F_WF  F_WR)        on: DssVector size: 3;

  auxiliar:    #(pi into_grad delta_rad) on: DssScalar;
               #(delta_vec)             on: DssCardan size: 3; ③
               #(F_WFL F_WRL)           on: DssScalar;
               #(dir)                   on: DssVector size: 3;

  auxiliarEquation:
    pi         := 4 * 1 arcTan;
    into_grad  := 1 / pi * 180;
    delta_rad  := delta * pi / 180.0;

    delta_vec at: 1 put: 0;
    delta_vec at: 2 put: 0;
    delta_vec at: 3 put: delta_rad;

    F_WFL := c_WFL * (delta_rad - (((v_WF at: 2)/(v_WF at: 1))
             arcTan));
    F_WRL := c_WRL * (((v_WR at: 2)/(v_WR at: 1)) arcTan);

    dir := #(0 1 0);
④
  outputEquation:
    F_WF :=  delta_vec asTransfMatrix asTransposed * F_WFL * dir; ⑤
    F_WR :=  F_WRL * dir negated;
end.
```

Fig. 4: The wheel forces calculation description

the size 3. ③ There are special classes representing variables useful for the calculation of values well known in multibody dynamics, e. g. the class *DssEuler* for euler angles, *DssCardan* for cardan angles, and *DssFrame* for record-like objects with the components position, velocity, orientation and angular velocity and their accompanying algebra, i. e. the transformation or the kinematic matrices.
④ Because of the fact that objects of the class *DssMathBe* are meant to represent nonlinear state-space systems a keyword is provided to formulate the state-space equations (not necessary for this model). For the declaration of the states all variable classes can be used, such as vectors, matrices, scalars, and record-like elements with their accompanying algebra. ⑤ In the output equation section the outputs are calculated in dependence of the parameters, inputs, auxiliars, and states. Here the special algebra introduced for objects representing cardan angles is used to yield the transformation matrix, transpose it and multiply it by the lateral wheel force of the front wheel.

Elements of the class *DssMechatronicHcs*: The rigid body and the wheel forces block are combined with the lateral dynamics model of the car. The interface of assembly objects which are instances of the class *DssMechatronicHcs* may contain the same elements as the single basic elements. The ports of *DssMechatronicHcs* elements are used to declare which subsystem ports are visible on the next hierarchical level. In this model three different port categories are visible on the next hierarchical level: MC_APF, APF and APR which are ports of the class *DssMbsAtp*, delta which is a scalar input and T_APF which represents a transformation matrix of the dimension 3.

The ports are connected to the designated subsystem ports via

```
DssMechatronicHcs named: lateralDynamicsType.
  port:     #(MC APF APR) on: DssMbsAtp;
  input:    #(delta)      on: DssScalar;
  output: #(T_APF)        on: DssTransformationMatrix size:9;

  auxiliar: #(c_WFL c_WRL) on: DssScalar;

  auxiliarEquation:
    c_WFL := 130000;
    c_WRL := 130000;

  partTypes:    #(carbodyType wheelForcesType);

  partDefinition:
    wheelForces is: wheelForcesType
              on: #(c_WFL => c_WFL
                    c_WRL => c_WRL);
    carbody     is: carbodyType;

  inputCondition:
    wheelForces at: delta     connectTo: delta;

  coupleCondition:
    wheelForces at: v_WF connectTo: carbody     at: v_WF;
    wheelForces at: v_WR connectTo: carbody     at: v_WR;
    carbody     at: F_WF connectTo: wheelForces at: F_WF;
    carbody     at: F_WR connectTo: wheelForces at: F_WR;

  wireCondition:
    MC  connectTo: carbody at: MC;
    APF connectTo: carbody at: APF;
    APR connectTo: carbody at: APR;

  outputCondition:
    T_APF connectTo: carbody at: T_APF;
end.
```

Fig. 5: The lateral dynamics model

parameter (not used in this example), input, output and wire conditions. Connections between the different parts of the system can be made by two different connection mechanisms: assigning the value of an output to an input of a system which has been introduced by the keyword: *coupleCondition:* or using a topological coupling interconnecting e. g. multibody ports used as shown in Fig. 8 and discussed in more detail in the next section.

2.3 Articulated Vehicles - Advantages of Topological Couplings

If the aim is to explore a system of vehicles, then two simple mechanisms help in the modelling: parametrized copies and a simple connection mechanism between the models. As an example, we use the model of the articulated vehicles, shown in Fig. 6, consisting of two lateral dynamics models and a coupling bar between them. The coupling bar is connected to the lateral dynamics model via rotational joints and can be exchanged by a single spring-damper element for the following analysis.

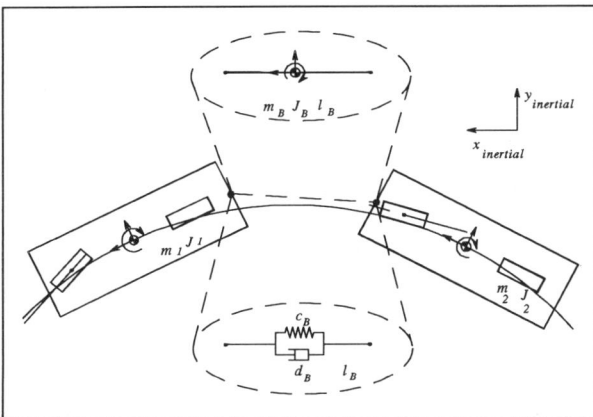

Fig. 6: Model of articulated vehicles

In Objective-DSS, this basic idea of the connection of ports via couple elements is formalized for the description of elements of the class *DssMechatronicHcs*. Couple elements are elements which have no ports themselves but interconnect two ports. They are of a physical or subject-oriented nature, just as joints in multibody systems or pipes in hydraulic systems. Joints in Objective-DSS are objects containing information about their degrees of freedom.

```
DssJoint named: rotJointType.
   degreeOfFreedom: #(PhiZ);
   orientation: #(0);
   angVelocity: #(0);
   dynamicJoint;
   stiffnessOfDynamicJoint: #(1e6 1e6 0.0 0.0 0.0 0.0);
   dampingOfDynamicJoint:   #(1e2 1e2 0.0 0.0 0.0 0.0);
end.
```

Fig. 7: Model of a Rotational Joint

Fig. 7 shows the description of a rotational joint used in the model of the articulated vehicles. It is also possible to declare the kinematic values of the joints, such as relative displacements and velocities. Furthermore the realization of the joints can be described as explained in more detail in the next section. Another basic element used for the articulated vehicles is an environment model of the multibody system of the class *DssMultibodyEnvironment*. It is used to declare environmental values, such as the direction of the gravitation, attachment points fixed in the inertial system, and excitations acting on the attachment points.

Multiple copies: If a model is sufficiently parametrized, it is advantageous to have multiple copies of a description element differing only in their parametrization. Objective-DSS provides this powerful feature for all modelling elements. Models are declared in the section introduced by the keyword *partTypes:* for basic elements and hierarchical sy-

```
DssMechatronicHcs named: singleTrackedCar.
  input:     #(draggingCarDelta)          on: DssScalar;
  output:    #(followingCarSteeringAngle) on: DssScalar;

  auxiliar: #(K_Steering) on: DssScalar;
  auxiliarEquation: K_Steering := 1;

  partTypes:    #( icsType lateralDynamicsType couplingBarType steeringAngleControllerType);
  coupleTypes: #( planarJointType rotJointType );

  partDefinition:
    ics                      is: icsType;
    draggingCar              is: lateralDynamics;
    followingCar             is: lateralDynamics;
    couplingBar              is: couplingBarType;
    steeringAngleController is: steeringAngleControllerType on: #(K_Steering => K_Steering);

  coupleDefinition:
    planarJoint is: planarJointType;
    rot1Joint   is: rotJointType;
    rot2Joint   is: rotJointType;

  inputCondition:
    draggingCar at: delta     connectTo: draggingCarDelta;

  connectCondition:
    planarJoint from: ics            at: ICS connectTo: draggingCar  at: MC;
    rot1Joint   from: draggingCar at: APR connectTo: couplingBar  at: P1;
    rot2Joint   from: couplingBar at: P2  connectTo: followingCar at: APF;

  coupleCondition:
    steeringAngleController at: T_10     connectTo: followingCar               at: T_APF;
    steeringAngleController at: T_20     connectTo: couplingBar                at: T_P2;
    followingCar            at: delta    connectTo: steeringAngleController at: delta_21;

  outputCondition:
    followingCarSteeringAngle connectTo: steeringAngleController at: delta_21;
end.
```

Fig. 8: The Objective-DSS model of the articulated vehicles

stems and *coupleTypes:* for couple elements. Multiple copies of the description elements are declared in the section introduced by the keyword *partDefinition:* for the parts and *coupleDefinition:* for the couple elements.

Topological couplings: Multibody systems representing open- and closed-loop structures connected in any hierarchical order can be built up in the section introduced by the keyword *connectCondition:*. The second connect condition in Fig. 8, for instance, is the connection between the dragging car (attachment point APR) and the coupling bar (attachment point P1); they are connected by rot1Joint representing a rotational joint introduced in the couple-definition section. Given only the connection structure of the multibody system, the user may yet concentrate on the modelling and the related questions and need not be bothered with vagabonding negative signs in a mathematical description.

2.4 Working with Objective-DSS Models

There are three ways of working with Objective-DSS models: the interactive assembly in a topology editor which allows the graphical connection and description of Objective-DSS models; the direct editing of models by means of a simple text editor; and the generation of models from other tools, e.g. multibody system tools. During interactive modelling, the model is checked syntactically and the topological connections resp. input/output connections are checked semantically.

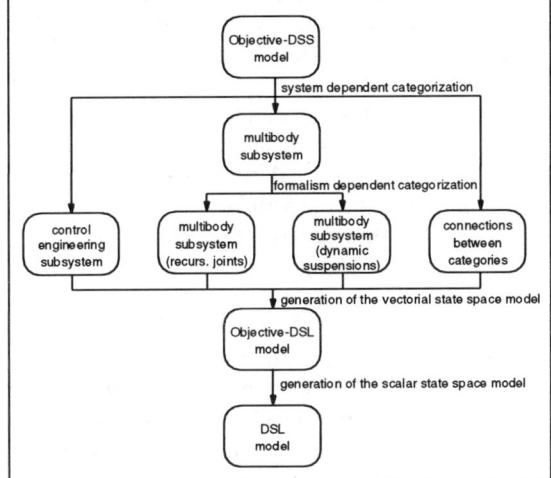

Fig. 9: The mathematical model generation

After loading and linking the system, different categories can be derived from the entire system. They depend either on the generation formalism or on the system. System-dependent categories are e. g. multibody systems and control engineering systems. The derivation of the system category called multibody system allows to implement specialized functions to calculate the absolute values of position, orientation, velocity, and angular velocity of the system in dependence of the relative joint values.

Likewise, one can implement a new behaviour (functions) into every system category. The generation of the mathematical model of mechatronic systems is another example of the system-dependent categorization. Derivation of the different subsystems (multibody systems, control engineering systems, etc.) and generation of the mathematical model for these topological substructures allows integration of different multibody system formalisms. Formalisms for multibody systems are the recursive formulation introduced by [9] and adapted to modular hierarchical block diagrams [10] resp. the formulation of dynamic suspensions in state-space description [11]. Because of the three-dimensional nature of the multibody systems model an automatic dimension reduction is implemented for two-dimensional models using dynamic suspensions. This leads to simpler models which are faster in numerical evaluation.

The problems arising of the formulation of multibody systems in state-space representation are solved by the introduction of mechanical couple elements. Because of the physical/subject-oriented nature of the connections (couple elements) it is easy to alter the structure of the model and the modelling depth. Exchanging the model of the mass in the lateral dynamics model for a more detailed one yields a model of the same interface. Yet the model has a different implementation as well as a highly different kinematical structure; this leads to a different mathematical model.

The topology editor used for the interactive graphical assembly of mechatronic systems allows the administration of the system via a database repository. Templates can be used for interactive system assembly. The topological structure of the system is displayed in the editor. In the special case of linear or nonlinear state-space representation, the topological structure is identical with the hierarchical block diagram. Furthermore the generation of the mathematical model can be derived from the Objective-DSS description in two steps. The first step is the generation of Objective-DSL, a fully symbolical vectorial state-space representation, which uses hierarchical block diagrams for aggregated structures. The second step is the transformation of Objective-DSL into DSL, a scalar state-space representation form. The DSL representation of the system allows the use of a variety of CAMeL tools, developed at the MLaP and discussed in chapter 6.

3 Development of Control Strategies for Lateral Dynamics

In the first modelling step as described in chapter 2 only the degrees of freedom of level motion are considered. This means that out of the six degrees of freedom only three (longitudinal, lateral, and rotational motion) are examined. Due to the separation of lateral and longitudinal dynamics, they are divided into two degrees of freedom for the lateral and one for the longitudinal dynamics. To obtain this separation the longitudinal velocity has to be set constant for the analysis of the lateral dynamic behaviour, and the steering angle has to be zero in the modelling of the longitudinal dynamics.

Now one concept for controlling the lateral dynamics of a car will be used as an example. In order to find out a control rule the behaviour of the purely mechanical components has to be analysed.

3.1 Analysis

First of all, the behaviour of the car-body and of the camera system is analysed.

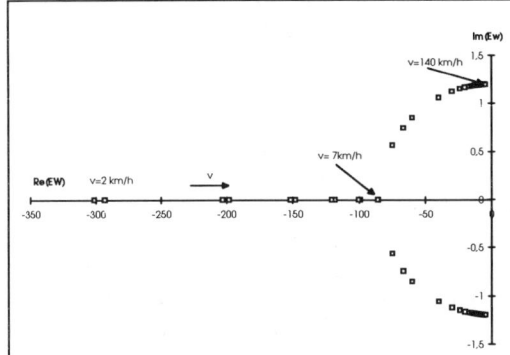

To yield more information on the dynamical behaviour of a single car the equations of motion are linearized at the zero steering angle [12]. With the CAMeL analysis tool LinTool [13],[16] the eigenvalues are calculated for different velocities. Fig. 10 shows these results:

Obviously the eigenvalues depend on the longitudinal velocity. From a limit speed up, the non-oscillatory behaviour changes into an oscillatory one. Higher velocities lead to a decrease in damping and eigenfrequencies.

Fig.10: Eigenvalues of the single car

To allow a closer look at the system, the transfer function from steering angle to lateral motion (of a single car) has been calculated from the linearised equations:

$$G(s) = \frac{Y}{\delta} = \frac{vC_F(s^2 I_{zz}v + (C_R x_{WR}^2 + x_{WF}x_{WR}C_R)s + v(C_R x_{WH} + C_F x_{WF}))}{\left(\begin{array}{c} s^2(mv^2 I_{zz}s^2 + (mv(C_F x_{WF}^2 + C_R x_{WR}^2) + vI_{zz}(C_R + C_F))s + \\ mv^2(C_R x_{WR} - C_F x_{WF}) + (\sqrt{C_F C_R}x_{WF} + \sqrt{C_F C_R}x_{WR})^2) \end{array}\right)} \tag{5}$$

This function has been formulated with car-specific parameters (mass, inertia, geometrical values, tyre stiffness) and depends on the longitudinal velocity. It has the form of a general transfer element of second order, combined with a double integrator.

$$G(s) = K \frac{(T_Z^2 s^2 + 2d_Z T_Z s + 1)}{s^2(T_N^2 s^2 + 2d_N T_N s + 1)} \tag{6}$$

In a second analysis and synthesis step the articulated vehicle have been considered. The y-component of the tow bar force F_{APF} was formulated as disturbance directly effecting the body mass (see fig. 14). The disturbance transfer function has been calculated analytically (not detailled here):

$$G_{Disturbance} = \frac{Y}{F_{APF}}$$ (7)

Together with the complete transfer function of the closed loop:

$$G_{Reference} = \frac{Y_{Actual}}{Y_{Reference}} = \frac{(T_z s + 1)K}{T_n s^3 + s^2 + KT_z s + K}$$ (8)

it is possible to calculate a disturbance compensation as shown in the next subchapter.

To enable the car to follow another running in front, a CCD camera system with an image-processing system has been designed and manufactured. This system supplies longitudinal and lateral distance and the relative angle between the two cars, evaluating the video signal. These values are calculated from the respective position and angles of three marks fixed on the backplane and the roof of the car in front. Furthermore, the velocity is determined by a measurement of the rotational velocity of the non-driven wheel. The dynamical behaviour of the sensor system has been elaborated by means of experiments (see chapter 5). For this purpose, synthetic driving manoeuvres were emulated on the testbed and measured both by means of the testbed sensors and the CCD camera sensor. In this way, the CCD system´s measurement delay of 40 - 60 milliseconds was traced, as was the discretization error of this sensor type, resulting from the limited resolution of the pictures.

These effects have to be taken into consideration in the design of a controller because they have an undamping influence on the closed-loop behaviour.

3.2 Synthesis

The transfer function (see equation 5) shows a velocity-dependent resonance compensation. Therefore, a compensator structure has been chosen as control strategy. The compensator parameters are adapted in dependence of the velocity. Thus the behaviour of the system is reduced to that of a double integrator. For this reduced system a real PD-controller is used (see Fig. 11):

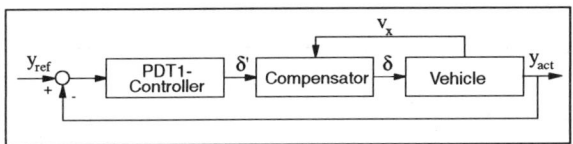

Fig.11: Closed loop of lateral dynamics

The parameter values of the PD-controller were determined by numerical optimization.

As optimization criterion the minimization of the quadratic error area between reference and actual position in y-direction (taken from ISO double lane change manouvre) was chosen. This optimization was performed using the CAMeL optimizer OPIDEX [14].

In the next evaluation step, the time delay of measurement has to be taken into account. Therefore the time gap was modeled by a Padé approximation and integrated into the controller structure.

To outline the influence of the delay time of image processing on the control quality the controller parameters have not been modified in a first design step. This delay has an

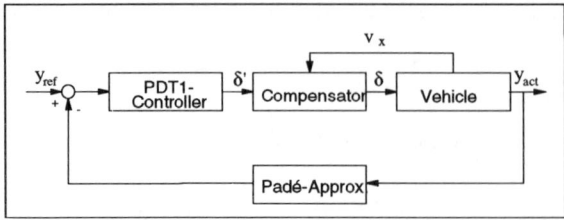

undamping influence on the closed loop behaviour. The phase decrease of a delay is about 57 Degree by 3 Hz.

For this new topology (see Fig. 12) the parameters were modified with the same method.

Fig.12: Closed loop with sensor dynamics

Fig.13 shows the simulation of an ISO double-lane change manoeuvre as the reference for the leading car. The maximum lateral displacements (shown in the second plot) between reference car are shown for three variants: 1. without Padé-block, 2. with Padé-block but without adapted parameters, and 3. with Padé-block and optimized parameters. The last variant shows a maximum displacement of 0.06 m. Furthermore the stationary accuracy of this control strategy can be recognized

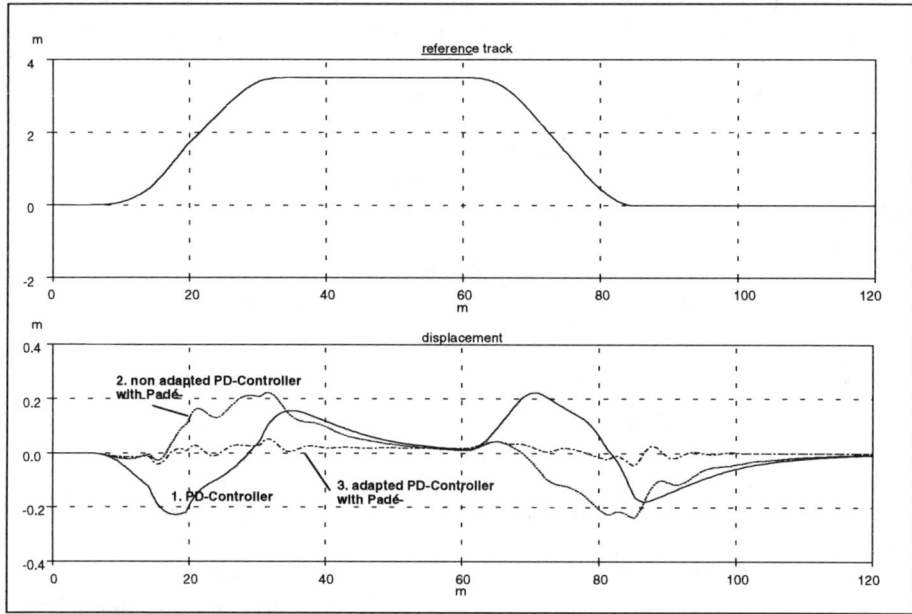

Fig. 13: ISO double lane change manouvre

Finally it will be shown how the lateral dynamics controller, if extended by a disturbance compensator, can also be used for the articulated vehicles coupled mechanically. (7) and (8) yield the necessary transfer function:

$$G_{Distcomp} = \frac{G_{Disturbance}}{G_{Reference}} \qquad (9)$$

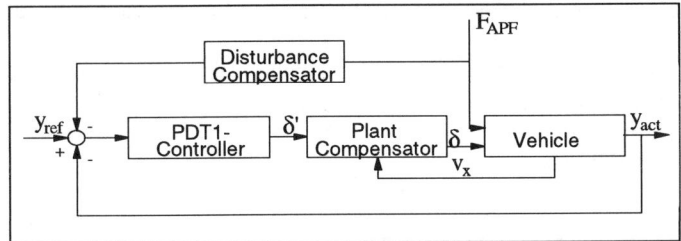

Fig.14 clarifies the structure of the lateral-dynamics closed loop of the articulated vehicles coupled mechanically.

Fig.14: Lateral dynamics closed loops for the articulated vehicles

4 Development of Control Strategies for Longitudinal Dynamics

4.1 Modelling

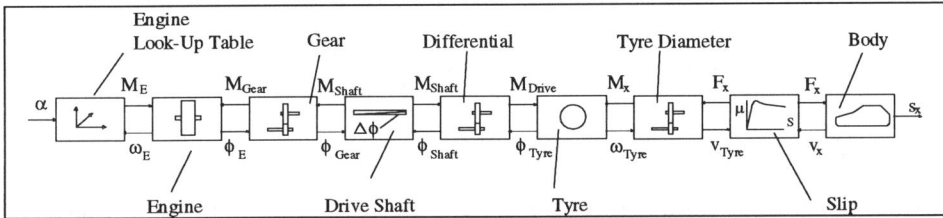

Fig. 15: Physical model of longitudinal dynamics

Fig. 15 shows the physical substitute model of the drive train in the applied modelling depth. 3 masses are coupled by the elements gear, shaft elasticity, and differential (between engine and tyre) resp. slip model (between tyre and chassis).

This approach is adjusted to the frequency range of the longitudinal control, in particular to the distance control (up to 5 Hz). A further criterion for this modelling is the availability of parameters.

The resistance forces effecting the chassis (i.e. drag, rolling, and hill driving resistance) are not displayed. Yet they are considered in the model.

For the driven case the throttle angle α forms the control input of the passive structure. A 3D-engine-look-up-table represents the stationary behaviour of the internal-combustion engine. Transition processes modelled by PT_1/PT_2- or delay blocks can extend the model (not detailled here). The actuator input of braking and the corresponding action chain are not considered here. Therefore the deceleration is limited by the engines drag torque ($M_{Eng} < 0$).

4.2 Analysis

Analogous to the lateral dynamics procedure the passive mechanical structure is analyzed in a linear way. Therefore the nonlinear model must be linearized. The first step is a manual linearization, that means the user has to substitute the nonlinearities directly. The nonlinear slip is replaced by a linear spring which corresponds to the initial gradient of the nonlinear origin. Drag resistance is calculated for the operating velocity and considered as a constant offset. The hill driving resistance is neglected. A proportional factor replaces the 3D-engine-look-up-table.

For these linearizations eigenvalues and frequency responses for the relevant transfer paths have been calculated.

In the next analysis step the linearized models are coupled symbolically to determine the transfer functions for disturbance and reference quantities. These symbolic models build the basis for the analytical determination of controller parameters resp. compensator and adaption structures (not detailled here).

4.3 Synthesis

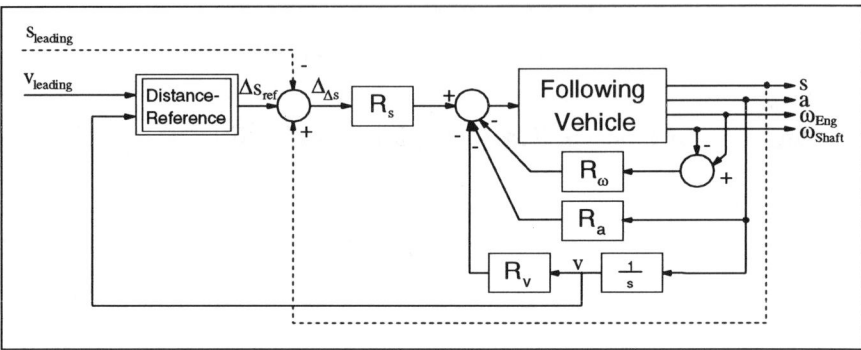

Fig.16: Longitudinal control concept

The selected strategy for the distance control of two cars adjusted to the available sensors is shown in Fig. 16. The dashed signal pathes mark the modelling of the distance sensor. This sensor consists of the image processing of the pictures delivered by a CCD-camera. The sensor also provides the lateral displacement and relative angle measurement. Furthermore there are the longitudinal acceleration and two shaft revolutions available.

For the entire control loop a symbolic model is developed. Therefore one can adjust the controllers parameters to meet a desired dynamic behaviour. This adjustment depends on the preconditions and the plant parameters.

The real distance control is formulated in relative quantities between the two cars ($\Delta_{\Delta s}$-control). Instead of the simple P-controller (shown in Fig. 16), one can apply controllers with differential effect to reach a phase lifting in the desired frequency range.

The approach for the underlying control loops is a state-space feedback which uses relative revolution, velocity, and acceleration. By means of these active inputs most of the drive train oscillations are controlled.

The velocity signal is determined by integration of the acceleration measurement signal. The reference distance can be adapted in dependence on this velocity.

5 Experimental Works

5.1 Laboratory Experiments

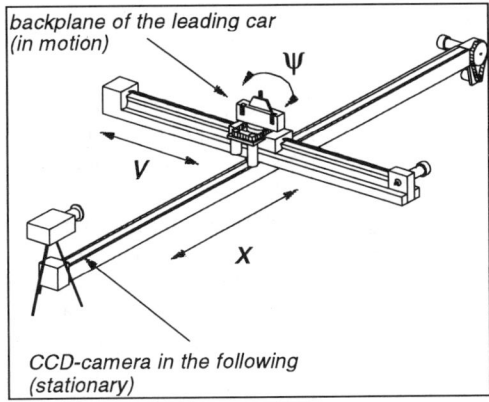

backplane of the leading car
(in motion)

ψ

V

X

CCD-camera in the following
(stationary)

Fig.17: Principle of laboratory works

To test the elected sensor system as well as the designed control algorithms in a safe way a testbed has been developed at the MLaP. This testbed consists of a three-axis mechanism which represents the three level degrees of freedom. This testbed visualizes the simulated relative position of the two cars, the second car following automatically. The actual relative position is measured by the camera system and then these data are used as input into the simulation. The simulation models employed may either be simple or complex car models. The natural limit to the complexity is the calculation in real-time.

5.2 Field Tests

After being evaluated and tested, the control strategies in computer simulations and on the testbed are verified by means of outdoor experiments. For this purpose two cars have been set up at the institute of a project partner (Prof. Schiehlen, University of Stuttgart) [8]. One of them has got a steering actuator; at present, control strategies for lateral dynamics are being tested on these cars. The longitudinal control has not yet been tested because the actuators required for influencing the longitudinal motion have until now been missing. At the moment the longitudinal coupling is being realized by a coupling bar.

In the outdoor experiments performed, a high correspondence between the simulated and the experimental results was detected.

6 Computer-Aided Mechatronics Laboratory (CAMeL)

This chapter introduces the design environment CAMeL (Computer-Aided Mechatronics Laboratory) which supports the entire cycle in the design of mechatronic systems. Modern CACSD (Computer-Aided Control System Design) tools have to be open and easily combinable with other existing tools designated to perform special design tasks. CAMeL is a process-based development framework for control systems. Along with an internal toolset for defined purposes, CAMeL has the possibility to use other, i. e. commercial design environments to enhance its performance [13],[14].

Fig. 18 shows the different system description languages (O-DSS, O-DSL, DSL, and DSC), arranged in chronological usage order along the systematic design cycle. The available CAMeL tools supporting the respective design steps are also displayed[15].

6.1 Internal Toolset

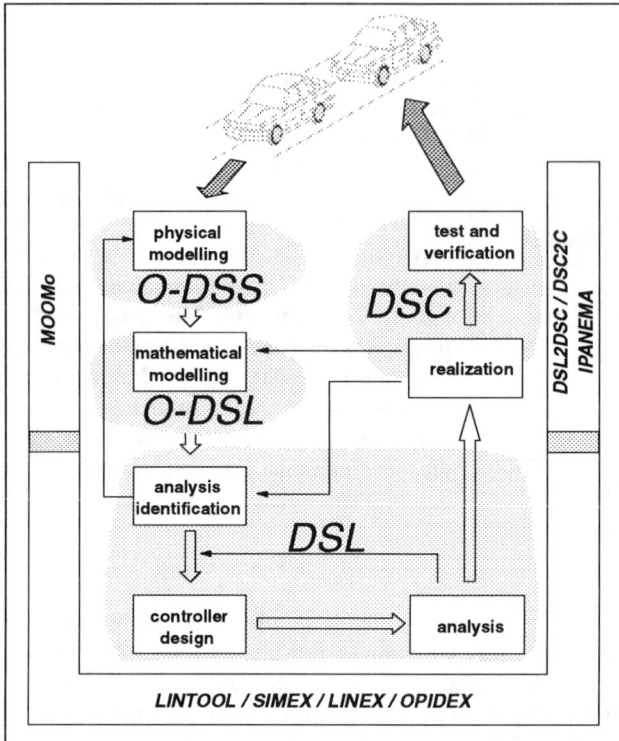

Fig. 18: CAMeL design environment

Packages treating numerics, graphics, control engineering, system engineering, and optimization form the internal toolset of CAMeL.

The object-oriented system formulations Objective-DSS (O-DSS) resp. Objective-DSL (O-DSL) support the modelling on the physical resp. the mathematical level. DSL forms the common system description of the analysis and synthesis phase [16]. This description has been developed first at MLaP. It forms the basis of all analysis and design tools. This toolset comprises software components for nonlinear simulation, symbolic linearization as well as a wide range of linear analysis and different optimization tools for synthesis and identification tasks. Most of these programs work off-line, not under real-time aspects. Some novel components in the CAMeL environment add real-time features (distributed, transputer-based simulation environment and corresponding controller realization hardware and software). They are based on DSC (Dynamic System Code), an abstract, machine related but still machine-independent notation of the systems evaluation order. Furthermore the code segments are marked which can be calculated in parallel.

All of these components have a graphical user interface in addition to the features related to the process-oriented user interaction principle.

A novel module produced directly within the context of this project allows processing the linguistic description of fuzzy systems. A specific input syntax to cover the fuzzy sets, the control basis and the various operators is transformed into the block-oriented input/output formulation of DSL [1].

6.2 Interfaces to External Tools

To enhance the performance of CAMeL, external commercial tools can be linked into the CAMeL design process. Thus they join their properties with CAMeL's. It provides interfaces to command-line-driven programs like MATLAB® and MAPLE® and to some internal programs of the MLaP outside the internal toolset. The interface is based on a communication package that provides encapsulated support to bi-directional UNIX®

pipes or to MS-Windows® DDE mechanisms. The exchanged ASCII data follow a simple protocol. The complete communication support is encapsulated by means of a so-called interface process that does not perform internal computation on the data it is supplied with, but transfers them to the external process, which again sends its results back; these are then transmitted to the next CAMeL process in the design cycle.

7 Conclusions

This contribution outlines the special requirements of the mechatronics design cycle on methods and design tools using a practice relevant example from vehicle dynamics.

The demands of an integrated system description of the heterogenous systems components are clarified. The solution approach suggested by the MLaP, the object-oriented formulation of the submodels and their couplings, have been presented.

Furthermore the analysis and synthesis of lateral and longitudinal control were shown for the vehicle convoy driving example. Robust controllers have been developed extending the simple state-space controllers by compensators and adaption mechanisms. These extensions have been worked out using computer algebra software.

The need for an open design environment, which allows tool coupling of internal methods with external commercial software components is elaborated. The MLaPs approach, the process-oriented software environment CAMeL is introduced.

8 References

[1]   Integration of distributed systems of mechatronics with special consideration of real-time aspects (In German: Integration verteilter Systeme der Mechatronik mit besonderer Berücksichtigung des Echtzeitverhaltens). Project Report, MLaP, Paderborn, 1994.

[2]   HAHN, M., Physical Modelling of Mechatronic Systems According to Object-Oriented Principles, Preprints of 1st MATHMOD, IMACS Symposium on Mathematical Modelling, Vienna, Austria, 1994.

[3]   HAHN, M.; SCHLÜTER, F.: A Physical Model Description and Generation Technique for Hybrid Mechanical-Hydraulic Structures, Tampere Int. Conf. on Machine Automation ICMA'94, Tampere, Finland, 1994.

[4]   RIEKERT, P.; SCHUNCK, T. E.: On Vehicle Mechanics of Cars with Rubber Tyres (In German: Zur Fahrzeugmechanik des gummibereiften Kraftfahrzeugs). Ingenieur Archiv 11 (1940).

[5]   PACEJKA, H. B. (ed.): Tyre Models for Vehicle System Dynamics. Vehicle System Dynamics, Vol. 21, Supplement 1993, Lisse, Netherlands, 1993.

[6]   Mechatronic Function Modules - Concepts and Realization. Internal Report, MLaP, Paderborn, 1994.

[7]   RICHERT, J.; SEUSS, J.: Integration of Distributed Mechatronic Systems, ICRAM'95, Int. Conf. on Recent Advances in Mechatronics, Istanbul, Turkey, 1995.

[8]   RÜKGAUER, A.; PETERSEN, U.; SCHIEHLEN, W.: Mechatronic Steering of a Convoy Vehicle. MEROCON '95, 3rd Int. Conf. on Mechatronics and Robotics, Paderborn, Germany, October 4-6, 1995.

[9]   BAE, D.-S.; HAUG, E.: A Recursive Formulation for Constrained Mechanical System Dynamics, Part I: Open Loop Systems, Mech. Struct. & Mach. 15 (1987), pp. 359-382.

[10]   JUNKER, F.; HAHN, M.: Systematic Modelling of Mechanical Parts in Mechatronic Systems, IMACS-SAS 1995, 5th Int. IMACS Symp. on System Analysis and Simulation, Berlin, 1995.

[11]   HENTSCHEL, M.: A Vehicle Model Library as a Basis for Mechatronic Systems (in German: Eine Fahrzeugmodellbibliothek als Basis für mechatronische Systeme). Diss., University of Paderborn, MLaP, 1995.

[12]   SEUSS, J.: Automatic Towing - Synthesis of Lateral-Dynamics Control (In German: Mechatronisches Folgefahren - Synthese der Querdynamikregelung). Internal Report, MLaP, Paderborn, 1994.

[13]   RICHERT, J.; RUTZ, R.: CAMeL - An Open CACSD Envirionment. IEEE Control Systems Magazine, April 1995.

[14]   CAMeL - Computer-Aided Mechatronics Laboratory - User's Guide, Mechatronics Laboratory Paderborn, Paderborn, Germany, 1994.

[15]   RICHERT, J.; HAHN, M.; DSS - DSL - DSC: The Three Levels of a Model Description Language for Mechatronic Systems, Tampere Int. Conf. on Machine Automation, Tampere, Finland, 1994.

[16]   SCHRÖER, J.: A Model Description Language for Simulation and Optimization of Linear and Linearized Hierarchical Systems (in German: Eine Modellbeschreibungssprache zur Simulation und Optimierung von nichtlinearen und linearisierten hierarchischen Systemen), VDI Press, Series 20, No. 128, Düsseldorf 1994.

Dipl.-Ing. Martin Hahn, Dipl.-Ing. Jobst Richert, Dipl.-Ing. Jürgen Seuss,
Mechatronics Laboratory Paderborn (MLaP), University of Paderborn,
Pohlweg 55, D - 33098 Paderborn, Germany,
phone: ++49-5251-60{2416, 2417, 2423}, fax: ++49-5251-603550,
Email: {hahn, rich, seuss}@mlap.uni-paderborn.de

Mechatronic Steering of a Convoy Vehicle

Andreas Rükgauer, Uwe Petersen, Werner Schiehlen
Institute B of Mechanics at the University of Stuttgart, Germany

Abstract: The development of a mechatronic vehicle convoy, without a driver in the rear vehicle, is discussed. The vehicles are linked together by a rigid tow bar. Throughout the modeling process, great attention is put on efficiency considerations vs. the model complexity. Two different approaches for the coupling by the tow bar are discussed. Topological considerations are performed for the system description. A model of the steering plant is introduced as well as a driver model apath description. A controller is supplied for the steering along with physically motivated control strategies for the vehicle positioning. An experimental setup is shown, consisting of a steering actuator and a special rigid tow bar with position encoders. The position measurement by an optical CCD sensor is discussed and ist implications on the control robustness are discussed. First experimental tests with steady state maneuvers are performed.

1 Introduction

In this paper the behavior of two vehicles in convoy is investigated. This work comprises of system modeling, analysis, control, and experimental setup. Thus it consists of all phases of the mechatronic design process. That is why it is embedded as a benchmark in a surrounding compound project led by 5 universities [1]. In contrary to other publications on this topic, see e.g. Dreyer et al. [2], Shladover [3], Trächtler et al. [4], and Narendran and Hedrick [5], the goal is not to develop a new product but rather to study concepts used with mechatronic modeling and design.

At the current stage of development, the convoy vehicles are linked together by a rigid tow bar, thus reducing the complexity to lateral dynamics only. The driver of the rear car is replaced by a steering actuator, sensors and controllers. The above aspects are discussed in the following and cover different goals, among them:

- Development of dynamic models,
- study of the complexity needed for dynamical models,
- setup and verification of control laws for the mechatronic steering and
- experimental setup and test of two vehicles in convoy.

2 Vehicle Modeling

As the two vehicles are linked together by a rigid tow bar, the complexity of the dynamic models needed for simulation studies and controller development is reduced in a great way. For such models, the multibody approach (MBS) is most suitable [6].

The models supplied are to be used for different tasks: For reference simulations, controller development, and real−time simulations in hardware−in−the−loop environments. So the question of the appropriate model−complexity raises. For this reason, different models are introduced and then compared with each other.

2.1 Single Track Models

The simplest vehicle models for lateral motion allow plane motion only, thus neglecting heave, roll, and pitch. As the complexity is reasonably low, these models can be used with modern methods of controller development easily.

2.1.1 Model of Riekert and Schunck

First, a simple lateral vehicle model on the basis of two single track models, Riekert and Schunck [7], connected by a rigid tow bar is implemented, see Figure 1. Both wheels on each axis of the single track model are assumed to be aligned in the middle, this results in a great simplification of the systems geometric properties. The tire forces are implemented using a linear characteristic. The equations of motion for this model are derived using the multibody system formalism NEWEUL, see e.g. Kortüm and Sharp [8]. The platoon velocity v_x is considered a system parameter, not a degree of freedom. This simple vehicle model is used for both, studying the cause of certain dynamical characteristics and designing lateral control laws for the following vehicle.

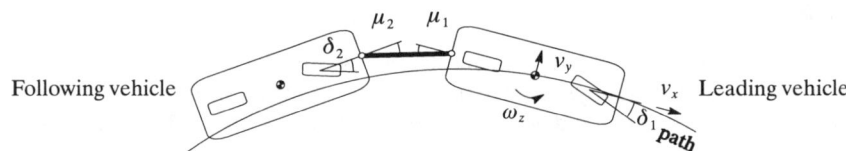

Figure 1: Single track model of the convoy vehicle system

The equations of motion of a single vehicle as stated by Riekert and Schunck are linear. When generating the equations of motion of the platoon, the symbolic equations generated by NEWEUL are usually nonlinear and read as

$$M(y,t)\,\dot{z} + k(y,z,t) = q_e(y,z,t) \text{ and } \dot{y} = \dot{y}(y,z,t) \tag{1}$$

with the vectors y and z of generalized coordinates and velocities, see Popp and Schiehlen [9]. The mass matrix is described by M, k denotes the circular and gyroscopic forces, q_e contains the applied forces. For the analysis in state space with contemporary tools, these equation must be linearized to the form

$$\dot{x} = A(t)\,x + B(t)\,u(t), \tag{2}$$

with the state vector $x = [y,z]^T$, where the state matrix can be found as

$$A = \begin{bmatrix} 0 & \left.\dfrac{\partial \dot{y}}{\partial z}\right|_{traj} \\ -M|_{traj}^{-1}\left(\left.\dfrac{\partial (M\dot{z})}{\partial y}\right|_{traj} + \left.\dfrac{\partial k}{\partial y}\right|_{traj} - \left.\dfrac{\partial q}{\partial y}\right|_{traj}\right) & -M|_{traj}^{-1}\left(\left.\dfrac{\partial k}{\partial z}\right|_{traj} - \left.\dfrac{\partial q}{\partial z}\right|_{traj}\right) \end{bmatrix} \tag{3}$$

by symbolical linearization along arbitrary trajectories x_{traj}. For the above vehicle platoon the state vector and the input vector u are given as

$$x = \begin{bmatrix} \mu_1 & \mu_2 & v_y & \omega_z & \dot{\mu}_1 & \dot{\mu}_2 \end{bmatrix}^T \text{ and } u = \begin{bmatrix} \delta_1 & \delta_2 \end{bmatrix}^T, \tag{4}$$

where v_y denotes lateral velocity of the leading vehicle with respect to the moving vehicle coordinate system and ω_z the yaw rate. The angles between the tow bar and the vehicles are denoted by μ_1 and μ_2. The input vector is composed of the steering angle δ_1 of the leading vehicle and the steering angle δ_2 of the following vehicle.

2.1.2 Subsystem Techniques

Usually the dynamical model of a vehicle platoon is derived from the model of a single vehicle. When generating the equations of motion based upon the geometric description with a multibody formalism, the system reduction by the tow bar is performed by a projection into the motion manifold as with D'Alembert's principle, Schiehlen [10].

When building the dynamic equations of the platoon based upon the equations of motion of a single vehicle, as for instance from a model–library, this is no longer possible, rather, the full set of equations of motion has to be generated completely new. In this case, an approach is likely to reduce the equations of motion after their derivation. Two strategies are imaginable: Using an integrator capable of handling such constraints or reformulating the constraints in a proper way. Here the second way is chosen: The force–coupling approach reads as

$$M(y,t)\,\ddot{y} + k(y,z,t) = q_e(y,z,t) + G^T(y,t)\,\lambda(y,z)\,,\, g(y,t) = 0 \text{ and } G = \frac{\partial g}{\partial y^T} \tag{5}$$

with the constraint g and the penalty λ as applied force. For the vehicle convoy, the constraint can be found by

$$g = |\,r_2 - r_1\,| - l_0 = 0\,, \tag{6}$$

where r_1 resp. r_2 describe vectors to the hitch points of the tow bar as shown in Figure 2. As penalty function a spring–damper combination is chosen:

$$\lambda = K\,g + D\,\dot{g}\,. \tag{7}$$

	Minimal description	Force coupling			
		$D = 10^1$		$D = 10^3$	
Value		$K = 10^3$	$K = 10^5$	$K = 10^3$	$K = 10^5$
Size of the equations of motion (FORTRAN)	126.13 kB	43.51 kB			
Simulation time with lane change [17] (22 m/s, 8 s)	8.1 s	4.21 s	7.26 s	4.28 s	4.13 s
Simulation time with curve (2 m/s, 20 s)	42.8 s	28.59 s	28.98 s	28.65 s	28.95 s

Table 1: Size of equations and simulation times with force coupling

As shown in Table 1, the force coupling not only leads to new models quite easily, but is also very efficient in numerical simulations. However, finding good parameters for the penalty functions is not always easy. In addition, the resulting system description is strictly spoken not minimal but rather consists of artificial coordinates with high frequency components. For this reason, description (1) with the minimal set of coordinates is better for the use with control tools.

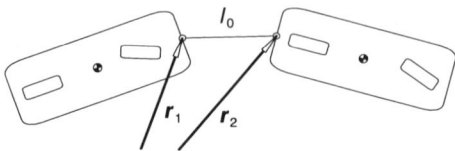

Figure 2:   Hitch point vectors for the constraint definition

2.2 Spatial Convoy Model

For the verification of control laws and for reference simulations, a rather complex, three−dimensional convoy model is needed.

2.2.1 Structure

The complex convoy model is again derived from a single car model. The spatial vehicle model is generated based on the plane model, as shown in Figure 3, by introducing a suspension for each of the wheels. The convoy model comprising of two such vehicles is described by 19 degrees of freedom and consists of 19 rigid bodies. Six degrees of freedom are related to the absolute position of the first car, these are x, y, z and the angles ψ, θ and ϕ. Another 5 degrees of freedom describe the revolute joints of the tow bar. The position of each wheel with respect to the vehicle body is described by w_i, $i = 1(1)8$, and the rotation of the wheels is introduced by a rheonomic constraint.

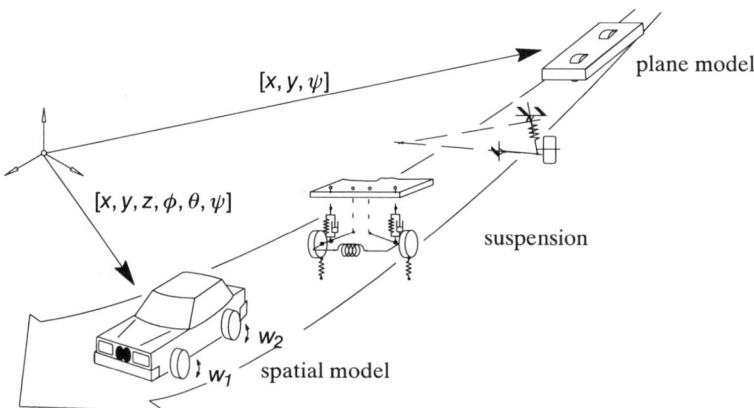

Figure 3:   From a plane model to the spatial vehicle model

2.2.2 Comparison with the Simple Model

To compare the simple model with the three dimensional one, numerical simulations using a single vehicle are performed. The vehicles run along a curve at constant diameter, at a velocity of 23 m/s. The trajectories of both vehicles are shown in Figure 4.

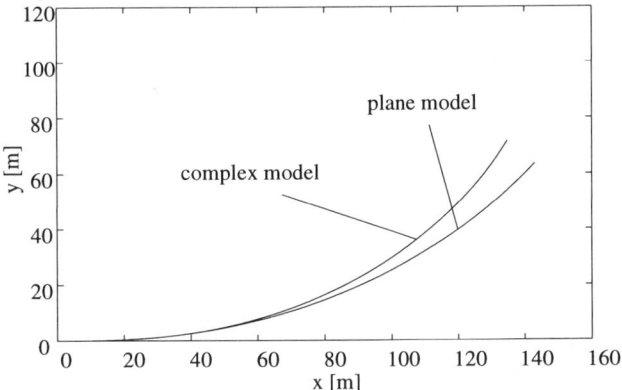

Figure 4: Comparison of plane and complex vehicle model

The complex model goes along a curve of smaller diameter at the same steering angle input. This is due to the nonlinear behavior of the spatial model. Because of the high lateral accelarations, the vehicle chassis begins to roll. As a reaction on this rolling, the car suspension performs slight alignment reactions for each of the wheels. For this reason, the tire forces are different and the car runs on a different path.

Summing up and considering the simulation cycles which are 40 times as long for the complex model, the spatial model is needed only for maneouvers at high lateral accelerations.

2.2.3 Topological Considerations

When using the dynamic models also in real−time simulations, much effort has to be spent to the efficiency of the model description. Again, the MBS approach is most appropriate, because of its high degree of system reduction. However, one must be careful about the geometric topology resulting from the degrees of freedom chosen when gererating the equations of motion.

For the above system description the topology is given by a chain, see Figure 5 (left). With this topology the dynamical expressions for the first car are quite simple, whereas those of the rear car are rather complicated, due to the $O(n^3)$ complexity of the equations of such systems with n being the number of joint coordinates in the chain.

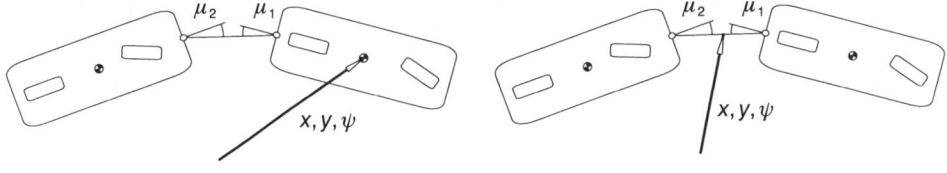

Chain Structure   Tree Structure

Figure 5: Different Topological Structures for the Vehicle Platoon

A better approach is shown in Figure 5 (right), where a tree structure is used. Here the complexity is distributed equaly on both vehicles. Table 2 shows the comparison of equation size and simulation cycle times for both approaches. The equations of motion for the tree structure are much shorter than those of the chain structure. For this reason, the simulation cycles are as much as 14% less time consuming.

Value	Chain structure	Tree structure
Size of equations of motion (FORTRAN)	1.332 MB	1.022 MB
Simulation time with lane change [17] (22 m/s, 8 s)	516.57 s	442.68 s
Simulation time with curve (23 m/s, 5 s)	349.34 s	297.3 s

Table 2: Size of equations and simulation times for chain− and tree structure

2.3 Tire Models

The characteristics of a real vehicle tire is nonlinear with respect to the slip angle. It is, however, as shown in Figure 6, reasonably linear for small slip angles but runs into saturation for larger angles.

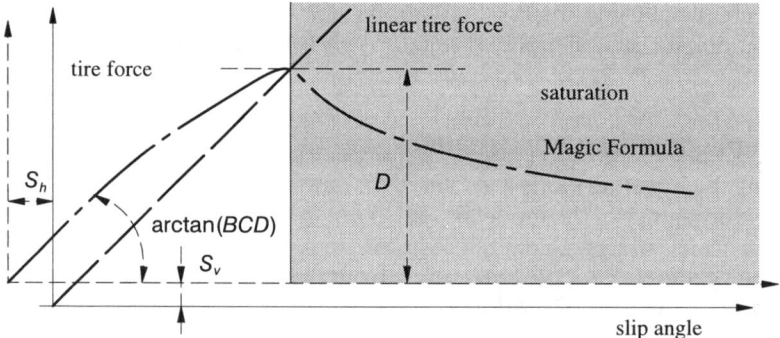

Figure 6: Nonlinear Behavior of the Vehicle Tire

A very simple tire model can be introduced by a linear function on the slip angle. For simulations at high lateral accelarations, a more complex model is introduced: Pacejka's Magic Formula [11] allows to adjust the tire force as function of the slip angle with only 7 parameters which are again functions of 16 basic parameters. Comparing the tire models by simulation studies one can find that the linear tire performs well for low and intermediate lateral accelerations. Only for maneuvers at very hig lateral accelerations a complex tire model is nesseccary.

3 Steering Plant

For the development of a steering controller and for full scale simulations, dynamic models of the hydraulic steering of the vehicle are to be supplied as well as a model of the behavior of the steering actuator.

3.1 Simple Steering Model

The behaviour of the hydraulic power steering is determined by a ball and nut steering unit [15]. For efficient simulations, a very simple model of the system behaviour must be found. The most simple approach is given by the combination of a mechanical PT2−block and a nonlinear input amplifier, as shown in Figure 7. The nonlinear amplifier supplies hydraulic power due to the steering torque M_l. Mechanical vibration of the steering wheel angle ϕ_l is introduced by a combination of mass m, damper D and spring C.

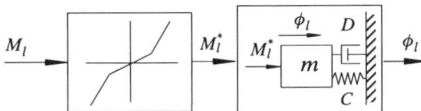

Figure 7: Simple model of the steering plant

3.2 Complex Steering Model

Important properties of a hydraulic power steering are play, friction, and the hydraulic flow with compressible characteristics. To incorporate these effects, a more elaborate steering model is introduced, see Figure 8.

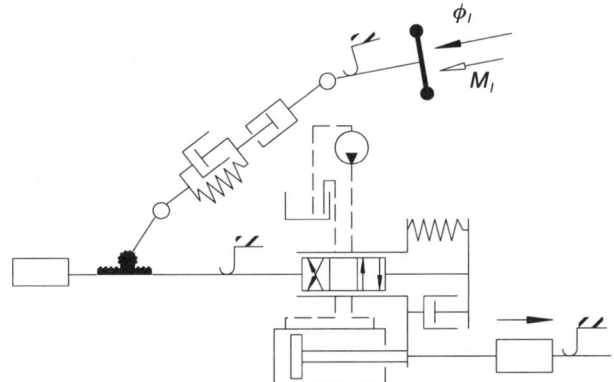

Figure 8: Complex model of the steering plant

This model is described by 3 mechanical and 2 hydraulic degrees of freedom. However, a few problems arise with this model: It turnes out to be very difficult to assess the correct parameters for the model. In addition, it is found that the shown three−land−four−way spool valve [12] is not accurate enough to act in a realistic way. Effects as self−exitation from road vibrations and out−of−phase reactions at high steering frequencies are still not considered in this model. For this reason, the complex model must be extended by 2 additional flow chambers in the hydraulic circuit [13].

3.3 Steering Actuator

The steering actuator is modeled as a DC motor at constant flux. Most common, an electric actuator is accompanied by an electronic torque control unit. The implementation of actuator and electronic control is shown in Figure 9. The actuator generates the steering torque M_l due to a control voltage U_{in}. Input voltage limits are included in this model. Figure 9 shows the the steering model and the steering actuator with control.

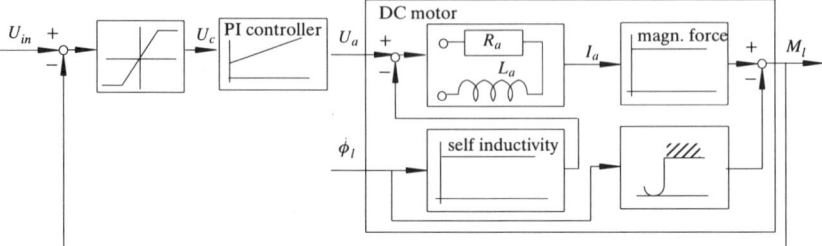

Figure 9:   Model of the controlled steering actuator

4 Driver Model, Path Generation and Control

To be able to simulate maneuvers along arbitrary paths, both, a path description and driver models are needed for the current project. Controllers are needed for steering as well as for vehicle positioning.

4.1 Driver Model and Path Generation

Input to the simulation model of the vehicle system is the steering angle of the leading vehicle δ_1. For more realistic maneuvers it is necessary to have the leading vehicle follow a certain given path, i.e. the steering angle has to be calculated by a driver model. In this research the double layered driver model by Donges [16] is chosen.

Common roads consist of straight lines, curves at constant diameters, and clothoides. Based on these components, an interface to a path description is implemented to store the desired path on a C^2 steady spline and use it within the simulation environment.

The path generation is performed in a preprocessing stage before the actual simulation starts. Note that although a C^2 steadiness is given by the splines, this is still not smooth enough for common integration schemes [14] and therefore long simulation times must be expected with the path generator.

4.2 Controller Synthesis

The lateral control of the vehicles comprises two levels: the relative position of the rear vehicle is calculated on a superior level in order to follow the towing vehicle autonomously. On a subordinated steering level, a compensator cares for the compliance of the steering plant with the desired steering wheel angle provided by the superior vehicle control.

4.2.1 Steering Control

A standard PID compensator is chosen for the steering control. The parameters of the PID controller are found by pole placement. The amplitude of the resulting system drops beyond 1 Hz which is within the range of the lateral dynamics of vehicles, see Figure 10. One must expect a big impact of the steering system on the vehicle control.

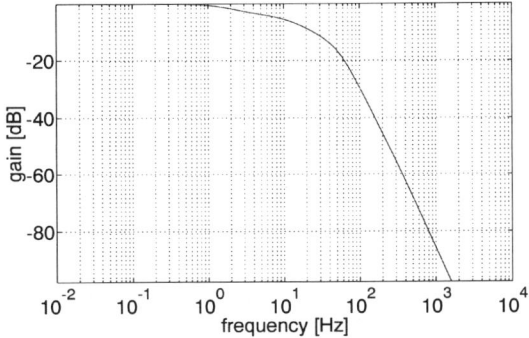

Figure 10: Frequency response of the controlled steering plant

4.2.2 Vehicle Control

For the vehicle control, tow contradictory goals must be dealt with: Offtrack and stability. The offtracking is shown in Figure 11. Here it can be studied, that for low velocities (left) the rear vehicle tends to smaller turning radii then these of the first vehicle. For higher velocities (right) the rear vehicle follows radii beyond those of the first vehicle.

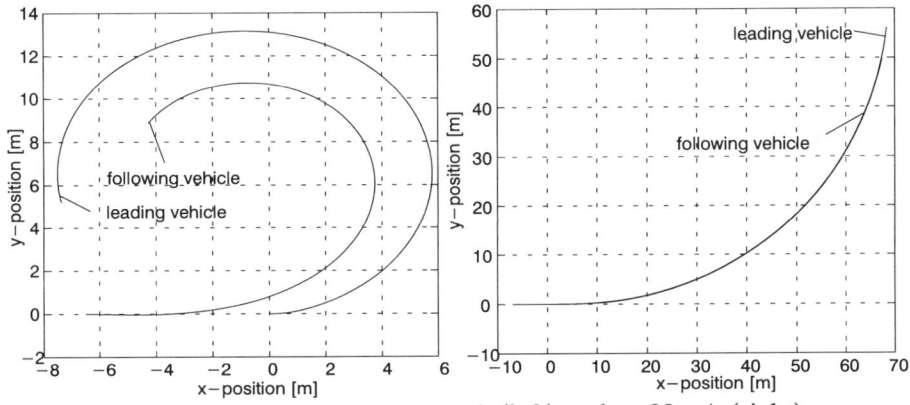

Figure 11: Cornering at v=1 m/s (left) and v=20 m/s (right)

Two principles are proposed for the vehicle control, both beeing based on physically motivated ideas, requiring minimum sensor information and being easy to implement. The first one, named Trailer Principle, sets the steering angle δ_2 of the rear vehicle exactly to the angle μ_2 between the rear vehicle and the tow bar, Figure 12 (left), thus the front wheels are always aligned parallel to the tow bar. This is similar to how a normal two−axis trailer pulled by a tractor is steered. Within the meaning of control theory, the heading angle μ_2 is fed back with a gain of one.

The second principle, named Stiffening Principle, requires a vanishing heading angle μ_2 as shown in Figure 12 (right). In the case of traveling on a straight lane the Stiffe-

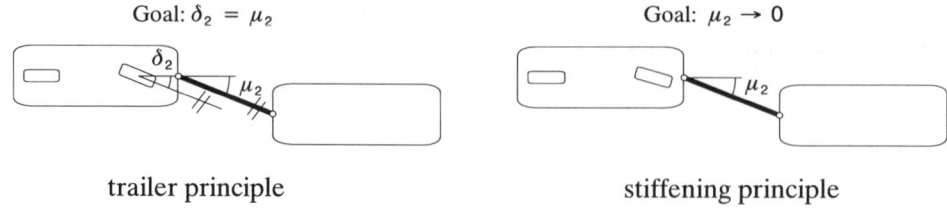

Goal: $\delta_2 = \mu_2$ Goal: $\mu_2 \to 0$

trailer principle   stiffening principle

Figure 12: Vehicle Control Concepts

ning Principle results in vanishing of lateral deviation of the following vehicle which is a function of the heading angle only. The Stiffening Principle also results in aligning the tow bar to the longitudinal axis of the leading vehicle. The implementation of the Stiffening Principle also uses feedback of μ_2. A PD controller is designed for this task by pole placement using the single track model of the convoy. Contrary to the Trailer Principle, the proportional part is chosen much bigger than one. If taken too big, the system can easily become unstable.

To examine stability of the convoy, the eigenvalues for Trailer Principle and Stiffening Principle are calculated for a range of velocities from 1 to 40 m/s, Figure 13. For both principles, the poles tend to move to the right for higher velocities. Stability problems do not occur at low velocities, but it is found that the Stiffening Principle results in a much more stable system at higher velocities than the Trailer Principle. When examining the stability with the steering system incorporated, the latter one even shows instability for higher velocities. The Stiffening Principle is almost unaffected by the steering dynamics.

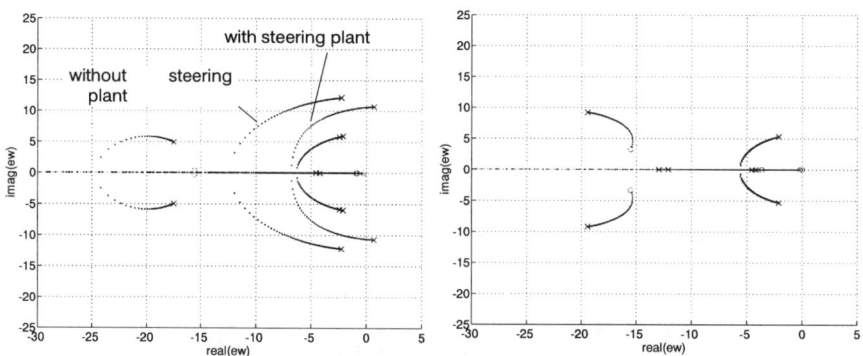

Figure 13: Eigenvalues with Trailer Principle (left)
and Stiffening Principle (right)

4.2.3 Rating of the Vehicle Control

To study transient behaviour, a severe lane change maneuver, where the leading vehicle has to change lanes twice, see Fancher et al. [17], is carried out at a high speed, Figure 14. As expected from the stability analysis, the Stiffening Principle performs better. Contrary to low speed, the Trailer Principle displays more lateral deviation and is closer to instability.

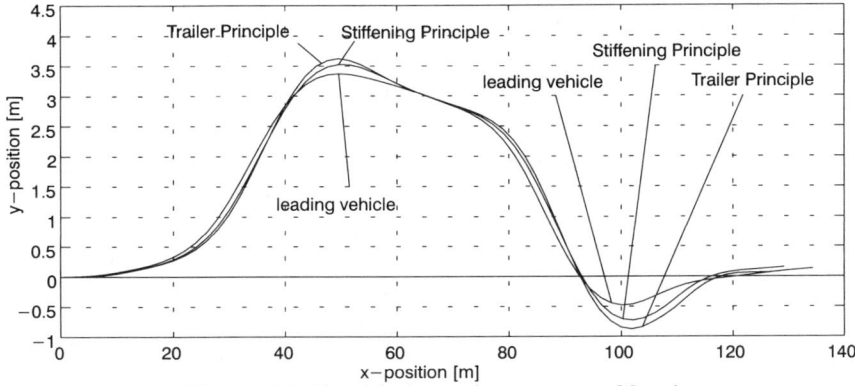

Figure 14: Double lane change at v=20 m/s

5 Experimental Setup

After having verified the controllers designed in numerical simulations, two BMW passenger cars are equipped with a tow bar and a steering actuator to run road tests. For the control strategies proposed above, the vehicle control must be fed with the rear tow bar angle μ_2. As it is likely to implement further steering strategies with regard to μ_1 in the future, this angle is to be measured, too. Further, the towing force is measured.

5.1 Tow Bar with Position Sensors and Steering Actuator

Tow bars commercially available show play. As force sensors are quite sensitive to sudden peaks, play must be avoided. This leads to a cardan design of the tow bar joints as shown in Figure 15.

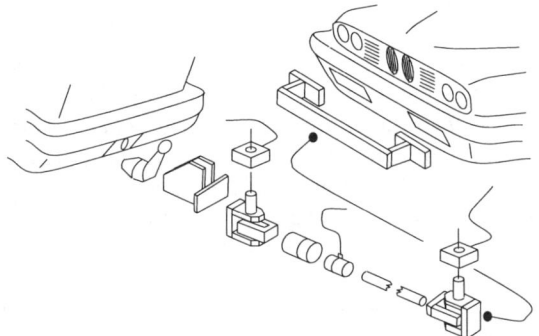

Figure 15: Towed vehicle with tow bar and sensors

This picture shows the modular approach for the tow bar. To reinforce the vehicle chassis, a welded beam as shown must be incorporated. For the measurement of the angles, encoders are placed on the joint axle. The towing force is measured using a piezo ceramic charge−sensor.

For the steering, a brushless DC motor is chosen. This housingless motor resides inside of a special construction replacing the normal steering wheel.

5.2 Other Sensor Input

The position sensors on the tow bar are to be replaced by an optical sensor in a further stage of development. This optical sensor is supplied by a project partner and consists of a CCD camera with frame grabber and driver software.

Special problems with this kind of sensor technique are accuracy and time delay. The sensor and software in the rear car searches for special markers mounted on the first car. As these markers have finite size, and the sensor has limited resolution, the sensor system inherently has a poor accuracy. In addition, the time delay is caused by the time needed for software image processing and analysis. The functionality of the optical sensor in the vehicle convoy must therefore be proven by numerical simulations.

To check the robustness of the above vehicle controllers on sensor problems, a special module is added to the simulation environment. With this sensor module it is possible to simulate arbitrary time delay, even with stepsize controlled, multiorder multistep integration schemes.

The simulation results show that the admissible convoy velocities depend directly on the quality of the sensor signal. For the Trailer Principle it is found that noise up to 10% is allowable in the measured signal. With increasing time delay, the admissible velocity drops exponentially and for a delay of 75ms the highest velocity possible is 10m/s.

The Stiffening Principle is much more crucial on sensor errors. It is found that it is most sensitive on noise. However, slight time delays are acceptable. But for a delay of 25 ms the highest velocity possible already drops to 6m/s.

6 Experimental Results

To date, safety strategies have not been incorporated into the experimental setup yet. For this reason, steady state maneuvers at low speed are performed solely.

In this first stage, simulation results are verified through a series of circular turns at walking speed. Having reached steady state behaviour, the vehicles are stopped and their position with respect to the center of the turning circle is measured in additon to the data mentioned above. Simulation results could be confirmed well, see Table 3. In both cases the second vehicle follows the leading vehicle with a smaller turning radius, where the Trailer Principle is closer to the path of the leading vehicle than the Stiffening Principle.

Variable	Simulation		Experiment	
	Trailer Principle	Stiffening Principle	Trailer Principle	Stiffening Principle
r_{lead}[m]	8.55	8.66	8.52	8.62
r_{follow}[m]	7.57	7.23	7.52	7.08
δr[m]	0.98	1.43	1.00	1.54

Table 3: Turning radii for narrow circle

Although the Stiffening Principle promises much better results then the Trailer Principle, regarding the numerical analysis, is is found to be quite pretentious in terms of data to be fed with. From the eigenvalue plots of Figure 13 right, rather stiff behaviour must be expected for the Stiffening Principle. Fed with the discrete data from the encoders, the physical system turns out to be on the edge to instability, which has to be considered when adjusting its control parameters.

7 Conclusions

The development of a mechatronic vehicle convoy, without a driver in the rear vehicle, is discussed. The vehicles are linked together by a rigid tow bar. This work covers all stages of the mechatronic design process and is therefore well suited as benchmark for a compund project on mechatronic system design.

The modeling, control, and experimental setup of the convoy lateral dynamics are described. Throughout the modeling process, great attention is put on efficiency considerations vs. the model complexity. The convoy models are derived based on single vehicles. Two different approaches for the coupling by the are discussed. The coupling by applied forces is quite efficient, whereas the minimal description gives equations better suited for the controller synthesis. Topological considerations are performed for the system description. A chain structure of the dynamic models must be avoided. A model of the steering plant, including steering actuator and hydraulic power steering, is introduced.

To simulate along arbitrary vehicle paths, a driver model is incorporated into the simulation package as well as apath description. A controller is supplied for the steering. In addition, physically motivated control strategies for the vehicle positioning are proposed. The Stiffening Principle performes in simulations quite well.

An experimental setup is shown, consisting of a steering actuator and a special rigid tow bar with position encoders. In a next stage, an optical CCD sensor is to be introduced for the detection of the convoy position. In simulations it turnes out that the Stiffening Principle struggles with the time delay and noise produced by the CCD sensor.

First experimental tests with steady state maneuvers show a good system performance. In the future, more tests will be carried out, including sensor data from the CCD camera. Also, the whole system, simulation and experimental setup, will serve as a test rigg for control strategies supplied by project partners.

8 References

[1] Färber, G.; Lückel, J.; Müller, P.C.; Pfeiffer, F.; Schiehlen. W.: Integration verteilter Systeme der Mechatronik mit besonderer Berücksichtigung des Echtzeitverhaltens. Work report for the time from Jan. 1, 1993 to June 30. 1994. Paderborn: Universität−GH, FB−10, Automatisierungstechnik, Prof.Dr.−Ing. J. Lückel, 1994.

[2] Dreyer, W.; Hoppe, P.; Jacob, U.; Maretzke, J.: ICAD − Intelligent Computer Aided Driving. Automobil−Industrie, Vol. 2, 1990.

[3] Shladover, S.E.: Longitudinal Control of Automotive Vehicles in Close−Formation Platoons. Journal of Dynamic Systems, Measurement and Control, Vol. 113, 1991.

[4]   Trächtler, A.; Lohmann, B.; Struck, G.: Regelung der Querbewegung eines ka-
      merageführten Straßenfahrzeugs. Automatisierungstechnik, Vol. 40,
      1992.

[5]   Narendran, V.K.; Hedrick J.K.: Autonomous Lateral Control of Vehicles in an
      Automated Highway System. Vehicle System Dynamics, Vol. 23, 1994.

[6]   Schiehlen, W. (Editor): Multibody Systems Handbook. Berlin, Heidelberg, ...:
      Springer Verlag, 1990.

[7]   Riekert, P.; Schunck, T.E.: Zur Fahrmechanik des gummibereiften Kraftfahr-
      zeugs. Ing.−Arch. 11, 1940.

[8]   Kortüm, W.; Sharp, R.S.: Multibody Computer Codes in Vehicle System Dyna-
      mics. Lisse, Netherlands: Swets & Zeitlinger, 1993.

[9]   Popp, U.; Schiehlen W.: Fahrzeugdynamik. Stuttgart: Teubner, 1993.

[10] Schiehlen, W.: Technische Dynamik. Stuttgart: Teubner Verlag, 1986.

[11] Pacejka, H.; Bakker, E.: The Magic Formula Tyre Model. In: Tyre Models for
      Vehicle Dynamic Analysis, Ed. Pacejka, H.B., Amsterdam/Lisse: Swets &
      Zeitlinger, 1993.

[12] Merritt, H.E.: Hydraulic Control Systems. New York: John Wiley and Sons,
      1967.

[13] Dürr, R.: Modellierung und Simulation eines Servolenksystems. Stuttgart: In-
      stitut B für Mechanik, Diplomarbeit DIPL−52, 1995.

[14] Shampine, L.F.; Gordon, M.K.: Computer Solution of ordinary differential
      equations. San Francisco: W.H. Freeman and Company, 1975.

[15] N.N.: ZF Kugelmutter−Umlauf Hydrolenkungen. Schwäbisch − Gmünd:
      Zahnrad fabrik Friedrichshafen AG, 1991.

[16] Donges, E.: Ein regelungstechnisches Zwei−Ebenen−Modell des menschli-
      chen Lenkverhaltens im Kraftfahrzeug. Zeitschrift für Verkehrssicher-
      heit, 24/3, 1978.

[17] Fancher, P.; Segel, L.; Bernard J.; Ervin, R.: Test Procedures for Studying Vehi-
      cle Dynamics in Lane−Change Maneuvers. SAE−Paper 760351, 1976.

Andreas Rükgauer, Uwe Petersen, Prof. Dr.−Ing. Werner Schiehlen
Institute B of Mechanics at the University of Stuttgart
Pfaffenwaldring 9, D−70550 Stuttgart
Tel.: (+49) 0711 685 6956
Fax: (+49) 0711 685 6400
Email: ru@mechb1.fertigungstechnik.uni−stuttgart.de

IX. General Aspects in Mechatronics

T. Fukuda, K. Shimojima
Fuzzy-Neuro-GA Based Intelligent Control

T. Raste, P. C. Müller
Modeling and Control of Mechatronic Systems by Decentralized Descriptor Systems

W. Brockherde, D. Hammerschmidt, B. J. Hosticka
Silicon Microsystems for Mechatronic Applications

Fuzzy-Neuro-GA based Intelligent Control

Toshio Fukuda and Koji Shimojima
Dept. of Micro System Engineering, Nagoya University
Furo-cho, Chikusa-ku, Nagoya 464-01, Japan

Abstract - This paper introduces some intelligent techniques, such as fuzzy theory, neural networks, genetic algorithms, and artificial intelligence, and its application for the hierarchical control system. The control system is classified into three levels: 1)learning level, 2)skill level, and 3)adaptation level. In the learning level, symbols are reasoned logically to control strategies. The skill level produces control references along with the control strategies and sensory information on environments. The adaptation level controls robots and machines while adapting to their environments which include uncertainties. For realization of these levels and connection among them, artificial intelligence, neural networks, fuzzy logic, and genetic algorithms are applied to the hierarchical control system while integrating and synthesizing themselves. To be intelligent, the hierarchical control system learns various experiences both in top-down manner and bottom-up manner. Thus, the hierarchical control scheme is effective for intelligent robotics and mechatronics.

I. Introduction

Many robotic systems are used in various fields and places in these days. In order to use robotic systems widely, the robots have to have some intelligence. The intelligent robots have to carry out tasks in various environments by themselves like human beings. They have to determine their own actions under uncertain environments based on their own sensory information. In advance, human operators can give the robots some of their knowledge and skills to some extent in top-down manner. However, when the robots perform tasks in an uncertain or inconsiderable environment, the knowledge may not be useful or meaningful. In this case, the robots have to adapt themselves to their environments and acquire new knowledge by themselves through learning. This process proceeds in bottom-up manner.

This paper introduces a control scheme for intelligent robotic systems, which this paper refers to as the *hierarchical intelligent control*. The hierarchical intelligent control consists of three levels: adaptation level, skill level and learning level. This scheme has two characteristics with respect to learning process: top-down approach and bottom-up approach. To link three levels and have such characteristics for knowledge acquisition, the scheme uses artificial intelligence (AI), fuzzy logic, neural networks (NN) and genetic algorithm (GA) [1 - 3]. Each technique has advantages and disadvantages. In order to overcome the disadvantages, this paper introduces integration and synthesis techniques of them. Those are key techniques for intelligent control of systems in robotics and mechatronics.

This paper describes advantages and disadvantages of each techniques in the section second. The section third explains integration and synthesis techniques of them to overcome the disadvantages (see Fig. 1). This paper also shows the skill acquisition scheme. These are key techniques for intelligent control of systems in robotics and mechatronics.

II. Artificial Intelligence, Fuzzy, Neural Network and Genetic Algorithm

A. Artificial Intelligence

In fields of intelligent information processing such as symbolic or language processing, conventional AI techniques have been used to manufacture knowledge-based systems as expert systems. As applications, knowledge-based reasoning system were used for diagnosis, design, and control. For the control, there were some examples of symbolic control which uses symbolic reasoning mechanisms. The AI has good performance with respect to symbolic manipulation. However, it is difficult to classify sensed data in order to transform numerical data set into symbolic data set for understanding process state. The signals are classified by using 'if - then rules' as shown in Fig. 2.

B. Fuzzy sets

The fuzzy logic is characterized as extension of binary crisp logic. Each fuzzy rule has an antecedent, or if, part containing several preconditions, and a consequent, or then, part which prescribes the value. The fuzzy set is a class in which transition from membership to non-membership is gradual rather than abrupt as shown in Fig. 3. Crisp sets allow only full membership or no membership at all, whereas fuzzy sets allow partial membership. In other words, an element may partially belong to a set. One identifies the main parameters and determines a term set which is the right level of granularity for describing the values of each linguistic variable. For example, a term set including linguistic values such as {Small (S), Medium Small (MS), Medium Big (MB), Big (B)} may be used.

Because of the partial matching attribute of fuzzy rules and the fact that the preconditions of rules do overlap, more than one fuzzy rule can fire at a time. The methodology which is used in determining which value should be taken as the result of the firing of several rules can be referred to as conflict resolution. Traditionally, fuzzy logic uses a minimum operator. In the simplified fuzzy logic, however, multipliers are used instead of the minimum operator. Defuzzification procedure is also simple. However, since the fuzzy set does not have learning capability, it is difficult for human operator to tune the rules from data set.

C. Neural network

Neural network, a model of the brain, artificially connects many nonlinear neuron model and processes information in a parallel distributed manner. The neural network has many characteristics such as nonlinear mapping, parallel processing, learning, and self-organization. It is applied to pattern recognition, control and so on. The neural network which consists of three layers (input/output layers and one hidden layer) are able to express any functions while using enough hidden units. The neural network produces transformative rules from empirical training sets through learning. To train the neural network, the back-propagation is used. However, the mapping rules in the network is not visible and is difficult to understand as shown in Fig. 4. Moreover, the convergence of learning is very slow and not guaranteed. To overcome those problems, some structured neural networks are proposed. The next section describes the networks as synthesis techniques.

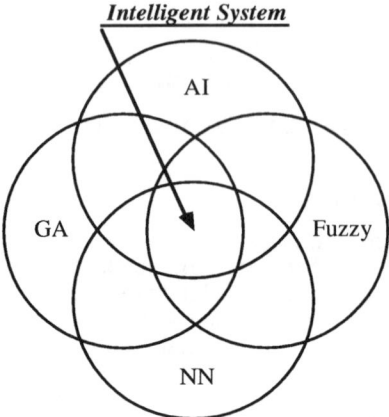

Fig.1 Synthesis of Fuzzy, Neural Networks, Artificial Intelligence, and Genetic Algorithms for intelligent System

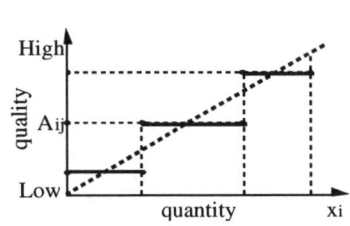

(a) Signal transformation from a set of numerals to a set of symbols by 'if ... then' rules.

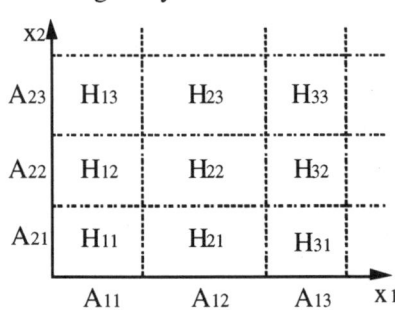

If $A_{1s} \wedge A_{2t}$ then H_{st}

H_{st}: Symbol or CF (certain factor)

(b) Partitioned areas in lattice

Fig. 2 Classification by 'if ... then ~' rules

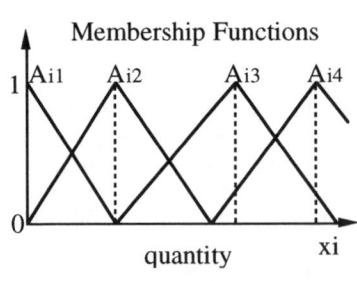

(a) Membership functions

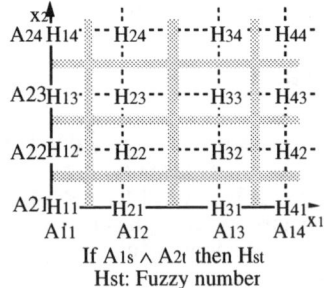

If $A_{1s} \wedge A_{2t}$ then H_{st}

H_{st}: Fuzzy number

(b) Gradually partitioned area in lattice

Fig. 3 Classification by fuzzy logic with membership functions

D. Genetic Algorithm

GA is one of search algorithms based on the mechanics of natural selection and natural genetics. It is not a gradient search technique. It combines survival of the fittest among string structures with a structured yet randomized information exchange to form a search algorithm with some of the innovative flair of human problem solving. An occasional new part is tried for good measure. While randomized, GAs are no simple random walk. They efficiently exploit historical information to speculate on new search points with expected improved performance.

GAs have traditionally three operations to abstract and rigorously explain the adaptive process of natural systems as follows: (1) Selection operation, (2) Crossover operation, (3) Mutation operation. Figure 5 shows a flow chart of the GA. The selection process is an operation to select the survivals in a set of candidate strings. In this process, the fitness value is calculated for each candidate string by using the fitness function which depends on a goal for searching problems. According to the fitness value, the selection rate is determined for the present candidate strings, and the survival is selected in any rate depending on the selection rate. The crossover process is a reform operation for the survival candidates. In natural system, a set of creatures creates a new set of the next generation by crossing among the creatures. In the same way, the crossover process is performed by exchanging pieces of strings using information of old strings. The pieces are crossed in couples of strings selected randomly. The mutation process is held to escape the local minima in search space in the artificial genetic approach. The calculation is stopped when the generation is up.

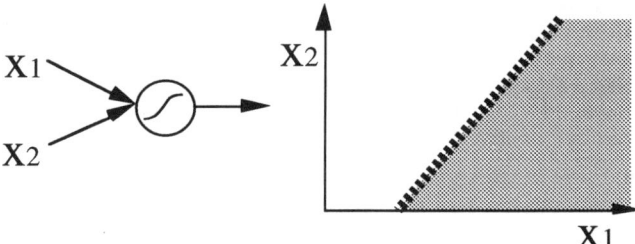

(a) Signal transformation at a neuron using a nonlinear function (i.e. sigmoid function)

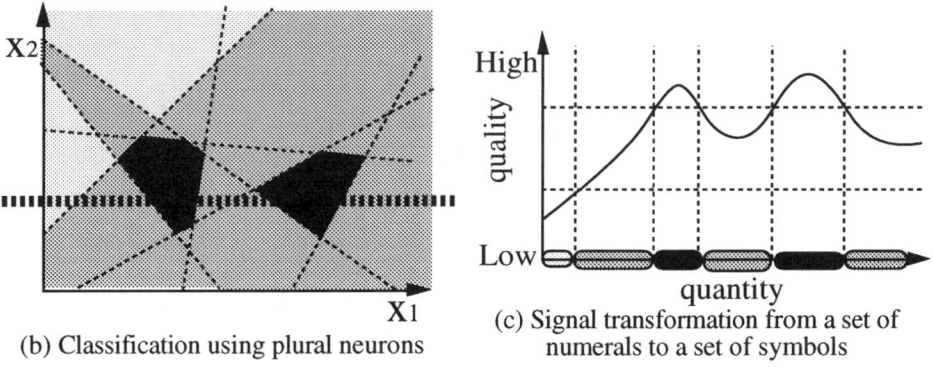

(b) Classification using plural neurons

(c) Signal transformation from a set of numerals to a set of symbols

Fig. 4 Classification by neural network with nonlinear functions

III. Integration and Synthesis of Neural Network, Fuzzy Logic and Genetic Algorithm

As described in the previous section, the AI, fuzzy logic and neural network have similar performance with respect to signal transformation, though methods of them are different. Each method has merits and demerits. Table I is the comparison of them. To overcome their demerits, some integration and synthesis techniques of them and GA have been proposed. This section explains these techniques which are indispensable to construct the hierarchical intelligent control architecture.

The fuzzy logic and the neural networks can be used as preprocessors of the AI. They transform numerical data set to symbolic data set. To give the rules for transformation, human operators easily determine rules of the fuzzy logic. However, when the number of input parameters increases, determination of the rules becomes laborious for the human operators. In this case, the neural networks are useful. While showing data sets of input/output to the neural network, it learns them and works as a transformative function. Drawbacks of the neural network are that the human operator can not give their knowledge beforehand nor understand the acquired rules. Moreover, the convergence of the learning is very slow and the neural network can not learn new patterns incrementally. To solve those problems, some kind of structured neural networks are investigated.

The fuzzy neural network is a combined neural network with the fuzzy logic. Figure 6 shows an example of the fuzzy neural networks. Human operators are able to give their knowledge in the fuzzy neural network by means of membership functions. The membership functions are modified through learning process as fine tuning. After the learning, the human operators can understand the acquired rules in the network. With respect to the convergence of the learning, the fuzzy neural network is faster than the conventional neural network. For multiple input parameters, the hierarchical fuzzy neural network is available [4, 11]. However, it is difficult to optimize the structure of the hierarchical fuzzy neural network.

On the other hand, the fuzzy logic is used as a critic for improvement of convergence of learning of the neural network [7]. In this case, the fuzzy logic determines the learning step depending on the state of convergence.

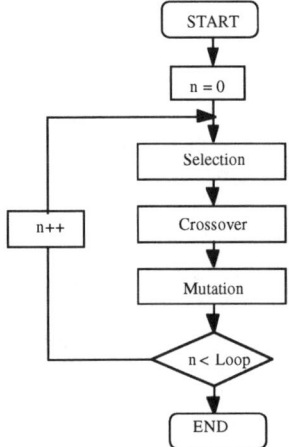

Fig. 5 Flow chart of the Genetic Algorithm

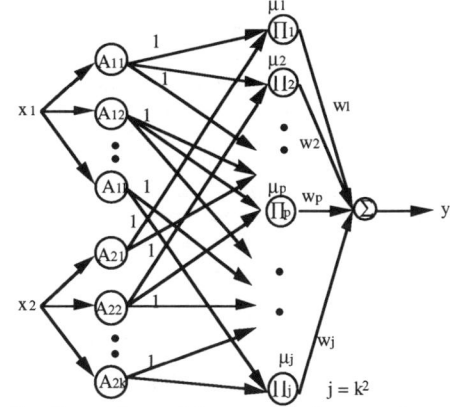

(a) Configuration of fuzzy neural network

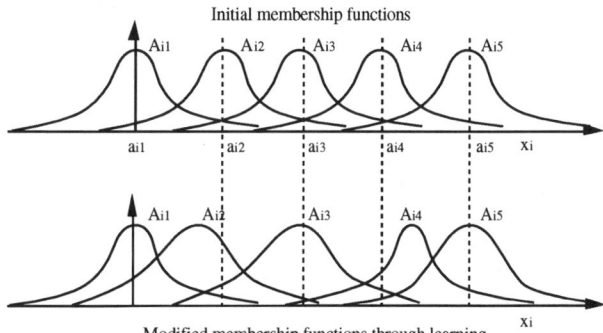

(b) Gaussian basis functions for membership functions in the fuzzy neural network

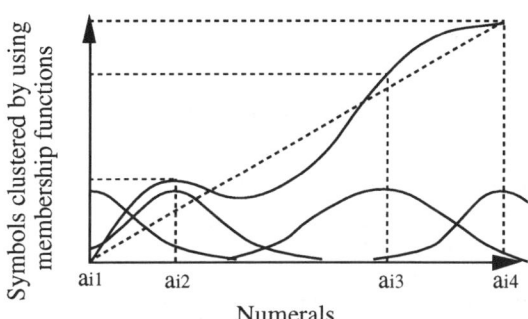

(c) Signal transformation from sets of numerals to sets of symbols or numerals

Fig. 6 Fuzzy neural network

The neural network with radial basis functions is also the structured one [5, 12]. It has potential to learn more quick and easier than the neural network with the sigmoid functions (Fig. 7) [5]. For incremental learning, the Adaptive Resonance Theory (ART) model has been proposed as Fig. 8. It has a two-layered structure. It learns patterns one by one incrementally. That is, it can correct errors by learning new patterns without old patterns. However, the ART model has a problem of bad classification ability. For example, the ART model cannot classify the two patterns shown in Fig. 9, though the RBF neural network can. The Neural network based on Distance between Patterns (NDP), shown in Fig. 10, has the abilities of incremental learning and classification [5, 18, 19]. The NDP learns categories of patterns one by one. It increases neurons of the output layer using the incremental learning algorithm. It uses the radial basis function at the output layer. Therefore, it can classify the patterns shown in Fig. 9. Depending on aims, human operators should give the neural network efficient structure if they have experiences. Or else, heuristic approach for structure optimization is necessary.

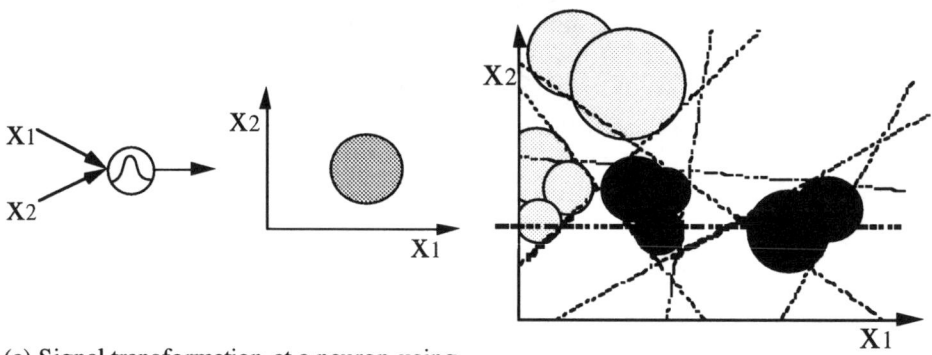

(a) Signal transformation at a neuron using a nonlinear function (i.e. radial basis function)

(b) Classification using plural neurons

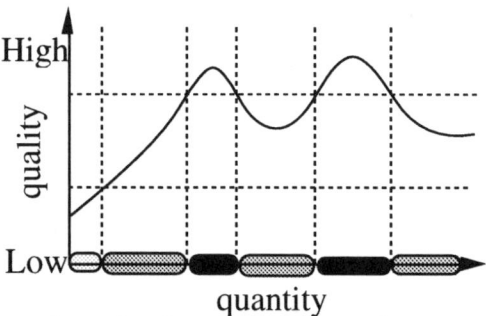

(c) Signal transformation from a set of numerals to a set of symbols

Fig. 7 Classification by a neural network with radial basis functions as a structured neural network

The GA is a powerful tool for structure optimization of the fuzzy logic and the neural networks (Figs. 11-13) [9, 13, 14, 20]. Particularly, the GA is powerful to optimize the hierarchical fuzzy neural network [20]. On the other hand, the fuzzy logic and the neural network can be a evaluation function for the GA [10]. It is difficult to define evaluation functions for complex optimization problems. However, while using the fuzzy logic or the neural network, human operators can transfer their criterion. Those are the complicated reinforce learning technique because they do not use teaching signal but obtain desirable states while manipulating a lot of parameters at the same time. The

Genetic Programming which is one of applications of the GA and manipulates symbols can produce new rules or knowledge for the AI [15].

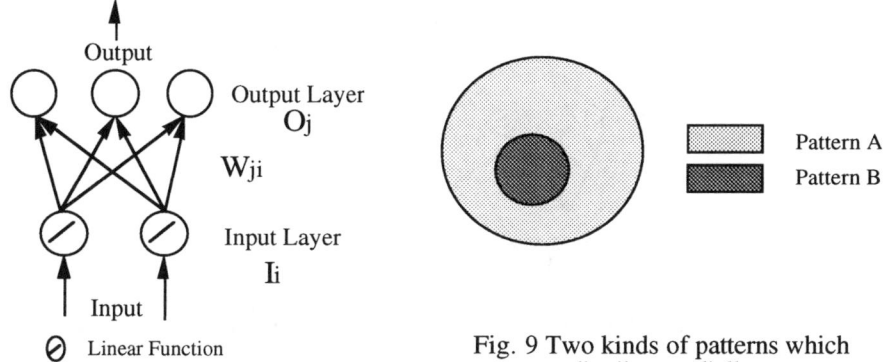

Linear Function
Fig. 8 Structure of Adaptive Resonance
Theory Model

Fig. 9 Two kinds of patterns which
distribute radially

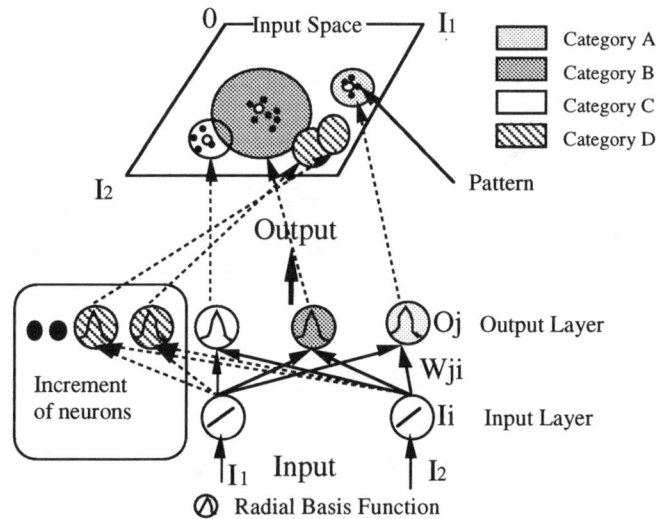

Radial Basis Function
Linear Function
Fig. 10 Structure of Neural network based on Distance between Patterns

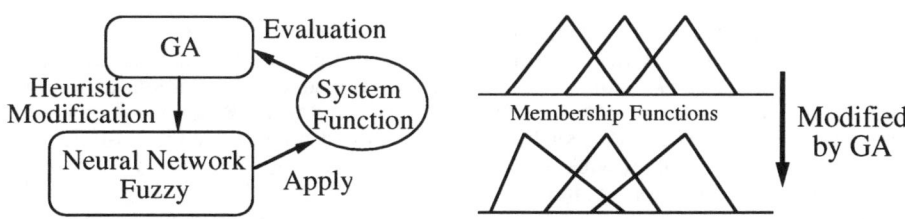

Fig. 11 Structure optimization of fuzzy or
neural network by genetic algorithm

Fig. 12 Structure optimization and learning
of fuzzy logic by genetic algorithm

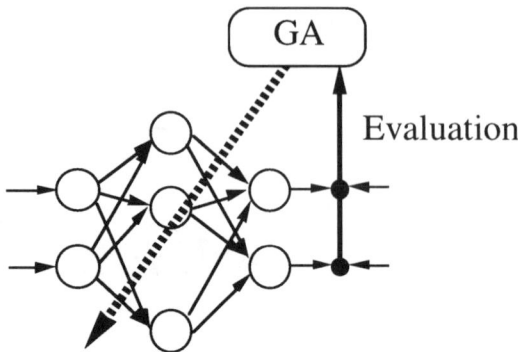

Evaluation

Fig. 13 Structure optimization and learning of neural network by genetic algorithm

Table I Comparison of Neural Network, Fuzzy Logic, AI, and Genetic Algorithm

	Math Model	Learning Data	Operator Knowledge	Real Time	Knowledge Representation	Nonlinearity	Optimization
Control Theory	O	×	△	O	×	×	×
Neural Network	×	O	×	O	×	O	◖
Fuzzy	◖	×	O	O	△	O	×
AI	△	×	O	×	O	△	×
GA	×	O	×	△	×	O	O

IV. Behavior Acquisition by Reinforcement Learning

In this section, we describe the behavior acquisition scheme based on a reinforcement learning[21, 22]. A reinforcement learning[23, 24] is an unsupervised learning method to learn control policies, or skills only from a scalar performance index without any explicit teacher which shows how to control a system at each moment. When we humans and animals perform complex motions such as walking, they can be divided into fundamental units of actions. For instance, walking can be divided into motions such as "an action to sustain the body'', "an action to stretch out a free leg'', "an action to swing arms'', "an action to control pitching, rolling and yawing'', "an action to find obstacles'', and so on. In this study, a 'behavior' means such a fundamental unit of an action function. After these required behaviors are learned well, a complex motion "walking'' can be realized with a sequential and/or parallel combination of them.

Figure 14 shows an essential structure of a motion learning agent which is consisting of functions of three levels: a planner at the highest level, behavior agents in the middle level, and a sensor data integrator and an action combinator in the lowest level. Each function can be described as follows:

[Planner]:Planner is a high level controller that plans how to activate sequentially or simultaneously which behaviors at which states by which parameters considering coordination of their actions. It receives the state information and sends activation signals and behavioral parameters to behavior agents and combination orders to a combinator.

[Behavior agents]:Each behavior agent receives some behavioral parameters and some state variables as its inputs, and it outputs some control variables. The behavioral parameters tell how it should behave.

For example, when we walk, we can change strides. In this case, a length of a stride can be a behavioral parameter. Each behavior agent may have different sets of inputs and outputs. Generally, inputs are abstract state variables, which are integrated from raw sensor data, and actions are abstract control variables, which will be combined and translated into direct actuator inputs.

[Integrator]: Integrator provides state variables in useful forms by integrating raw information from sensors.

[Combinator]: Combinator combines actions from behavior agents according to the planner's order and yields control inputs for actuators.

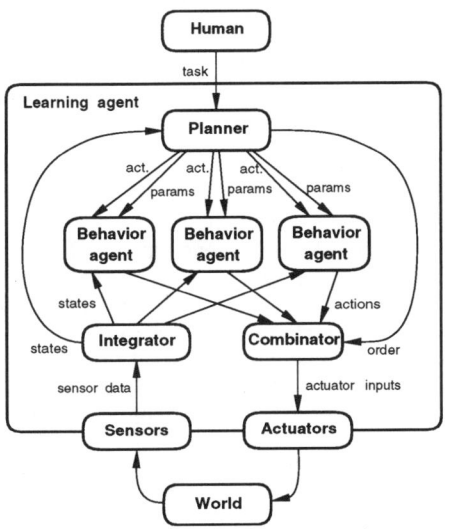

Fig. 14 Structure of a motion learning agent

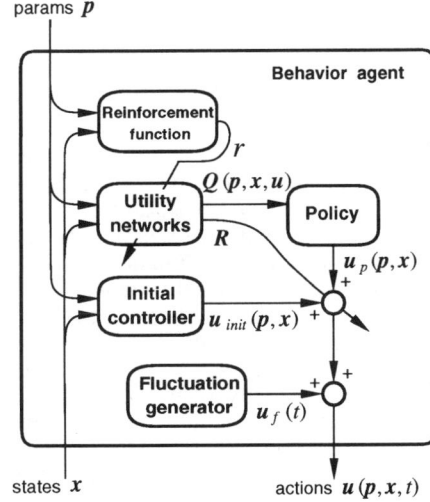

Fig. 15 Structure of each behavior agent

The behavior agent which we propose has a structure shown in Fig. 15. Here *params* are behavioral parameters given by a planner and states are integrated state variables provided by an integrator. Utility networks represent utility functions (Q-values in the Q-learning) by connectionist networks and the policy determines the best actions according to the utility values. The reinforcement function is assumed to be given by a human or a planner.

The goal of learning is to acquire the behavior to reactively output optimal actions according to the parameters and the states. All of the parameters, states and actions are real values, and this scheme can be implemented to dynamic motion learning of real robots [21, 22].

V. Hierarchical Intelligent Control

The hierarchical intelligent control scheme comprises three levels: a learning level, a skill level, and an adaptation level as shown in Fig. 16 [4, 8]. Therefore, there are three feed-back loops. The learning level is based on the expert system for a reasoning mechanism and has a hierarchical structure: recognition and planning to develop control strategies. The recognition level uses neural networks and fuzzy logic combined with the neural network as nodes of a decision tree. In the case of the neural network, inputs are numeric quantity sensed by some sensors, while outputs are symbolic quality which indicates process states. The structured neural network for incremental learning is effective to memorize new patterns [5]. In the case of the fuzzy neural network, inputs and outputs are numeric quantities and the fuzzy neural network clusters input signals by using membership functions. That is, the fuzzy neural network transforms numerical quantity into symbolic quality by using membership functions. Both the neural network and the fuzzy neural network are trained with the training data sets of a-priori knowledge obtained from human experts. As a result, the neural network and the fuzzy neural network can transform various sensed data from numerical quantities to symbolic qualities, and perform sensor fusion and production of meta-knowledge at the learning level. The important information is sensed actively on using the knowledge base. The sensors of vision, weight, force, touch, acoustic, and others can be used as nodes of decision tree for recognition of the environment.

Then, the planning level reasons symbolically for strategic plans or schedules of robotic motion, such as task, path, trajectory, force, and other planning in conjunction with the knowledge base. The system can include another common sense for robotic motion. The GA optimizes control strategies for robotic motion heuristically [6, 15]. The GA also optimizes structures of neural network and fuzzy logic connecting each levels. Thus, the learning level reasons unknown facts from a-priori knowledge and sensory information. Then, the learning level produces control strategies for skill level and adaptation level in a feed-forward manner. Following the control strategy, the learning level selects initial data set for a servo controller at the adaptation level from a data base which maintains some gains and initial values of interconnection weights of the neural network in the servo controller. Moreover, the recent sensed information from the skill level and the adaptation level updates the learning level through long-term learning process with human instruction. Therefore, knowledge at the learning level is given by human operator in top-down manner and acquired by heuristics of the skill level and the adaptation level in bottom-up manner.

In the same task and different environments, it is necessary to change control references depending on the environment for the servo controller at the adaptation level. At the skill level, the fuzzy neural network is used for specific tasks following the control strategy produced at the learning level in order to generate appropriate control references. Input signals into the fuzzy neural network are numerical values sensed by some specific sensors and some symbols which indicate the control strategy produced at the learning level. Output of the fuzzy neural network is the control reference for the servo controller at the adaptation level. This output is based on the skill extracted from human experts through learning training sets obtained from them. At the same moment, the fuzzy neural network clusters the input signals in the shape of membership functions. These membership functions are used as the symbolic information for the learning level.

In the adaptation level, a neural network in the servo controller adjusts control law to current status of dynamic process [7, 16]. Particularly, compensation for non-linearity of the system and uncertainties included in the environment must be dealt with by the neural network. Thus, the neural network in the adaptation process works more rapidly than that in the learning process. It is shown that the neural network-based controller, the Neural Servo Controller, is effective to the nonlinear dynamic control with uncertainties such as force control of a robotic manipulator. Eventually, the neural networks and the fuzzy neural networks connect neuromorphic control with symbolic control for hierarchical intelligent control while combining human skills.

The hierarchical intelligent control is applied not only to a single robot, but also to multi-agent robot system. If there is no interaction between robots, each robot has to work optimally for its purpose, so that the total task should be achieved optimally. That is, each robot should work selfishly. Or else conflicts among the robots might occur when using a public resource. The competition may cause collisions and deadlock states among the robots in a local area. In order to avoid competition, it is necessary for the robots to communicate and to coordinate among themselves. The coordination among the robots is as important as selfishness. The GAs are applied hierarchically to balance selfishness with coordination for efficient motion planning [6]. When multiple robots works independently as decentralized system, the learning capability of the robots is indispensable for evolution of the system [17].

As results, integration and synthesis of AI, Fuzzy Logic, Neural Network and GA are important for intelligent system, depending on their characteristics. Hierarchical intelligent control using these techniques is effective to control intelligent systems in robotics and mechatronics.

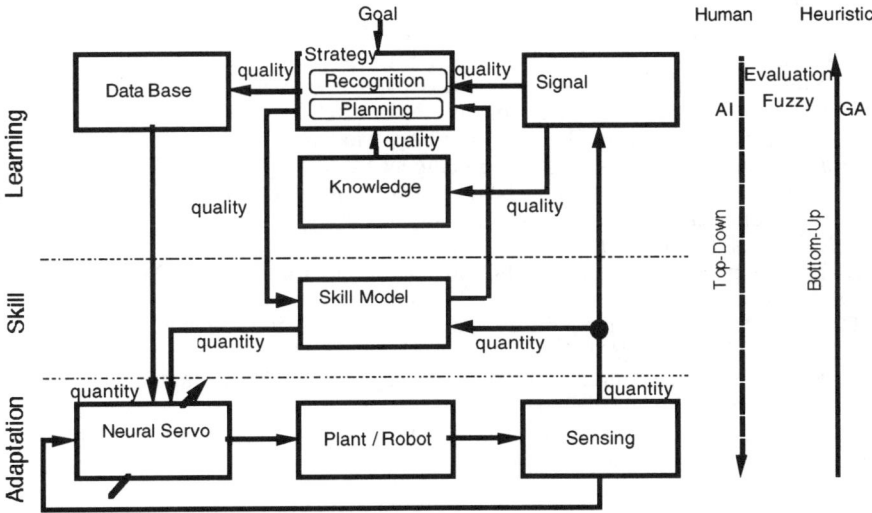

Fig. 16 Hierarchical intelligent control system

VI. Conclusions

In this paper, some intelligent techniques, such as fuzzy logic, neural networks, genetic algorithms, and artificial intelligent are introduced and its application for a hierarchical intelligent control of robotic systems are described. Integration and synthesis techniques of AI, fuzzy, neural network and GA make the robot system be intelligent, and the hierarchical control system has both top-down and bottom-up learning abilities while integrating and synthesizing those techniques. The reinforcement learning as the skill acquisition technique also introduced. For these schemes, robotic systems would have the flexibility for the environmental changes or unknown environment.

References

[1] T. Fukuda and T. Shibata, Theory and Applications for Neural Networks for Industrial Control Systems, IEEE Trans. on Industrial Electronics, Vol. 39, No. 6, pp. 472-489 (1992)

[2] L. A. Zadeh, Fuzzy Sets, Information and Control, Vol. 8, pp. 228, (1965)

[3] D. E. Goldberg, Genetic Algorithms in Search, Optimization, and Machine Learning, Addison Welsey (1989)

[4] T. Shibata and T. Fukuda, Skill Based Control by using Fuzzy Neural Network for Hierarchical Intelligent Control, Proc. of IJCNN'92 - Baltimore, Vol. 2, pp. 81-86 (1992)

[5] T. Fukuda, S. Shiotani, F. Arai, A New Neuron Model for Additional Learning, Proc. of IJCNN92-Baltimore, Vol. 1, pp. 938-943, (1992)

[6] T. Shibata, T. Fukuda, K. Kosuge, F. Arai, Selfish and Coordinative Planning for Multiple Mobile Robots by Genetic Algorithm, Proc. of the 31st IEEE Conf. on Decision and Control, Tucson, Vol. 3, pp. 2686-2691 (1992)

[7] T. Fukuda, T. Shibata, M. Tokita, T. Mitsuoka, Neuromorphic Control - Adaptation and Learning, IEEE Trans. on Industrial Electronics, Vol. 39, No. 6, pp. 497-503 (1992)

[8] T. Shibata and T. Fukuda, Hierarchical Intelligent Control of Robotic Motion, Trans. on NN (1992) (in Press)

[9] T. Fukuda, H. Ishigami, F. Arai, T. Shibata, Structure Optimization of Fuzzy Neural Network using Genetic Algorithm, Proc. of IFSA (1993)

[10] T. Shibata and T. Fukuda, Fuzzy Critic for Robotic Motion Planning by Genetic Algorithm in Hierarchical Intelligent Control, Proc. of IJCNN'93-Nagoya (1993)

[11] H. Ichihashi, Learning in Hierarchical Fuzzy Models by Conjugate Gradient Method using Backpropagation Errors, Proc. of Intelligent System Symp., pp. 235-240 (1991)

[12] T. Parisini, R. Zoppoli, Radial basis function and multilayered feedforward neural networks for optimal control of nonlinear stochastic systems, Proc. of Int'l Conf. on Neural Networks, pp. 1853-1858 (1993)

[13] D. E. Goldberg, Genetic Algorithm in Search, Optimization and Machine Learning, Addison Wesley (1989)

[14] C. L. Karr, E. J. Gentry, Fuzzy Control of pH Using Genetic Algorithm, IEEE Trans. on Fuzzy Systems, Vol. 1, No. 1, pp. 46-53 (1993)

[15] J. Koza, Genetic Programming on the Programming of Computers by means of Natural Selection, MIT Press (1992)

[16] D. A. Sofge (Ed.), Handbook of Intelligent Control - Neural, Fuzzy, and Adaptive Approaches, Van Nostrand Reinhold (1992)

[17] T. Shibata and T. Fukuda, Coordinative Behavior by Genetic Algorithm and Fuzzy in Evolutionary Multi-Agent System, Proc. of IEEE Int'l Conf. on Robotics and Automation, Vol. 1, pp. 760-765 (1993)

[18] S. Shiotani, T. Fukuda, T. Shibata, Recognition System by Neural Network for Incremental Learning, Proc. of the IEEE/RSJ Int'l Conf. on Intelligent Robotics and Systems, pp. 1729-1735 (1993)

[19] S. Shiotani, T. Fukuda, T. Shibata, An Architecture of Neural Network for Incremental Learning, Neurocomputing (1993) (unpublished)

[20] H. Ishigami, Y. Hasegawa, T. Fukuda, T. Shibata, Automatic Generation of Hierarchical Structure of Fuzzy Inference by Genetic Algorithm, Proc. of Int'l Conf. on Neural Networks '94 (1994) (in Press)

[21] F. Saitoo T. Fukuda, Learning Architecture for Real Robotics System-Extension of Connectionist Q-Learning for Continuous Robot Control Domain, Proc. of Int'l Conf. on Robotics and Automation, Vol.1, pp. 27-32, (1994)

[22] F. Saitoo T. Fukuda, Two-Link-Robot Brachiation with Connectionist Q-Learning, Proc. of 3rd Int'l Conf. on Adaptive Behavior (From Animals to Animats 3), pp. 309-314, (1994)

[23] J.H. Connell, S. Mahadevan, Robot Learning, Kluwer Academic Publishers, (1993)

[24] A.G.Barto, R.S. Sutton, C.W. Anderson, Neurolike adaptive elements that can solve difficult learning control problems, IEEE Transactions on Systems, Man, and Cybernetics, SMC-13(5), pp.834-846, (1983)

Authors:

Professor Toshio Fukuda
Dr. Koji Shimojima
Dept. of Micro System Engineering, Nagoya University
Furo-cho, Chikusa-ku, Nagoya 464-01, Japan
Tel: +81-52-789-4478
Fax:+81-52-789-3909
E-mail: fukuda@mein.nagoya-u.c.jp

Modeling and Control of Mechatronic Systems by Decentralized Descriptor Systems

Thomas Raste and Peter C. Müller
Safety Control Engineering
University of Wuppertal, Germany

Abstract: Mechatronic systems in descriptor form arise very naturally and conveniently when various subsystems are composed to a complete overall system. Usually the descriptor model is transformed into state-space form. The alternative approach presented in this paper is based directly on the descriptor form. Because of severe requirements concerning real-time and safety aspects, the mechatronic system is separated into interconnected subsystems. The control concept is based on compensating these interconnections by local feedback including decentralized observers. Controlling the motion of a vehicle convoy demonstrates the design procedure.

1 Introduction

The theory of large-scale systems gives the control engineer powerful methods at hand he needs to design controllers for systems with complex structures [1]. Mechatronic systems are a representative class of those systems which is expressed on process level by high dimensionality with manifold interactions [1]. Usually this high dimensionality is counteractive against on-line real-time control with a centralized control structure. Therefore distributed control structures have been well established in mechatronic system design.

The basic concept is decomposing the whole process into less complex subprocesses and specifying the interactions between them. Multilevel/multilayer hierarchical control structures compress the rate of information from lower to higher level components. Computation on higher levels is based on condensed information. Real-time computers (RTC) control only a local subprocess. Special purpose computers (SPC) are provided for coordination of the RTC's to reach a global process goal or they fulfil fault-tolerance or diagnosis functions. To avoid a costly point to point cable connection higher level components are connected with a high-speed communication network. In recent years intelligent actors and sensors (IA/IS) came into use, i.e. these components have inherent communication capacities. In fig. 1 an example of a complex mechatronic system is shown. Fig. 2 is the corresponding block-diagram system representation used in control design.

---

[1]This work was supported by the VW-Foundation under Grant I/67 977

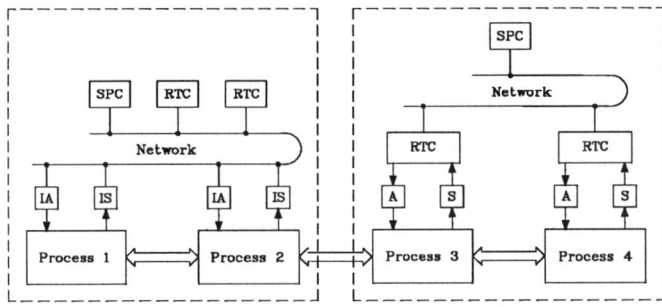

Figure 1: Complex mechatronic system

For technical or economical reasons information exchange between controller networks sometimes is impossible. The dashed boxes in fig. 1 illustrate this "information constraint". Although a complete decentralized control design is much more difficult than a centralized or hierachical one the design often leads to a higher degree of safety or reliability. To illustrate this effect an enlargement of the second controlled subsystem is seen in the lower part of fig. 2. If the network is down coordination between the controllers of subsystem 2 through the dashed paths is impossible and a decentralized subsystem remains. Provided a decentralized control design procedure, subsystem 2 is tolerant to the network breakdown, i.e. the local controllers can stabilize the subsystem. If network breakdown happens to subsystem 1, this subsystem will loose control and may become unstable.

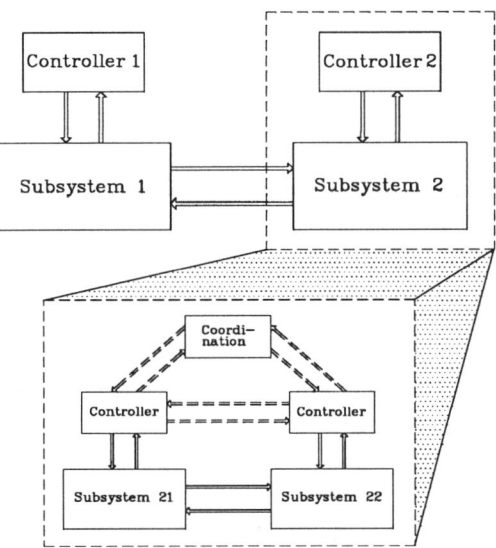

Figure 2: System representation for control design

2 Modeling of Decentralized Descriptor Systems

Usually the dynamic behaviour of each isolated subsystem, i.e. subsystem without interactions, can be described by a mathematical model in the linear case such as

$$\dot{\mathbf{x}}_i = \mathbf{A}_i\, \mathbf{x}_i + \mathbf{B}_i\, \mathbf{u}_i, \tag{1a}$$

$$\mathbf{y}_i = \mathbf{C}_i\, \mathbf{x}_i, \qquad i = 1, \ldots, N, \tag{1b}$$

in the state space with state vector $\mathbf{x}_i \in \mathcal{R}^{n_i}$, input vector $\mathbf{u}_i \in \mathcal{R}^{q_i}$, measured output vector $\mathbf{y}_i \in \mathcal{R}^{m_i}$ of each subsystem $i = 1, \ldots, N$. The constant matrices $\mathbf{A}_i$, $\mathbf{B}_i$ and $\mathbf{C}_i$ are of appropriate dimensions. Composing the subsystems to a complete mechatronic system additional couplings will appear. The couplings and dynamical interactions may be couplings by springs and dampers or kinematic constraints such as joints restricting the freedom of motion. All over a system of differential and algebraic equations has been received which is called a decentralized descriptor system

$$\dot{\mathbf{x}}_i = \mathbf{A}_i\, \mathbf{x}_i + \mathbf{B}_i\, \mathbf{u}_i + \sum_{\substack{j=1 \\ j \neq i}}^{N} \mathbf{A}_{ij}\, \mathbf{x}_j + \mathbf{L}_i\, \boldsymbol{\lambda}_i, \tag{2a}$$

$$\mathbf{0} = \sum_{j=1}^{N} \mathbf{N}_{ij}\, \mathbf{x}_j, \tag{2b}$$

$$\mathbf{y}_i = \mathbf{C}_{xi}\, \mathbf{x}_i + \sum_{\substack{j=1 \\ j \neq i}}^{N} \mathbf{C}_{xij}\, \mathbf{x}_j + \mathbf{C}_{\lambda i}\, \boldsymbol{\lambda}_i \qquad i = 1, \ldots, N. \tag{2c}$$

The dynamic interactions are described by the terms $\mathbf{A}_{ij}$ and the kinematic constraints are represented by the additional algebraic equations (2b) and by some Lagrange multipliers $\boldsymbol{\lambda}_i \in \mathcal{R}^{p_i}$. The terms $\mathbf{C}_{xij}$ yield e.g. from "relative" measurements between two subsystems. For convenience all constraint forces effecting subsystem i caused by kinematic constraints between subsystem i and other subsystems j are composed into

$$\mathbf{L}_i\, \boldsymbol{\lambda}_i = \sum_{\substack{j=1 \\ j \neq i}}^{N} \mathbf{L}_{ij}\, \boldsymbol{\lambda}_{ij}, \qquad i = 1, \ldots, N. \tag{3}$$

Mechanical parts of a mechatronic system can be modeled due to the multibody system method [2]. Fig. 3 illustrates the decomposition of a multibody system into interconnected subsystems. In this example eq. (3) leads to

$$\mathbf{L}_1\, \boldsymbol{\lambda}_1 = \begin{bmatrix} \mathbf{L}_{12} & \mathbf{L}_{13} \end{bmatrix} \begin{bmatrix} \boldsymbol{\lambda}_{12} \\ \boldsymbol{\lambda}_{13} \end{bmatrix}, \quad \mathbf{L}_2\, \boldsymbol{\lambda}_2 = \mathbf{L}_{21}\, \boldsymbol{\lambda}_{21}, \quad \mathbf{L}_3\, \boldsymbol{\lambda}_3 = \mathbf{L}_{31}\, \boldsymbol{\lambda}_{31}. \tag{4}$$

If there are no constraints between subsystem i and a certain subsystem j it is $\mathbf{N}_{ij} = \mathbf{0}$ and therefore $\mathbf{L}_{ij} = \mathbf{0}$. In a similar manner it is $\mathbf{A}_{ij} = \mathbf{0}$ if there is no dynamic interaction. For the control design procedure introduced in the next chapter all interaction terms of the decentralized descriptor system (2) are composed into the expressions given in eq. (5) with interconnection variables $\mathbf{d}_i \in \mathcal{R}^{n_{di}}$. It is assumed that there are no "zero-columns" in the matrices of eqns. (3) and (5).

$$\mathbf{X}_{Ai}\,\mathbf{d}_i = \sum_{\substack{j=1 \\ j\neq i}}^{N} \mathbf{A}_{ij}\,\mathbf{x}_j, \quad \mathbf{X}_{Ni}\,\mathbf{d}_i = \sum_{\substack{j=1 \\ j\neq i}}^{N} \mathbf{N}_{ij}\,\mathbf{x}_j, \quad \mathbf{Y}_i\,\mathbf{d}_i = \sum_{\substack{j=1 \\ j\neq i}}^{N} \mathbf{C}_{xij}\,\mathbf{x}_j. \tag{5}$$

Augmenting system (2) with an equation for interesting variables $\mathbf{z}_i \in \mathcal{R}^{n_{zi}}$ to be controlled and considering (5) leads to

$$\begin{bmatrix} \mathbf{I}_{ni} & 0 \\ 0 & 0 \end{bmatrix} \begin{bmatrix} \dot{\mathbf{x}}_i \\ \dot{\boldsymbol{\lambda}}_i \end{bmatrix} = \begin{bmatrix} \mathbf{A}_i & \mathbf{L}_i \\ \mathbf{N}_{ii} & 0 \end{bmatrix} \begin{bmatrix} \mathbf{x}_i \\ \boldsymbol{\lambda}_i \end{bmatrix} + \begin{bmatrix} \mathbf{B}_i \\ 0 \end{bmatrix} \mathbf{u}_i + \begin{bmatrix} \mathbf{X}_{Ai} \\ \mathbf{X}_{Ni} \end{bmatrix} \mathbf{d}_i, \tag{6a}$$

$$\mathbf{y}_i = \begin{bmatrix} \mathbf{C}_{xi} & \mathbf{C}_{\lambda i} \end{bmatrix} \begin{bmatrix} \mathbf{x}_i \\ \boldsymbol{\lambda}_i \end{bmatrix} + \mathbf{Y}_i\,\mathbf{d}_i, \tag{6b}$$

$$\mathbf{z}_i = \begin{bmatrix} \mathbf{H}_{xi} & \mathbf{H}_{\lambda i} \end{bmatrix} \begin{bmatrix} \mathbf{x}_i \\ \boldsymbol{\lambda}_i \end{bmatrix} + \mathbf{Z}_i\,\mathbf{d}_i, \qquad i = 1,\dots,N. \tag{6c}$$

It is assumed that the matrices $\mathbf{N}_{ii}$ have full rank, i.e.

$$\operatorname{rank}\mathbf{N}_{ii} = p_i, \qquad i = 1,\dots,N. \tag{7}$$

The physical principle "actio=reactio" implies that some pairs of Lagrange multipliers $\boldsymbol{\lambda}_{ij}$ are equal, e.g. in fig. 3 the pairs $\lambda_{12} = \lambda_{21}$ and $\lambda_{13} = \lambda_{31}$. A remarkable consequence is the fact that the global constraint equation matrix $\mathbf{N}$ has never full rank,

$$\operatorname{rank}\mathbf{N} = \operatorname{rank} \begin{bmatrix} \mathbf{N}_{11} & \cdots & \mathbf{N}_{1N} \\ \vdots & \ddots & \vdots \\ \mathbf{N}_{N1} & \cdots & \mathbf{N}_{NN} \end{bmatrix} < \sum_{i=1}^{N} p_i. \tag{8}$$

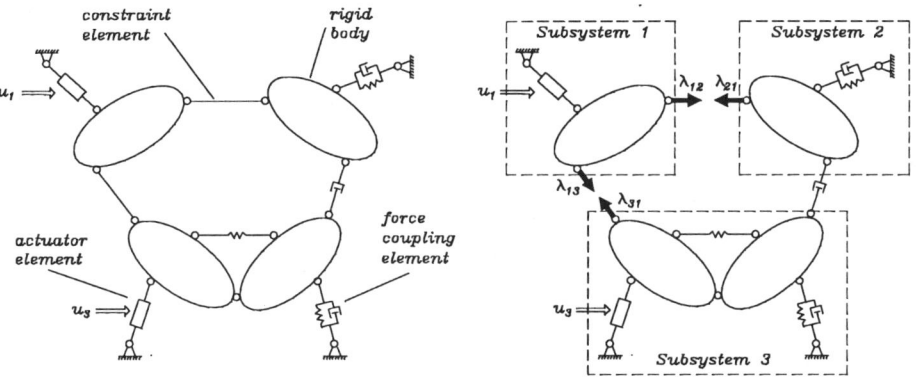

Figure 3: Multibody system with typical elements and decomposition into interconnected subsystems

3 Control of Decentralized Descriptor Systems

The design procedure is similar to disturbance rejection [3] or compensation of non-linearities [4] for normal state-space systems. The control aim is to design a local feedback such that the closed loop subsystem is asymptotically stable and influence of the interconnection variables d_i is asymptotically compensated. This implies that the interesting variables z_i are controlled to zero independently of d_i:

$$z_i \to 0 \quad \text{for} \quad t \to \infty \tag{9}$$

The design strategy is separated into two steps. First an estimate $\hat{d}_i$ will be constructed by the use of decentralized observers. In a second step local feedback to compensate d_i will be designed. The result is in general a local dynamic output feedback. In the special case when x_i, λ_i and d_i are measurable no observer is needed and the local feedback is constant.

3.1 Design of Decentralized Observers

The design of decentralized observers is based on an approximation of the time behaviour of d_i by the output of a homogeneous linear time invariant system with state $w_i \in \mathcal{R}^{r_i}$,

$$d_i \approx D_i w_i, \quad \dot{w}_i = W_i w_i. \tag{10}$$

Substitution of d_i in (6) with (10) leads to $i = 1, \ldots, N$ extended subsystems

$$\underbrace{\begin{bmatrix} I_{ni} & 0 & 0 \\ 0 & I_{ri} & 0 \\ 0 & 0 & 0 \end{bmatrix}}_{E_e} \begin{bmatrix} \dot{x}_i \\ \dot{w}_i \\ \dot{\lambda}_i \end{bmatrix} = \underbrace{\begin{bmatrix} A_i & X_{Ai}D_i & L_i \\ 0 & W_i & 0 \\ N_{ii} & X_{Ni}D_i & 0 \end{bmatrix}}_{A_e} \begin{bmatrix} x_i \\ w_i \\ \lambda_i \end{bmatrix} + \underbrace{\begin{bmatrix} B_i \\ 0 \\ 0 \end{bmatrix}}_{B_e} u_i, \tag{11a}$$

$$y_i = \underbrace{\begin{bmatrix} C_{xi} & Y_i D_i & C_{\lambda i} \end{bmatrix}}_{C_e} \begin{bmatrix} x_i \\ w_i \\ \lambda_i \end{bmatrix}. \tag{11b}$$

A decentralized observer in state-space form for system (11) can be realized if the R-observability condition

$$\text{rank} \begin{bmatrix} sE_e - A_e \\ C_e \end{bmatrix} = n_i + p_i + r_i \quad \text{for all finite } s \in \mathcal{C} \tag{12}$$

and the I-observability condition

$$\text{rank} \begin{bmatrix} E_e & A_e \\ 0 & E_e \\ 0 & C_e \end{bmatrix} = n_i + p_i + \text{rank } E_e \tag{13}$$

is fulfilled. Making use of the special structure of (11) the observer is with $\boldsymbol{\xi}_i = \begin{bmatrix} \hat{\mathbf{x}}_i^T & \hat{\mathbf{w}}_i^T \end{bmatrix}^T$ of the following form

$$\dot{\boldsymbol{\xi}}_i = (\bar{\mathbf{A}}_i - \bar{\mathbf{L}}_i \bar{\mathbf{C}}_i)\, \boldsymbol{\xi}_i + \bar{\mathbf{B}}_i\, \mathbf{u}_i + \bar{\mathbf{G}}_i\, \mathbf{y}_i, \tag{14a}$$

$$\hat{\boldsymbol{\lambda}}_i = \bar{\mathbf{G}}_{1i}\, \mathbf{y}_i - \bar{\mathbf{G}}_{2i}\, \boldsymbol{\xi}_i, \tag{14b}$$

All matrices of the observer (14) except $\bar{\mathbf{L}}_i$ and parts of $\bar{\mathbf{G}}_i$ are build from the matrices in (11) by certain transformations described in [5]. The observer gain matrix $\bar{\mathbf{L}}_i$ can be determined using design methods for normal state-space systems, e.g. linear quadratic methods. It is worth pointing out that conditions (12) and (13) are sufficient but not necessary for observer design. R-observability, which is similar to observability of normal state-space systems, can be weakened to R-detectability. I-observability implies measurement of the constraint forces $\boldsymbol{\lambda}_i$. Weakening this condition leads to a varied palette of alternative observers up to singular observers in descriptor form [6].

3.2   Compensation of Interconnections

The result of this second design step is a local feedback

$$\mathbf{u}_i = -\mathbf{K}_{xi}\, \hat{\mathbf{x}}_i - \mathbf{K}_{\lambda i}\, \hat{\boldsymbol{\lambda}}_i - \mathbf{K}_{wi}\, \hat{\mathbf{w}}_i. \tag{15}$$

The gain matrices $\mathbf{K}_{xi}$, $\mathbf{K}_{\lambda i}$ can be determined for each local subsystem

$$\underbrace{\begin{bmatrix} \mathbf{I}_{ni} & 0 \\ 0 & 0 \end{bmatrix}}_{\mathbf{E}_c} \begin{bmatrix} \dot{\mathbf{x}}_i \\ \dot{\boldsymbol{\lambda}}_i \end{bmatrix} = \underbrace{\begin{bmatrix} \mathbf{A}_i & \mathbf{L}_i \\ \mathbf{N}_{ii} & 0 \end{bmatrix}}_{\mathbf{A}_c} \begin{bmatrix} \mathbf{x}_i \\ \boldsymbol{\lambda}_i \end{bmatrix} + \underbrace{\begin{bmatrix} \mathbf{B}_i \\ 0 \end{bmatrix}}_{\mathbf{B}_c} \mathbf{u}_i \tag{16}$$

if R-controllability is satisfied, i.e.

$$\text{Rank}\begin{bmatrix} s\mathbf{E}_c - \mathbf{A}_c & \mathbf{B}_c \end{bmatrix} = n_i + p_i \quad \text{for all finite } s \in \mathcal{C}. \tag{17}$$

Because no interconnection variables occur in (16), well known methods from centralized controller design for causal and regular descriptor systems can be applied [7]. Causality means that no time derivatives of $\mathbf{u}_i$ are involved in the time response of the system. This property is fulfilled for multibody systems without direct input $\mathbf{u}_i$ into the constraint equation (2b). Regularity of (16) is given by assumption (7). It should be noted that for multibody systems no feedback of $\hat{\boldsymbol{\lambda}}_i$ is necessary, i.e. $\mathbf{K}_{\lambda i} = \mathbf{0}$, and therefore estimate (14b) is superfluous.

The determination of gain matrix $\mathbf{K}_{wi}$ is based on the idea that assumption (9) is fulfilled for sufficiently large t and that there is a stationary behaviour

$$\begin{bmatrix} \mathbf{x}_{i\_stat} \\ \boldsymbol{\lambda}_{i\_stat} \end{bmatrix} = \begin{bmatrix} \mathbf{M}_{1i} \\ \mathbf{M}_{2i} \end{bmatrix} \mathbf{w}_{i\_stat}, \qquad \mathbf{z}_{i\_stat} = \mathbf{0}. \tag{18}$$

This concept leads to the following linear equations for the unknown matrices $\mathbf{M}_{1i}$, $\mathbf{M}_{2i}$ and $\mathbf{K}_{wi}$

$$
\begin{bmatrix} \mathbf{A}_i - \mathbf{B}_i\mathbf{K}_{xi} & \mathbf{L}_i - \mathbf{B}_i\mathbf{K}_{\lambda i} \\ \mathbf{N}_{ii} & 0 \end{bmatrix} \begin{bmatrix} \mathbf{M}_{1i} \\ \mathbf{M}_{2i} \end{bmatrix} - \begin{bmatrix} \mathbf{I}_{ni} & 0 \\ 0 & 0 \end{bmatrix} \begin{bmatrix} \mathbf{M}_{1i} \\ \mathbf{M}_{2i} \end{bmatrix} \mathbf{W}_i - \begin{bmatrix} \mathbf{B}_i \\ 0 \end{bmatrix} \mathbf{K}_{wi} = \begin{bmatrix} -\mathbf{X}_{Ai}\mathbf{D}_i \\ -\mathbf{X}_{Ni}\mathbf{D}_i \end{bmatrix}
$$

$$
\begin{bmatrix} \mathbf{H}_{xi} & \mathbf{H}_{\lambda i} \end{bmatrix} \begin{bmatrix} \mathbf{M}_{1i} \\ \mathbf{M}_{2i} \end{bmatrix} = -\mathbf{Z}_i\mathbf{D}_i \quad (19)
$$

Solving (19) simplifies essentially if the fictitious model (10) is realized by n_{di} integrators

$$
r_i = n_{di}, \quad \mathbf{D}_i = \mathbf{I}_{n_{di}}, \quad \mathbf{W}_i = 0. \tag{20}
$$

Equation (19) modified with (20) can be written as

$$
\underbrace{\begin{bmatrix} \mathbf{A}_i - \mathbf{B}_i\mathbf{K}_{xi} & \mathbf{L}_i - \mathbf{B}_i\mathbf{K}_{\lambda i} & -\mathbf{B}_i \\ \mathbf{N}_{ii} & 0 & 0 \\ \mathbf{H}_{xi} & \mathbf{H}_{\lambda i} & 0 \end{bmatrix}}_{\tilde{\mathbf{A}}} \begin{bmatrix} \mathbf{M}_{1i} \\ \mathbf{M}_{2i} \\ \mathbf{K}_{wi} \end{bmatrix} = \underbrace{\begin{bmatrix} -\mathbf{X}_{Ai} \\ -\mathbf{X}_{Ni} \\ -\mathbf{Z}_i \end{bmatrix}}_{\tilde{\mathbf{B}}} \tag{21}
$$

Compensation of the interconnection variables is possible if

$$
\text{rank } \tilde{\mathbf{A}} = \text{rank } \begin{bmatrix} \tilde{\mathbf{A}} & \tilde{\mathbf{B}} \end{bmatrix} \tag{22}
$$

Some remarks to the introduced design procedure are important. First, no separation principle holds for the closed loop subsystem with decentralized observer. The design procedure uses only local isolated subsystem (16) and extended subsystem (11) with approximations of the interconnection terms. On the other hand stability has to be checked for each controlled subsystem (2), (14) and (15) with the real interconnection terms. Like in normal decentralized state-space systems the difference between approximated and real interconnection terms leads to violation of the separation principle [8]. Stability check on the basis of the controlled isolated subsystems (14), (15) and (16) is only possible utilizing the sufficient but not necessary criterion of connective stability, which is not part of this paper.

Second remark concerns the modes of motion of the decentralized global overall system, i.e. the augmentation of all $i = 1, \ldots, N$ systems (2). For example it may be possible that although every subsystem is completely controllable and observable via $\mathbf{u}_i$ and $\mathbf{y}_i$, i.e. conditions (17) and (12) are true, and additionally the global overall system is controllable and observable too, there are some modes which cannot be controlled or observed by a decentralized controller. Those "decentralized fixed modes" are the result of the structural constraints on the decentralized control law. If there is an unstable decentralized fixed mode no decentralized control is applicable [9].

4 Application: Vehicle Convoy

To illustrate the above design procedure the results will be applied to an actual problem out of the fields of road vehicles. The aim is to control the lateral and longitudinal motion of a convoy of two passenger cars connected together with a tow-bar. More precisely, only the second following car has to be controlled automatically while the first car is steered by a human driver. Each car is modeled as a rigid body with three planar degrees of freedom. One of the basic simplifications is neglection of changes in the wheel loads. The model is sometimes called "single track model" because the wheels on one axle are represented by one wheel in the middle of the axle [10], cf. fig. 4.

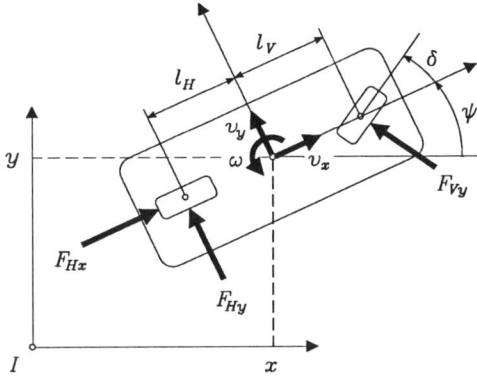

Figure 4: Single track model

The dynamics of the car is described in a chassis coordinate system fixed in the center of gravity with velocity components v_x, v_y and yaw velocity ω. The position of the car is described in the inertial fixed coordinate system I with coordinates x, y and yaw angular ψ. Control input is the steer angle δ of the front wheel. The equations of motions are given with mass m including mass of wheels and moment of inertia J_z by

$$
\begin{aligned}
\dot{x} &= v_x \cos \psi - v_y \sin \psi \\
\dot{y} &= v_x \sin \psi + v_y \cos \psi \\
\dot{\psi} &= \omega \\
m\dot{v}_x - mv_y\omega &= F_{Hx} - F_{Vy} \sin \delta \\
m\dot{v}_y + mv_x\omega &= F_{Hy} + F_{yV} \cos \delta \\
J_z \dot{\omega} &= F_{Vy}l_V \cos \delta - F_{Hy}l_H
\end{aligned}
\tag{23}
$$

The longitudinal tire friction force and the lateral tire sideforces are assumed to be linear functions of the slip s_x, i.e. the relative velocity of the tire to the ground, and the side slip angles α_V, α_H respectively

$$F_{Hx} = C_x\, s_x, \qquad s_x = \frac{v_{Rx} - \omega_R\, r}{v_{Rx}}$$

$$F_{Vy} = C_V\, \alpha_V, \qquad \alpha_V = -\arctan\frac{-v_x\,\sin\delta + (l_V\,\omega + v_y)\,\cos\delta}{v_x\,\cos\delta + (l_V\,\omega + v_y)\,\sin\delta} \tag{24}$$

$$F_{Hy} = C_H\, \alpha_H, \qquad \alpha_H = -\arctan\frac{v_y - l_H\,\omega}{v_x}.$$

The drive train, see fig. 5, consists of simple models for engine and driven wheels. Considering only small accelerations braking torques are excluded.

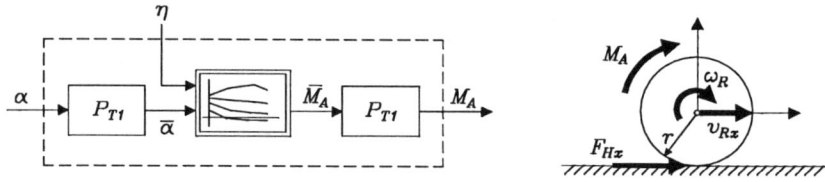

Figure 5: Drive train model

Control input of the engine is the throttle valve position α measured in angular units. The engine torque M_A is specified with α and the engine speed η by a stationary characteristic field. Unsteady processes like throttle actuator dynamics and time lag of engine torque are considered by first order systems. Transmission of gear-box and differentials is summed up into the gear ratio i_G. The drive train model yield with angular velocity ω_R, moment of inertia J_R and constant radius r of the wheel

$$\begin{aligned}
J_R\,\dot{\omega}_R &= i_G\, M_A - F_{Hx}\, r, \\
\dot{M}_A &= -\tfrac{1}{T_M}\, M_A + \tfrac{K_M}{T_M}\, \bar{M}_A(\eta, \bar{\alpha}), \\
\dot{\bar{\alpha}} &= -\tfrac{1}{T_\alpha}\, \bar{\alpha} + \tfrac{K_\alpha}{T_\alpha}\, \alpha.
\end{aligned} \tag{25}$$

The tow-bar connecting the two cars to a convoy is assumed to be massless and of constant length l_0. Fig. 6 shows the geometric description of the vehicle convoy with difference vector $\mathbf{r}_{12}$ evaluated in the inertial fixed coordinate system

$$\mathbf{r}_{12}^I = \begin{bmatrix} x_2 + d_{x2}\cos\psi_2 - d_{y2}\sin\psi_2 + d_{x1}\cos\psi_1 + d_{y1}\sin\psi_1 - x_1 \\ y_2 + d_{x2}\sin\psi_2 + d_{y2}\cos\psi_2 + d_{x1}\sin\psi_1 - d_{y1}\cos\psi_1 - y_1 \\ 0 \end{bmatrix}. \tag{26}$$

Summing the position coordinates of both cars into $\mathbf{p} = [x_1\, y_1\, \psi_1\, x_2\, y_2\, \psi_2]^T$ leads to the algebraic constraint equation and constraint forces respectively

$$\mathbf{f}(\mathbf{p}) = |\mathbf{r}_{12}^I| - l_0 = 0, \qquad \mathbf{F}(\mathbf{p})^T\boldsymbol{\lambda} = \left(\frac{\partial\mathbf{f}}{\partial\mathbf{p}^T}\right)^T \boldsymbol{\lambda} \tag{27}$$

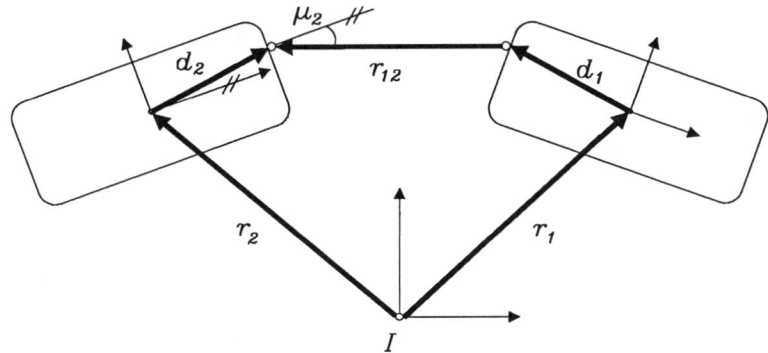

Figure 6: Geometric description of the vehicle convoy

The decentralized subsystems of the vehicle convoy are derived from the equations of motion and the kinematic interconnection terms by linearizing (23), (25) and (27) about a nominal trajectory. The nominal trajectory is given by a straight on drive of the vehicle convoy with constant velocity v_s. In this special case all matrices of the linearized model are constant. Because only car 2 has to be controlled automatically the drive train of car 1 is neglected and the "human driver" is simplified to a given steering signal δ_1. In the following all state, input and output variables describe the small deviation from the nominal trajectory. Subsystem 1 is given by

$$\mathbf{x}_1 = \begin{bmatrix} x_1 & y_1 & \psi_1 & | & v_{x1} & v_{y1} & \omega_1 \end{bmatrix}^T, \quad \mathbf{u}_1 = \delta_1$$

$$\mathbf{A}_1 = \begin{bmatrix} \mathbf{H}_{p1} & \mathbf{H}_{v1} \\ -\mathbf{M}_1^{-1}\mathbf{Q}_1 & -\mathbf{M}_1^{-1}\mathbf{P}_1 \end{bmatrix}, \quad \mathbf{B}_1 = \begin{bmatrix} \mathbf{0} \\ \mathbf{M}_1^{-1}\mathbf{S}_1 \end{bmatrix}, \quad \mathbf{L}_1 = \begin{bmatrix} \mathbf{0} \\ \mathbf{M}_1^{-1}\mathbf{F}_1^T \end{bmatrix}, \tag{28}$$

$$\mathbf{N}_{11} = \begin{bmatrix} \mathbf{F}_1 & \mathbf{0} \end{bmatrix}, \quad \mathbf{N}_{12} = \begin{bmatrix} \mathbf{F}_2 & \mathbf{0} \end{bmatrix},$$

with constant matrices and system parameters

$$\mathbf{M}_1 = \begin{bmatrix} m_1 & & \\ & m_1 & \\ & & J_{z1} \end{bmatrix}, \quad \mathbf{H}_{p1} = \begin{bmatrix} 0 & 0 & 0 \\ 0 & 0 & v_s \\ 0 & 0 & 0 \end{bmatrix}, \quad \begin{array}{l} \mathbf{H}_{v1} = \mathbf{I}_3 \\ \\ \mathbf{Q}_1 = \mathbf{0} \end{array}$$

$$\mathbf{P}_1 = \begin{bmatrix} 0 & 0 & 0 \\ 0 & \frac{c_{V1}+c_{H1}}{v_s} & m_1 v_s + \frac{c_{V1}l_{V1}-c_{H1}l_{H1}}{v_s} \\ 0 & \frac{c_{V1}l_{V1}-c_{H1}l_{H1}}{v_s} & \frac{c_{V1}l_{V1}^2+c_{H1}l_{H1}^2}{v_s} \end{bmatrix}, \quad \mathbf{S}_1 = \begin{bmatrix} 0 \\ c_{V1} \\ c_{V1}l_{V1} \end{bmatrix} \tag{29}$$

$$\mathbf{F}_1 = \frac{1}{l_0}\begin{bmatrix} \bar{l} & \Delta d & -(d_{x1}\Delta d + d_{y1}\bar{l}) \end{bmatrix}, \quad \Delta d = d_{y1} - d_{y2}$$

$$\mathbf{F}_2 = \frac{1}{l_0}\begin{bmatrix} -\bar{l} & -\Delta d & -(d_{x2}\Delta d + d_{y2}\bar{l}) \end{bmatrix}, \quad \bar{l} = \sqrt{l_0^2 - \Delta d^2}$$

Subsystem 2 includes the drive train. System state, input and matrices yield as

$$\mathbf{x}_2 = \begin{bmatrix} x_2 & y_2 & \psi_2 \mid v_{x2} & v_{y2} & \omega_2 \mid \omega_{R2} & M_{A2} & \bar\alpha_2 \end{bmatrix}^T, \quad \mathbf{u}_2 = \begin{bmatrix} \delta_2 & \alpha_2 \end{bmatrix}^T,$$

$$\mathbf{A}_2 = \begin{bmatrix} \mathbf{H}_{p2} & \mathbf{H}_{v2} & \mathbf{0} \\ -\mathbf{M}_2^{-1}\mathbf{Q}_2 & -\mathbf{M}_2^{-1}\mathbf{P}_2 & \mathbf{M}_2^{-1}\mathbf{R}_{s2} \\ \mathbf{R}_{p2} & \mathbf{R}_{v2} & \mathbf{R}_2 \end{bmatrix}, \quad \mathbf{B}_2 = \begin{bmatrix} \mathbf{0} \\ \mathbf{M}_2^{-1}\mathbf{S}_2 \\ \mathbf{S}_{s2} \end{bmatrix}, \quad \mathbf{L}_2 = \begin{bmatrix} \mathbf{0} \\ \mathbf{M}_2^{-1}\mathbf{F}_2^T \\ \mathbf{0} \end{bmatrix}, \quad (30)$$

$$\mathbf{N}_{21} = \begin{bmatrix} \mathbf{F}_1 & 0 & 0 \end{bmatrix}, \quad \mathbf{N}_{22} = \begin{bmatrix} \mathbf{F}_2 & 0 & 0 \end{bmatrix}$$

System parameters of subsystem 2 are given with the following matrices

$$\mathbf{M}_2 = \begin{bmatrix} m_2 & & \\ & m_2 & \\ & & J_{z2} \end{bmatrix}, \quad \mathbf{H}_{p2} = \begin{bmatrix} 0 & 0 & 0 \\ 0 & 0 & v_s \\ 0 & 0 & 0 \end{bmatrix}, \quad \mathbf{H}_{v2} = \mathbf{I}_3,$$

$$\mathbf{P}_2 = \begin{bmatrix} \frac{C_{x2}}{v_s} & 0 & 0 \\ 0 & \frac{C_{V2}+C_{H2}}{v_s} & m_2 v_s + \frac{C_{V2}l_{V2}-C_{H2}l_{H2}}{v_s} \\ 0 & \frac{C_{V2}l_{V2}-C_{H2}l_{H2}}{v_s} & \frac{C_{V2}l_{V2}^2+C_{H2}l_{H2}^2}{v_s} \end{bmatrix}, \quad \mathbf{Q}_2 = 0$$

$$(31)$$

$$\mathbf{R}_{s2} = \begin{bmatrix} -\frac{C_{x2}r_2}{v_s} & 0 & 0 \\ 0 & 0 & 0 \\ 0 & 0 & 0 \end{bmatrix}, \quad \mathbf{S}_2 = \begin{bmatrix} 0 & 0 \\ C_{V2} & 0 \\ C_{V2}l_{V2} & 0 \end{bmatrix}, \quad \mathbf{S}_{s2} = \begin{bmatrix} 0 & 0 \\ 0 & 0 \\ 0 & \frac{K_\alpha}{T_\alpha} \end{bmatrix},$$

$$\mathbf{R}_2 = \begin{bmatrix} -\frac{C_{x2}r_2^2}{J_{R2}v_s} & \frac{i_G}{J_{R2}} & 0 \\ \frac{30}{\pi}\frac{K_M}{T_M}i_G M_\eta & -\frac{1}{T_M} & \frac{K_M}{T_M}M_\alpha \\ 0 & 0 & -\frac{1}{T_\alpha} \end{bmatrix}, \quad \mathbf{R}_{p2} = 0, \quad \mathbf{R}_{v2} = \begin{bmatrix} -\frac{C_{x2}r_2}{J_{R2}v_s} & 0 & 0 \\ 0 & 0 & 0 \\ 0 & 0 & 0 \end{bmatrix}$$

The interesting variable to be controlled is the meassured angle μ_2 between tow-bar and car 2, see fig. 6. Besides μ_2 further measurements are necessary to fulfil observability condition (12). The output equation of subsystem 2 is according to (2c)

$$\mathbf{y}_2 = \begin{bmatrix} \mu_2 \\ x_2 \\ y_2 \\ x_1 \\ \lambda_2 \end{bmatrix} = \underbrace{\begin{bmatrix} c_1 & c_2 & c_3 \\ 1 & 0 & 0 \\ 0 & 1 & 0 \\ 0 & 0 & 0 \\ 0 & 0 & 0 \end{bmatrix}}_{\mathbf{C}_{x2}} 0 \; 0 \Big| \mathbf{x}_2 + \underbrace{\begin{bmatrix} c_4 & c_5 & c_6 \\ 0 & 0 & 0 \\ 0 & 0 & 0 \\ 1 & 0 & 0 \\ 0 & 0 & 0 \end{bmatrix}}_{\mathbf{C}_{x21}} 0 \Big| \mathbf{x}_1 + \underbrace{\begin{bmatrix} 0 \\ 0 \\ 0 \\ 0 \\ 1 \end{bmatrix}}_{\mathbf{C}_{\lambda2}} \lambda_2 \quad (32)$$

The interaction matrices used in (5) are

$$\mathbf{X}_{A2} = 0, \quad \mathbf{X}_{N2} = \mathbf{F}_1. \quad (33)$$

Matrix $\mathbf{Y}_2$ consists of the first three columns of $\mathbf{C}_{x21}$. All elements c_i in (32) result from the linearization process as

$$c_1 = \tfrac{1}{l_0^2}\Delta d, \qquad c_2 = -\tfrac{1}{l_0^2}\,\bar{l}, \qquad c_3 = \tfrac{1}{l_0^2}\left(d_{x1}\bar{l} + d_{y1}\Delta d\right),$$

$$c_4 = -c_1, \qquad c_5 = -c_2, \qquad c_6 = \tfrac{1}{l_0^2}\left(l_0^2 + d_{x2}\bar{l} - d_{y2}\Delta d\right). \tag{34}$$

The interconnection variables of subsystem 2 are summarized in the following vector

$$\mathbf{d}_2 = \begin{bmatrix} x_1 & y_1 & \psi_1 \end{bmatrix}^T. \tag{35}$$

Although position x_1 is measured a ficticious model with three integrators

$$\mathbf{D}_2 = \mathbf{I}_3, \qquad \mathbf{W}_2 = \mathbf{0} \tag{36}$$

is chosen. An observer of form (14a) and a local feedback $\mathbf{K}_{x2}$, ($\mathbf{K}_{\lambda 2} = \mathbf{0}$), has been designed for subsystem 2. It is worth pointing out that without feedback of $\hat{\mathbf{w}}_2$ the vehicle convoy is unstable. To determine feedback gain matrix $\mathbf{K}_{w2}$ linear equation (21) is solved with

$$\mathbf{Z}_2 = \begin{bmatrix} c_4 & c_5 & c_6 \end{bmatrix}, \qquad \mathbf{H}_{\lambda 2} = \mathbf{0}$$

$$\mathbf{H}_{x2} = \begin{bmatrix} c_1 & c_2 & c_3 & 0 & 0 & 0 & 0 & 0 \end{bmatrix}. \tag{37}$$

Fig. 8 shows the results of the controller design for a selected driving manoeuvre. The manoeuvre is a lane change with speed $v_s = 10\text{m/s}$ of the vehicle convoy initiated by a given input function δ_2 which is represented by the top curve in fig. 8. A plot of the trajectories of the center of gravity of both cars demonstrates that the controller is working well taking small geometric deviations between the trajectories of car 1 and car 2 as a criterion. To achieve a realistic view of the manoeuvre the nominal trajectory is added to the deviations in this plot. Fig. 7 demonstrates that although the integrator model (36) is a coarse approximation the interconnection variables are suitably estimated.

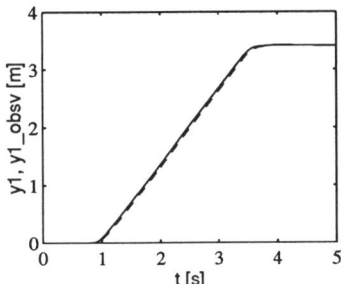 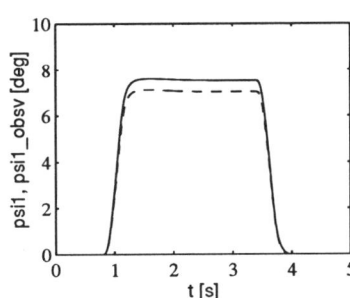

Figure 7: Time history of interconnection variables y_1, ψ_1 (——) and estimation $\hat{y}_1$, $\hat{\psi}_1$ (- - -) during the lane change manoeuvre

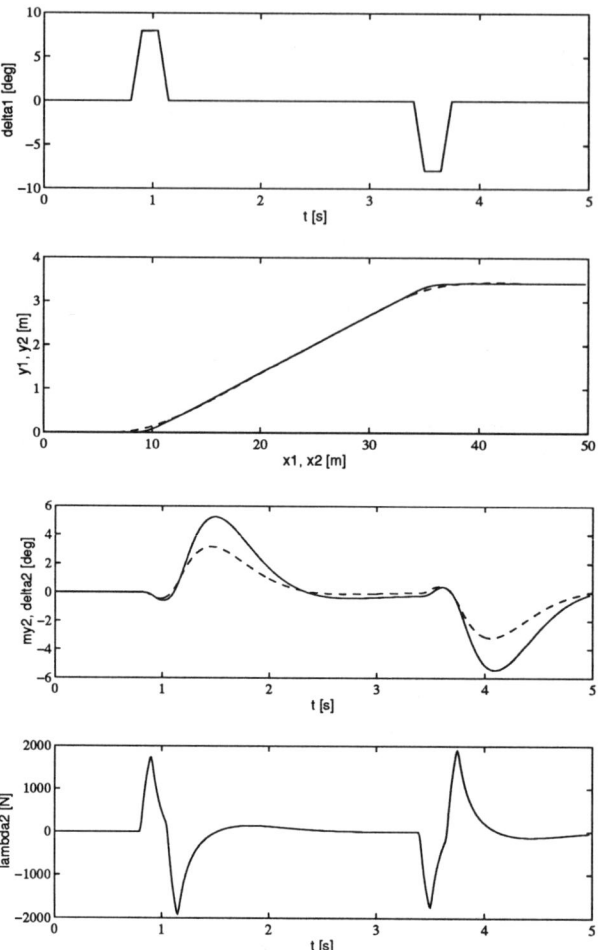

Figure 8: From top to bottom: input function δ_1, trajectories of the center of gravity of car 1 (———) and car 2 (- - -), interesting variable μ_2 (———) and input δ_2 (- - -), constraint force $\lambda_2 = \lambda_1$ for a lane change manoeuvre of the vehicle convoy

5   Conclusions

The introduced design procedure facinates by efficiency and easy use. An advantage of the approach is that the design is based only on the decentralized interconnected subsystems. A limitation is providing equations for the interesting controlled variables which fulfil at the same time eq. (22). Beyond that stability of the overall system has to be checked carefully which is done at this state of research by eigenvalue analysis of the overall system. Future research will integrate the concept of connective stability into the design procedure. The application of the control design to the vehicle convoy has shown the feasibility of the approach. Limiting factor for realization are the measurement requirements. Measuring absolute positions will not be realized for economical reasons. Alternative studies with more suitable vehicle convoy models in descriptor form and with different observers are going on.

6 References

[1] Lunze, J.: Feedback Control of Large-Scale Systems. Prentice Hall, New York, 1992.

[2] Schäfer, P.; Schiehlen, W.: High-Dynamic Test Bed for Mechatronic Vehicle Suspensions. Proceedings of the 2nd Conference on Mechatronics and Robotics'93, Duisburg/Moers, Germany, Sept 27-29, 1993, pp. 367-382.

[3] Müller, P.C.; Lückel, J.: Zur Theorie der Störgrößenaufschaltung in linearen Mehrgrößenregelsystemen. Regelungstechnik 25, 1977, pp. 54-59.

[4] Müller, P.C.: Estimation and Compensation of Nonlinearities. Proceedings of the Asian Control Conference, Tokyo, July 27-30, 1994, pp. 641-644.

[5] Hou, M.: Descriptor Systems: Observers and Fault Diagnosis. Ph.D Dissertation, Safety Control Engineering, University of Wuppertal, Germany, 1994.

[6] Dai, L.: Singular Control Systems. Lecture Notes in Control and Information Sciences, 118, Springer, Berlin, 1989.

[7] Schüpphaus, R.: Regelungstechnische Analyse und Synthese von Mehrkörpersystemen in Deskriptorform. Fortschritt-Berichte VDI, Reihe 8, Nr. 478, VDI-Verlag, Düsseldorf, 1995.

[8] Litz, L.: Dezentrale Regelung. Oldenburg, München, 1983.

[9] Chang, T.N.; Davison, E.J.: Decentralized Control for Descriptor Type Systems. Proceedings of the 25th IEEE Conference on Decision and Control (CDC), 1986, pp.1176-1181.

[10] Lugner, P.: Horizontal Motion of Automobiles. Theoretical and Practical Investigations. In: Dynamics of High-Speed Vehicles. Schiehlen, W. (Ed.). CISM Courses and Lectures No. 274, Springer, New York, 1982, pp. 83-146.

Prof. Dr. Peter C. Müller, Dipl.-Ing. Thomas Raste
Safety Control Engineering, University of Wuppertal
Gaußstr. 20
D-42097 Wuppertal, Germany

Tel.: ++49 202 439-2017 / -2336
Fax: ++49 202 439-2901
Email: mueller@wrcd1.urz.uni-wuppertal.de
         raste@wrcd1.urz.uni-wuppertal.de

SILICON MICROSYSTEMS
FOR MECHATRONIC APPLICATIONS

Werner Brockherde, Dirk Hammerschmidt, and Bedrich J. Hosticka
Fraunhofer Institute of Microelectronic Circuits and Systems,
Duisburg, Germany

Abstract: While modern microelectronic fabrication processes allow monolithic integration of millions and millions of electronic devices on tiny silicon chips and thus form a technological base for implementation of high-performance electronic systems, they are also capable of realizing sensing and actuating functions. Although this may require some minor changes in standard processing and thus increase the fabrication costs, the unique combination of sensors, actuators, and electronic circuits can provide implementation of miniature microsystems featuring unprecedented functionality. Their principles, potentials, and applications in mechatronic systems are the central topic of this contribution.

1 Introduction

Modern microelectronic technologies are capable of integrating huge amount of transistors on a single silicon chip and their products have penetrated wide range of applications [1]. Their adaption and subsequent application to silicon micromachining allow a monolithic cointegration of sensors, actuators, and electronic circuits on silicon chips and thus create an enormous potential for realization of new generation of powerful silicon microsystems [2]. Any penalty paid occassionally in reduced performance of monolithic sensors and actuators due to necessary processing trade-offs, when compared to hybrid implementations using optimized devices, is more than offset by almost unlimited capabilities provided by on-chip electronic signal processing [3]. The key question is here whether we succeed in developing global system approaches towards microsystem design, fabrication, assembling, packaging, and testing and utilize effectively and economically their market potential [4].

2 Principles of Silicon Microsystems

Single-crystal silicon is arguably one of the best materials for microsystem technologies. While its electrical properties are well known and widely used in modern microelectronic circuits, it also offers some excellent mechanical properties [2]. Thus it surpasses stainless steel in yield-strength ratio and lacks mechanical hysteresis. Though difficult to machine using cutting tools, it can be shaped using processing techniques known from microelectronic fabrication, e.g. by chemical etching [5]. Other materials used in standard processing, such as silicon dioxide, silicon nitride, aluminium, and polycrystalline silicon ("polysilicon"), can also be micromachined. There are two major techniques that can be employed: bulk and surface micromachining. Bulk micromachining uses the entire mass of the silicon chip, while surface micromachining works only the surface. These micromachining technologies have already produced a variety of tiny micromechanical devices, such as beams, cantilevers, springs, bridges, diaphragms, gears, micromotors, valves, pumps, etc. The catalogue of applications includes among others suspended or moving silicon structures that can be used for implementation of pressure and acceleration sensors. Bulk micromachining is becoming an established technology for manufacturing of silicon nozzles, e.g. for ink jet or fuel injection. Surface micromachining has found its application in fabrication of micromirror arrays developed for all-digital HDTV ("high-definition television") projection displays. Another application example are microgrippers which can be used in microsurgery and are actuated by electrostatic forces. All these devices greatly benefit from presence of microelectronic circuits cofabricated on the same silicon substrate.

A prominent example of class of devices that benefit from microelectronic functionality at its utmost is represented by silicon-based sensors [3,4]. Such sensors utilize the fact that silicon is sensitive to light, temperature, and magnetic fields. They can be even made sensitive to chemical substances using additional deposited materials. Any of these sensors, either alone or combined with others to form multisensors or sensor arrays, greatly gains from signal processing capabilities provided by on-chip electronic signal processing. Add-on functions, such as sensor readout, amplification, data compression, noise, offset and drift reduction, linearization, calibration, compensation of nonideal effects, communication interfacing, sensor control, etc., can serve to yield high-performance low-cost data acquisition components. Also, short on-chip sensor-electronics interconnects minimize noise pick-up and reduce parasitic capacitive loading of the sensor, thus improving sensor performance.

Monolithic actuators benefit of course owing to on-chip electronics as well, as their potential can be greatly enhanced by including control and supervisory functions. Although high-power and high-voltage capabilities may again require some additional processing steps, any price increase must be considered with respect to increased functionality and widened application range.

3 Potentials of Silicon Microsystems

During the discussion on principles of microsystems in the last section it has become clear that the miniature size is not the only significant feature of silicon microsystems. Their even more outstanding property is the capability to provide combined mechanical, sensing, actuating, and electronic functions merged in a tiny silicon device. This enables consequent functional optimization and interface tailoring, and also allows development of new approaches to signal acquisition and processing, calibration, assembling, packaging, and testing. In case of sensor arrays and multisensors substantial cost saving are feasible due to sharing of electronics hardware. On-chip signal storage and software programming not only enable implementation of advanced signal processing algorithms but also of novel concepts, such as adaptive processing, user-defined flexibility, self-test, self-calibration, sensor and actuator fusion, etc.

Other principles that can be applied to future microsystem concepts are that of "fuzzy logic", "fuzzy control", and "neural networks" [6]. Especially the latter harbour a high potential due to their "learning" capabilities. These can lead to inclusion of new advanced features, such as statistical signal processing, user-defined functionality with zero-programming, and self-diagnosis.

3 Applications of Silicon Microsystems

Due to their small size, high functionality, and capability to interface electronic, electrical, and mechanical "worlds", silicon microsystems can be deemed to represent almost ideal micro-mechatronical building blocks [7]. Microsystems can serve as autonomous multifunctional subsystem components with build-in sensory and actuatory interfaces. It can be envisioned that due to their incorporated "intelligence" such components can be readily installed in large systems, and should be easy to use and operate. This will significantly simplify overall system assembly, operation, diagnosis, maintenance, and repair and, hence, have a profound impact on construction of future mechatronic systems. Also, novel components will stimulate new applications and, thus, contribute to creation of new markets. As an example, "smart" vision sensors can perform image compression and recognition, and evaluate textures in quality inspection for manufacturing or enable collision avoidance in vehicle guidance and robot control. This can result in yielding new products, such as intelligent inspection microdevices for preventive maintenance, sensory-guided miniature robots and microsurgery devices, or portable orientation and navigation systems for blind persons.

4 Dependable Microsystems

As microsystems start invading new applications, the users will face new problems. One of the most critical issues is microsystem reliability in safety-critical applications, e. g. in plant control, transportation, and health-care. Faults, errors, and failures can have tragic consequences, as they can cause not only substantial material damage, but also bodily harm or even death. While standard product quality assurance programs carried out by manufacturers cover component testing and reliability monitoring as far as MTBF (i. e. mean-time-between failures) are concerned, the problems of faults and failures occuring during microsystem operation remain matter of the user and are largely unsolved. E. g., on-line failure detection and repair of a microsystem operated in real-time control system represent still a major problem.

As classical approaches, such as fault-tolerant computer systems, rely usually on multiple component redundancy and are thus prohibitively expensive, new concepts will be required for competitive products. Very often programatic solutions can defuse the issue: e. g., if safety is of paramount importance in a high-performance sensor application and a temporary mild performance degradation is admissible in case of sensor failure, a downgraded redundancy can be applied. Then, a secondary low-performance low-cost sensor may be used as a back-up, if the main high-performance high-cost sensor has been detected as faulty. Nevertheless, the detection requires continuous sensor monitoring and evaluation of its operation, and thus a local "intelligent" self-test, evaluation, and control will be needed for both sensors and accompanying electronic circuits.

Another solution can be employed in case of piezo- or magnetoresistive sensors: while these are usually realized as full resistive bridges (i. e. 4 resistors), in case of a faulty resistor the remainig devices can be used and operated as a half resistive bridge.

5 Examples of Silicon Sensor Microsystems

Although silicon microsystems are still in their infancy, first such systems have already been reported [2]. In this section we will present three examples from the smart sensor area.

A first example is a magnetic-field sensor system integrated in CMOS technology [8]. Additional processing steps have been included that are necessary for depositing of a magnetoresistive permalloy microbridge acting as sensor. The system also contains temperature compensation circuitry, programmable readout electronics, reference voltage bias, and clock generation (Fig. 1a). The maximum magnetic-field sensitivity of the system is 88.2 mV / (A/m), integral full-scale linearity is better than 0.1 %, and the temperature sensitivity is below 260 ppm/$^\circ$C in the range between - 50 $^\circ$C and + 100 $^\circ$C. The chip area is 29.6 mm^2 (see Fig. 1b).

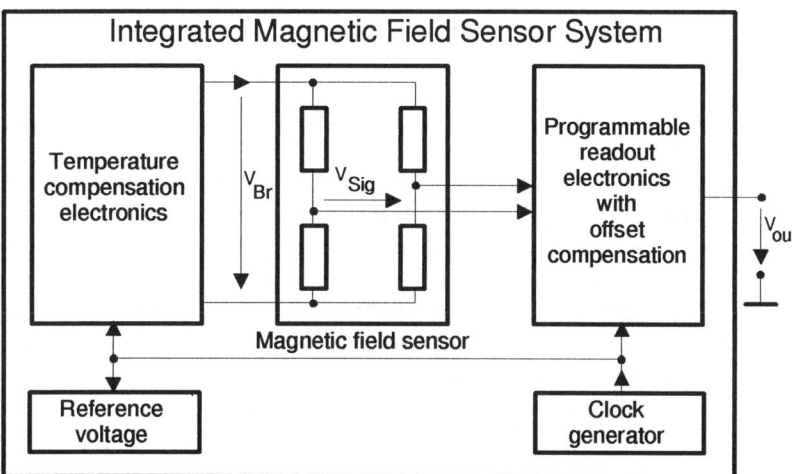

Figure 1a: Block diagram of the single-chip magnetic-field sensor system

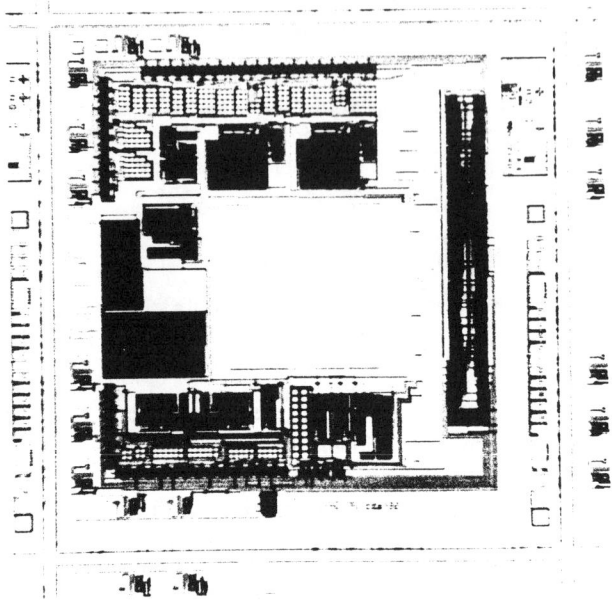

Figure 1b: Chip microphotograph of the CMOS magnetic-field sensor system

Another example which shows how silicon pressure sensor and electronics can yield a smart transducer component is depicted in Fig. 2a. The sensor itself is based on silicon diaphragm formed by anisotropically etched cavity and four piezoresistors [9]. The accompanying CMOS electronics corrects the sensor offset and 1st and 2nd order temperature sensitivity dependence of the sensor in the temperature range between - 40 ^{o}C and + 120 ^{o}C. Parameters for digitally programmable analog CMOS readout and calibration on-chips electronics are stored either in a 30-bit static shift register or permanently with fusible links as a nonvolatile storage (see Fig. 2b). The pressure range is 0.5 - 2 bar and the chip area is 29 mm^2.

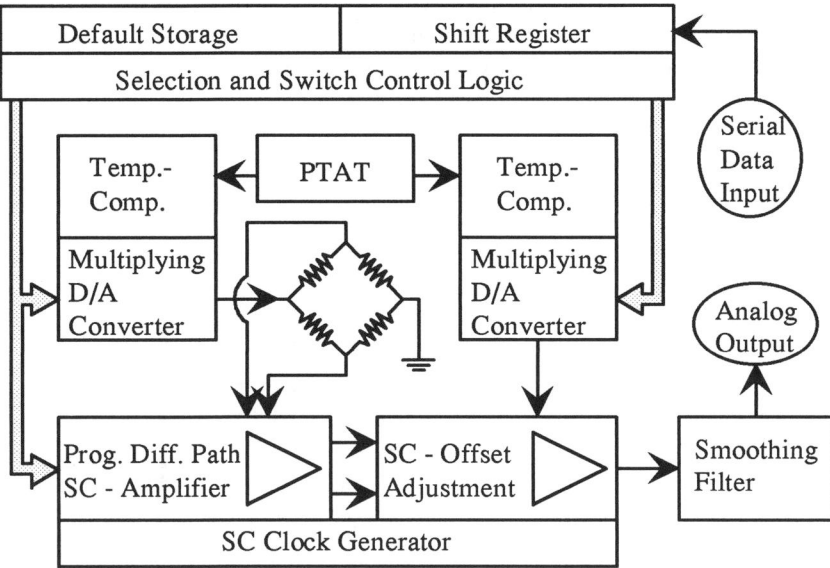

Figure 2a: Block diagram of the piezoresistive presssure sensor system

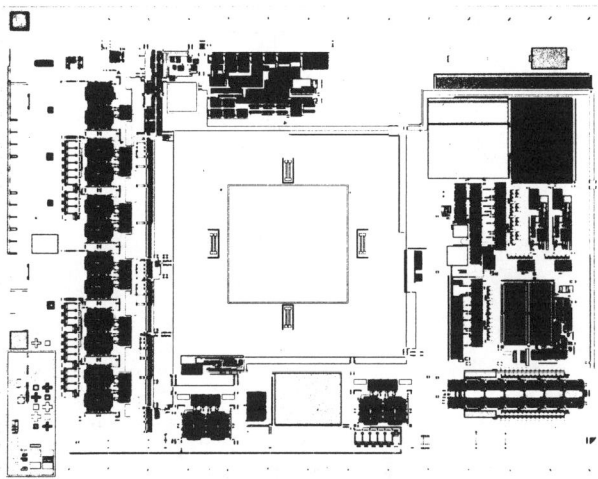

Figure 2b: Chip microphotograph of the CMOS piezoresistive pressure sensor with on-chip programming and calibration

A final example of a smart microsensor is shown in Fig. 3a. This represents a block diagram of a linear image sensor array with programmable on-chip signal processing, which also has been fabricated in CMOS technology [10]. The sensor is based on active pixel sensing principle which utilizes high light sensitivity of a standard PMOS transistor located in a floating n-well.

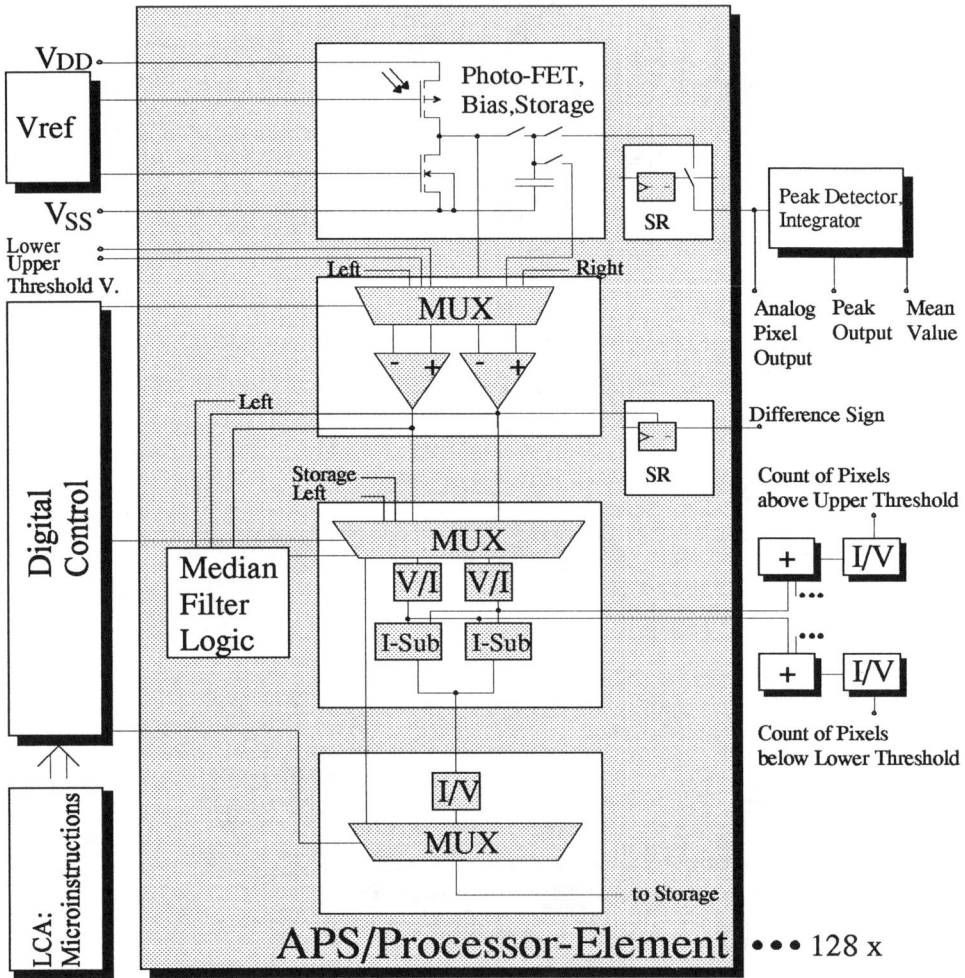

Figure 3a: Block diagram of the CMOS linear image sensor array

In contrast to classical CCD image sensors this technique features not only a much higher sensitivity and dynamic range, but also absence of blooming effects, possibility to design various sensor array and readout geometries, and compatibility with mainstream silicon technologies. This last property is of extreme importance, as it makes possible to realize complex analog, digital, or mixed signal processing on the same silicon chip as the image sensor. Also, since this concept enables parallel sensor access, it can implement very efficiently those image processing algorithms that are based on processing of neighbouring sensors and thus contribute significantly to data reduction while maintaining low system cost.

The chip (see Fig. 3b) contains 128 sensor elements arranged in a line. Each sensor has its own bias and signal processing circuitry which allows both, local and global, signal processing operations.

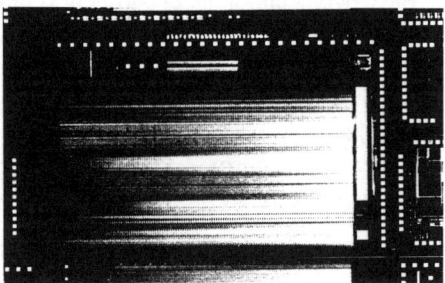

Figure 3b: Chip photomicrograph

The chip can perform the following functions on its sensor pixels:

- serial image readout,
- serial detection of the peak-level pixel signal and its position,
- serial averaging over all pixels in an image,
- parallel time differencing of an image on pixel by pixel basis,
- parallel spatial differencing of all neighbouring pixels in an image,
- parallel median filtering, and
- parallel maximum and minimum signal thresholding.

The chip has been fabricated in a 2μm double metal n-well CMOS process and its area is 54 mm^2 though the sensor array occupies only 0.3 mm^2. The device operates at a clock frequency of 200 kHz and its single power supply range is 5 - 12 Volts. The chip operates at least from -35 up to 120°C and exhibits a maximum temperature pixel drift of about 8 mV/K at the output.

While the examples shown so far have covered the areas of magnetic-field, pressure, and photo sensors, any silicon-compatible sensor can be made "intelligent" in the same way as the above examples, e. g. infrared [11] or chemical sensors [12].

6 Summary

Silicon microsystem technologies allow an almost perfect "marriage" of micromechanical, sensory, actuatory, and electronic signal processing functions on tiny monolithic chips. Thanks to high functionality and performance offered by silicon microelectronics, silicon microsystems can yield components for mechatronic systems with unique features and thus change profoundly design manufacturing, assembly, operation, diagnosis, maintenance, and repair of such systems.

7 Acknowledgement

The authors appreciate the contributions of Michael Schanz to the development of the image sensor and of Andreas Sprotte and Rolf Buckhorst to the magnetic-field sensor. Finally, the support of the FhG-IMS technology department is gratefully acknowledged.

8 References

[1] H. Komiya, "Future technological and economic prospects for VLSI", Digest of Techn. Papers IEEE Int. Solid-State Circ. Conf. (ISSCC '93), San Francisco (Cal.), pp. 16-19, Febr. 1993.

[2] J. Bryzek, K. Petersen, and W. Mc Culley, "Micromachines on the march", IEEE Spectrum, vol. 31, no. 5, pp. 20-31, May 1994.

[3] B. J. Hosticka, "Circuit and system design for silicon microsensors", Proc. IEEE Int. Symp. on Circuits and Systems (ISCAS '92), May 10-13, 1992, San Diego (Cal.), vol. 4, pp. 1824-1827, May 1992.

[4] K. D. Wise and N. Najafi, "The coming opportunities in microsensor systems", Dig. of Techn. Papers IEEE TRANSDUCERS '91, San Francisco (Cal.), pp. 2-7, June 1991.

[5] H. Guckel, "Micromechanisms fabrication: A challenge in micromechanics and micro-electronics", Digest of Techn. Papers IEEE Int. Solid-State Circ. Conf. (ISSCC '92), San Francisco (Cal.), pp. 14-17, Febr. 1992.

[6] D. L. Reilly and L. N. Cooper, "An overview of neural networks: Early models to real world systems", in An Introduction to Neural and Electronic Networks (Eds. S. F. Zornetzer, I. L. Davis, and C. Lan), pp. 227-248, Academic Press, 1990.

[7] B. Hosticka, "Potential of microelectronics for mechatronics", Proc. 2nd Conf. on Mechatronics and Robotics, Duisburg/Moers (Germany), Sept. 27-29, 1993, pp. 555-565.

[8] A. Sprotte, R. Buckhorst, W. Brockherde, Bedrich J. Hosticka, and D. Bosch, "CMOS magnetic-field sensor system", IEEE Journal of Solid-State Circuits, vol. 29, no. 8, pp. 1002-1005, Aug. 1994.

[9] D. Hammerschmidt, F. V. Schnatz, W. Brockherde, and B. J. Hosticka, "A CMOS piezoresistive pressure sensor with on-chip programming and calibration", Digest of Techn. Papers IEEE Int. Solid-State Circ. Conf. (ISSCC '93), San Francisco (Cal.), pp. 128-129, Febr. 1993.

[10] M. Schanz, W. Brockherde, B.J. Hosticka, and R. Klinke, "A CMOS linear image sensor array with on-chip programmable signal processing", Digest of Technical Papers, 21st European Solid-State Circuits Conference (ESSCIRC'95), Lille (France), Sept. 1995.

[11] R. Jähne, W. Budde, R. Gottfried-Gottfried, H. Kück, and M. Müller, "Monolithic infrared sensor system with programmable readout electronics," Digest of Technical Papers, 20th European Solid-State Circuits Conference (ESSCIRC'94), Ulm (Germany), pp. 292-295, Sept. 1994.

[12] W. Mokwa, "Silicon technologies for sensor fabrication", in Chemical Sensor Technology (Ed.S. Yamauchi), vol. 4, pp43-62, Kodansha Ltd., Tokyo (Japan), 1992.

Dipl.-Ing. Werner Brockherde
Fraunhofer Institute of Microelectronic Circuits and Systems
Finkenstrasse 61, D-47057 Duisburg, Germany
Phone: +49-203-3783-230
Fax: +49-203-3783-266
e-mail: brockherde@ims.fhg.de

Dipl.-Ing. Dirk Hammerschmidt
Fraunhofer Institute of Microelectronic Circuits and Systems
Finkenstrasse 61, D-47057 Duisburg, Germany
Phone: +49-203-3783-207
Fax: +49-203-3783-266
e-mail: hammer@ims.fhg.de

Prof. Bedrich J. Hosticka, Ph. D.
Fraunhofer Institute of Microelectronic Circuits and Systems
Finkenstrasse 61, D-47057 Duisburg, Germany
Phone: +49-203-3783-222
Fax: +49-203-3783-266

X. Real-Time Processing

E. G. M. Holweg, G. Honderd, W. Jongkind
Transputer Software Tool for Robot Control Applications

U. Honekamp, R. Stolpe
Design and Application of a Distributed Simulation- and Runtime-Platform for Mechatronic Systems in the Field of Robot Control

J. Pfefferl
MERKUR: A Real-Time and Fault-Tolerant Communication System for Mechatronic Applications

Transputer Software Tool for Robot Control Applications

E.G.M. Holweg, G. Honderd and W. Jongkind,
Control Laboratory,
Department of Electrical Engineering,
Delft University of Technology,
P.O. Box 5031, 2600 GA Delft,
The Netherlands,
Tel.: +31 (0)15 785119,
FAX +31 (0)15 626738,
Email: e.g.m.holweg@et.tudelft.nl

Abstract: For the implementation of a primitive based robot control system a Software Tool has been developed which runs on a transputer network. The Transputer Software Tool provides the user with a standard I/O protocol and a Virtual Robot concept to connect different robots to the control system. On top of this I/O level standardized joint and cartesian controllers can be used to perform simple tasks. The open architecture of the Transputer Software Tool allows the user to add their own task specific primitives or to use other controllers than supplied. Using TCP/IP socket communication programs that execute on a different hardware platform can easily be connected to the transputer network creating a complex control architecture.

1 Introduction

1.1 Primitive control

Within the framework of the robotic project Teleman-18 research has been carried out at the Control Laboratory of the Department of Electrical Engineering in order to design and develop a dexterous robotic gripper [6]. This gripper has to be able to execute tasks in areas which can be described as dangerous for human, e.g. the hot area of a nuclear plant or a chemical polluted area.

For the benefit of this dexterous gripper a control system has to be made which will operate as a man-machine interface and besides execute the given instructions without intervention of the operator. Complex operator tasks such as "Grasp Object" are too difficult to execute at once by a general control system.

One common strategy is to divide the complex task into several smaller tasks, which can be called primitives [1] [2]. These primitives can be executed more easily by the control system. Using the primitives a complex task can be created and executed by building a hierarchy of primitives. E.g. the task "Grasp Object" can be executed using "Open Hand", "Move to Object" and a "Close Hand" primitive.

In general primitives can be described as small processes which execute a specific task without any interference by the user. Whenever necessary it can make use of other primitives, creating an hierarchical structure. To optimize such a structure a primitive must include the possibility to be manipulated in such a manner that it executes within specified precision and speed. The following constraints can be imposed on the primitives.

- Primitives have to operate autonomous with the exception of serious errors,
- As far as the first requirement allows a primitive should provide a certain amount of feedback, especially after finishing its task,
- A primitive must register its parent to be able to pass on messages, errors or other data,
- A primitive can complete its task if all the primitives it started are finished,
- Each primitive has only knowledge within its own scope. E.g. the joint controllers know everything about the joints, but nothing about a group of joints or a finger.

The computer system on which the primitive system will be implemented has to be real-time, multi-tasking and allow a-synchronous communication between processes.

1.2 Implementation

For the implementation of the primitive control system a transputer network has been used [3] [4] [5]. For a general overview of a transputer see Fig. 1. A transputer is a high performance microprocessors that support parallel processing through on-chip hardware. They can be connected together by their serial links called edges and can be used as building blocks for complex parallel processing systems. Parallel processes on a single transputer can communicate through a system of internal links, called channels. On the level of software development there is no difference between the external links, the edges and internal links the channels. This makes the topology of the network easy to adapt to current needs. The communication between processes is synchronous.

On each transputer a highly efficient built-in run-time scheduler is present for processes running in parallel on that transputer. Processes waiting for channel I/O or a timer consume no CPU resources. The context switching time can be as little as 1 µs. The communication links between processors operate concurrently with the processing unit and can transfer data simultaneously on all links without intervention of the CPU.

For the implementation of the Transputer Software Tool so called T805 transputers are used. These transputers run with a 35 MHz clock-cycle, an on-chip floating point processor and 2 Mb of main memory.

The primitive control system is directly connected to the robot manipulator and gripper. The lowest level within the primitive hierarchy are the actual joint controllers. It is important that these primitives are executed in real-time for proper control of the gripper and manipulator. Higher in the hierarchy, the need for real-time behaviour will be less

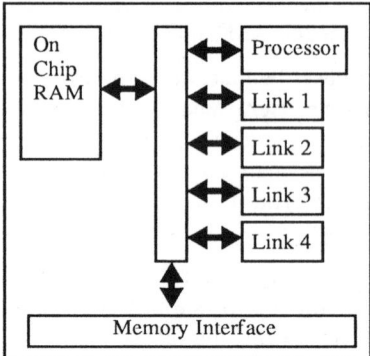

Figure 1:   Schematic view of a transputer.

demanding. This will reflect on the used sampling rate of these primitives. In general one can state that the sampling rate will increase when traversing through the hierarchy of the primitives.

2   The real-time kernel

2.1   The communication model

The primitive processes run in real-time, therefore communication between the primitives is of vital importance. Whenever a primitive sends a message it must be certain that within a specific time interval the message will be delivered, guaranteeing its real-time behaviour. It stands clear that every primitive can not be connected to any other primitive directly, a more structured model is needed. In this model two types of channels are present. High priority channels and low priority channels. Messages send with low priority will try to avoid the high priority channels whenever possible in order to increase the throughput on the high priority channels which are intended for time critical messages. The following approach has been used:

A router process is placed on each individual transputer within the network. The edge connections are used to connect all router processes in the desired topology. The router process is in charge of all primitives running on its transputer. Each primitive will be connected to this router by means of a message queue implemented using a common memory block. The router decides when a primitive should be started or stopped. E.g. it can start a primitive when it is not running and a message for it arrives. It can stop a primitive if it has been idle for a specific duration. The router process is aware of the topology of the network and which primitive is running on which transputer. Using this topology the router can calculate the shortest route from one primitive or transputer to another. In Fig. 2 a simple example is depicted. In this example the network consists of three transputers and five primitives or processes. The three router processes are connected to each other by means of edge channels. Between transputer A and B a special edge is being used, a high-priority channel to be used for high priority messages.

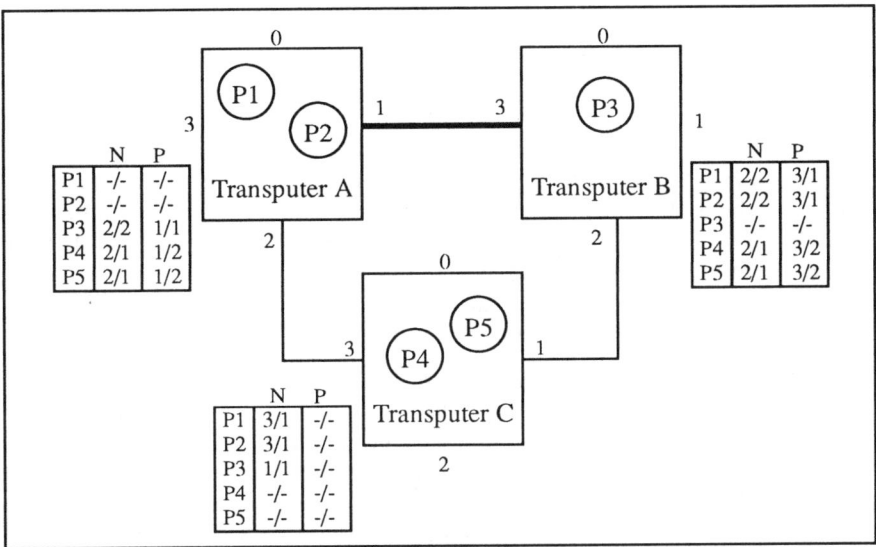

Figure 2:   A schematic overview of the routing tables.

Using the topology of the network the router process can create a look-up table or routing table in which the shortest routes between primitives are stored. The three routing tables for this particular network of three transputers are depicted in Fig. 2. For each process in the network two entries are available. If both entries are empty it indicates that the process runs on the local transputer, messages can be send directly. The first entry is used if the message is of low priority. The second entry is used if the message should use high priority channels. In each entry the first number indicates the edge over which the message has to be send out. The second number in the entry indicates the number of hops it has to made before it will reach the goal transputer. If only high priority channels are available for a low priority message the router will send the message over a high priority channel. A system warning will be generated to state this event. If a high priority edge is not available for a high priority message the router will use a low priority edge. This will also generate a system warning. In both cases the user probably didn't create a sensible network configuration. For the network depicted in Fig. 2 high priority messages should only send between processes P1, P2 and P3.

To make proper use of this routing table, messages send between different primitives are constructed out of four different fields, see Fig. 3.

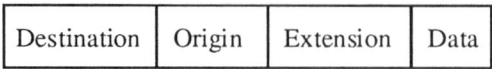

| Destination | Origin | Extension | Data |

Figure 3:   Message lay-out.

- Destination field: containing the identifier of the primitive process to which the message should be delivered. An unique integer number is used as identifier.
- Origin field: containing the identifier of the sending primitive process. This can be used for returning messages to sender.
- Extension field: used for future extensions.
- Data field: the actual data to be transferred.

Whenever a primitive decides to send a message to another primitive it sends the message to the local router. The router uses the look-up table to determine if the primitive is running locally or on a remote transputer. If the primitive is running locally the message is put directly in the input queue of the process. Otherwise the message is send over the indicated edge channel. This process will be repeated after arrival on the next transputer until finally the goal transputer and primitive has been reached.

This communication scheme is a-synchronous and differs from the synchronous transputer channel communication. To avoid communication problems the router processes are divided into a send daemon, a receive daemon and the actual router process. A schematic overview of this is depicted in Fig. 4. For every edge connected a send and receive daemon pair is present. For each edge their is a separate output queue. As long as there are messages in the output queue the send daemon will try to send them over the edge. If the output queue is empty the send daemon will be idle. The receive daemon will also be idle except when a message is received on an edge. After reception of the message it will be put in one single input queue. The router seen in Fig. 4 will check this queue and use the local routing table to see if messages has to be delivered locally or to another transputer. If so the message will be put in one of the output queues otherwise in the input queue of a process.

2.2 Initialization

During run-time each router process has to know the location of all processes available on the transputer network. To gain a real-time character it is of vital importance that messages between processes will be send using the shortest route available. A special start-up sequence is introduced which is being executed after the network is loaded to acquire exactly this information.

Initially every router process creates a first instance of the routing table which holds all local processes. Each process starts initially with 0 hops. A hop is defined as the number

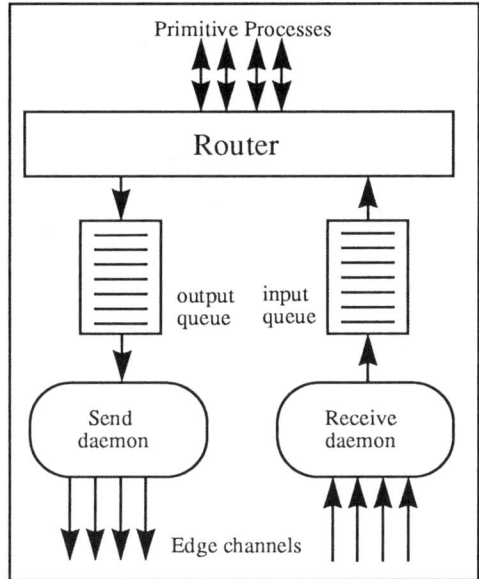

Figure 4: Detailed view of the router
process.

of transputers a message has to travel to reach its destination. For each process two elements are present in the table. One for sending a message using low-priority and one for high priority. At the start of the first iteration the routing table will be send over all connected edges. All router processes will receive the routing table of their neighbour routers. Each received routing table is updated by increasing all hops in the table by one. All processes in the updated routing table will now be matched against all processes in the local table. Whenever a process in the incoming table does not exist in the local table it is copied together with the edge number over which it received the routing table.

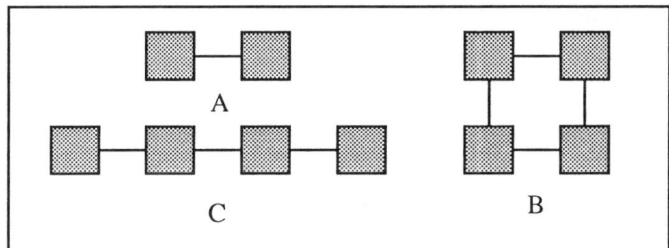

Figure 5: Different type of network topologies.

If the routing table was received over a high-priority edge the second element in the table will be checked, otherwise only the first element. If a process in the received routing table has fewer hops than in the local routing table, the local entry will be replaced together with the edge number over which it received the routing table. Otherwise nothing happens. This iteration has to be repeated as many times as necessary for each router process to receive all other routing tables.

The problem however is to determine the maximum number of iterations needed to complete all routing tables. If a network topology is used seen in Fig. 5 (A) after one iteration both routing tables will be complete. Two iterations are needed if topology (B) is used. The worst case network topology is depicted in Fig. 5 (C). It can easily be seen that in general N-1 iterations are needed for a network with consists of N transputers.

2.3 Routing Kernel Interface

For interaction between the operator and the processes running on the transputer network a special interface has been developed. This interface behaves exactly as a transputer

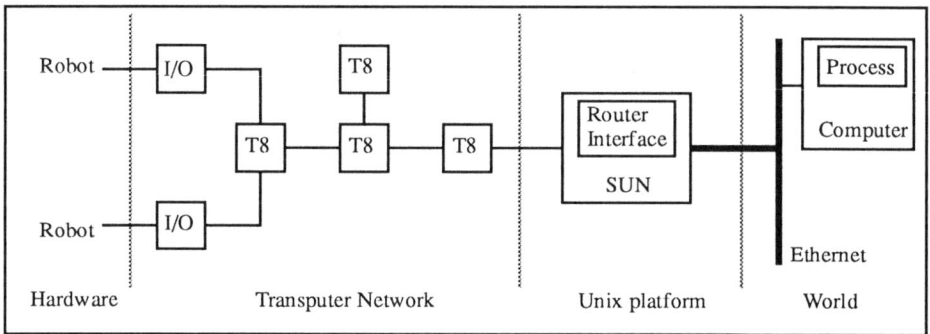

Figure 6: Overview of the complete routing kernel and primitive system.

router process with this difference that it can perform I/O functions which are not available on the transputer network. For each I/O function a virtual process is available within the interface. The I/O functions can be accessed from the transputer network by sending messages to the virtual processes. I/O functions which are implemented are:

- File I/O,
- Data logging,
- Graphical Time displays,
- Graphical Phase plots,
- Graphical 3D plots and
- User defined control panels.

Operator functions are available to bootstrap and stop the transputer network kernel and to stop, start and send messages to processes on the network. Graphical tracing of the throughput and other important kernel information is available for each router in the network. With this information the operator can visually check the status of the transputer network.

The routing interface as mentioned before acts as a routing process itself. On one side it is connected to the router processes of the transputer network. On the other side it is connected to the ethernet. Any given process which has access to this ethernet can connect to the interface using a TCP/IP connection. After a connection is made the process runs together with all processes on the transputer network. Using this open technology it is possible to connect software applications that run on different hardware platforms to the real-time transputer kernel.

3 The control layers

3.1 Low level and I/O control scheme and joint controllers

In this section the generalized I/O system and the joint controllers will be discussed. In order to avoid the problem that for each robot which is connected to the transputer system new I/O programs and controllers have to be developed a more generalized approach has been used. The general scheme is given in Fig. 7.

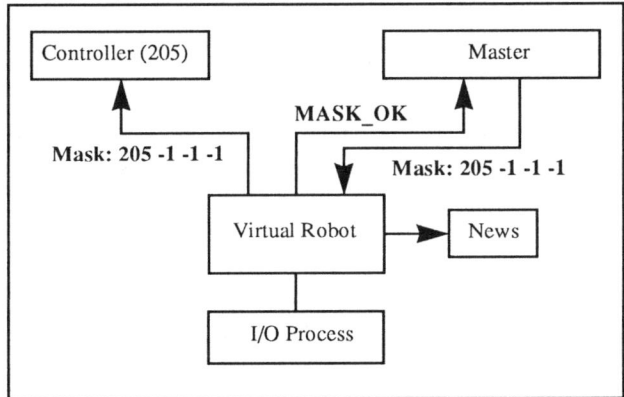

Figure 7: Schematic overview of the I/O system and the controllers.

The generalized I/O system consists of three processes. First the real I/O process, which takes care of all I/O actions with the robot hardware. Second the Virtual Robot process acts as an interface for all processes in the network. The News process is used to supply the network with sensor information about the robot such as joint positions and force information.

The Virtual Robot is master over the I/O process and is the only process allowed in the network to give commands to the I/O process. The Virtual Robot also takes care of timing and safety aspects. If a process decides to control one or more joints of a robot it has to send a mask to the Virtual Robot requesting the specific joints. For each joint one element is available in the mask. Each element holds a process id of the controller process which will control that joint. If the mask is successful the Virtual Robot will reply with a MASK_OK message to the Master process. The controller process will receive the current mask and will receive at the start of each sample the current joint values of the robot. It has to respond within the same sample with new control values. If too much underruns or overruns occur the controller process will be disconnected from the Virtual Robot and the Virtual Robot will reply with a JOINT_ERROR message to the Master and controller process.

If a joint is already controlled by another process the request will fail and the Virtual Robot will reply with a MASK_ERROR. If the number in the mask is -1 the current joint status is unchanged. If the number is 0 the joint will be closed and the connected controller process will be disconnected. This operation can only be performed by the same process that either sent the mask or is currently controlling that joint.

As mentioned the News process is used to supply the network with sensor information. Only the joint controllers which are directly connected to the Virtual Robot will receive the joint information immediately. Other processes which require also this information can subscribe themselves as a client to the News process. The client process can state which information is required and at what interval it needs the data. The News process will send each interval the required data to the client process. Multiple processes can subscribe to the News process and subscribe for different sensor information at specific intervals. By using the News process overhead in the Virtual Robot is being minimised.

Within the Transputer Software Tool several types of joint controllers are present. Currently the following are implemented:

- Joint position controller,
- Joint velocity controller and
- Joint torque controller

All controllers are implemented according to the safety aspects discussed in this paragraph and can all be used to control different robots.

3.2 Cartesian controllers

The Cartesian controller is placed one level higher than the joint controllers. The connection scheme of the cartesian controller is depicted in Fig. 8. In this scheme the cartesian controller will be the master of the joint controller process. It can decide to start either a joint velocity or joint torque controller by sending an appropriate mask to the Virtual Robot process. The cartesian position of the TCP of the robot is received by subscribing to the News process. The inverse or transpose Jacobean is available using the Jacobean process. This process works in principle the same as the Virtual Robot. For each robot the Jacobean process can calculate the inverse or transpose Jacobean. The cartesian process or any other process who needs to use this Jacobean can send input vectors and will receive output vectors in return. In this fashion the cartesian controller

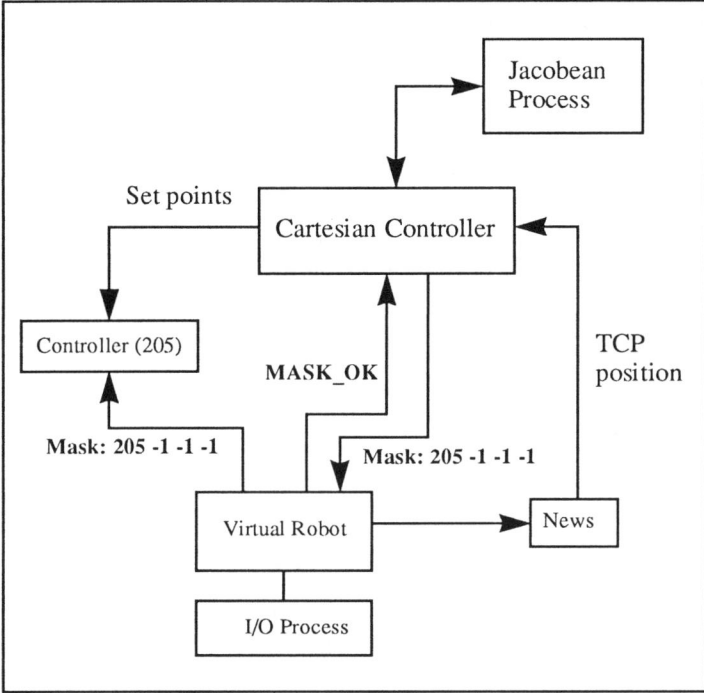

Figure 8:   Topology of the Cartesian controller.

can control different robots by selecting the appropriate Virtual Robot, controller processes and the Jacobean process.

3.3   Collision detection system

Within the transputer network a process is available which takes care of the safety of the attached robot systems. This collision detection system checks periodically if a robot is in collision with itself or with obstacles in its environment.

The collision detection system can serve multiple clients and multiple robots comparable to the News process. Robots can be attached to another in order to actually mount a gripper on the TCP of a manipulator robot. The client has to supply the collision detection system with the Denavith Hartenberg notation of the robot, the dimensions of each link of the robot, static objects present within the robot environment and the sample rate in which the collision checks have to be performed.

For every client, each sample time new joint positions of the robot will be retrieved by registering itself to the News process. Using the joint values, the DH parameters and the link dimensions a cartesian representation of the robot is being generated using spheres to model the link volumes. The obstacles in the robot environment are also modelled using spheres. This representation allows fast collision checks between objects. After the robot has been completely modelled in cartesian space each joint is checked against other joints to check for an internal collision. After the internal collision check the complete robot will be checked against the obstacles in the robot environment. If a collision is detected the system will stop the robot and inform the operator and client process of this event. A safety margin is implemented by increasing the diameter of the spheres used to model the robot and obstacles allowing a collision to be detected before it actually will happen. If the safety margin is chosen larger, fewer checks have to be performed or the robot can move faster with the same number of checks per second.

4 Experiments and Results

4.1 General setup

At the Control Laboratory of the Department of Electrical Engineering of the Delft University of Technology several robots are present. Each robot is connected to the transputer hardware. For each robot a separate I/O process had to be developed. After the development of one Virtual Robot another Virtual Robot could easily be created for a new robot. After the development of the I/O and Virtual Robot processes the joint and cartesian controllers could be used to control the robots.

Because there is no limitation in the number of robots to be controlled by a single transputer application it is possible to create in a simple way a master-slave system using two robots. This is of particular interest for the development of a bi-lateral force reflection data glove to operate the dexterous gripper as discussed briefly in 1.1 and 4.2.

4.2 Collision detection system

The collision detection system has been tested for the OctoVera manipulator. This is a highly redundant 6 D.O.F. robot. This robot has been developed to work in a steam-generator of a nuclear power plant, therefore it moves relatively slow. Using a sampling rate of 100 ms. the collision detection system proved to be fast enough to prevent internal collisions and collisions with obstacles.

The system has also be tested on the HandyMan dexterous gripper [6]. This gripper has three independent fingers with each 3 D.O.F. and an active palm of 2 D.O.F. Within the collision detection system the gripper is modelled as four independent robots which are connected to the TCP of the OctoVera manipulator. As stated before the OctoVera manipulator is checked every 100 ms. The gripper is checked every 175 ms.

4.3 Autonomous control set-up

In Fig. 6 a schematic overview of the real-time kernel, the router interface and a 3D-CAD system LinC is depicted. This set-up is used for autonomous control of the

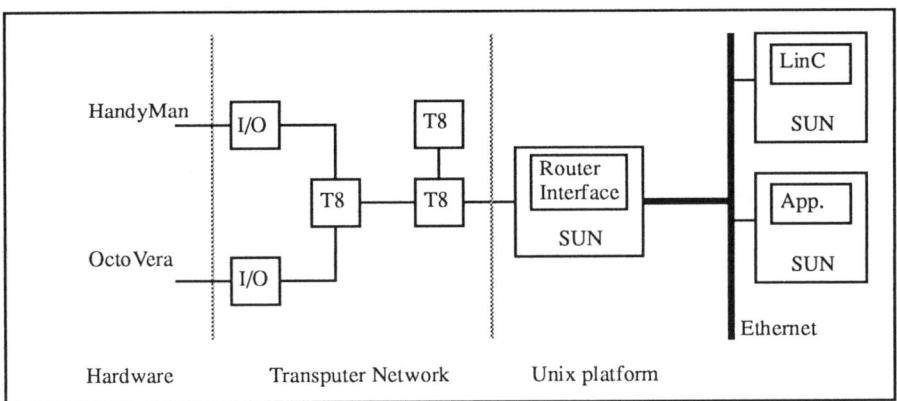

Figure 9: Connection of 3D-CAD system to real-time kernel.

OctoVera manipulator and the HandyMan dexterous gripper. LinC is used to display in 3D models of the robot work-cell. Using LinC it is possible to calculate cartesian robot motions, obstacle free paths and test these paths in simulation. Because LinC is a task independent program another program is needed to make a connection between LinC and the Transputer Software Tool. This program is depicted in Fig. 9 as the Application process. This process will connect to LinC and to the Router Interface using TCP/IP socket communication in order to act as a gateway between the two programs.

LinC is able to send path information to the controllers running on the transputer network. Also processes on the transputer network can send joint information back to LinC in order to show the robot movements on the display. Using this setup it is possible to update LinC every 100 ms with new joint information and render the pictures every 300 ms. This result depends strongly on the complexity of the rendered models and the graphical speed of the machine. In our case the images are rendered using a SparcStation 20 with a ZX graphical frame buffer.

5 Conclusions

The control system is fast enough to control a robot with 4 D.O.F. using a sampling rate of 2 ms. The number of robots controlled by the same application does not influence the speed, because the controller, I/O and Virtual Robot processes run in parallel on different transputers.

The Router processes proved to be very reliable and easy to use. Within the primitives knowledge about the communication is reduced to a minimum allowing the programmer to concentrate fully on the controller design. The open architecture of the Transputer Software Tool allows simple adding of new robots to the software, a new I/O process has to be developed and a Virtual Robot adapted. By adding new primitives using the existing layers it is possible to create more sophisticated and complex tasks. E.g. creating complex arm movements can be implemented using the joint and cartesian controllers.

Applications running on different hardware platforms can easily be connected to the real-time system by means of TCP/IP socket communication allowing the user to build a complex control architecture.

The collision detection system works well for relative slow moving robots. The time needed for a check depends on the D.O.F. of the robot and the number of obstacles present in the environment. But most time during a check is needed to create a cartesian representation of the robot and obstacles. In general one can state that the collision system works well during fine manipulation and for slow moving robots such as the dexterous gripper discussed in 4.2. For faster moving robots another approach has to be used. In stead of checking the robot and obstacles at fixed time intervals it should predict using the current speed of the robot and the obstacles in which period it has to check again to prevent collision. By doing so the robot can move fast in obstacle free areas and slow in close proximity of obstacles.

6   References

[1]   Speeter, T.H., "Primitive based control of the Utah/M.I.T. Dextrous Hand", IEEE International Conference on Robotics and Automation, April 1991.

[2]   Speeter, T.H., "Control of the Utah/M.I.T. Dextrous Hand: Hardware and Software Hierarchy", Journal of Robotic Systems, 1991.

[3]   Holweg, E.G.M.: "Real-time Transputer Kernel", IEEE Sprann, Lille, France, Sept. 27-29, 1994, pp. 1-10.

[4]   Gomes, M.: "De ontwikkeling van een robotbesturing gebaseerd op primitieven", Master Thesis in Dutch, Delft University of Technology, Jan., 1993.

[5]   Berger, J.S.: "Transputer Operating-Systeem voor robotbesturingen", Master Thesis in Dutch, Delft University of Technology, Dec., 1994.

[6]   Jongkind, W.: "Dextrous Gripping in a hazardous environment", PhD, Delft University of Technology, Jun., 1993.

[7]   Feiner, S., Salesin D., and Banchoff T., DAIL: "A diagrammatic animation language", IEEE Computer and Animation, vol 2, September 1982.

[8]   Korein, J.U., Maier G.E., Taylor R.H. and Durfee L.F., "A Configurable System for Automation Programming and Control", IEEE Internationl Conference on Robotics and Automation, 1986

[9]   Doll, T.J. and Schneebeli H.-L, The Karlsruhe Hand, "ARW on Redundant Robots", Salo, Italy, june 1988.

[10] Wöhlke, G. and Braun Th., An Action-Based Process Model for Dextrous Multifingered Hands, Journal of Intelligent and Robotic Systems 4, 1991

Design and Application of a Distributed Simulation- and Runtime-Platform for Mechatronic Systems in the Field of Robot Control

Uwe Honekamp, Ralf Stolpe,
Mechatronics Laboratory Paderborn (MLaP),
University of Paderborn, Germany

Abstract

A basic feature of mechatronic design is the decomposition of a mechatronic system into smaller subsystems. Furthermore each of these subsystems can be equipped with a local information processing unit. This paper deals with a novel approach for supporting real-time information processing in mechatronic systems. We discuss some aspects of implementation and illustrate the application of the approach by means of position and distance control of a 3-dof revolute arm robot. For this system a C-code-generation out of a special description language and the distributed simulation on a multiprocessor hardware are explained.

1 Introduction

Usually the structure of a mechatronic system (e. g. an industrial robot) is very complex. In order to achieve handability of such systems one approach of mechatronic design leads to a decomposition of the entire mechatronic system into smaller subsystems.

Furthermore each of these subsystems can be equipped with a local information processing unit built of hard- and software. Thus, the employment of a parallel computer for the implementation of the a. m. subsystems seems to be highly recommended. Both the implementation of the local information processing units and their arrangement in a communication network can be supported by suitable software tools.

The subsystems (including information processing) can be tested separately by means of the hardware-in-the-loop simulation technique [1]. For this purpose the subsystem is embedded in a realistic environment which interacts with the subsystem as if the real operation environment was present (e. g. employment of a hydraulic test stand for hardware-in-the-loop-simulation of an active suspension strut [2]).

The approach for supporting distributed hardware-in-the-loop-simulation and realization of mechatronic systems presented in this paper is called **IPANEMA** (*Integration Platform for Networked Mechatronic Applications*).

IPANEMA is a strategy for organizing parallel processes in a mechatronic system as well as software support for creating distributed real-time information-processing components for mechatronic systems. The software support comes as tools and an object library which contains C-functions that can be called by the information processing part of the mechatronic application. Due to space limitations the software part of IPANEMA will not be discussed in detail in this paper.

2 IPANEMA

2.1 Conception

In order to achieve portability and well-defined interfaces it is always recommended to project a software application following the multilayer model [e. g. 3]. IPANEMA there-fore implements a software layer between a real-time operating system (if there is any, at least there must be OS-like services) and the "informational content" of the mechatronic application (Figure 1).

To be sure: the intention for introducing this special software layer is encapsulation of OS-services from the application and vice versa. Additionally IPANEMA refines OS-services in a "closer-to-mechatronic-application" manner (e. g. communication resources are delivered including a suitable protocol for data transmission in mechatronic systems).

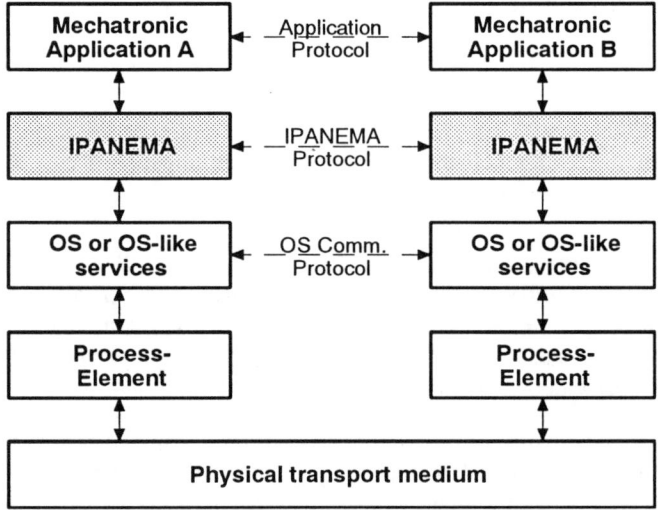

Figure 1: Software layer model

IPANEMA has been projected and designed for use in early design phases of a mecha-tronic product (i. e. operation in research laboratory). Its services are:

- *Organization of application processes*: the number and kind of IPANEMA processes for a specific mechatronic application is determined by the IPANEMA conception.

Furthermore a special process mapping strategy must be taken into account for implementations on certain hardware platforms (see chapter 2.3.1).

- *Deliverance of communication resources*: the communication among processes is channel-based.

- *Coupling of a technical process* (i. e. the physical part of the mechatronic system): this is a prerequisite for hardware-in-the-loop-experiments.

- *Resources for user-friendly communication*: during the experiment the application must be able to communicate with a control center on a host computer.

- *Administration of the application*: during the experiment system parameters might have to be changed. Also the experiment can be stopped and restarted.

- *Data-logging*: values of system variables can be sampled during a specified time duration for documentation and visualization purposes.

- *Supervision of real-time conditions*: for hardware-in-the-loop-experiments real-time conditions must be supervised in order to avoid system destabilization.

Being an approach for supporting distributed real-time information-processing IPANEMA makes use of the **process** construct as a basic conception to build the information processing part of the mechatronic system.

Additionally the processes used to build an IPANEMA application can be considered as **objects** in the sense of the object-orientation paradigm [4] (Figure 2). Therefore a class scheme can be found to structure the process classes of IPANEMA.

2.2 Architecture

First of all processes are needed that implement the "informational content" of the mechatronic system. Normally differential equations (and suitable solvers) are used to describe and/or determine the dynamical behaviour of the system. Additionally state machines can be used to implement any event driven behaviour.

Processes concerned with computation tasks are called **calculators** (Figure 2). Note that calculator-processes are not limited to simulation purposes although simulation is the only application of calculator-processes implemented by now.

The information channels between calculator-processes are established (interface to the OS) depending on the particular application and are shared among calculator- (and adaptor-, see below) processes. Both synchronuous ("once every sampling period") and asynchronuous messages can be transmitted.

In order to interface a technical process (i. e. the physical part of the mechatronic system) for use in a hardware-in-the-loop-simulation a sufficient resource for process coupling must be provided. For this purpose another class of IPANEMA-process has been invented. According to its functionality it is called **adaptor** class (Figure 2) and its instances have to meet hard real-time conditions.

Unlike calculator-processes the implementation of adaptor-processes strongly depends on the underlaying hardware and process periphery. Therefore a port of adaptor-processes to another hardware might be expensive.

For use in a hardware-in-the-loop-simulation calculator- and adaptor-processes have to operate under hard real-time conditions. Usually the load of the individual processors is very high. Therefore it seems to be attractive to separate non-essential services from the calculator-processes in order to save computation time. The non-essential services can be implemented in another class of process which instances can run in parallel with calculator-processes.

Thus, calculator processes are supported by so-called **assistants** (Figure 2). Each calculator-process gets use of its own (exclusively assigned) assistant process. The range of services of an assistant process includes:

- administration of variables (e. g. parameters of the mechatronic system);

- sampling of certain variable values every computation interval and storing the data for visualization and documentation purposes;

- supervision of real-time conditions (e. g. was the duration of the last computation interval greater than the maximum?).

Furthermore, the set of assistant-processes must be able to communicate to the control center running on the host computer in order to change system parameters, start and stop the computation and so on. For this purpose another class of IPANEMA process has been introduced.

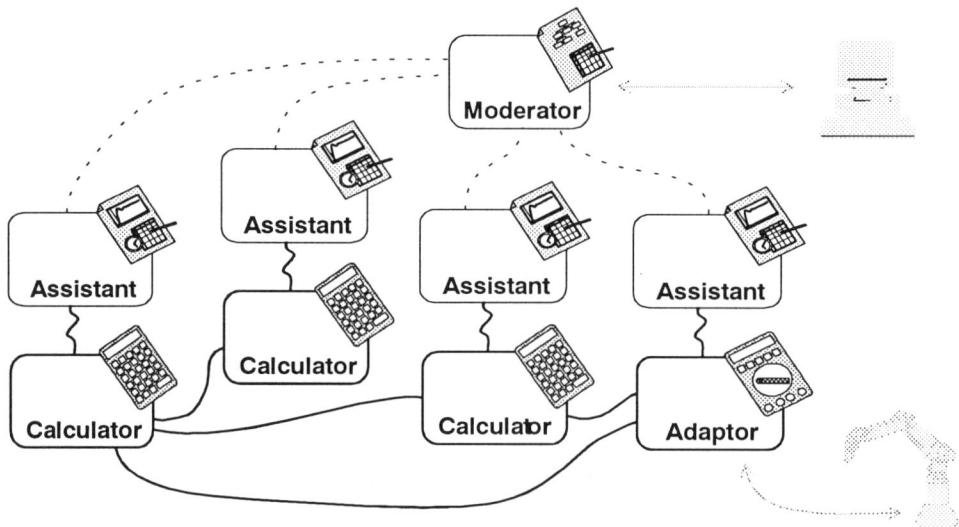

Figure 2: Example of an IPANEMA application

The **moderator** acts as a kind of gateway between the set of assistant processes and the control center. Any application of IPANEMA gets only one moderator-process. The process provides as its main service net transparency. For example: The engineer at the control center would not have to know the "physical location" of a parameter variable on the network in order to change the parameter value. Rather the moderator interprets the request and forwards it to the target process that contents the appropriate variable.

For certain applications the presence of assistant- and moderator-processes is not mandatory. If no user-interaction with the application is necessary, these non-essential processes could be left out.

2.3 Current Implementations

2.3.1 Hardware Platforms

Up to now IPANEMA has been implemented on two different hardware platforms. The original implementation has been done on a transputerbased multiprocessor network [5]. By adding the Transputer Interface Board (TIB) [6] which contains an interface for coupling a technical process to the network this version can be used for hardware-in-the-loop-simulation under hard real-time conditions.

For the implementation of the real-time-version of IPANEMA several process priorities are necessary. The microcoded scheduler of the transputer offers only two priorities, so a special topology for process mapping on certain processors had to be found to overcome this problem. Additionally a set of router processes had to be implemented in order to provide transparent communication for the IPANEMA-processes.

The second implementation is based on a SPARC workstation running the SunOS operating system. The transputer model (based on CSP [7]) seems to be the sufficient communication model even for implementation under UNIX. Therefore, a socket-based communication library has been created in order to provide transputer-like communication resources.

2.3.2 From Model Description to Calculator-Processes

During the last years three stages of model description and their respective description languages have been developed at MLaP [8]. Furthermore, several software tools have been invented as part of the *Computer-Aided Mechatronic Laboratory* (CAMeL) in order to support the engineer in his work of modelling, analyzing, simulating and optimizing mechatronic systems based on these languages [9,10].

The first level covers the discipline-specific[1] description of the different parts of the mechatronic system and their interconnections. Each technical discipline has its specific way of system description. The language *Dynamic System Structure* (DSS) [8] is able to integrate all these specific description formalisms into a method for modelling purposes of entire mechatronic systems.

---

[1] Mechatronic systems are usually built of mechanical, electrical, electronical, hydraulical and information-processing components.

In order to analyze, simulate and optimize a mechatronic system by means of the CAMeL tools the system description has to be abstracted in a mathematical formalism, a system of ordinary differential equations (ODE) of order 1. The description language of this level is *Dynamic System Language* (**DSL**) [8].

The third level offers a machine- and calculation-oriented system description on the basis of the description language *Dynamic System Code* (**DSC**) [8]. DSC is not intended to be directly edited by the user, rather DSC descriptions are automatically generated from DSL descriptions.

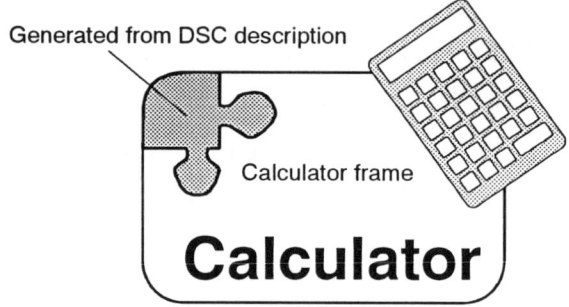

Figure 3: Generated and static parts of the calculator process

And now, a DSC description can be processed by a load distribution tool in order to prepare an implementation of the system on a multiprocessor hardware. The result of the load distribution is another DSC-file, but improved by the information for the multiprocessor implementation.

The improved DSC description is then compiled to a set of files written in ANSI-C including some extensions for parallel computing. For each calculator process in the IPANEMA-application one C-file derived from the DSC description is generated. The calculator is not only built of the generated C-file, but also the application-independent *program frame* has to be added in order to get an entire calculator process (Figure 3).

2.3.3 From Periphery Description to Adaptor Processes

Depending on the formal interface between the information processing part and the physical part of the mechatronic system special hardware circuits have to be added to the application in order to create a suitable process coupling. While this process is application-specific a software tool has been designed and implemented for an automatic creation of adaptor processes.

Analog to model description a specific language called *Hardware Mapping Language* (**HML**) has been invented and implemented for the TIB in combination with industry-standard periphery hardware [11]. The Tool HML2C can be used to generate an entire adaptor process on the basis of the description in HML.

3 Application of IPANEMA to Robot Control

3.1 Position and distance control of a 3-dof revolute arm robot

The following chapter deals with the example of the 3-degree-of-freedom (3-dof) revolute arm robot with hybrid position and distance control (Figure 4). All three joints of the robot have parallel axes. Therefore the robot's movement is limited to the x/y plane. The control problem is to move the robot with a flanged sensor along a surface with steps at a given distance and a constant orientation.

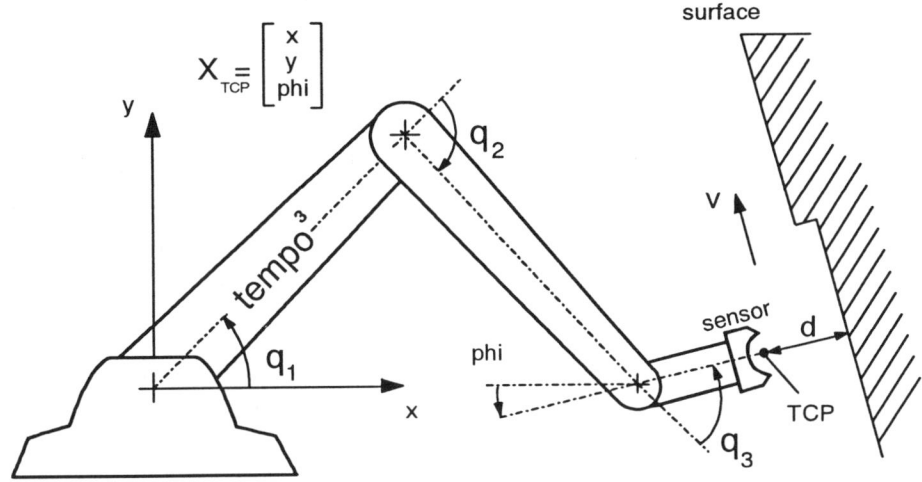

Figure 4: 3-dof robot with environment

The position controller causes the tool frame with the Tool Center Point (TCP) of the robot to follow a planed position trajectory. The distance controller checks the desired distance (d) of the tool frame from the surface. For this purpose, the generated trajectory has to be altered so that the distance to the surface always remains the given one.

Figure 5 shows a block diagram of the hybrid joint controller [12]. The block *trajectory generation* computes the trajectory which describes the planed motion in the x, y coordinate and the angle of the tool frame phi $X = [x,y,phi]$. In addition to position, velocity and acceleration with a trapezoidal profile over the path are calculated. With these cartesian values the *inverse kinematics* computes the desired joint angles (q1, q2, q3), joint velocities, and joint accelerations.

The mapping between the current joint values of the rigid body model of the robot (*rigid model*) and the cartesian space, i.e. the cartesian position and orientation value X, $\dot{X}$, is done by the block *direct kinematics*. The *selection matrix 1* and *selection matrix 2* are switches which set the control mode to be used with each degree of freedom in the base frame [13]. The selection matrix 2 fades out the generated trajectory values in the direction in which the distance controller has to be active.

The selection matrix 1 fades the feedback of the current position and velocity in. This leads to a replacement of the generated values (trajectory generation) with the current values (direct kinematic) in the selected direction. A specific movement of the robot in this direction is now possible, because the new position will always be faded in the generated trajectory values. The distance controller takes over the movement in the selected direction in order to keep the desired distance to the surface.

The controller used is partitioned into two parts, a model-based portion and a servo portion. The parameters of the system appear only in the model-based portion whereas the servo portion is independent of these parameters. The model-based portion decouples the equations of motion over the *mass matrix* of the rigid body model [13].

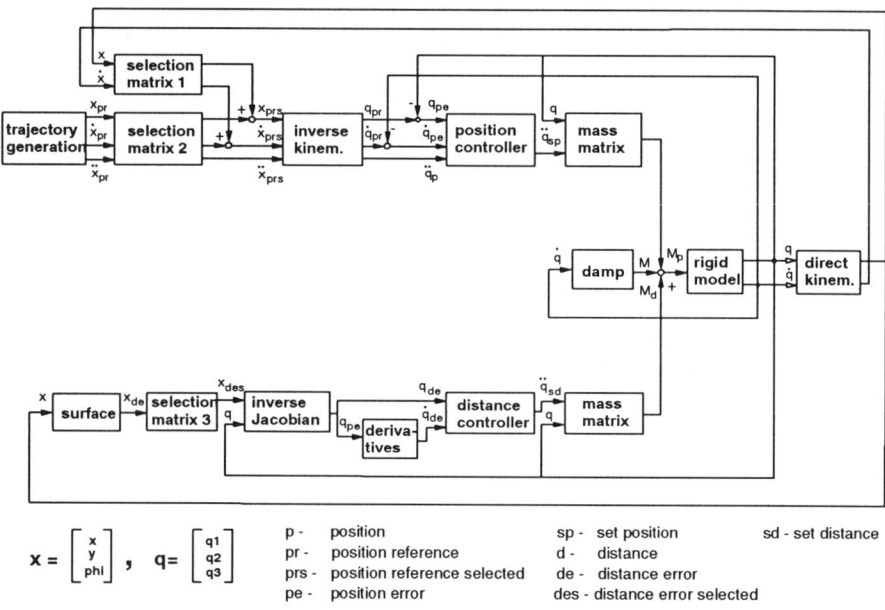

Figure 5: Scheme of the hybrid joint control

This and the *damp* of the model-based portion reduces the system so that it appears to consist of three independent unit masses. The trajectory following resp. the observance of the distance can be controlled by means of a simple PD-controller. Inputs into the block mass matrix are the amplified position / velocity errors of the joints and the generated joint accelerations.

For the distance control the *inverse Jacobian* block maps the cartesian error onto the joint- angle errors. This requires the cartesian error always to be small. Both the error and its derivatives are inputs to the PD-controller. The computed acceleration is also sent to the mass matrix block.

The description of the surface with its steps and the calculation of the distance error to the TCP can be found in the *surface* block.

3.2 DSL basic and couple systems

The entire system is built in DSL. As shown in chapter 2.3.2 DSL is a block-oriented, hierarchical description language. On the basis of an explicit input-output formulation of state-space systems DSL comprises the mathematical description of the models. The state-space formulation consists of nonlinear differential equations for the derivatives of the states (state equations) and a nonlinear algebraic equation for the outputs (output equations).

```
BASIC SYSTEM   DT (
     PARAMETER: k :=1.0, f :=200.0;
     INPUT     : u_q;
     OUTPUT    : y_qp ) IS
     STATE     : x;
     AUXILIAR  : w;
           w := 8*atan(1)*f;
     STATE_EQUATION:
           x':= -w*x + w*u_q;
     OUTPUT_EQUATION:
           y_qp := k*x';
  END DT;
```

Figure 6: Example of a DSL basic system

Two modelling elements of DSL are the *basic system* and the *coupled system*. The derivation block of the distance control, Figure 6, is a simple example of a basic system in DSL. Coupled systems are used to interconnect different basic systems. In this way it is possible to connect a basic system on the next hierachical level.

The described control system consists of DSL basic systems as shown in Figure 5. With a view of the implementation of the system on a multiprocessor hardware some basic systems were combined to coupled systems. For the coupling of the basic systems the functionality and the computing time of each system was taken into account. It is important to pay attention to the replacement of the robot model and the modelled environment with the real system for the realization (see chapter 4). The scheme of the coupled systems is shown in Figure 7.

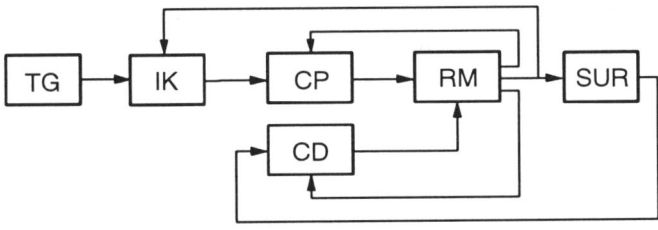

TG   - *trajectory generation*
IK   - *selection matrix 1, 2, inverse kinematics*
CP   - *position controller, mass matrix, damp*
CD   - *inverse Jacobian, derivatives, distance controller, mass matrix*
RM   - *rigid model, direct kinematics*
SUR - *selection matrix 3, surface*

Figure 7: Scheme of the coupled systems

3.3 C-code-generation from DSL systems

For the C-code-generation a set of tools [8] is provided. The tool *DSL* generates out of the discussed DSL basic and coupled systems a description based on Dynamic System Code (DSC). This is also a block-oriented description language for mechatronic systems based on an explicit input-output formulation of state-space-systems (see chapter 2.3.2). For every basic system a separate DSC module has to be generated.

The binder *DSCBIND* combines the coupled system with the associated basic systems to one DSC module. From every present DSC module the translator *DSC2C* generates an ANSI-C source file. This file consists of a fixed number of computing functions with given interfaces (see Figure 8).

The initialization and the setting of the parameters of the system belong to the computing functions just as the initialization of the state vector. The calculation of the derivatives of the states could be found in the eval() function, where the derivatives are functions of the states, the inputs, and the time, as shown in the interface below. The output() function computes the coupled system outputs, which are functions of time, the states, and the inputs. The function update() performs other actions at the major integration time step, e. g. updating the discrete states.

```
static void init_chan()
static void set_p(REAL *p)
static void init_p(REAL *p)
static void init_x(REAL *x)

static void eval(REAL *dx, REAL *x, REAL *u, REAL t)
static void outputs(REAL *y, REAL *x, REAL *u, REAL t)
static void update()
```

Figure 8: Computing functions of the generated Code

For these functions the user can choose float or double for the REAL type. To simulate with the reduced accuracy could be important for the realization because of the decreased computing time of the system. The init_chan() function is necessary to initialize the channels for the distributed simulation. These channels were used for the data exchange between the coupled systems shown in Figure 7.

```
static SimModel M = {"DT",
                     2, 1, 1, 1,
                     same_x_u_t,
                     init_chan,
                     set_p, init_p, init_x,
                     eval, outputs, update};
```

Figure 9: Data structure SimModel

Information about every generated coupled system will be maintained in a data structure called the SimModel. The data structure SimModel provides the number of the parameters, inputs, outputs and states and the names of the discussed functions. The name of the system, here the basic system derivatives from Figure 6, could be found as the first entry in the data structure.

3.4 Application of IPANEMA

The described system can be simulated by the use of the runtime-platform IPANEMA on a transputer network. The main motive is to test the basic systems (*trajectory genera-tion, inverse kinematics* ..) of the generated coupled systems and to determine the dy-namic behavior of the modelled entire system. For the presented six subsystems, Figure 7, ANSI-C code were generated in the mentioned way.

The generated computing functions are embed in the calculators, Figure 3. The com-puting functions for the evaluation of the right sides of the state equations are called as part of the numerical integration function. The calculation of the output equation and the numerical integration function are called cyclic in the calculator program frame. As an example an excerpt of the calculator program frame of the robot model is shown in Figure 10.

```
M->outputs(y, x, u, t);
send(OUT_CHANNELS, y);
integrate(x, u, M, &S, t, h, I);
M->update();
recv(IN_CHANNELS, u);
```

Figure 10: Excerpt of the calculator program frame

The send() communication function sends the outputs of the model to the other subsys-tems. The current joint angles / velocities and the cartesian positions / velocities belong to the sent outputs. With these outputs the subsystems compute the values for the next time step. After the integration the recv() communication function receives the setting moments which are calculated by the controllers. The new outputs are calculated with these inputs and so on. The modules were compiled by the C-compiler of the INMOS-C-Toolsets [14] and linked with the library IPANEMA.lib.

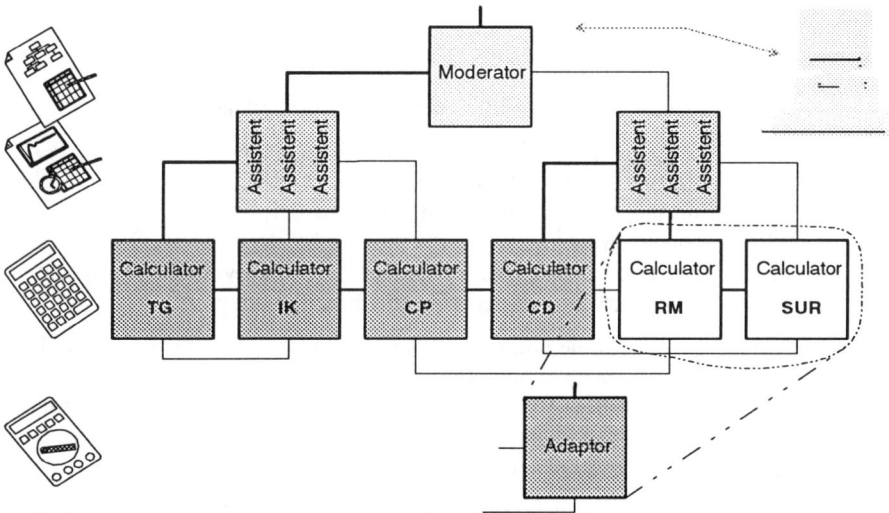

Figure 11: Processortopology of the entire system

After that a bootable file can be built. This includes the logical placement of the de-scribed processes on the available processors of the parallel computer. The in this man-ner prepared bootable file could be downloaded onto a transputer network. The simula-tion of the entire system was done on the IMS B008 transputer board with nine proces-sors. The distribution of the described calculators, assistants and the moderator on the processor hardware is shown in Figure 11. The coupled system from Figure 7 forms the level of the calculators. Three assistents which have direct access to the associated calcu-lators act parallel on one processor.

The in such a manner built hard- and software were used for the simulation of the sys-tem. The assistent knows the names and the units of all the system values such as the outputs, inputs, states and parameters of the associated subsystem. The user can choose under the control center the system values which have to be logged during the running time. The assistent takes the chosen simulated values and books them in a special data structure.

The moderator transmits the saved simulation data to the control center, where they are displayed or saved to file. All the required informations about the application are col-lected from the moderator and sent to the control center.

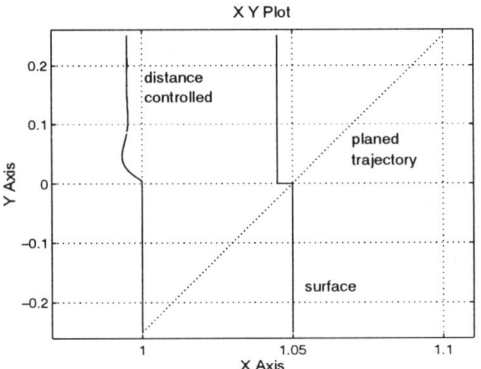

Figure 12: Planed and distance controlled trajectory

For the simulation the following settings were done. The planned trajectory is described by point P1 = [1.0, -0.25] and P2 = [1.1,0.25] in the x/y plane. For the angle phi zero degree and for the surface a vertical line with a step of 0.5 cm for y = 0 are specified. The desired distance between the TCP and the surface has to be 5 cm. Therefore the x-coordinate has to be extracted from the generated trajectory values. These values are substituted by the current cartesian values from the direct kinematics.

The used integrator was a simple euler algorithm. Figure 11 shows the executed trajec-tory of the robot without distance control (-.-) and with distance control (-). These data were traced during the distributed simulation (see Figure 10). The described simulation was done on a Unix SPARC workstation too. For this purpose the created socket-based communication library was used (see chapter 2.3.1) .

For the realization the robot model and the modelled surface, in Figure 11 circled with dotted lines, can be substituted by the adaptor process, discussed in chapter 2.3.3. For example the robot model computes no longer the joint angles / velocities but the adaptor provides the required values.

The task of the adaptor is to couple the periphery hardware with the computing processes, to limit and to scale the process data, and to check the real-time condition of the system. The validated algorithms are now ready to be ported onto a real-time hardware for further tests. The calculation interval could be down to 1 ms.

For the realization the six degree of freedom system **tempo** (**t**est **m**anipulator **p**aderborn) [15] will be reduced to a three dof system. The first generation of the robot control software for the **tempo** was implemented as a modular basis for linear controller implementation and test [16]. The next generation of the robot control software will be based on IPANEMA. The runtime-platform offers a wide range for simulation and realization of complex linear and nonlinear control systems.

4 Conclusions

As written earlier, IPANEMA is a strategy for organizing processes in order to implement distributed information processing in mechatronic systems as well as software to support that task. In particular a software library with C-functions is provided completed by software tools for automatic code generation.

As an application the example of a position and distance control of a 3-dof revolute arm robot was discussed. The C-code-generation out of DSL for this system and the distributed simulation was presented. Therefore an explicit distribution of the coupled system on the multiprocessor hardware was done.

Although IPANEMA is still under construction we reached a stage of implementation that can be employed for tests on our various laboratory systems.

5 References

[1] Lückel, J.: *The Concept of Mechatronic Function Modules applied to Compound Active Suspension Systems*, Symposium: "Research Issues in Automotive Integrated Chassis Control Systems", International Symposium for Vehicle System Dynamics, Herbertov, CSFR, 1992.

[2] Castiglioni, G.; Jäker, K.-P.; Lückel, J.; Rutz, R.: *Active Vehicle Suspension with an Active Vibration Absorber*, in: Proceedings of the International Symposium on Advanced Vehicle Control AVEC '92, Yokohama, September 14-17, ed. Society of Automotive Engineers of Japan, Inc, 1992.

[3] Tanenbaum, A. S.: *Computer Networks*, Second Edition, Prentice-Hall, 1988.

[4] Booch, G.: *Object-Oriented Design with Applications,* Benjamin/Cummings Publishing Company, Inc., Redwood City, California, 1991.

[5] Sczyrba, Ch.: *Entwurf und Implementierung einer portablen, echtzeitfähigen Laufzeitplattform zur Hardware-in-the-Loop-Simulation mechatronischer Systeme*, Diplomarbeit, University of Paderborn, Mechatronics Laboratory Paderborn, 1995.

[6] Gaedtke, Th.: *Entwicklung eines Transputer-Interface-Boards (TIB) als neuer zentraler Baustein für eine industriell gefertigte Prozeßperipherie*, Final Report for the ZIT-Project "Modulsystem zur digitalen Simulation", MLaP, 1990.

[7] Hoare, C. A. R.: *Communicating Sequential Processes*, Communications of the ACM (1978), Volume 21, Number 8, August 1978.

[8] Richert, J.; Hahn, M.: *DSS - DSL - DSC. The Three Levels of a Model Description Language for Mechatronic Systems*, ICMA '94 Mechatronics Spells Profitability, Tampere, Finland, 1994.

[9] Jäker, K.-P.; Klingebiel, P.; Lefarth, U.; Lückel, J.; Richert, J.; Rutz, R.: *Tool integration by way of a Computer Aided Mechatronic Laboratory (CAMeL)*, 5th IFAC/IMACS Symposium on CADCS 91, Swansea, UK, 1991.

[10] Hahn, M.; Richert, J.; Seuss, J.: *Mechatronic Object-Oriented Modelling and Control Strategies for Vehicle Convoy Driving*, 3rd Conference on Mechatronics and Robotics, October 4-6, Paderborn, Germany, 1995.

[11] dSpace GmbH: *DSP-CITpro Hardware*, Paderborn, Germany, 1991.

[12] Hedrich, A.: *Aufbau einer Simulationsumgebung zur Auslegung von hybriden Kraft/Lage und Abstandsregelungen für elastische Knickramroboter*, Studienarbeit, University of Paderborn, Mechatronics Laboratory Paderborn, 1995.

[13] Craig, J. J.: *Introductions to Robotics, Mechanics and control*, Second Edition, Addison-Wesley Publishing Company, Inc., 1989.

[14] Inmos Ltd., *ANSI C Toolset - Reference Manual*, Doc. Nr. 72 TDS 346 01, UK, 1992.

[15] Schütte, H.; Moritz, W.; Neumann, R.; Wittler, G.: *Practical Realization of Mechatronics in Robotics*, Preprints of the Third International symposium on Experimental Robotics, Kyoto, Japan, 1993.

[16] Stolpe, R.; Schütte, H.; Honekamp, U.: *Eine Mehrprozessor-Robotersteuerung als modulare Reglerimplementierungs- und -testumgebung*, 6. Transputer Anwender Treffen TAT´94, Aachen, Germany, 1994.

Uwe Honekamp, Ralf Stolpe
University of Paderborn
Mechatronics Laboratory Paderborn (MLaP)
D-33095 Paderborn
Tel: ++49/(0)5251/60-2420 (U. Honekamp) -3857 (R. Stolpe)
Fax: ++49/(0)5251/60-3550
E-Mail: {hone, stol}@mlap.uni-paderborn.de

MERKUR: A real–time and fault tolerant communication system for mechatronic applications

Pfefferl Johann

Department of Process Control Computers, Prof. Dr. G. Färber
Technical University of Munich, Germany

Abstract: A mandatory condition for mechatronic systems to be realized in a unified manner is a communication system interconnecting the different, spatially distributed subsystems. In the last few years so called fieldbusses have been proposed to fulfill this task. A new communication concept presented in this paper provides the means to integrate sensors, actuators and computers to a complete mechatronic system. The requirements are analyzed and discussed in this paper. The article also explains, how these requirements are taken into account to realize MERKUR. The last part outlines the main aspects of the hardware implementation.

1 Introduction

Distributed real–time computer systems are replacing conventional centralized control techniques especially in this field of mechatronic systems, e.g. process control systems or hardware–in–the–loop simulations. Typical distributed systems are characterized by a controlled object (the mechanical system) and a control system (the computers). These two elements are connected via sensor based and actuator based interfaces. The control system accepts data from the sensors, processes them and outputs the results to the controlled object via the actuators. The aim of this action is to affect the dynamic mechanical object in such a way, that its behavior is optimized.

The general information flow between the two main components is illustrated in Figure 1. This process of merging mechanical parts with sensors, actuators, computer and software to an integrated system is known as mechatronics.

In practice, Figure 1 describes only a global view of the system. Typically, the mechanical system is spatially distributed. It consists of multiple subsystems and its dynamics model is represented by several degrees of freedom. The situation of the control system is similar. Due to the complexity of today's mechanical systems and due to the increasing iteration rates of the control loops, the computational loads of the algorithms are extremely high.

For that reason, most control units do not "fit" within a single computer module. Instead, many require multiple Central Processing Units (CPUs) to handle the mas-

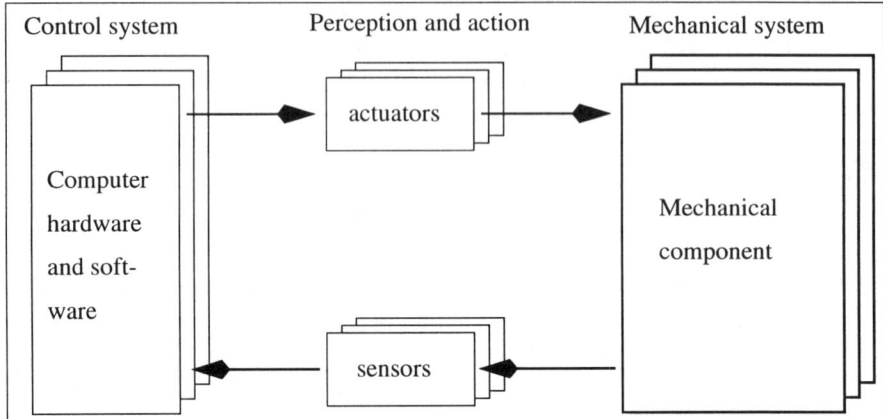

Figure 1: A general mechatronic system

sive load of the algorithms. These distributed approaches of realizing the mechanical and the control system have presented some problems in themselves. Probably the most demanding of these problems has been the question how to interconnect all these distributed components, including the sensors and actuators, to form an unit as displayed in Figure 1.

This paper will describe a ring–structured communication system based on a deterministic channel access protocol. First of all the requirements of the system will be analyzed and compared to existing concepts. The next part of the article will outline the main architectural principles which have been followed in the design of **MERKUR** (**ME**chatronic **R**edundant **K**[**C**]ommunication **U**nder **R**eal–Time Conditions). Finally, the paper will take a look at the hardware structure of the implemented system.

2 Architectural characteristics of MERKUR

2.1 Requirements and existing concepts

As already mentioned, the fundamental processing work of the control stations consists of the following sequence. First of all, the actual state of the mechanical system is perceived by reading new values of the connected sensors, then the new actions that should take place are calculated and finally the new control values are transmitted via the actuators to the controlled system.

This general processing sequence is essential for mechatronic systems, but not exclusively found there. It requires a suitable communication system for transporting the input and output information of the control loop from respectively to the controlled object. It is extremely important to avoid inconsistencies between the internal states of the control system and the controlled object. The control system has to respond to an external stimulus from the environment within an interval dictated by the object, called response time. This time must be guaranteed on the one hand by the control, on the other hand by the communication system. In a multitude of mechatronic applications typical sample rates are in the order of 1 to 100 milliseconds. In addition, the communication system must support a kind of fault tolerance, timeliness, maintainability and extensibility.

In detail, the communication system has to offer the following criteria to satisfy the requirements of the mechatronic system:

- The number of participants, consisting of sensors, actuators and computer nodes, is typically ≤ 100.

- The distributed components reside within a radius of about 50 meters.

- The main load on the communication media is caused by periodic messages transporting sensor and actuator values.

- The sample rate of the control loops are in the range of 10 to 1000 Hz.

- A typical resolution of 16 bits for the data types has to be supported.

- Jitter in data transfer can only be tolerated when its order is much smaller than the sample period.

- Under certain circumstances, sporadic communication with predictable small latency should be possible. Such a situation can occur when switching to an emergency mode is necessary.

- Because of the critical environmental conditions, in which the communication system will operate, the system has to be insensitive to electromagnetic disturbance.

- Integration respectively disintegration of communication nodes during operation is a desired feature, but not of utmost importance.

- A simple interface to the application layer should allow an efficient access to the distributed data.

These requirements are the results of a study made by the interdisciplinary research project [14].

MERKUR isn't the first concept for a bus system interconnecting distributed systems. Many other concepts for so called fieldbusses like "PROFIBUS" [2, 5], "Interbus–S" [8, 3], "CAN" [13, 6], "ASI" [12], "SERCOS" [15] and the real–time network "SCRAMNET" [4] exist. All these implementations were examined and compared with regard to the important items mentioned above. It's far beyond the scope of this article to discuss this study in all its details here.

As a short conclusion, we can establish, that none of the existing fieldbus concepts accomplish all the mechatronic requirements listed above which are necessary for coupling multiple computer nodes and sensor/actuator units. Many of them can't guarantee the timing constraints because of too small transmission rates of the communication medium or too much overhead in the communication protocol. An exception represents the "SCRAMNET" implementation, whose design philosophy is to realize a distributed real–time simulation environment consisting of multiple workstations. This design allows a node delay of 1μsecs. The actual concept of MERKUR combines many aspects found in the different realizations and some new ones especially with respect to a fast, deterministic and reliable communication system.

2.2 Topology and transmission medium

The node interconnection topology is perhaps the most common way to describe local network architectures. The MERKUR net topology can be characterized as a ring structure. Each node possesses 4 ports. The two neighbors are connected via 2 ports to this node (Figure 2). This structure has been chosen for the following reasons.

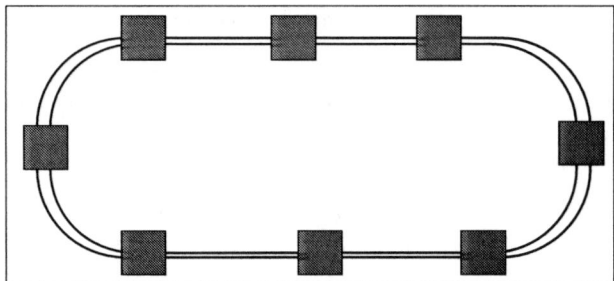

Figure 2: MERKUR net topology

First of all, the choice of the physical medium influences the whole system design. Due to the increasing importance of optical fibers, MERKUR is based on a low cost plastic optical fiber. This fiber is the only supported transmission medium. By employing this fiber, we can profit by different advantages: fibers minimize the electromagnetic disturbance caused by the environment. Also the aspired high bandwidth of 10 to 50 MBit/sec can be realized with few effort over a length of about 20 meters.

One disadvantage of optical fibers is manifested in the fact that only point–to–point connections between two participants can be built with low expense. For that reason, the chosen topology of the communication system is a ring structure as found in many other systems based on optoelectronic transmission techniques.

2.3 Deterministic information distribution

Each real–time system has to provide the specified timely service to its environment (see section 2.1). To meet this requirement, two fundamentally different methods exist: the event–triggered approach and the time–triggered approach [10].

In an event driven system, a significant event in the environment or in the computer triggers the start of a corresponding system action — for example the activation of a special task on a node. In such systems the communication happens only on demand.

A time driven system is characterized by the fact, that the moment when a particular message is passed over the communication system is predefined and therefore known a priori. Because of this property, the system behavior concerning the information transport is totally fixed. Data transmission happens permanently, even when no task needs the data at the moment.

These two main concepts are compared comprehensively by Kopetz [10, 11]. By analyzing the desired requirements of the mechatronic communication architecture as described in section 2.1 and by comparing them with the theses in [10], the only possible realization consists of a time based system architecture. The following considerations will emphasize the decision for a time orientated concept.

To satisfy the demand of timeliness in all imaginable situations like peak load, the system performance must not degrade with variations in the frequency of external stimuli or due to message congestion on the real–time bus [9]. The medium access delay time of the bus must be independent of the communication traffic on it. To realize and especially to prove this behavior by event–triggered concepts is much more complicated than by time–triggered approaches. Therefore in MERKUR a TDMA (Time Division Multiple Access) strategy provides a deterministic, load independent and collision free procedure for medium access like other existing real–time busses designed for special purposes as MARS [9] or SERCOS [15].

2.4 Data communication

Distributing information over the MERKUR ring is both simple and fast. The engineer of the control system software only has to know, that the system–wide common data is represented as a contiguous dataset. The application program can access this dataset by linking a start address to the beginning of the shared–memory block. The operating system can assist this access by a special service routine.

Once this is accomplished, data communication can take place. Each time a new variable value is written to the shared–memory, it is automatically updated in all other nodes on the network ring.

Refer to Figure 3 for an overview of the just mentioned process. The CPU ”wri-

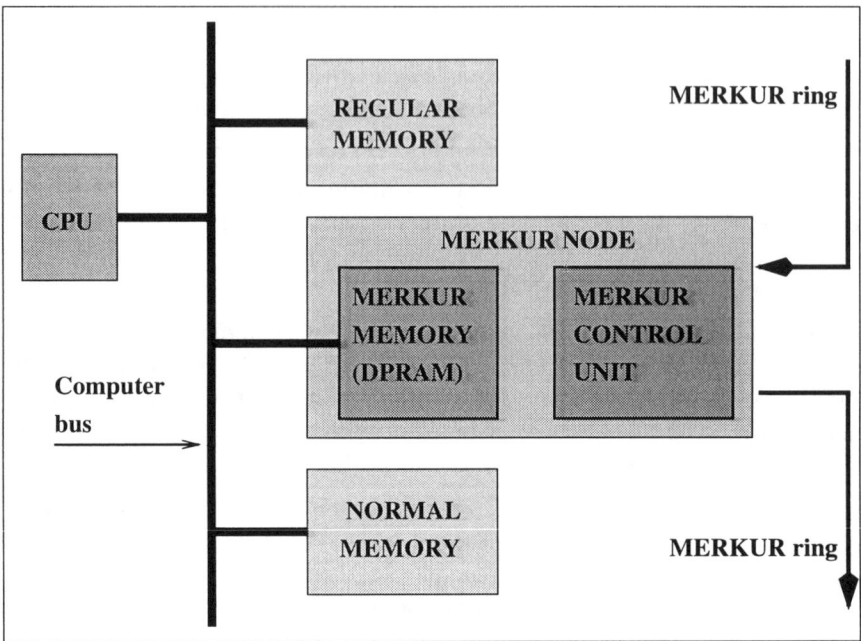

Figure 3: MERKUR data access

tes" a dataword of 16 bit width to a memory location, which physically resides on the MERKUR card, by simply executing an assignment statement like *speed = actual_speed* or a system call such as *WriteDataValue(Speed, actual_speed)*. This formulation is well known from most high–level programming languages. The MERKUR memory, also named real–time data base, "looks" to the CPU as all other memory existing in the computer system, because it is mapped in the normal address space of the host computer.

All the other work concerning the transmission of the data is managed transparently by the MERKUR electronics. Every other component connected to the ring places the 16-bit message automatically in its own local memory at the same relative physical address as the producer node during the next communication cycle. This cycle is started periodically.

The key features of this technique are the simplicity, the speed of data exchange, the access at any time and the unified and easy structure of the software.

2.5 Aperiodic communication

Beside the normal data traffic over the communication channel, a mechanism that handles reactions on sporadic, external events and exceptions is required. For treating this type of message, the basic tool of the system engineer is to utilize hardware interrupts. Because of the decentralized system concept, these interrupts must be communicated among nodes efficiently, fast, and "deterministically".

This fact is also taken into consideration by the design of MERKUR. Every node has the possibility to generate two different interrupts whose meanings can be defined freely by the application designer. In addition to this, all nodes have the ability to generate a so called failsafe interrupt, whose meaning can't be modified.

The failsafe interrupt is used, if a serious failure occurs, either in the control system or the controlled object. If such a situation happens, a further continuation of the operation of the object is not possible or does no longer make sense. The system must shut down and stop in a controlled predetermined manner. Similar to normal exceptions, the activation of failsafe can be done by every node integrated in the communication scheme.

Up to now only the possibility of generating aperiodic events has been discussed. The mechanism, how all these abilities are implemented in MERKUR, is described in section 2.7.

2.6 Fault tolerance

The ring topology naturally includes an uncertainty, because a break of one transmission line stops the complete information transfer of the entire system.

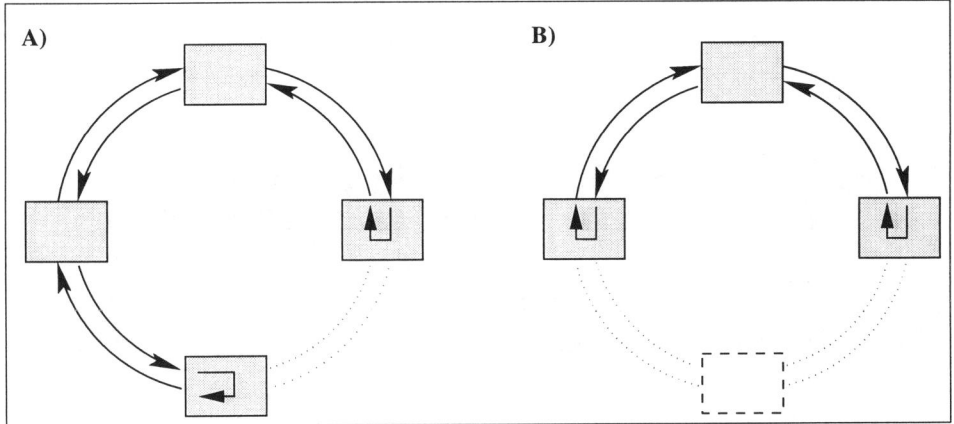

Figure 4: Failure situations: A) channel failure B) node failure

Therefore, the communication topology of MERKUR consists of an antiparallel duplex–ring (see Figure 2). This design feature covers permanent transmission line faults by using the secondary ring segment for bypassing the data flow at the error location as shown in Figure 4. With this method it is also possible to bypass a defective node. When the ring is already operating in one of the two states illustrated in Figure 4, a repeated occurrence of a similar fault forces the system to enter a failsafe state. This step is absolutely necessary because this additional event would divide the remaining ring into two independent ring systems. Such a constellation does no longer represent a valid configuration.

Another fault situation occurs, when data transport fails because of mainly transient failures. This is the case, when a bit toggles falsely during communication. This kind of error is handled by the protocol mechanism.

2.7 Efficient protocol

In the protocol, driven on the communication system, each participant is represented at least by one 16–bit slot. These slots of the TDMA frame are arranged successively according to their slot number i. This number is also used by the MERKUR control unit as an offset for addressing the real–time data base (see section 2.4). The start of a communication cycle is marked by the header packet "CS" as illustrated in Figure 5.

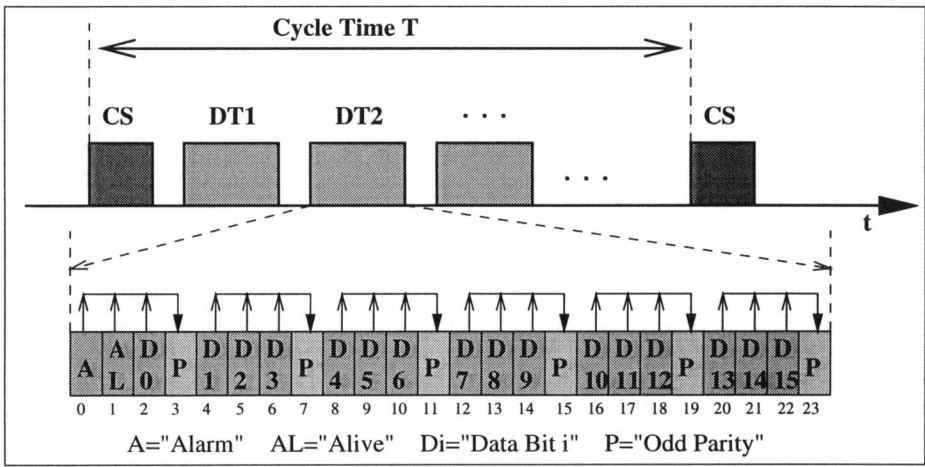

Figure 5: TDMA protocol of MERKUR: CS Communication synchronization slot, DTi Data telegram slot i

This form of protocol allows an implicit addressing of the different nodes inside one communication cycle whereby the efficiency of the protocol increases and the overhead of sending explicit addresses is avoided. When a message is detected faulty — the odd parity bits indicate this — it is rejected and replaced by the message of the following cycle. This is done as fast as possible and avoids overhead of message acknowledge and message repeating mechanisms.

Furthermore, every instant of the control system can be configured in such a way that it checks the actuators for being alive and operating correctly. The method for doing this is the "Alive"–bit in the data telegram. All bad actuators are not able to toggle this bit. For example this is the case, when an actuator is powered down.

Finally, the protocol has to handle reactions on sporadic, external events and exceptions. As mentioned in section 2.5, every node possesses the possibility to generate two universal and one failsafe interrupt. Their transmission is done by replacing the normal data packet. First of all, the "Alarm"–bit is set in the own slot. Then the alarm reason is encoded in the data field (D0 ... D15). Again, the address of the node generating the interrupt is given implicit. This method of overriding implies, that one data message gets lost. To avoid the loss of more than one message, these two different information types are alternated in worst case situations.

3 Node architecture

3.1 Types of participants

A mechatronic system is composed of a passive mechanical system and computer stations. These two parts are linked together by the sensors and actuators. The communication involves only the sensors, actuators and computers. Therefore, MERKUR offers two basic types of nodes. The first type is named "slave" and allows to connect interface components (sensors, actuators). Its functionality is limited to simple IO operation. A slave occupies exactly one slot of the TDMA frame (see Figure 5).

More than one slot can only be assigned to the so called "master" nodes which are part of the control system. Such a node is always realized as an extra board, mounted in a computer as shown in Figure 3. These master nodes are the second type of nodes.

3.2 Slave node

The slave node is arranged in multiple functional modules. One of the most important units is the shift register in the middle of Figure 6. Its function is to read or write values from or to the mechanical objects depending on the chosen operating mode of the node (sensor or actuator). This register reflects also the whole organization of the ring as a large distributed shift register. Every transmitted information passes this register unit. This method is one of several to realize medium access [7]. The slot counter supports the selection of the slot, which corresponds to the node, by comparing the actual counter value with the node address adjusted by DIP switches.

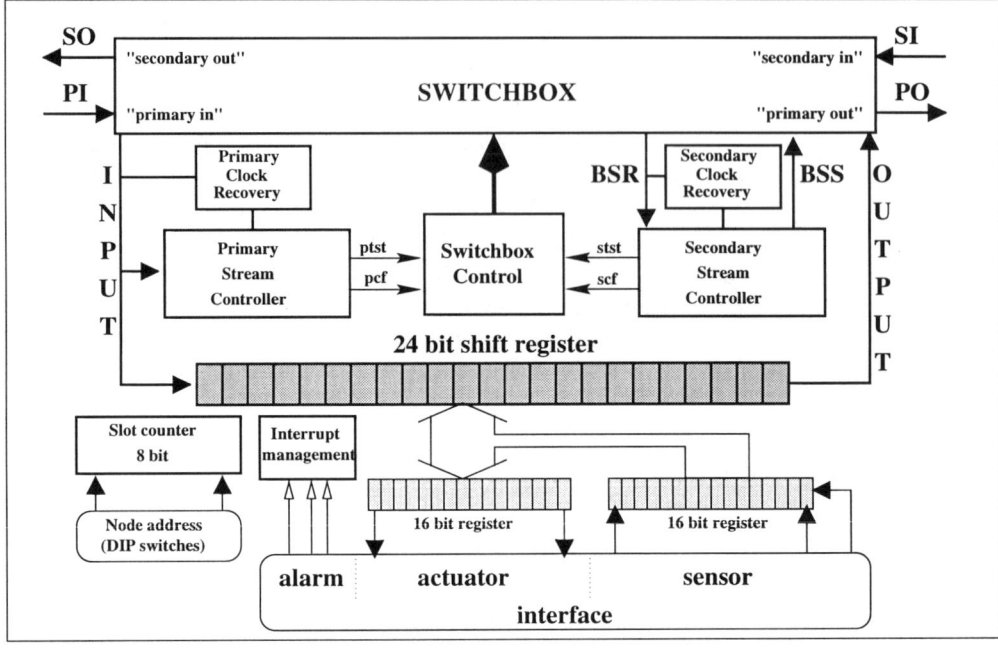

Figure 6: Functional block diagram of a slave node

Another essential part is the switchbox. Its task is to attach the physical inputs
PI and SI (respectively outputs PO and SO) to the logical inputs $INPUT$ and
BSR (respectively outputs $OUTPUT$ and BSS). The primary stream controller
is utilized to supervise the logical input $INPUT$ for a channel failure (signal pcf)
or a test signal ($ptst$), which serves for checking an already broken line. The same
work is done by the secondary part for the input BSR (signals $stst$ and scf). Refer
to Table 1 for an overview of the mapping process for all imaginable channel faults.
In normal operating mode, the signals BSR and BSS are used to handle the data
streams on the secondary ring segments.

channel failure type	$INPUT$	$OUTPUT$	BSR	BSS
all 4 ports ok	PI	PO	SI	SO
SO or PI defective	SI	PO	PI	SO
SI or PO defective	PI	SO	SI	PO

Table 1: Channel mapping from physical to logical ports of the switchbox: BSR bit
stream receiver, BSS bit stream sender

This concept implements a local fault management for the data channels, because
no central station is needed to handle such situations. Every node is checking and
reacting for itself, but all master nodes are informed of the new ring configuration

by sending an alarm packet after the fault detection. This mechanism allows a fast rearrangement of the ring in the case of a line break. The detailed implementation is described in the masterthesis [1].

3.3 Master node

The master node integrates the general aspects of a slave with the extended functionality necessary for a master node. One of these features is the interconnection between MERKUR and the computer. For that reason, the master possesses an AT bus interface. The actual design is implemented for a personal computer (PC) as master station. The AT bus allows to access two different memory blocks on the MERKUR board. On the one hand, the normal data memory already described in section 2.4 is located on the master board. On the other hand, a special configuration memory is accessible.

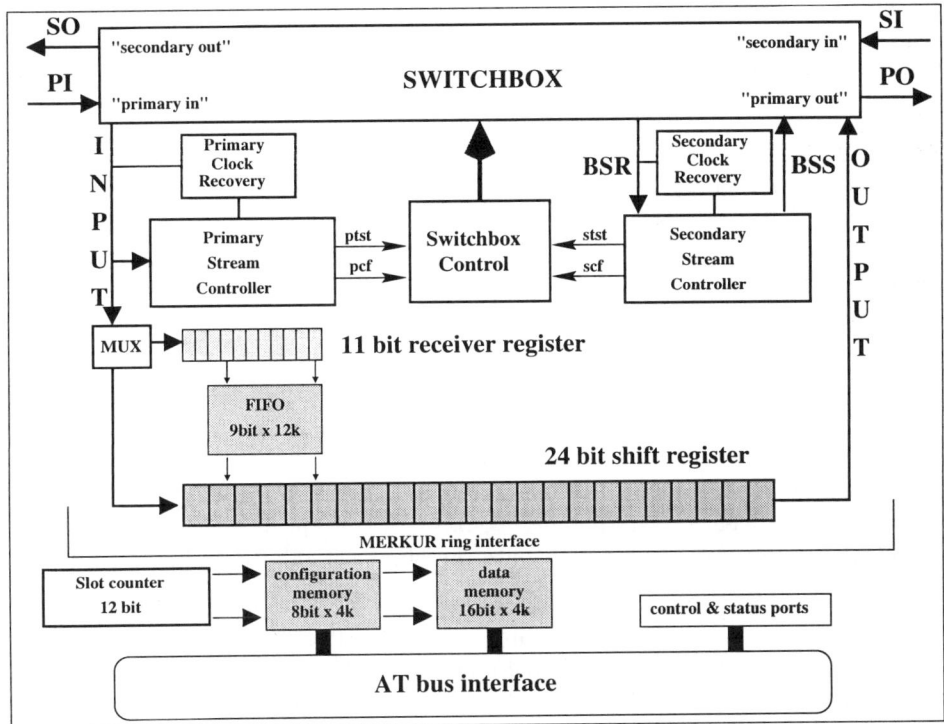

Figure 7: Functional block diagram of a master node

The meaning of that memory block is to describe every TDMA slot (maximum number is 4096) more precisely by one byte, so that the hardware knows how to handle every slot. The 8 bit width dataword encodes the following attributes of one slot: the first bit marks the last valid slot of the whole frame, another one determines, whether the slot is an actuator, which must be served by the node. The third and fourth bit declare, if the "Alive" bit is enabled or ignored and if a slot belongs to the local or a foreign host.

To evaluate the contents of this SRAM–based configuration memory for each slot during execution, it is necessary to read the value at the beginning of the outgoing message slot. Before the ring is started by setting a bit in the control port, it must be guaranteed, that the configuration memory is initialized correctly. A slot counter is used for addressing both memory types.

As illustrated in the Figures 6 and 7, every node uses a clock recovery unit to extract the clock from the data stream. The ring structure must possess one node, which generates this clock base. Exactly one master must be selected by setting a bit in the control port to do this job. Such a master is named "special master".

As mentioned above, a master can also possess more than one slot. In a system with n nodes the shift registers can only store n datawords. A FIFO buffer is activated exclusively by the special master to hold the rest of the distributed information. This FIFO also compensates minimal timing discrepancies between the incoming and outgoing data stream clocks.

Finally, the special master initiates also the operation of the secondary ring. This work includes the generation of a special test pattern for the output BSS (bit stream sender), the corresponding clock base and the evaluation of the received pattern at port BSR (bit stream receiver). During normal operation (no channel faults exist), all nodes communicate over the primary ring segments and the secondary part is only tested for a fault.

4 Conclusion

The design of distributed mechatronic systems requires a suitable communication system. In this paper a real–time interconnection concept has been presented, that provides all services needed to distribute the necessary information.

For proving the correct operation of the described ideas in practice, the MERKUR system has been implemented as a prototype consisting of two slave modules and one master node. The last mentioned one is realized as a PC based system. The aspired bandwidth of 20 MBit/sec is not reached at the moment. Actually a transmission rate of 4 MBit/sec is working. This lower rate is caused by the applied fiber optic transmission devices and the provisional board layout used to implement the prototypes.

The whole interpretation of the communication protocol is done by hardware without a microcontroller. The slave modules consist of two CPLDs (complex programmable logic devices). The master uses five of these chips. A further improvement in system behavior is expected, when the message protocol format is modified in such a way, that the encoding of the aperiodic information is separated from the normal data. This can be done by extending the slot format with several additional bits. This extension avoids the loss of one data message in the case of an alarm.

Beside the hardware, a run–time system is needed to allow a unified integration of the application software. This run–time system serves as an interface between the hardware and the user software. The architecture of the run–time software is developed at the moment. Results of this design process will be presented in following papers.

5 Remarks

The work presented is sponsored by the Volkswagen–Stiftung, Hannover, Germany, as part of the interdisciplinary research project "Integration of distributed mechatronic systems with special regard to real–time behavior".

References

[1] Wolfgang Baldauf. Implementierung eines echtzeitfähigen und fehlertoleranten Sensor/Aktor Kommunikationssystems für mechatronische Anwendungen. Diplomarbeit, Lehrstuhl für Prozeßrechner, Technische Universität München, 12 1994.

[2] Klaus Bender, editor. *Profibus. Der Feldbus für die Automation.* Hanser, 1990.

[3] R. Bent. Interbus–S: Offene Kommunikation für Sensoren/Aktoren. Technical description, Phoenix Contact, Blomberg, Germany, 1992.

[4] Tom Bohman. Shared–memory computing architectures for real–time simulation — simplicity and elegance. Technical description, SYSTRAN Corporation, Dayton, Ohio, 1993.

[5] J. Ehrenberg, E.-J. Heins, P. Leymann, and W. Schumacher. Automatisierungspraxis: Profibus Anwendungen, Produkte, Trends. *Elektronik plus*, 1, 1993.

[6] Konrad Etschberger, editor. *CAN. Controller–Area–Network. Grundlagen, Protokolle, Bausteine, Anwendungen.* Hanser, 1994.

[7] Georg Färber. Feldbus–Technik heute und morgen. *Automatisierungstechnische Praxis*, 36, 1994.

[8] Bernhard Jünger. Profibus contra Interbus–S. Ein aktueller Vergleich. *Elektronik*, 21, 1994.

[9] H. Kopetz, A. Damm, Ch. Koza, M. Mulazzani, W. Schwabl, Ch. Senft, and R. Zainlinger. Distributed fault–tolerant real–time systems: The MARS approach. Research Report Nr. 4/88, Institut für technische Informatik, Technical University of Vienna, 1988.

[10] Hermann Kopetz. Event–triggered versus time–triggered real–time systems. Research Report Nr. 8/91, Institut für technische Informatik, Technical University of Vienna, 1991.

[11] Hermann Kopetz. Should responsive systems be event–triggered or time–triggered? *IEICE Trans. on Electronics, Inst. of Electronics, Information and Comm. Engineers, Tokyo Japan*, E76–C(11), 11 1993.

[12] Werner Kriesel and Otto W. Madelung, editors. *ASI. Das Aktuator–Sensor–Interface für die Automation.* Hanser, 1994.

[13] Wolfhard Lawrenz, editor. *CAN. Controller–Area–Network. Grundlagen und Praxis.* Hüthig, 1994.

[14] J. Richert, A. Rükgauer, U. Petersen, V. Hadwich, T. Raste, K. Gresser, and J. Pfefferl. Integration verteilter Systeme der Mechatronik mit besonderer Berücksichtigung des Echtzeitverhaltens. Interdisciplinary Research Report Az.: I/67975-9, Uni–GH Paderborn, 1994.

[15] Heribert Winkler. SERCOS Interface auf dem Weg zum Standard. *Elektronik*, 6:116–124, 1991.

Dipl.–Ing. Pfefferl Johann
Lehrstuhl für Prozeßrechner, Prof. Dr.–Ing. G. Färber
Technische Universität München
Arcisstr. 21, 80333 München, Germany
Tel +49–89–2105 3557
Fax +49–89–2105 3555
Email: pfefferl@lpr.e–technik.tu–muenchen.de

Optimization of the Inertial and Acceleration Characteristics of Non-Redundant Manipulators

Oussama Khatib and Alan Bowling
Robotics Laboratory
Department of Computer Science, Stanford University
Stanford, CA, USA

Abstract

This article investigates the problem of manipulator design for increased dynamic performance. Optimization techniques are used to determine the design parameters that improve manipulator performance. The dynamic performance of a manipulator is characterized by the inertial and acceleration properties of the end-effector. Our studies of inertial and acceleration properties have provided separate descriptions of the characteristics associated with linear and angular motions. This allows a more physically meaningful interpretation of these properties. The article presents these models, discusses the design optimization criteria, and formulates the optimization problem. The approach is illustrated in the design parameter selection of a parallel mechanism.

1 Introduction

The research presented here is concerned with increasing dynamic performance of manipulators; increasing their ability to move very quickly and apply forces in any arbitrary direction within the workspace. Different measures have been proposed to characterize manipulator performance. These measures are derived from different considerations including analysis of workspace characteristics, kinematic performance, structural integrity, dexterity, controllability, and dynamic performance. These include the manipulability measure [1] and the dynamic manipulability ellipsoid [2], the generalized inertia ellipsoid [3], the ellipsoid of gyration [4], the Jacobian condition numbers [5], the acceleration parallelepiped [6], acceleration sets [7], and the ellipsoid expansion model [8].

The inertial properties as perceived at the end-effector have a significant effect on the dynamic performance of a manipulator, especially in tasks where the end-effector comes in contact with the environment. We have recently [9] shown that the inertial properties of a manipulator can be decomposed into end-effector mass and inertia properties, affecting linear and angular motions. This allows a separate, more meaningful description of the end-effector inertial properties.

However, the inertial properties alone do not describe the full dynamic behavior at the end-effector. The dynamic behavior, although influenced by the inertial properties, is also influenced by gravity, centrifugal and Coriolis forces, and actuator capacity. In addition to the inertial properties, the isotropic acceleration, the maximum acceleration achievable in or about all directions, is used to characterize the dynamic behavior of the manipulator system. The acceleration capability governs the quickness of response of a robotic system to commands. The acceleration characteristics have also been decomposed [8] into two separate descriptions associated with linear and angular motions. The goal of the optimization is to obtain a design with the smallest, most isotropic, and most uniform inertial characteristics, and the largest, most isotropic, and most uniform acceleration capability at the end-effector.

We first present the models for the characterization of the inertial and acceleration properties. This is followed by a discussion of the optimization scheme. Results of the application of this scheme to the design of a parallel mechanism will be given.

2 End-Effector Inertial Properties

Discussion of the end-effector inertial properties begins with the operational space form of the equations of motion,

$$A(\mathbf{q})\ddot{\mathbf{q}} + \mathbf{b}(\dot{\mathbf{q}}, \mathbf{q}) + \mathbf{g}(\mathbf{q}) = \tau \tag{1}$$

In equation (1) $\mathbf{q}$ is the vector of n joint coordinates, $A(\mathbf{q})$ is the joint space mass matrix, $\mathbf{b}(\dot{\mathbf{q}}, \mathbf{q})$ is the vector of centrifugal and Coriolis forces, $\mathbf{g}(\mathbf{q})$ is the gravity vector, and τ is the vector of joint torques.

The above equation can be transformed into operational space, projected to the end-effector, using the following relationships;

$$\vartheta \triangleq \begin{bmatrix} \mathbf{v} \\ \omega \end{bmatrix} = J_0(\mathbf{q})\dot{\mathbf{q}}. \tag{2}$$

and

$$\tau = J_0(\mathbf{q})^T \mathbf{F} \tag{3}$$

where $\mathbf{v}$ and ω are the end-effector linear and angular velocities, $J_0(\mathbf{q})$ is the basic Jacobian matrix, and $\mathbf{F}$ is the vector of forces applied at the end-effector.

The inertial properties as perceived at the end-effector are contained in the kinetic energy matrix, $\Lambda(\mathbf{q})$, from the operational space form of the equations of motion,

$$\Lambda(\mathbf{q})\dot{\vartheta} + \mu(\dot{\mathbf{q}}, \mathbf{q}) + \mathbf{p}(\mathbf{q}) = \mathbf{F}. \tag{4}$$

In equation (4) $\mu(\mathbf{q}, \dot{\mathbf{q}})$ is the centrifugal and Coriolis force vector, and $\mathbf{p}(\mathbf{q})$ is the gravity force vector, (for notational simplicity $(\mathbf{q})$ and $(\mathbf{q}, \dot{\mathbf{q}})$ will be omitted from subsequent equations).

The Λ matrix can be written in terms of joint space components;

$$\Lambda = (J_0 A^{-1} J_0^T)^{-1}. \tag{5}$$

The eigenvalues and eigenvectors of Λ usually represent some mixture of mass and inertial properties which are not very meaningful. A decomposition of Λ into separate mass and inertia properties is much more physically meaningful. Omitting the details, which can be found in [9], this decomposition is accomplished as follows. The Jacobian is rewritten as

$$\vartheta \triangleq \begin{bmatrix} \mathbf{v} \\ \omega \end{bmatrix} = J_0 \dot{\mathbf{q}} = \begin{bmatrix} J_\mathbf{v} \\ J_\omega \end{bmatrix} \dot{\mathbf{q}}. \tag{6}$$

where $J_\mathbf{v}$ is a rectangular matrix which transforms joint velocities into linear velocities and J_ω does the same for angular velocities. Using equation (6) the matrices $\Lambda_\mathbf{v}$, representing the mass properties, and Λ_ω representing the inertia properties, can be formulated as,

$$\Lambda_\mathbf{v} = (J_\mathbf{v} A^{-1} J_\mathbf{v}^T)^{-1} \tag{7}$$

and

$$\Lambda_\omega = (J_\omega A^{-1} J_\omega^T)^{-1}. \tag{8}$$

Part of the goal in the optimization is to achieve the smallest, most isotropic, and most uniform inertial properties. Indicators for these properties can be derived from $\Lambda_\mathbf{v}$ and Λ_ω and are used in the cost function of the optimization. This will be discussed further in section 4.2.

3 End-Effector Acceleration Characteristics

The end-effector acceleration can be characterized by the isotropic acceleration. The isotropic acceleration is easily determined using the *ellipsoid expansion model*. This model completely describes the effect of inertial properties, gravity, centrifugal and Coriolis forces, and actuator capacity on end-effector accelerations. The formulation of the *ellipsoid expansion model* is presented below.

3.1 Torque/Acceleration Relationship

The development of this model begins with the operational space form of the equations of motion given in equation (4). Using the relationship between joint torques and end-effector forces, equations (3), and (4), yields,

$$J_0^T (\Lambda \dot{\vartheta} + \mu + \mathbf{p}) = \tau. \tag{9}$$

The bounds on τ, which are determined by the maximum torque of the actuators, can be written as

$$-\tau_{bound} \leq \tau \leq \tau_{bound}. \tag{10}$$

To normalize the bounds on τ, we introduce the diagonal matrix N with components $N_{ii} = \frac{1}{\tau_{bound_i}}$. Now using equations (9) and (10) yields,

$$-\mathbf{1} \leq NJ_0^T(\Lambda\dot{\vartheta} + \mu + \mathbf{p}) \leq \mathbf{1}; \tag{11}$$

where $\mathbf{1}$ is a vector of length n with each element equal to one. Using equation (2) the above equation can be rewritten as,

$$\tau_{lower} \leq [E_{\mathbf{v}}\ E_\omega]\begin{bmatrix}\dot{\mathbf{v}}\\\dot{\omega}\end{bmatrix} + \begin{bmatrix}\vartheta^T M_1\vartheta\\\vdots\\\vartheta^T M_n\vartheta\end{bmatrix} \leq \tau_{upper}; \tag{12}$$

where M_i are symmetric matrices and,

$$[E_{\mathbf{v}}\ E_\omega] = NJ_0^T\Lambda; \tag{13}$$

$$\begin{bmatrix}\vartheta^T M_1\vartheta\\\vdots\\\vartheta^T M_n\vartheta\end{bmatrix} = NJ_0^T\mu; \tag{14}$$

$$\tau_{upper} = \mathbf{1} - NJ_0^T\mathbf{p}; \tag{15}$$

$$\tau_{lower} = -\mathbf{1} - NJ_0^T\mathbf{p}. \tag{16}$$

Finally, the governing equation for this analysis is

$$\tau_{lower} \leq E_{\mathbf{v}}\dot{\mathbf{v}} + E_\omega\dot{\omega} + \begin{bmatrix}\vartheta^T M_1\vartheta\\\vdots\\\vartheta^T M_n\vartheta\end{bmatrix} \leq \tau_{upper} \tag{17}$$

In equation (17) the separation of linear and angular accelerations is motivated by the need to analyze each of them independently. The matrices $E_{\mathbf{v}}$ and E_ω are used to characterize the mapping from acceleration to torques. If the mappings are isotropic the acceleration capability will also be as isotropic and uniform as possible. This is due to the fact that $E_{\mathbf{v}}$ is normalized by the torque bounds. The condition numbers of $E_{\mathbf{v}}$ and E_ω, $\kappa(E_{\mathbf{v}})$ and $\kappa(E_\omega)$, are used in the cost function of the optimization. It is also desirable to decrease the amount of torque required for a given acceleration. So the norms of $E_{\mathbf{v}}$ and E_ω, $\|E_{\mathbf{v}}\|$ and $\|E_\omega\|$, are both used in the cost function to minimize the magnification of a given acceleration into required torque.

3.2 Analysis

The basic approach used to analyze equation (17) is to visualize each component of the equation as a geometric object. This allows a simple determination of the relationships between each component of the equation. The visualization process begins with the torque bounds which are visualized as an n-dimensional hypercube whose center is shifted from the origin by the gravity effect, i.e. $NJ_0^T\mathbf{p}$.

To visualize the isotropic accelerations first consider only end-effector linear acceleration in equation (17);

$$\tau_{lower} \leq E_{\mathbf{v}}\dot{\mathbf{v}} \leq \tau_{upper}. \tag{18}$$

Determining isotropic linear acceleration involves finding the maximum magnitude of $\dot{\mathbf{v}}$, achievable in every direction. This is idea can be visualized as a hyper-sphere with some radius a,

$$\dot{\mathbf{v}}^T\dot{\mathbf{v}} = a^2. \tag{19}$$

Since the bounds are being considered in torque space, the acceleration hyper-sphere must be mapped into torque space using the relationship;

$$\tau_{\mathbf{v}} = E_{\mathbf{v}}\dot{\mathbf{v}}.$$

It has been shown in [8] that the acceleration hyper-sphere can be transformed into a torque ellipsoid of dimension three or less of the form,(partly due to the fact that $E_{\mathbf{v}}$ is at most of rank three),

$$\tau_{\mathbf{v}}^T(E_{\mathbf{v}}E_{\mathbf{v}}^T)^+\tau_{\mathbf{v}} = a^2. \tag{20}$$

This ellipsoid is now mapped into the torque bounds. The isotropic acceleration is determined by expanding/contracting the ellipsoid equation (20), changing a, until it lies within and is tangent to one of the torque bounds. In Figure 1 this process is shown for a simple case and Figure 2 shows a more general case. The dashed ellipse in both figures corresponds to $a = 1$. Note that only the vectors associated with the tangent points, $2n$ points, need to be examined.

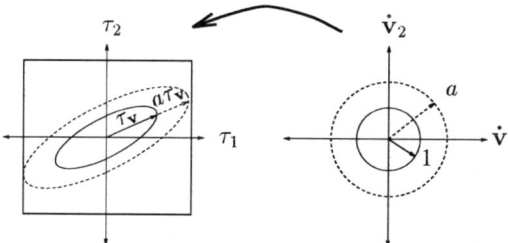 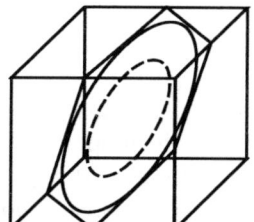

Figure 1: Ellipse $\tau_{\mathbf{v}}$.   Figure 2: Ellipsoid in 3D.

The same type of process can be performed on the $E_{\mathbf{v}}\dot{\omega}$ term and the resulting ellipsoid is added to the linear acceleration torque ellipsoid within the torque bounds. The centrifugal and Coriolis terms are analyzed by mapping isotropic linear and angular velocity hyper-spheres, $\mathbf{v}^T\mathbf{v} = c^2$ and $\omega^T\omega = d^2$, through those terms. The information from the resulting surface is then mapped into the torque bounds and added in with the other ellipsoids. Analysis of the centrifugal/Coriolis terms is difficult to perform analytically so an approximation is used as the result from this process.

3.3 Results

The information resulting from the above analysis is in the form of a set of $2n$ inequalities which give the relationship between isotropic end-effector accelerations, isotropic end-effector velocities, gravity, and actuator maximum torque capacity. Equation (17) can be rewritten as,

$$-\tau_{bound} \leq N^{-1}E_{\mathbf{v}}\dot{\mathbf{v}} + N^{-1}E_{\omega}\dot{\omega} + \begin{bmatrix} \tau_{bound_1}\vartheta^T M_1 \vartheta \\ \vdots \\ \tau_{bound_n}\vartheta^T M_n \vartheta \end{bmatrix} + \mathbf{g} \leq \tau_{bound}. \quad (21)$$

Using the above model each of the $2n$ resulting relationships has the form,

$$\beta_1\|\dot{\mathbf{v}}\| + \beta_2\|\dot{\omega}\| + \tau_c + \mathbf{g}_i \leq \tau_{bound_i} \quad (22)$$

where β_1 and β_2 are configuration dependent constants and τ_c represents the torque used compensating for centrifugal/Coriolis forces. τ_c has the form,

$$\tau_c = \|\mathbf{v}\|\sqrt{\gamma_1\|\mathbf{v}\|^2 + \gamma_2\|\mathbf{v}\|\|\omega\| + \gamma_3\|\omega\|^2} + \|\omega\|\sqrt{\gamma_4\|\mathbf{v}\|^2 + \gamma_5\|\mathbf{v}\|\|\omega\| + \gamma_6\|\omega\|^2}$$
$$+ \gamma_7\|\mathbf{v}\|^2 + \gamma_8\|\mathbf{v}\|\|\omega\| + \gamma_9\|\omega\|^2. \quad (23)$$

where γ_i are configuration dependent constants. Usually these equations are used to produce plots of surfaces and curves which describe the relationships between the variables. An example of the surface represented by the above set of equations is given in Figure 3 for the PUMA560.

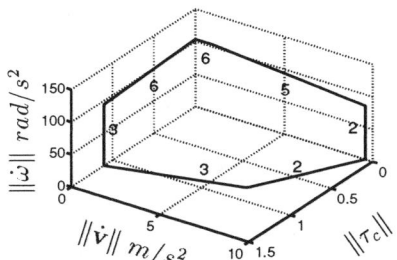

Figure 3: Isotropic Surface for PUMA560.

In Figure 3 the numbers labeling each line segment represent the actuator whose capacity limited the acceleration along that line segment. In Figure 3 only the line segments where the surface intersects the coordinate planes is shown. Since the surface is convex, (with planar facets), it is possible to specify its shape by giving a few points on the surface. From the above set of equations it is clear that if desired $\|\dot{\mathbf{v}}\|$ and $\|\dot{w}\|$, and desired $\|\mathbf{v}\|$ and $\|\omega\|$, are given for a configuration, it is easy to determine the maximum torque required of each actuator. In the optimization this information is provided as input to the process. This will be discussed further in sections 4.1 and 4.3.

4 Optimization Scheme

The goal of this optimization is to achieve, over the workspace, the smallest, most isotropic, and most uniform inertial properties and the largest, most isotropic, and most uniform acceleration properties at the manipulator's end-effector. In the following sections we will discuss the inputs to the optimization, the cost function, and the iteration process.

4.1 Inputs

In this optimization an attempt to improve the design at different configurations in the workspace is made. However, only a few representative configurations need be chosen for evaluation during the optimization.

Of course the parameters to be optimized must also be specified along with the constraints on these variables. In the example which follows a penalty function was added to the cost function to enforce the constraints.

4.2 Cost Function

The cost function should reflect the goals of the optimization as closely as possible. The stated goals can basically be achieved by reducing the norms and condition numbers of the mass and inertia matrices as low as possible, and by decreasing the norms and condition numbers of E_v and E_ω. Towards this end the cost function at a given configuration is defined as;

$$cost_i = w_1 \, \kappa(\Lambda_v) \; + \; w_2 \, \kappa(\Lambda_\omega) \; + \; w_3 \, \|E_v\| \; + \; w_4 \, \|E_\omega\| \tag{24}$$

where w_i is a weight, $\kappa(X)$ is the condition number of X, and $\|X\|$ is the norm of X. Equation (24) is evaluated at each specified configuration to obtain the total cost;

$$cost = \sum_i cost_i.$$

4.3 Iteration

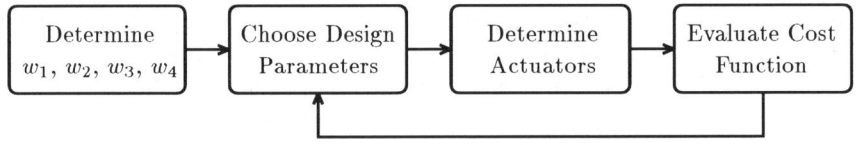

Figure 4: Optimization Flowchart

The steps in the optimization are shown in Figure 4. The only step requiring any further discussion is determining actuators. As was stated earlier in section 3.3 the maximum required torque for each actuator can be found using equation (22). The actuators are chosen so that the the torque capacity is greater than, yet nearest to, the required torque.

However, a new selection of actuators alters the inertial properties which in turn alter the coefficients in equation (22), possibly yielding a new required torque. Thus some iterations might be needed to obtain convergence between required torques and selected actuators.

5 Application

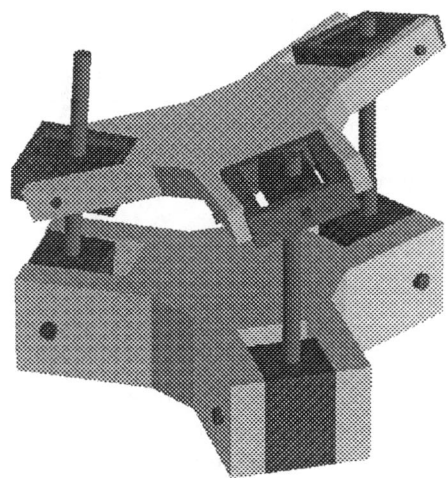

Figure 5: Parallel Mechanism.

The manipulator chosen for this example is the three-degree-of-freedom parallel mechanism shown in Figure 5. It consists of upper and lower plates connected by three identical ball screws. The ball screws are attached to the fixed lower plate by a one-degree of freedom passive joint, and to the top plate by a three-degree of freedom passive joint. The actuators are mounted to the ends of the ball screws beneath the lower plate.

Here it is only necessary to examine $\Lambda_{\mathbf{v}}$ and $E_{\mathbf{v}}$. The design parameters optimized were the mass of the upper plate and the pitch of the ball screw. Actuators were chosen from a discrete set. The desired performance was specified as the isotropic linear acceleration, $\|\dot{\mathbf{v}}\| = 40m/s^2$, at an isotropic velocity, $\|\mathbf{v}\| = 0$, for all configurations. The search was performed using a gradient search algorithm and the cost function of section 4.2. It is also important to mention that the final solution is highly dependent on the choice of weights, w_i, in equation 24.

In Figure 6 the mass and acceleration properties at five different configurations are shown. The mass properties are represented as ellipsoids whose projections are

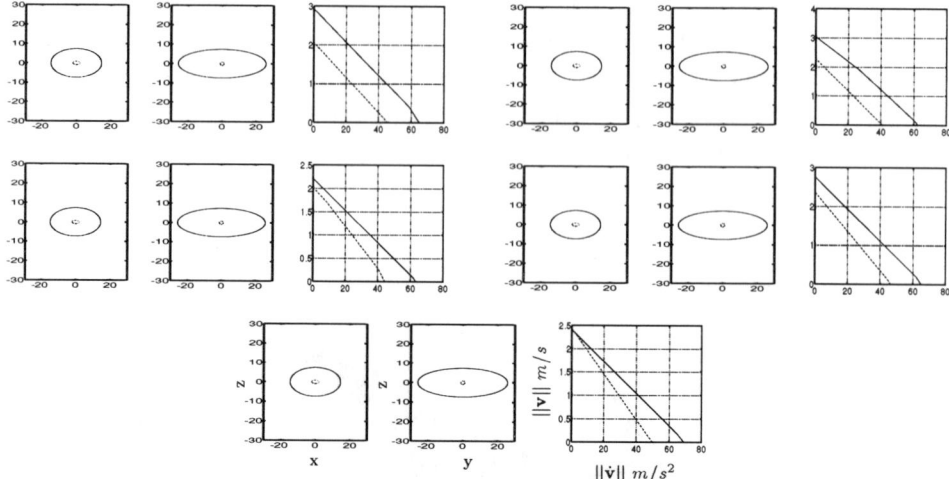

Figure 6: Manipulator Characteristics.

shown side by side. The coordinate axes are the eigenvectors of the matrix. To the right of the two ellipses is the plot of isotropic linear acceleration versus isotropic linear velocity at the same configuration. The solid lines indicate the characteristics of the initial design while dotted lines represent the optimized design.

Figure 6 shows the improvement of inertial properties with optimization. The average mass matrix norm for the five configurations decreased from 23 kg to 2 kg and the average condition number decreased from 2.7 to 2. However, it is also clear that some performance was sacrificed to improve the inertial properties. The average condition number of $E_{\mathbf{v}}$, $\kappa(E_{\mathbf{v}}) \approx 1.4$, did not change significantly, but the average norm, although very small in magnitude, increased from 3×10^{-7} to 5×10^{-5}.

Lastly note that the desired performance was met for every configuration. Achieving this required an increase in actuator capacity. The resulting optimized design has ball screws with a higher pitch, and a lighter upper plate than the original mechanism.

6 Conclusion

In this article we have presented a methodology for the optimization of the dynamic characteristics of manipulators. The formulation is based on our recent studies concerning end-effector inertial and acceleration properties. In particular, the decomposition of these characteristics into two separate models affecting linear and angular motions allow more physically meaningful descriptions of these properties. The effectiveness of this methodology has been illustrated in design of a three-degree-of freedom mini-manipulator system.

7 Acknowledgments

The financial support of NASA/JPL, Boeing, Hitachi Construction Machinery, and NSF, grant IRI-9320017, are gratefully acknowledged.

References

[1] Yoshikawa, T.: Manipulability of Robotic Mechanisms, The International Journal of Robotics Research, Vol. 4, No. 2, MIT Press, 1985.

[2] Yoshikawa, T.: Dynamic Manipulability of Robot Manipulators, Proceedings 1985 IEEE International Conference on Robotics and Automation, St. Louis, 1985, pp. 1033-1038.

[3] Asada, H.: Dynamic Analysis and Design of Robot Manipulators Using Inertia Ellipsoids, Proceedings of the IEEE International Conference on Robotics, Atlanta, March 1984.

[4] Hogan, N.: Impedance Control of Industrial Robots, Robotics Computer-Integrated Manufacturing vol.1, no. 1, pp. 97-113.

[5] Salisbury, J. K./Craig, J.: Articulated Hands: Force Control and Kinematic Issues, The International Journal of Robotics and Automation, vol. 1, No. 1, 1982.

[6] Khatib, O./Burdick, J.: Dynamic Optimization in Manipulator Design: The Operational Space Formulation, The International Journal of Robotics and Automation, vol.2, no.2, 1987, pp. 90-98.

[7] Kim, Y./Desa, S.: The Definition, Determination, and Characterization of Acceleration Sets for Spatial Manipulators, The International Journal of Robotics Research, vol. 12, no. 6, 1993, pp. 572-587.

[8] Bowling, A./Khatib, O.: Analysis of the Acceleration Characteristics of Non-Redundant Manipulators, Proceedings IEEE/RSJ International Conference on Intelligent Robots and Systems, Pittsburgh, August 1995.

[9] Khatib, O.: Inertial Properties in Robotic Manipulation: An Object-Level Framework, The International Journal of Robotics Research, vol. 13, no. 1, February 1995, pp. 19-36.

Professor Oussama Khatib
Computer Science Robotics Laboratory
Stanford University
MS: 4110 Cedar Hall
Stanford, CA, USA 94305
Tel: (415)723-9753
Fax: (415)725-1449
e-mail: ok@cs.stanford.edu

Alan Bowling
Computer Science Robotics Laboratory
Stanford University
MS: 4110 Cedar Hall
Stanford, CA, USA 94305
Tel: (415)725-8812
Fax: (415)725-1449
e-mail: bowling@flamingo.stanford.edu

A LASER BASED 3D CORRELATION PROCEDURE FOR THE EXECUTION OF A BIOMEDICAL TASK IN A ROBOTIZED CELL

Roberto CARACCIOLO, Francesco FANTON, Alessandro GASPARETTO and Aldo ROSSI

DIMEG - Department of Innovation in Mechanics and Management
University of Padova
Via Venezia 1 - 35131 Padova - ITALY
Tel: +39 49 828 6855 - Fax: +39 49 828 6816
e-mail: gaspare@hpdimeg.unipd.it

ABSTRACT

A 3D adjusting procedure is presented, able to find the correlation between a real object and its representation in a simulation environment. This procedure enables one to design and test complex tasks in the simulation environment and guarantees the correct execution in the real cell. The proposed algorithm is based on the surface matching theory and is designed to work whenever does not exist a one to one correspondence between the sets of points. Two types of matching are introduced, called *point to point* and *surface to point*. Both matching types are implemented and discussed. The results of some performance tests are presented. The proposed procedure is applied to a skull phantom whose CAD model come from combined CT and MR.

1. INTRODUCTION

Robotic simulators have been built in the last years, based on CAD system, that associate the typical CAD data structures with high quality images. The simulator user can off line generate operating sequences representing the robot movements and test the interactions of the robot with the parts inside the cell. In this way off line programming of robots becomes much easier and more efficient. As a matter of fact, the off line programming technique increases the productivity of a robotized cell, by avoiding that the robot be stopped for a long time, in order to be reprogrammed by means of teach-in operations.

The sequences generated by the user in the simulation environment should then be exported and executed in the real cell. A correct execution of these sequences is possible only if the correlation between the simulated and the real cell is known.

A 2-dimensional adjusting procedure, that finds the correlation in the case of a real object lying on a working table and its simulated model, has been described in [1]. This adjusting procedure, based on an infrared sensor which detects the position of the object, has been tested and applied to the field of automatic assembly.

A more general correlation procedure, which works in the 3-dimensional case, will be described in this paper. The approach to the problem is rather different: a laser sensor has been used instead of the infrared sensor, so that analog distance measurement in a longer range are now possible, and an algorithm has been developed, based on the surface matching theory instead of simpler 2-dimensional geometric considerations.

The paper is organized in six main sections. Firstly, an overview of the surface matching theory is presented. Then, a general purpose matching algorithm is described. Then the algorithm application to different type of matching is presented. Finally the results of some performance tests are shown and a biomedical application using the developed procedure is described.

2. SURFACE MATCHING THEORY

The state of the art in surface matching have been described with great detail in [4]; however, a brief overview will be given in the following.
The matching problem can be defined in two different ways.
The first type of matching is the so called *point to point* matching. It can be defined as follows.
Given two sets of points, representing the same surface in two different reference frames, find the rigid transformation (expressed by a rototranslation matrix) mapping one set of points into the other.
Such a transformation must have the following characteristics:
a)   *must be optimal with respect to some criterion (e.g. minimize either the maximum or the mean squared distance between corresponding points);*
b)   *must work for sets of points with different dimensions;*
c)   *must work if points in one set do not correspond exactly to points in the other;*
d)   *must work if the points are corrupted by noise.*
A second type of matching can be defined, called *surface to point* matching, defined as follows.
Given an analytical surface and a set of points obtained by sampling the corresponding real surface, find the rigid transformation rototranslating the analytical surface in order to minimize its distance from the set of points.
Such a transformation must have the following characteristics:
a)   *must be optimal with respect to some criterion;*
b)   *must work if some points are corrupted by noise.*
A comparison between this two types of matching was made by means of several tests. A more detailed discussion and the results are presented in Chapter 4 below.
A more formal approach to the matching problem is now given for the point to point case, because it will be shown in the following that the surface to point case can be brought to the former one.
Be X a set of points and $(R, \vec{t})$ a rototranslation defined by a rotation matrix R and a translation vector $\vec{t}$, let us call P the set of points obtained applying the rototranslation $(R, \vec{t})$ to the set X. It is simple to obtain the rototranslation matrix $(R, \vec{t})$ starting from the knowledge of X and P, if the one-to-one correspondence of the points of the two sets is known. The problem of determining the transformation $(R, \vec{t})$ becomes more difficult if the points of one set are affected by noise, in the sense that the relationship

$$\vec{x}_i = R \cdot \vec{p}_i + \vec{t} \tag{1}$$

does not hold for all pairs of points of X and P. In the above equation $\vec{x}_i$ and $\vec{p}_i$ are the coordinates of the i-th point ($i = 1...N$) of the sets X and P respectively.

In this case the problem becomes a minimization problem: it is required to find the matrix $\mathbf{R}$ and the vector $\vec{t}$ that minimize the sum of the errors

$$\vec{e}_i = \vec{x}_i - \mathbf{R} \cdot \vec{p}_i - \vec{t} \tag{2}$$

The general matching problem does not require any one-to-one correspondence between the points of X and the points of P. This assumption is required because in the general case the two sets of points come from different environments, such as the sampling of an analytical surface and the measurement of a real surface. This assumption implies that no rototranslation exists, which maps exactly every point of X into a point of P even in the case of zero noise.

Some authors have investigated the matching problem, applying their algorithms to specific cases.

If the one-to-one correspondence is known the matching problem can be solved using the methods proposed by Horn [10] and by Haralick [9] (discussed in [4]).

Another solution to the surface matching problem in the 3-dimensional case is given by Besl. He proposes a method, based on the Iterative Closest Point (ICP) algorithm, to match two 3-dimensional surfaces. This method reveals itself accurate and computationally efficient; furthermore, it works also if there is no one-to-one correspondence between the two sets of points representing the surfaces.

3. THE PROPOSED ALGORITHM

An algorithm has been developed, which correlates the analytical model of a surface with a surface descriptor extracted from the corresponding real object (e.g. a set of points measured on the surface of the object in the real world). This algorithm is a modification and an evolution of the Closest Point Algorithm proposed by Besl [2][3]. An important feature of this algorithm is that it is suitable to work in the case that there is no one-to-one correspondence between the points of the two sets.

Let $P = \{\vec{p}_i\}$ and $X = \{\vec{x}_i\}$ be the two sets of points to be matched.

Firstly, let us suppose that P and X have the same dimension ($N_x = N_p$). Then, the matching problem can be solved using Haralick's method (described in [4][9]), setting the initial conditions: $\mathbf{R}_0 = I_3$, $\vec{t}_0 = \vec{0}$ (so that $P_0 \equiv P$). We define the Q operator as the function that performs the registration between P and X, i.e. computes the optimal rototranslation matrix that matches P and X. So, for each iteration new values for $\mathbf{R}$ and $\vec{t}$ are obtained by applying the Q operator as follows:

$$(\mathbf{R}_k, \vec{t}_k, d_k) = Q(P_k, X) \quad k > 1 \tag{3}$$

where d_k is the mean squared error. The value of P_k is obtained applying the rotation $\mathbf{R}_{k-1}$ and the translation $\vec{t}_{k-1}$ to the whole set P_{k-1}, summarized by the formula:

$$P_k = \mathbf{R}_{k-1} P_{k-1} + \vec{t}_{k-1} \tag{4}$$

The iterations stop when the absolute value of the difference between two consecutive mean squared errors is lower than a fixed positive threshold τ:

$$|d_k - d_{k+1}| < \tau \tag{5}$$

Now, let us consider the more general problem of matching two sets of points with different dimensions. To solve this problem an iterative algorithm of the "closest point" type is used.
Let us suppose that the dimension of the set of the model points is greater than that of the set of data points ($N_x > N_p$), and let us call Y_k the set of the N_p points of X which are the closest to the points of P (i.e. are the "best correspondent points") in the k-th iteration; this defines, for each iteration, a new correspondence K. Let us call C the operator performing this computation:

$$Y_k = C(P_k, X) \tag{6}$$

Now the optimal rotation matrix R and the optimal translation vector $\vec{t}$ can be computed using the above defined Q operator applied to the Y_k set:

$$(\mathbf{R}_k, \vec{t}_k, d_k) = Q(P_k, Y_k) \tag{7}$$

The rototranslation $(\mathbf{R}_k, \vec{t}_k)$ thus computed is then applied to all the points of X, obtaining a new set P_{k+1} which is closer to the X set.
The C operator is now applied to the new set P_{k+1} in order to determine the new set Y_{k+1} of points closest to X.
The loop is iterated until the difference between the mean squared errors in two consecutive iterations is lower than a fixed positive threshold τ.

The convergence of the Closest Point algorithm to a local minimum has been demonstrated [2][3]. However, the convergence to the global minimum is not assured in the general case. A way to make the algorithm converge to the smallest local minimum is to start the algorithm choosing $\mathbf{R}_0$ in an adequate set of initial rotations, called "states", instead of choosing $\mathbf{R}_0 = \mathbf{I}_3$.

4. POINT TO POINT AND SURFACE TO POINT MATCHING

The matching algorithm described above requires two sets of points, namely the model and the data set. These sets of points are gotten by sampling a surface, either an analytical or a real one. Thus, if two sets of points are available, the matching algorithm can be applied directly. This situation corresponds to the *point to point* matching above described.
The situation is different when we attempt to use the surface description given in an analytical form (e.g. that given by a CAD system) without loss of information due to an *a priori* sampling of the surface itself. This situation corresponds to the *surface to point*

matching.

We can note that the *surface to point* matching requires the computation of the distance between the surface and each point of the set. In other words, this corresponds to calculate at every iteration a new set of points composed of the nearest points belonging to the surface.

Therefore, the two situation just described can correspond to a unique matching algorithm which enables the user to select whether he wishes to sample the surface:

1) just once, before the iterations start;

2) at each iteration, when the algorithm requires the set of nearest points.

Let us now consider that a set of points is fixed because it comes from the measurement of a real surface.

In the first case, also the second set of points in which the nearest point algorithm searches is fixed before the matching algorithm runs.

In the second case the second set of points is not fixed *a priori*, but it is computed at each iteration. In this way, the whole surface is used at each iteration, not just a sampled representation of it.

However, this latter method (*surface to point* matching) may exhibit a higher probability to converge to local minima. As we will see, being the computation time very different in the two cases, it is important to highlight the features of the two methods in order to choose which one is preferable, and under which conditions.

Some considerations on the complexity of the two methods are summarized in the following.

1) *Point to point* matching:

Given two sets of points X and P with $N_X \geq N_P$, first translate one set so as to overlap the centroids of the two sets of points; then find the set of points X' made of the points of the set X that are the nearest to the points of the set P. The dimensions of X' and P are now the same. It is important to note that the same point of X might correspond to different points of P. The complexity of the algorithm which searches the nearest points is proportional to the product of the numbers of points of the two sets. The computation of the closest points is made at every iteration.

2) *Surface to point* matching:

The algorithm proceeds as the *point to point* (calculating now the centroid of the surface) but at each step the set of nearest points X' is chosen using the function that gives the nearest point on the surface (i.e. a distance function), so the number of points that are taken into account each time is infinite. Every point of P is associated with the closest point on the surface. The complexity of the algorithm is proportional to the product between the number of points of P and the complexity of the function that gives the minimum point-surface distance.

As a general consideration, we can say that the algorithm that searches the nearest points to a given set of points is more complex in the second case: to have the same quality of representation of the surface, in the second case at least the same number of points as the P set is required. Thus, if the number of iterations required to find the matching between the two entities is the same, the second solution will be slower. This consideration is true even if the minimum point-surface distance is an internal function of the CAD system and its complexity is unknown.

5. GROSS AND FINE MATCHING TO AVOID LOCAL MINIMA

In the description of the theory, we stated that the matching algorithm may converge to local minima; to avoid it, Besl [2][3] introduced the concept of initial states. The matching procedure is repeated using different starting conditions, and the solution with the minimum matching error is chosen as the solution of the matching problem. The use of the initial states is powerful when the types of surface to be matched are unknown: as a matter of facts, the above described method (i.e. trying different starting conditions of the algorithm), works for any type of surface.

In those cases when the types of the surfaces to be matched are known, it is possible to apply a modified matching algorithm that takes into account the nature of the surfaces.

Generally, we can state that: *if a surface has a particularly irregular (with curvature changes) region it is preferable to use a two steps algorithm: in the first step a gross matching is performed using the irregular region of the surface only (local matching); in the second step a finer result is obtained using the matching algorithm on the rest of the surface, starting from the rototranslation just detected with the gross matching.*

The utility of the two steps method involves also the measurement procedure: if we know that a gross matching is made as first step, the irregular region of the surface can be measured with a lower precision, saving a lot of time. This is always the case of strongly irregular regions of the surface, where, to get an exact measure of the surface itself, an optical sensor has to be endowed with many degrees of freedom, to be able to follow the surface.

6. PERFORMANCE TESTS

To test the *point to point* matching algorithm we followed the steps described below: first of all we verified the convergence of the algorithm in the case of a perfect correspondence between the two sets (ideal case). Then we introduced a perturbation to the points of the P set: in this way we simulated the imprecision of measurement and the inaccuracy of the model. The perturbation was implemented by adding a pseudo-random, uniform, zero mean noise. Finally, we applied the algorithm to two sets of points representing the same surface but gotten using different methods of sampling.

Be X the model set and P the data set; to test the algorithm, we calculated P:

$$P = R_t \cdot X + N = X' + N, \tag{8}$$

where R_t is a rototranslation matrix and N the noise vector.

When the matching algorithm has been run, we obtain the rototraslation matrix R_t'.

In the ideal case (N=0) $R_t' = R_t^{-1}$.

In the other cases, in general $R_t' \neq R_t^{-1}$ and it is useful to calculate the Mean Square Error (MSE) to evaluate the global matching error.

Two kind of MSE can be defined now:

1) Using the just found matrix R_t', we can transform the starting set X, composed of N_x points, obtaining X'; the MSE_1 is evaluated by computing the distance between X' and P (see the fourth col. of Tab. 1):

$$MSE\_1 = \left(\Sigma \left(\left(P_x - X'_x \right)^2 + \left(P_y - X'_y \right)^2 + \left(P_z - X'_z \right)^2 \right) \Big/ N_x \right)^{1/2} \tag{9}$$

2) Using the matrix R_t', we can calculate $X'' = R_t' \cdot R_t \cdot X$. the MSE_2 is evaluated by computing the distance between X and X":

$$MSE\_2 = \left(\Sigma \left(\left(X_x - X''_x \right)^2 + \left(X_y - X''_y \right)^2 + \left(X_z - X''_z \right)^2 \right) \Big/ N_x \right)^{1/2} \tag{10}$$

We can note that this second index does not contain the error due to the added noise.
In this way by comparing the two defined MSE it is possible to evaluate how the measurement errors modify the behaviour of the matching algorithm.
Table 1 shows the results of these two tests. The model set X was of N_x=210 points and the data set P was of 60 points.
The data in the first row refer to the ideal case, in which the set of points P was just the set X after a rototranslation (T_x = 400 mm, T_y = 500 mm, T_z = -300 mm, a_x = 0.348 rad, b_y = 0.697 rad, g_z = 0.872 rad). In this case the algorithm converges after 9 iterations, with a null residual MSE.

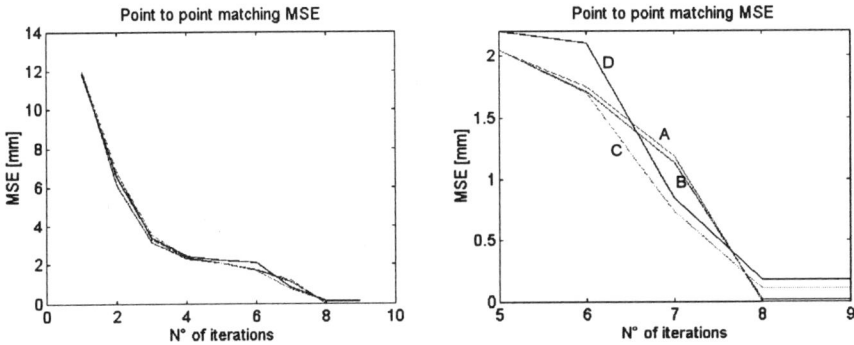

Figures 1 and 2 - *Point to point* matching MSE_2

The other three rows show the behaviour of the algorithm when the variance of the added noise increases. Figures 1 and 2 emphasise the trend of the MSE_2 versus the number of iterations in four test cases. Data describing these test cases are summarized in Table 1.

curve	nr. iter.	noise MSE	MSE_1	MSE_2	Translation			Rotation		
					X	Y	Z	α_x	β_y	γ_z
A	9	0.0		0.0	399.99	500.00	-300.00	0.3253	-0.7079	0.8656
B	9	0.0923	0.0903	0.0189	399.98	500.01	-300.00	0.3256	-0.7075	0.8658
C	9	0.4958	0.4841	0.1069	400.03	500.03	-299.97	0.3295	-0.7125	0.8659
D	9	0.9684	0.9518	0.1783	400.10	500.05	-300.09	0.3224	-0.7058	0.8660
-	9	0.0	-	-	399.98	500.07	-299.98	0.3247	-0.7131	0.8653

Table 1 - *Point to point* matching performance test results

Figure 2 is a zoom of the plot in Figure 1 that highlights the MSE_2 in the last steps of the algorithm. The MSE is of the same order of magnitude of the mean square error of the added noise. Of course the algorithm cannot correct the errors made during the measurements, that we simulate by adding the noise.

Another test was made by sampling the same surface with two different sampling steps. In this way no correspondence between the two sets exists (see the last row of Tab. 1).

The test on the *surface to point* matching algorithm was done using the same method and the same surface. MSE_1 and MSE_2 are defined as in (9) and (10), where the set X corresponds to the set Y_k of (6) and is recalculated at each step k of iteration. Both sets X and P contain 60 points. The results are shown on Table 2 and in Figures 3 and 4.

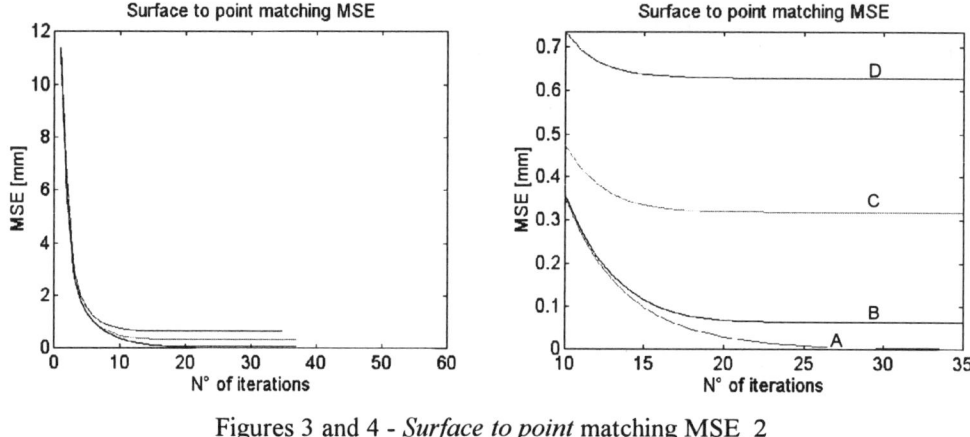

Figures 3 and 4 - *Surface to point* matching MSE_2

curve	nr. iter.	MSE noise	MSE_1	MSE_2	Translation			Rotation		
					X	Y	Z	α_x	β_y	γ_z
A	58	0.0		0.0000	399.99	500.00	-300.00	0.3253	-0.7079	0.8656
B	37	0.0925	0.0956	0.0625	400.00	499.99	-300.00	0.3258	-0.7079	0.8659
C	37	0.4963	0.5269	0.3179	400.08	499.98	-299.89	0.3285	-0.7111	0.8677
D	935	0.9768	0.9944	0.6292	400.10	499.85	-299.99	0.3218	-0.7150	0.8685

Table 2 - *Surface to point* matching performance test results

The comparison of the two methods shows that the *point to point* method converges quickly and with a smaller MSE than the *surface to point* one.

Therefore, it turns out that in all cases the *point to point* matching gives best results.

7. BIOMEDICAL APPLICATION

The case of a robotized surgical operation has been considered. A robotized cell has been built, containing a robot, a sliding working table acting as a bed for the patient, and a skull phantom representing the patient's head (see Fig. 5). A laser mounted on the end effector of the robot is used as a measurement system. The whole cell has been modelled in a CAD simulation environment, as shown in Figure 6. The CAD analytical representation of the phantom skull has been obtained by combining CT and MR results [13][14][15].

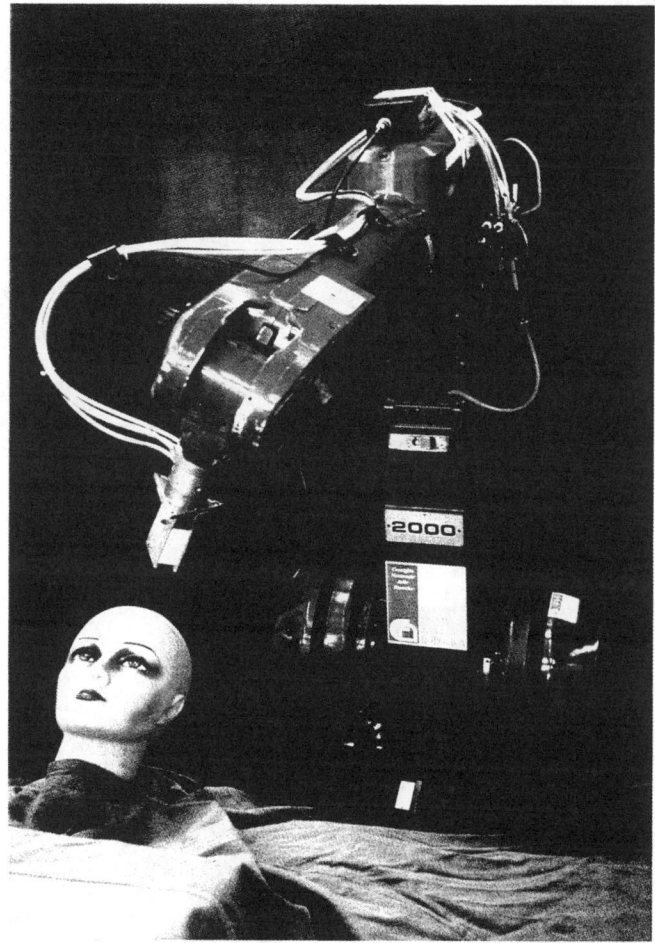

Figure 5 - The real cell with the robot and the skull phantom

The proposed algorithm has been used for calibrating the position of the head of the patient with respect to the robot reference frame. The output of this calibration procedure is a rototranslation matrix which is necessary to correlate the real cell with the simulation environment.

The goal of the experimentation was to realise a simple surgical operation using the robot. This operation consists in moving the robot, with high precision, along a trajectory designed by the surgeon in the simulation environment.

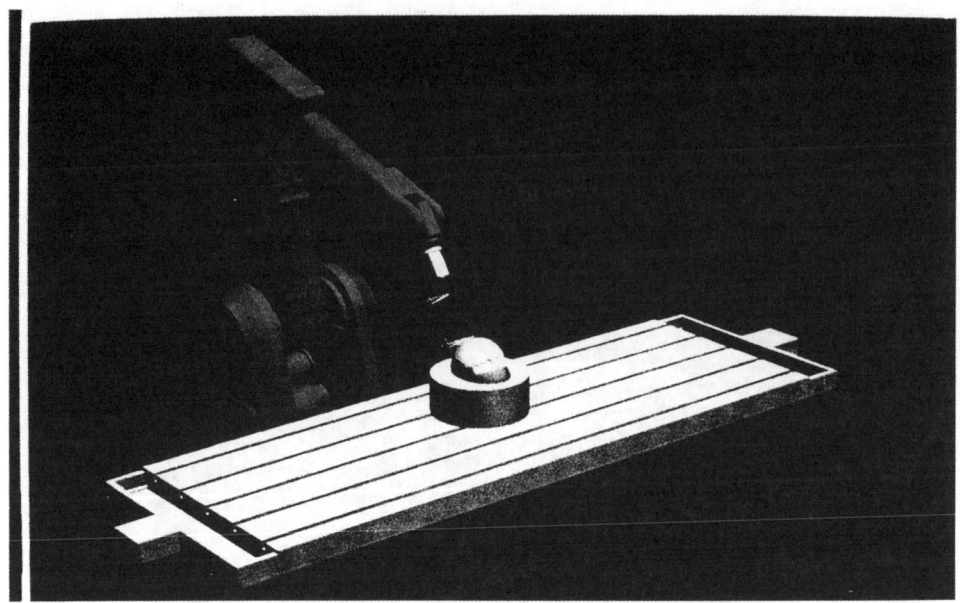

Figure 6 - The simulated cell

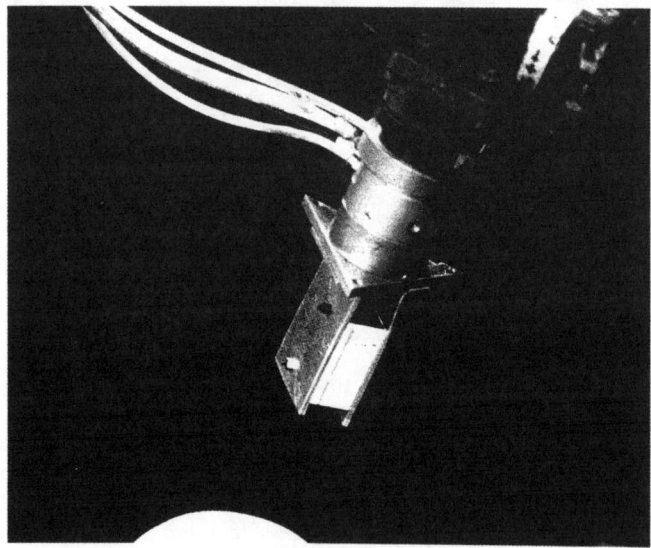

Figure 7 - The laser sensor mounted on the robot end effector

The matching algorithm described above was used as follows:

1) The patient is positioned on the operative bed, which is set in a fixed position with respect to the robot; the position of the patient is known with an approximation of some centimeters.

2) A laser sensor is mounted on the robot end effector and an automatic procedure is executed to measure the skull surface: the robot moves following a programmed trajectory, designed to avoid dangerous situations for the patient: whenever the control system detects an unforeseen situation the procedure is stopped.

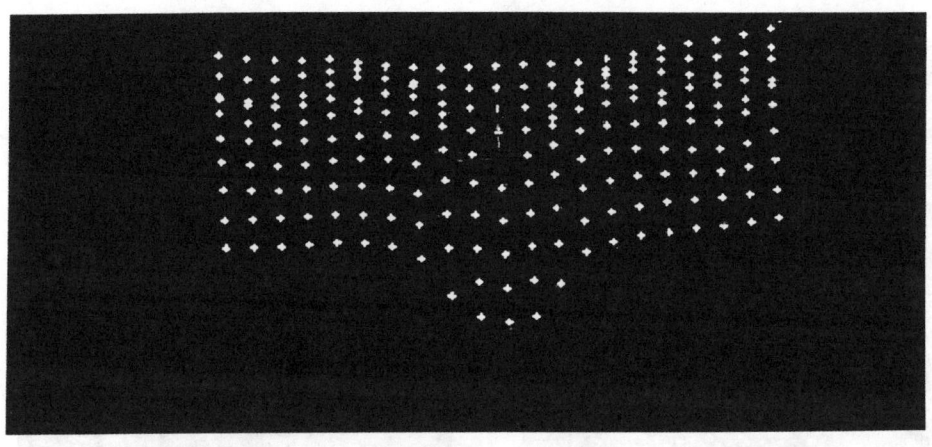

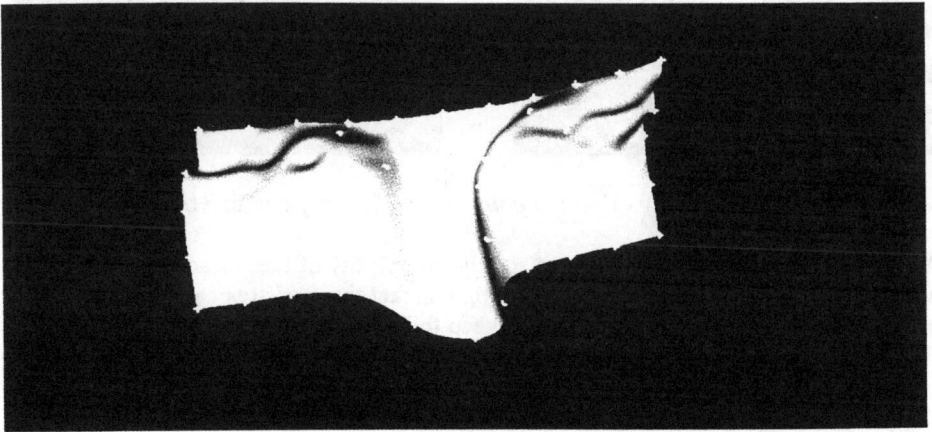

Figures 8 and 9 - *Point to point* and *surface to point* gross matching

The laser sensor works with high precision if the beam is normally reflected by the surface; the distance range ensuring the sensor linearity is 6±1 cm. So the measuring procedure has to be designed in order to ensure the correct position and orientation of the emitted beam. The measuring procedure starts from a point situated over the head of the patient, far enough to avoid collisions with the head itself. Then the robot moves toward the head until the laser reads a distance of 6 centimeters. A rectangular spiral is followed, modifying at each movement the orientation and the position of the tool. When the required number of points are gotten, the procedure stops, and the robot returns to its home position. The output of the procedure is a set of points taken over the skull surface.

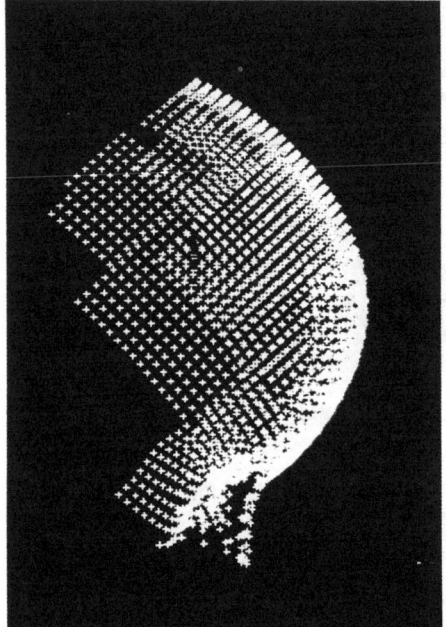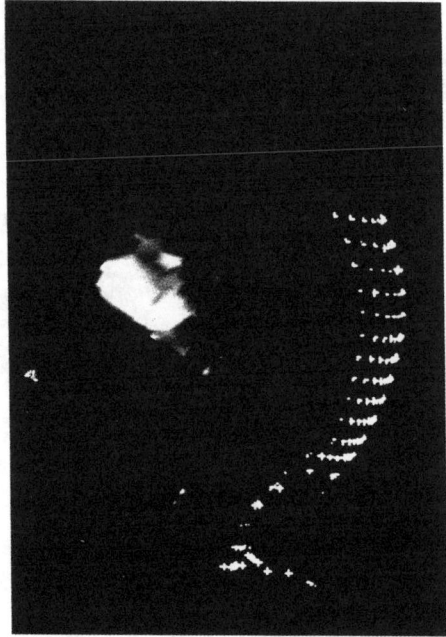

Figures 10 and 11 - *Point to point* and *surface to point* fine matching

As described above, due to the nature and the complexity of the surface of the skull, it is preferable to use a 2 steps matching algorithm; in fact the eyes area can be considered an irregular region. The robot is not able to keep the laser beam normal to the skull surface in the eyes region. So the measuring procedure executed by the robot takes a set of points over the eyes region with a lower precision (i.e. without verifying the perfect normality between the beam and the surface). We can note that sometimes the complexity of the surface would even get the robot to enter the skull to keep the normal direction; of course, this must be avoided.

Another set of points on the parietal and frontal bones on the skull is acquired with high precision to perform the second step of matching.

The integration of CT and MR results has been imported in a CAD environment and a complex analytical model has been built up. The matching algorithm works with the CAD modelled and the measured surfaces. Both *point to point* and *surface to point* matching are performed.

In details, the set of points gotten by edge extraction on the CT and MR were imported in the CAD system and a B-Surface of degree 3 in the *u* and *v* directions was built up.

3) The matching algorithm is applied to the eyes' set of points and to a set of point extract from the model of the same region (see Figs. 8 and 9).

Once the gross matching is done, a more accurate matching is performed between the second set of points and the corresponding model (see Figs. 10 and 11).

4) After the matching algorithm has run, the position of the patient is known with a precision of about 0.1 mm, same order of magnitude of the robot precision. It is now possible for the surgeon to supervise the automatic execution of the trajectory designed in the virtual environment.

8. CONCLUSION

A 3D correlation procedure for robotized application in structured environments has been presented. This procedure finds the correspondence between the simulated and the real cell, thus enabling off line programming of high precision tasks, to be executed on 3D objects of complex shape. The proposed algorithm is an application of the surface matching theory; it is *ad hoc* designed to work whenever a set of points comes from a measurement system and then it not corresponds a priori to the model data set. Two type of matching are defined and discussed, namely the *point to point* and the *surface to point* matching and the results of some performance tests are shown. Finally a neurosurgical application is presented.

9. ACKNOWLEDGEMENT

The authors thanks Dr. Ing. Flora Cavinato for its contribution.
This work was partially financed by the CNR (Italian National Council of Research) through the *Progetto Finalizzato Robotica* project.

10. REFERENCES

[1] S. Badocco, R. Caracciolo, M. Giovagnoni, A. Rossi, "An On-Line Adjusting System for Assembly Robots", Proc. of the Second International Symposium on Measurement and Control in Robotics, Tsukuba Science City, Japan, November 1992.

[2] P.J. Besl and N.D. McKay, "A Method for Registration of 3-D Shapes", IEEE Transactions on Pattern Analysis Intelligence, Vol. 14, n. 2, February 1992.

[3] P.J. Besl, "The Free-Form Surface Matching Problem", in Machine Vision for Three-dimensional Scenes, H. Freeman ed., C. E. New York Achademic, 1990.

[4] R. Caracciolo, F. Fanton and A. Gasparetto, "Surface Matching for Correlation of Virtual Models: Theory and Applications", Proc. of the NASA Workshop on Virtual Reality, Houston, December 1994

[5] F. Cavinato "Matching di superficie per robotica medica" Unpublished Undergraduate Dissertation, University of Padova, Italy

[6] P. Clarysse, D. Gibon et al., "A Computer-Assisted System for 3-D Frameless Localization in stereotaxic MRI", IEEE Transactions on Medical Imaging, vol.10, n.4, December 1991.

[7] R. Dann et al., "Evaluation of Elastic Matching System for Anatomic (CT, MR) and Functional (PET) Cerebral Images", Journal of Computer Assisted Tomography, 13(4): pagg. 603-611, July/August 1989.

[8] N.T.S. Evans, "Combining Imaging Techniques", Clin. Physiol. Meas., vol. 11, Suppl. A, pagg. 97-102, 1990.

[9] R.M. Haralick et al., "Pose Estimation from Corresponding Point Data", in Machine Vision for Inspection and Measurement, H. Freeman ed., C. E. New York Achademic, 1989.

[10] B.K.P. Horn, "Closed-form solution of absolute orientation using unit quaternion", J. Opt. Soc. Am. A, vol.4, n.4, April 1987.

[11] M.L. Kessler, S. Pitluck et al., "Integration of Multimodality Imaging Data for Radioterapy Treatment Planning", Int. J. Radiation Oncology Biol. Phys., vol.21, 1991, pp. 1653-1667.

[12] Y.S. Kwoh, J. Huo et al., "A Robot with Improved Absolute Positioning Accuracy for CT Guided Stereotactic Brain Surgery", IEEE Transaction on Biomed. Eng., vol.35, 153-160, 1988.

[13] C.A. Pelizzari, G.T.Y. Chen et al., "Accurate Three-Dimensional Registration of CT, PET, and/or MR Images of the brain", Journal of Computer Assisted Tomography, 13(1): 20-26, January/ February, 1989 Raven Press, Ltd., New York.

[14] C.A. Pelizzari, G.T.Y. Chen, "Image Correlation Techniques in Radiation Therapy Treatment Planning", Computerized Medical Imaging and Graphics, vol. 13, n. 3, 235:240, 1989, Printed in the U.S.A.

[15] T.M. Peters, J.A. Clark, et al., "Integrated Stereotaxic Imaging with CT, MR Imaging, and Digital Subtraction Angiography", Radiology , 1986; 161: 821-826.

Servodrive equipped linkages to generate exactly defined flexible and adaptable movements

Jörg Möckel, Maik Berger, Jürgen Schönherr
Institute of Engineering and Drive Technology
Chair of Mechanism Theory
Technological University of Chemnitz-Zwickau
09107 Chemnitz, Federal Republic of Germany

Abstract: A new direction of development emerges with the opportunity to combine linkages with servomotors. This combination adds the advantage of flexibility to the benefits of conventional linkages. A slave servodrive is used either to change the position of a frame point of the linkage or to vary the length of a moving part of the linkage. In this way it is possible to change the transmission function, to create harmonic output movements or to create completely new forms of coupler curves.

1 Introduction

The increased demands made to up-to-date drive systems result mainly from the rapidly developed degree of industrial automation and the increasing tendency to small-series production with quickly changing ranges of products and rising demands on flexibility. Even though in some cases it is enough to optimize mechanisms by improved synthesis methods, so the call for adaptable (flexible) mechanisms require a new approach, what cannot be realized any longer by using mechanical means only. The flexibility of a mechanism that is equipped with at least one controlled drive and in the mechatronical sense represents an optimal combination of both mechanical and electronic components in conjunction with elements of informatics is to a large degree determined by the interaction and efficiency of all components [1].

Interrupted movements, large strokes, wide angles of the swinging motion or motions with a strongly changing of velocity at the driven side as well as most of the path-generating movements can be generated by fixed linkages only by at least 6 moving parts. In contrast to this, a large variety of transmission functions and path-generating movements can already be generated by the use of five moving parts provided at least one drive is controlled [2]. Beyond this, an additional controlled drive provides cyclic variation of the mechanism, or the movement can be controlled such that any influences of masses or play as well as manufacturing tolerances and vibrations in the mechanisms are reduced or compensated.

A special attention is devoted to drive-controlled linkages used as path-generating mechanisms. They are especially suited to be used for tasks of handling and transportation at high cycle numbers, as they provide high stiffness in comparison with industrial robots having an open kinematic structure and getting a large part of their motion energy from a continuously rotating drive.

Such a five-bar mechanism has been installed as test mechanism at the Chair of Mechanism Theory at the Technological University of Chemnitz. The information on the design of the control as well as on the experiments refer to this test mechanism.

2 Servodrive equipped path-generating mechanisms

2.1 Master-slave control

Apart from various applications of planar linkages used as function generators, the latter gain more and more importance for special path-generating tasks. Simple kinematic structures are combined with efficient servodrives in order to widen the motion travel and open up new fields of application of planar path-generating mechanisms. To provide this, a servomotor (slave) working as servodrive is assigned to an almost continuously rotating main drive (master). They are linked one to another via the used control unit and the power-output unit. The control unit processes the signals provided from the incremental transducer of the main drive and controls the servodrive according to a given motion function which is stored in a Setpoint Table. Fig. 1 shows the block diagram of the master-slave control.

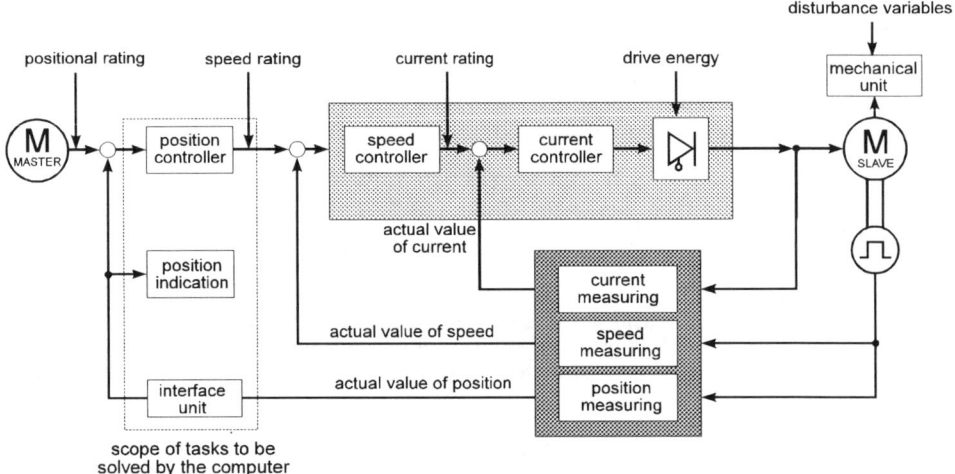

Fig. 1: Block diagram of the master-slave control

To position the servodrive according to the ratings determined in the Setpoint Table, a position-control loop is used. A speed and current controller is cascaded to the position controller in order to improve the control quality. A power amplifier provided with a 16-bit power processor controls speed and current as well as performs various monitoring functions. The control PC works as a proportional controller. According to the Setpoint Table, 1000 position rating per revolution of the main drive are provided to the servodrive. The position controller written completely in Assembler allows every rating to be controlled 50 to 70 times even at the maximum of the main drive speed. The large number of settings allows the servomotor to follow any travels within its limits.

A DAC transfers the regulated quantity determined by the position controller to the setpoint input (±10 V) of the speed controller in the drive unit. An RS 232 Interface allows the adaptation of the controlled quantities (PID) provided by the speed and current controller to the controlled system. A brushless d.c. motor with sinusoidal commutation is used as servomotor.

In contrast to the often described feature of function-generating mechanisms that an additional servodrive serves to perform preferably small motions with only a little torque [3], the path sections to be manipulated in conjunction with the kind of the positioning motion decisively determine the drive parameters. This may have the result that a motor having at least the same output with the main drive has to be used for starting the positioning motion.

2.2 Travel ranges of appropriate mechanisms

Already the extension of four-bar linkages, such as crank-rocker linkages to five-bar mechanisms with a degree of freedom F=2 considerably enlarges the range of paths that can be generated.

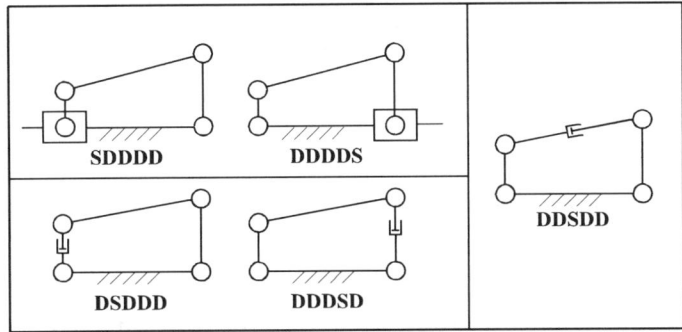

Fig. 2: Four-bar mechanism extended by a translating drive

The obligatory second driving motion can be either translatorical or rotatorical. Simple arrangements are always possible whenever both drives are born in the frame and act at separate points (DDDDS). In contrast to this, starting a positioning motion at a moved part or drive joint of the basic mechanism will cause either one of the two drives to be acting between the moved parts or a drive motion to be transmitted via a synchronously working clutch. Fig. 2 illustrates different arrangements of a additional drive. All arrangements starting out from a four-bar mechanism as basic mechanism. The upper configurations (SDDDD and DDDDS) show a translatorical positioning motion shifting the joints in the frame, and with the other arrangements the dimensions of the moved parts of the mechanism are changed.

Apart from the constructional variants, the selection of the mechanism is considerably determined by the effects of the positioning motion to the path curve to be manipulated. For a defined motion to be performed, a visual comparison can be carried out by representing the motion variants of the coupler point as a function of the arrangement of the translating drive. When doing this, the path curves of the path-generating coupler point under the effects of a set positioning motion can be recorded for any position. When representing all these curves over an entire drive period, a family of curves defining the travel range of the coupler point will be the result (see Fig. 3). If the required points are across this range, the mechanism used for the task is not suited, and another arrangement has to be used. In addition, it should be considered that any manipulation of the dimensions of the bars will change the transmission properties of the mechanism.

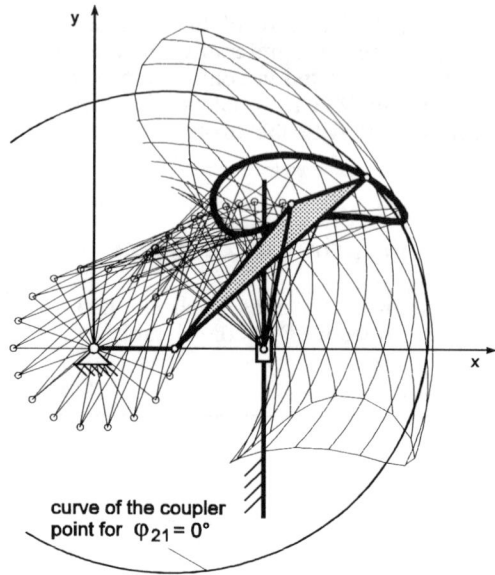

curve of the coupler
point for $\varphi_{21} = 0°$

Fig. 3: Curves and range of travel of the coupler point of a four-bar mechanism

2.3 Approximation of modified path-sections and the required positioning motions

Depending on the technological task to be performed, the sections of a path curve to be manipulated by a servodrive are described either point by point or by equations of a straight line or circle. There are two different approaches to calculate the positioning motion required for the manipulation.

In the first case, the curve sections defined either point by point or section by section and deviating from the path curve of the four-bar mechanism are analytically represented by the help of approximating methods and linked with the sections of the path-curve sections that have not been manipulated by means of transition curves. As approximation and transition functions polynomials, trigonometrical functions and splines are used [4]. This approach has the advantage that the evenness properties of the calculated path curve and its deviation from the given points or sections can be checked directly. However, the disadvantage is that the controlled variables of the related positioning function can generally be calculated only point by point from the determined kinematic conditions.

The other case does not have this disadvantage. The positions of the servodrive are directly determined from the manipulated points or sections of the path curve and then approximated to smooth the curve and continuously extended. This approach provides the opportunity to directly control those parameters of the positioning function such as maximum speed or maximum acceleration. However, the effects of the positioning function on the curve march can be checked only indirectly.

In many cases the path curve of the coupler point K can be approximated by appropriate selection of the coordinate system between the two link points x_A, y_A and x_E, y_E as an explicitly polynomial function of Q-th degree:

$$y(x) = \sum_{q=0}^{Q} c_q \cdot x^q, \quad x_A \leq x \leq x_E \tag{1}$$

The distance of the function $y(x)$ to the supporting points x_i, $\hat{y}_i$ r.m.s. can be minimized by the following equation of optimization, whereby the link-up conditions $y(x_{A/E}) = y_{A/E}$, $y'(x_{A/E}) = y'_{A/E}$ und $y''(x_{A/E}) = y''_{A/E}$ of 0th to 2nd order have to be observed:

$$L(c_q, \lambda_k) = \sum_{i=1}^{m} (\hat{y}_i - \sum_{q=0}^{Q} c_q \cdot x_i^q)^2 + \lambda_1 \cdot (\sum_{q=0}^{Q} c_q \cdot x_A^q - y_A)$$

$$+ \lambda_2 \cdot (\sum_{q=0}^{Q} c_q \cdot x_E^q - y_E) + \lambda_3 \cdot (\sum_{q=1}^{Q} q \cdot c_q \cdot x_A^{(q-1)} - y'_A)$$

$$+ \lambda_4 \cdot (\sum_{q=1}^{Q} q \cdot c_q \cdot x_E^{(q-1)} - y'_E) + \lambda_5 \cdot (\sum_{q=2}^{Q} q \cdot (q-1) \cdot c_q \cdot x_A^{(q-2)} - y''_A)$$

$$+ \lambda_6 \cdot (\sum_{q=2}^{Q} q \cdot (q-1) \cdot c_q \cdot x_E^{(q-2)} - y''_E) \rightarrow \text{minimum} . \tag{2}$$

The partial derivatives of the Lagrange function L from the coefficient c_q and the Lagrange multiplier λ_k which are set to zero result in a soluble system of linear equations. The calculation of the required positioning motion depends on the arrangement of the drives and is aimed at representing the individual position values as a function of the main drive and then optimally adapting them to the conditions and limits of the drive system to be used. The process of synthesis is determined mainly by how a subsequent optimization of the positioning function effects on the appearance and march of the path curve and its deviation from the settings within a given range of tolerance as well as what is the way how to realize compromise solutions.

The used control philosophy may require to eliminate the influence of those control parameters, such as the occurring delay between input and output values at the position controller in dependence of the transmission properties in the controlled system. Apart from the use of those electronic means, such as the acceleration precontrol, the correction of the reference input variable by the Laplace transformation with consideration of cycle time and step response will provide good results.

3 Adaptive correction of the reference motion

Between the incremental transducers providing the actual position values of the main and servodrives and the coupler point executing the desired reference motion there are the moving parts of the mechanism with all their technologically caused geometrical errors. These errors sum up with the plays of the joints and have their influence on shape and position of the real travel of the path. To record these negative effects and correct them, it is necessary to measure the real path curve of the coupler point while the mechanism is moving. The deviations of the rated from the measured path curve has to be deduced to the positioning motion with the help of a special mathematic algorithm. The correction values calculated by this method determine the positional settings of the servomotor for the subsequent motion cycles.

3.1 Measuring systems for even path curves

The path curve of the installed test mechanism may range over an area of max. 1 m² movement along the curve is possible up to 100 times per minute. To obtain a reasonable comparison between actual and rated form of the path curve, a sufficient number of positions has to be measured along the curve. The mathematical analysis requires one value per step, whereby one step corresponds to 2 .. 3° of a crank revolution. This means that with maximum speed of the main drive (100 rpm) at least 200 values per second have to be measured.

To measure the even path curve, both contact and non-contact measuring methods can be used. However, optical, non-contact measuring methods provide only a very little resolution. When the dot matrix of a CCD camera is divided into 512 x 512 pixels, the resolution with refer to the required large measuring range amounts to approx. 2 mm [5]. Contact measuring methods based on path transducers are considerably more exactly; their resolution is better than 0.1 mm. However, these transducers have to be integrated in the housing of the linkage. The transducer used in the test mechanism consists of two rope-operated path transducers. These rope-operated transducers use the absolute measuring principle, i. e. the reference points must be approached for calibration purposes only. The length of the rope can be polled over a measuring range of 2 meters up to 500 times per minute at a resolution of 0.1 mm and a max. error of ±0.05 mm.

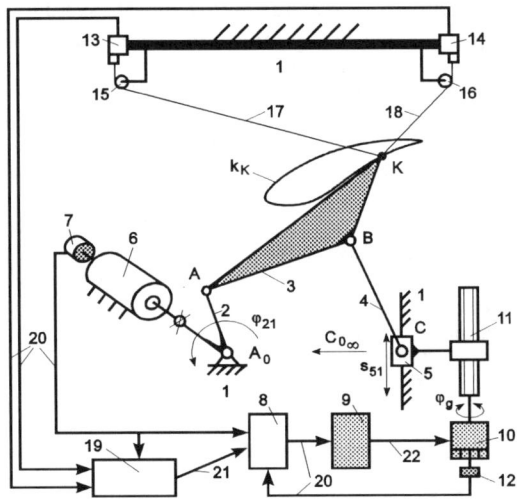

Fig. 4: Test mechanism arrangement

Fig. 4 shows a diagram of the arrangement of the test mechanism. The main motor (6) drives the crank (2) continuously. The positioning motion is generated by the servomotor (10) and provided to the linkage at the bearing point of the crank (C) via a pitch- line mechanism (11) with linear guide (5). The incremental transducers (7 and 12) installed on the motor shafts transmit the actual positions to the position controller (8). The rope-operated transducers (13 and 14) are allocated above the mechanism. The two ropes

(17 and 18, resp.) run over deflection pulleys (15 and 16); their ends are fixed at the coupler point K. While the mechanism is moving, both the lengths of the rope and the angle between the ropes change. The position of the coupler point in the coordinate system in the plane of the frame and the length of the ropes must be calculated at the same time. The measured rope lengths are the normal lines of the involute to the circle of the deflection pulleys. As a result of this, the position of the coupler point to be determined will be the intersection point of these involutes. The intersection point is calculated by an iteration method based on linearization. Starting out from an appropriate initial approximation, 2 ... 4 steps of iteration will be enough to calculate the intersection point. Logging of the measuring values, calculating of the positions as well as storing of the measuring values are carried out by the measuring PC (19) belonging to the next higher level. The measuring values are transmitted to the measuring PC by tackle lines via two parallel opto-decoupled 16-bit interfaces. A special timer module ensures the exact time basis and initiates the measuring process up to 100 times per second. While the measuring values are being logged, it is not possible to calculate the actual positions of the coupler point or the corrective function for the servodrive.

3.2 Calculating the corrective function

To calculate the corrective function for the positioning slide bar, the rated and actual path curves are compared one to another with consideration of the kinematics of the mechanisms, which has been used. The demands made on the algorithm of correction are determined by those aspects, such as
- the possibility of application for most different path forms,
- no limits for solution in singular points, and
- calculation of corrective motions with good evenness properties.
It should be taken into account that only corrections in the theoretically existing range of travel can be determined (see Fig. 5).

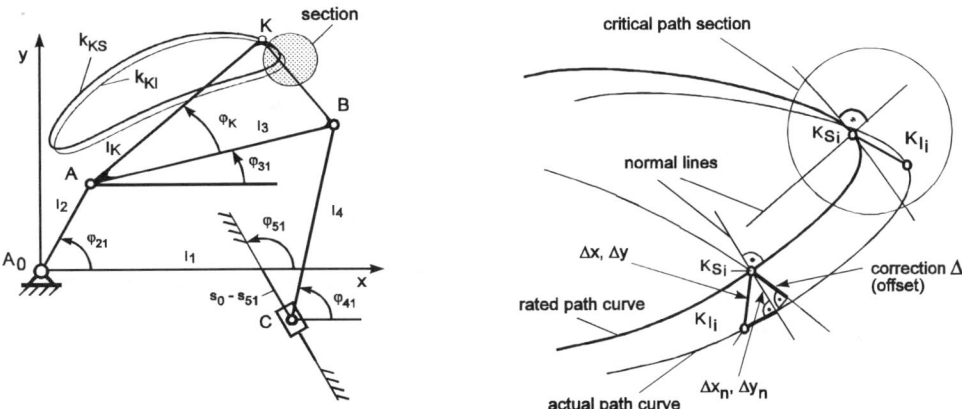

Fig. 5: Kinematic diagram of the mechanism

Another aspect is the method how the measuring value is acquired, whereby in principle two methods are possible: With drive-relevant measuring, the coordinates of the path curve are acquired by an angle transducer with refer to the positions of the drive crank and then directly assigned to them. When performing quick motions, errors may occur, since the required measuring time is not obtained or phase displacements occur due to delayed signal transmission. These errors can be avoided by measuring irrespective of the drive and depending on time. However, in this case the uneven distribution of the measuring points along the path curve due to different speed conditions may have negative effects.

For comparison of the two path curves they are scaled to the calculated curve length and approximated as a function of the parameters through a finite trigonometrical series [6]. Using this approach, the displacement of the phases of the two curves to one another can be determined quickly. After then, in the course of an iterative process, i.e. with the help of a linearized mathematical pattern for calculating the position of the coupler point K, a corrective way Δs_{51} for given drive positions are determined such that the differences between the two path curves are reduced to a minimum.

The distance vector between rated and actual curve in the normal direction in the point K_S of the rated curve can be approximated by projecting the distance vector $(\Delta x, \Delta y)$ of an adjacent point K_I of the actual curve to the normal line. When using γ_K as a rising angle of the normal line, the following is applicable:

$$\Delta x_n = \Delta x \cdot \cos^2\gamma_K + \Delta y \cdot \cos\gamma_K \cdot \sin\gamma_K \tag{3}$$

$$\Delta y_n = \Delta x \cdot \cos\gamma_K \cdot \sin\gamma_K + \Delta y \cdot \sin^2\gamma_K \tag{4}$$

Using the coordinates x_K, y_K of the point K_S, the drive angle φ_{21} and the direction angle φ_{31} of the coupling result from the compulsory conditions:

$$x_K = l_2 \cdot \cos\varphi_{21} + l_K \cdot \cos(\varphi_K + \varphi_{31}), \tag{5}$$

$$y_K = l_2 \cdot \sin\varphi_{21} + l_K \cdot \sin(\varphi_K + \varphi_{31}), \tag{6}$$

(see Fig. 5). The direction angle φ_{41} of the crank rocker as well as the drive coordinate s_{51} of the slide bar are calculated with the help of the following equations:

$$x_B = l_2 \cdot \cos\varphi_{21} + l_3 \cdot \cos\varphi_{31} = l_1 + (s_{51} - s_o) \cdot \cos\varphi_{51} + l_4 \cdot \cos\varphi_{41} \tag{7}$$

$$y_B = l_2 \cdot \sin\varphi_{21} + l_3 \cdot \sin\varphi_{31} = (s_{51} - s_o) \cdot \sin\varphi_{51} + l_4 \cdot \sin\varphi_{41} \tag{8}$$

If the last two equations with a fixed drive angle φ_{21} are differentiated from the drive coordinate s_{51} and solved from $\dfrac{\delta\varphi_{31}}{\delta s_{51}}$, the result is

$$\frac{\delta\varphi_{31}}{\delta s_{51}} = \frac{\cos(\varphi_{51} - \varphi_{41})}{l_3 \cdot \sin(\varphi_{41} - \varphi_{31})} . \tag{9}$$

Using this equation, the derivatives of the equations (5) and (6) from s_{51} will be

$$\frac{\delta x_K}{\delta s_{51}} = -i \cdot \sin(\varphi_K + \varphi_{31}) \tag{10}$$

$$\frac{\delta y_K}{\delta s_{51}} = i \cdot \cos(\varphi_K + \varphi_{31}), \tag{11}$$

whereby

$$i = \frac{l_3 \cdot \sin(\varphi_{41} - \varphi_{31})}{l_K \cdot \cos(\varphi_{51} - \varphi_{41})} \tag{12}$$

determines the transmission ratio from the servo drive 51 to the motion of the coupler point. When replacing the differential values in the equations above by finite displacements Δx_K, Δy_K and Δs_{51}, the correction of the vector component $(\Delta x_n, \Delta y_n)$, which is in the motion direction of the coupler point and manipulated by the servo drive, is calculated with the help of the following equation

$$\Delta = -\Delta x_n \cdot \sin(\varphi_K + \varphi_{31}) + \Delta y_n \cdot \cos(\varphi_K + \varphi_{31}) \tag{13}$$

whereby the linear displacement of the servodrive is approximated by

$$\Delta s_{51} = i \cdot \Delta. \tag{14}$$

Thus, the correction of the generated actual curve will reach as a maximum only the value of the normal deviation from rated and actual curves. So the corrective value Δ will not exceed all limits as with a complete error compensation near singular curve points.

In the last step of calculation, the corrective function Δs_{51}, that is now present point by point, must be summed up with the valid positioning function, and a new path-generating function must be determined with consideration of the characteristics of the controller. Because the path-generating function is approximated by a Fourier series, too, the curve of the path-generating function can be smoothed by selecting suitable harmonic components and in this way matched with the conditions of the drive system. The new deviation of the path curve that is then to be expected in any case can be improved later in the course of further correction cycles; however, with the same input parameters only a slight improvement is possible and mistakes in the calculation near singular points will sometimes have a more negative effect.

The corrected motion function for the servodrive is entered directly into the Setpoint Table of the position controller (control PC) and used with the subsequent revolutions of the main drive. The transmission is carried out through a parallel 16-bit data line.

3.3 Using the corrective method to generate an exact straight-line motion by means of a five-bar linkage

The motion example illustrates the opportunity of simulating a calculated straight-line section of a path curve of the test mechanism by using an suitable positioning function. The deviations of the real path march from an exact straight line are compensated by corrective functions. If the coupler point K of the test mechanism is desired to generate a horizontal straight-line section of a given position, which is crossed by the curves in alternating directions (see Fig. 6), the bearing point of the crank (C) allocated on the positioning bar (5) must move as illustrated in Fig. 7.

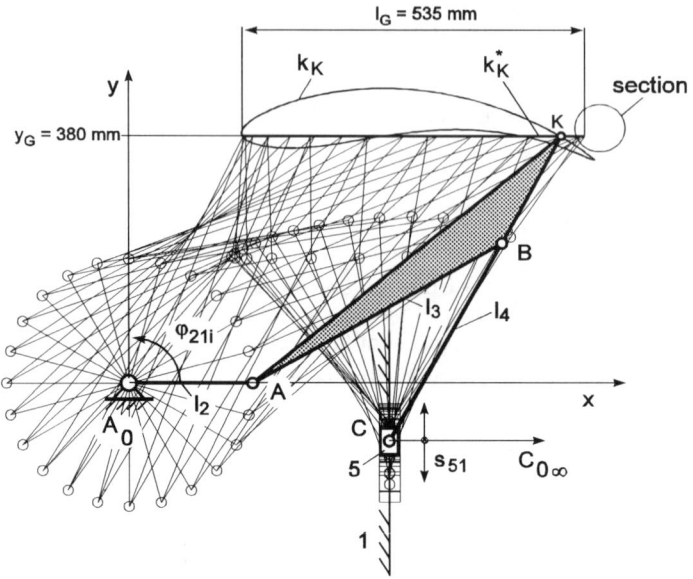

Fig. 6: Four-bar mechanism with path curve k_K and straight line section k_K^* generated by an additional positioning motion s_{51}

Because the required positioning function does not contain a locking position corresponding to the basic mechanism, the starting-up function has to be used. This function is intended to move the slide bar off its locking position during the first crank revolution such that a smooth and jerk-free transition into the actual positioning motion with constantly rising acceleration is possible, which is then permanently repeated. The starting-up function illustrated in Fig. 7 has been structured such that the transition to the required positioning function is carried out at the end of the first crank revolution. The motion functions were calculated for a main drive speed of 60 rpm.

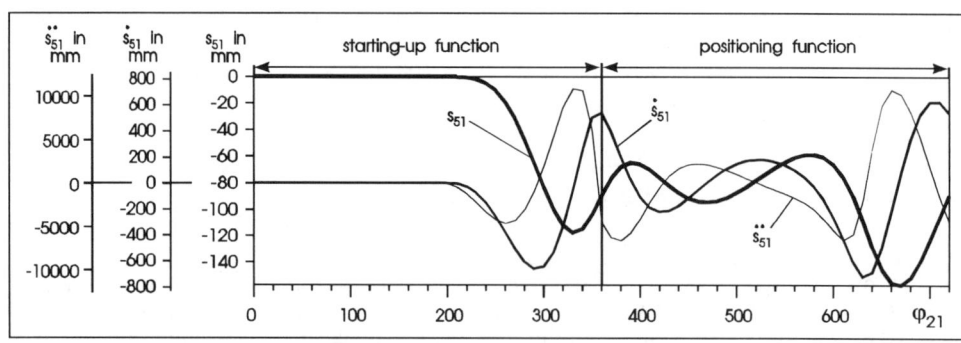

Fig. 7: Starting-up and positioning function for generating a straight-line motion

To demonstrate the corrective algorithm, the bar-relevant position of the coupler point K has intentionally been set to an incorrect value; as a consequence of this, it turned out that this was the main cause of the measured path deviations in the first uncorrected motion cycle. The maximum deviations were -5.3 mm (see Fig. 8).

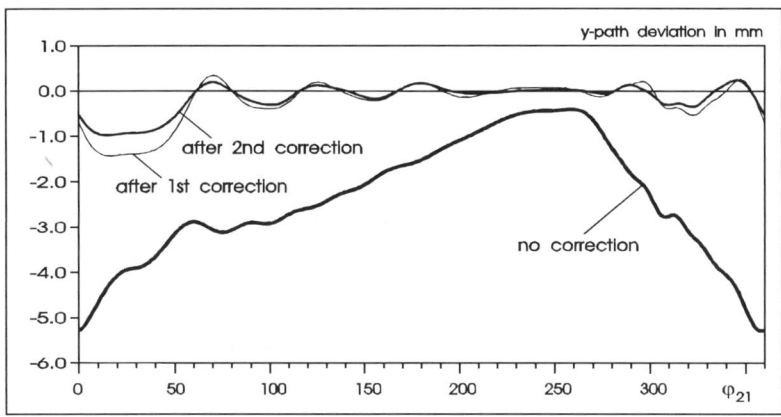

Fig. 8: Measured deviations of the actual path curves from the rated curve

These path deviations have been used to set a first corrective function. Because the real path curve is beneath the required straight line, the curve has positively to be corrected with consideration of the kinematic arrangement of the mechanism used (see Fig. 9). Any measuring values of the actual-path curve across the given range of travel of the rated curve cannot be compensated (see Fig. 10).

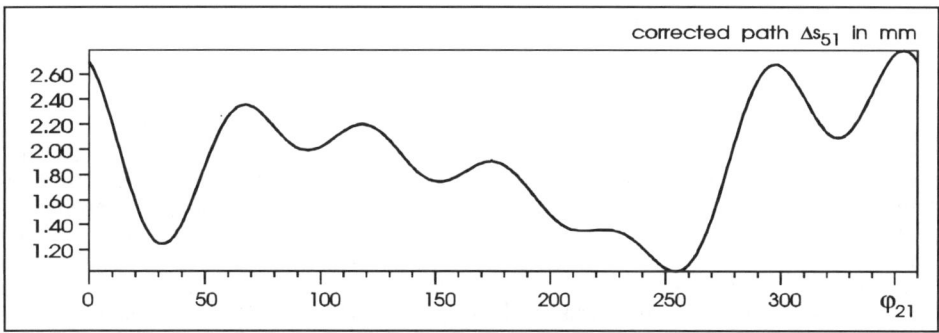

Fig. 9: 1st corrective function

Because the method of calculation is based on theoretically determined dimensions of the mechanism and aimed at projecting the actual curve to the respective rated curve at the limits of the travel range, i. e. near the singular positions of the mechanism, an error will occur that with this motion task will have an especially negative effect due to the new form of the path curve. If one tries to limit this error by exact simulation of the corrective function point by point, a heavily oscillating positioning motion with bad dynamic properties will be the result. This is the reason why it is tried to change the corrective motion by an appropriate number of harmonics such that, on the one hand, a sufficient number of corrective values is generated and, on the other hand, the peaks of the amplitudes in points where strong vibrations occur are weakened. That's why the corrective motion illustrated in Fig. 9 has been calculated with 6 harmonics.

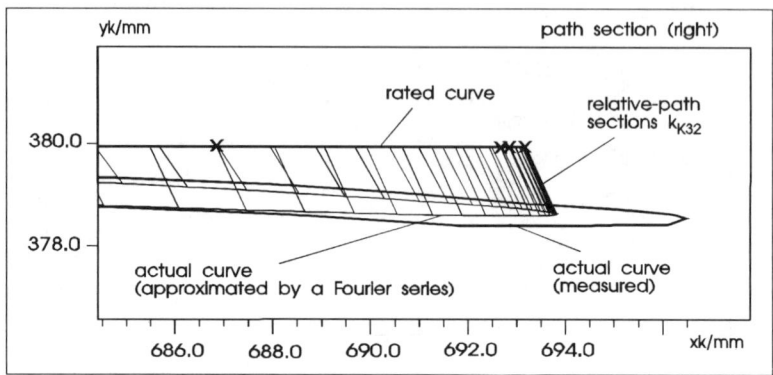

Fig. 10: Representation of path error for the right path section

The y-path deviations illustrated in Fig. 8 show that the dominating part of the path curve coincides with the rated curve to a large degree already after the 1st correction cycle. The deviations from the given path curve of max. 0.35 mm, which are across the range which cannot be manipulated, nearly reach the max. possible measuring accuracy. In the path sections at the boundaries of the travel range the existing mechanisms will not provide further improvement, as the degree of freedom of the mechanism is reduced to $F = 1$.

4 Conclusions

The combination of asynchronously transmitting mechanisms and controlled servodrives provides the opportunity to generate flexible path-generating motions. In addition, the controlled drive in conjunction with a transducer can be used for adaptive compensation of inaccuracies in the bar lengths as well as of the joint play or any deformations of the bars of the mechanism. The suggested solution is suited to be used in quickly-running conveyor lines or processing machines and can be considered in both applications as a cost-effective compromise between a full-flexible robotal device and a mechanism whose motion characteristics cannot be changed.

Reference:

[1] Isermann, R.: Integrierte mechanisch-elektronische Systeme.
          VDI-Z 135 (1993) Nr. 10, S. 64-69

[2] Fricke, A.: Erzeugung ungleichmäßiger Bewegungen durch Getriebe in Verbindung mit rechnergesteuerten Antrieben.
Dissertation B, TU Chemnitz, 1990

[3] Hammerschmidt, Chr.; Fricke, A.: Getriebe mit rechnergesteuerten Antrieben zur Erzeugung ungleichmäßiger Bewegungen.
VDI Berichte Nr. 958, S. 231-246, 1992

[4] Volmer, J. (Hrsg.): Getriebetechnik-Kurvengetriebe.
Berlin: Verlag Technik 1989

[5] Breuckmann, B.: Bildverarbeitung und optische Meßtechnik in der industriellen Praxis. Franzis-Verlag GmbH, München, 1993

[6] Schönherr, J.: Synthese von Punktführungsgetrieben und Qualitätskontrolle von Kurvenkörpern auf der Basis der Fourieranalyse.
Konstruktion 47 (1995), S. 69-73

Dipl.-Ing. Jörg Möckel,   (lecturer)
Dipl.-Ing. Maik Berger
Prof. Dr.-Ing. habil. Jürgen Schönherr,
Institut für Konstruktions und Antriebstechnik i. G.
Lehrstuhl Getriebetechnik
Technische Universität Chemnitz-Zwickau
Straße der Nationen 62
09111 Chemnitz

Tel.: (0371) 531 1254 / (0371) 531 1487
Fax.: (0371) 531 1239

Avoiding Singularity Problems of Manipulators with Redundant Kinematics by On-line Dynamic Trajectory Optimization

M. Schlemmer, R. Finsterwalder, G. Grübel

DLR Oberpfaffenhofen
Institute for Robotics and System Dynamics
D-82234 Wessling, Germany
e-mail: Maximilian.Schlemmer@dlr.de

Abstract

A method is addressed for on-line dynamics trajectory optimization of kinematically redundant manipulators. As an example, the considered task of trajectory planning is to teach interactively position and orientation of the tool-centerpoint frame, which is fixed in the manipulator hand. Thereby the manipulator has to autonomously preserve kinematic constraints such as moving obstacle avoidance, singularity avoidance, and physical joint limitations as well as dynamic constraints like box-constraints on joint velocities, accelerations and motor torques. The key idea is to transform the resulting overall motion planning problem into a timed series of point-to-point trajectory planning problems, which, in turn, may be formulated as parameter optimization problems, that can be efficiently solved in on-line mode by the numerical method of sequential quadratic programming. Since the approach does not require an inverse kinematics formulation it is feasible for manipulators with redundant kinematics.

Keywords: Manipulator trajectory optimization, On-line trajectory planning, Moving-obstacles avoidance, Kinematical Redundancy, Singularity avoidance, Ten-degrees-of-freedom manipulator, Interactivity by teaching.

1 Introduction

1.1 Previous work

Path planning [14],[12], [25] means purely geometric / kinematic based motion planning, whereas trajectory planning is more complex, because system dynamics are to be considered in addition to kinematics. In a general motion planning approach both geometric path planning requirements and dynamical constraints like bounds on velocities, accelerations and motor torques, as well as arbitrary performance indizes to be minimized, may be formulated as a so-called general trajectory planning

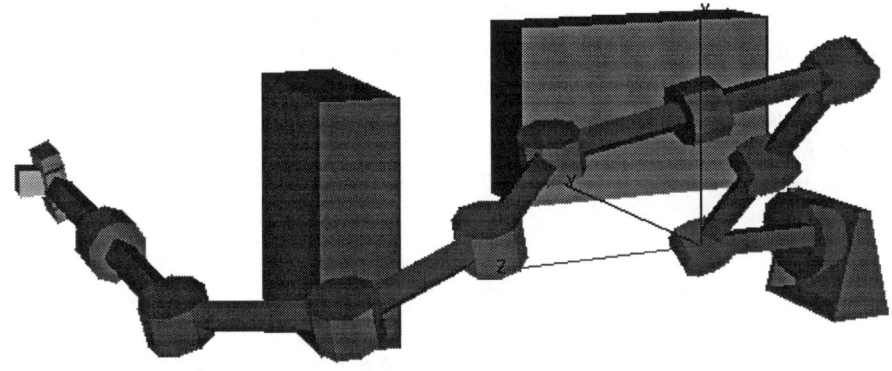

Figure 1: A ten-dof manipulator and obstacles

problem. A common solution strategy to this problem is to subdivide it into a geometric path planning problem and a one-dimensional velocity planning problem and then solving the two problems separately: after producing a collision-free geometric path a dynamically feasible velocity profile along this path is generated [20], [23], [22], [6]. This approach, however, may yield trajectories with a rather conservative dynamic behaviour, because the geometric path is not optimal with regard to system dynamics.

To solve the general motion planning problem commonly used approaches are based on optimality conditions of calculus of variations [2], [13] and solving the resulting boundary value problem by multiple shooting [4], [16] or solving a discretized optimal control problem as a nonlinear programming parameter optimization problem [24], [19], [21]. More heuristically motivated methods are based on dynamic programming techniques [1] or combinatorical algorithms [8].

However the field of applications of such approaches is typically restricted to offline trajectory planning problems, because these methods often exhibit exhaustive computational solution processes. Also, all geometrical information (e.g an evolving environment) neccessary to set up the motion planning problem is to be known in advance.

1.2 Task to solve

This paper should provide a contribution with regard to new approaches for on-line optimal trajectory generation in cases where on-line (i.e during motion execution) information of a changing environment is available by virtue of sensor data devices or via an interacting human operator. Position and/or orientation of the tool-centerpoint frame of a kinematically redundant ten-dof manipulator shall be controlled on-line by an operator. The three-dimensional work space can be populated by arbitrarly shaped, fixed and moving obstacles. The manipulator has to autonomously preserve explicitly formulated kinematic constraints such as moving obstacle avoidance, singularity avoidance, and box-constraints on joint positions as well as dynamic box-constraints on joint velocities, accelerations and motor torques. For that a feasible trajectory of the manipulator is to be computed on-line. Furthermore, the trajectory shall be optimized considering multi-performance criteria, such as power, energy consumption and time.

1.3 The trajectory planning approach

Basic ideas are addressed for on-line trajectory planning under explicitly formulated kinematic and dynamic constraints and multi-criteria optimization. The feasibility of these ideas has been demonstrated by simulations of a ten-degrees of freedom manipulator, which is on-line controlled by human interaction. The optimization up-date rate was less than 0.1 s (20 Mflops computing power). A main point of the approach is that the trajectory planning algorithm is independent of special kinematic and dynamic robot model representations. Therefore any kind of model, produced by some model building environment, can be used. For instance, the forward kinematic and dynamic model of the ten-dof manipulator according to figure 1 is modelled formally by the multibody description approach due to Otter [15], where the physical objects of bodies and joints are described by their local properties and an $O(n)$ symbolic formalism to compute the overall nonlinear system kinematics and dynamics is generated automatically by the object-oriented model building environment Dymola. In order to achieve fast solution processes during optimization, the trajectory planning formulation is realized in minimal coordinate parameter spaces. Collision avoidance and detection is performed simultanously in Cartesian space using a *forward* kinematic approach for implicitly casting Cartesian space obstacles and endeffector goal position/orientation in joint space. Therefore the algorithm is especially suited for motion planning of kinematically redundant manipulators, because no configuration space obstacles must be derived explicitly. The obstacles are described by a set of points that are arranged on the surfaces of the obstacles. The geometry of the robot itself is approximated by a set of (intersecting) spheres. A smooth distance function serves for collision avoidance. The workspace is partitioned into equally sized cells for reducing the number of obstacles to be considered at a given time instant. Self-collision between non-adjacent manipulator links is avoided by considering them as moving obstacles.

Especially for kinematically redundant manipulators, singularity avoidance is a major problem. Near to singularities the forward kinematic jacobian becomes ill-conditioned, resulting in high joint position displacements for executing rather small perscribed Cartesian motion increments. Our trajectory planning algorithm realizes singularity avoidance via dynamic constraints and minimal time motions. Hence, as an advantage, no special analysis of the kinematic jacobian has to be performed.

2 Trajectory planning

The trajectory planning problem is reduced to a series of variational problems to be solved in an iterative timeframe. Each variational problem can be split into two parts that can be computed in parallel. The two parts are:

1. **Goal Position Update:** The actual robot joint goal position is updated based on the actual six-dimensional Cartesian velocity increment, interatively demanded by the operator or calculated by a planning system based on sensor information.

   This is done on-line by successively solving closely related, smooth nonlinear parameter optimization problems using the numerical method of sequential quadratic programming. Box constraints on joint positions, a distance function formulation for collision avoidance, and singularity avoidance based on a kinematic jacobian approach, are set and define constraints within an optimization problem for generating feasible goal joint positions.

   The actually computed robot goal configuration is stored for the next trajectory optimization.

2. **Trajectory optimization**: The trajectory from the actual robot joint position to the actually computed new joint position goal (Goal Position Update) is computed in joint space with respect to kinematic and dynamic constraints as mentioned above.

The timing of the target updating process and of the trajectory optimization at a time instant t_{i-1} is organized according to figure 2:

1. The robot moves along the previously computed trajectory q_{i-1} at time t_{i-1}.

2. Depending on the required sample rate (e.g. 100 ms), the next time instant t_i is defined where the trajectory q_{i-1} should smoothly (twice-continuously differentiable) branch into the updated trajectory q_i.

3. At time t_{i-1} the actual joint position goal is updated (Goal Position Update) and this information is sent to the trajectory optimizer.

4. The trajectory q_i is computed as a function of the actual robot joint position and new joint position goal.

In the next section, it is shown how on-line trajectory optimization can be performed.

3 Discretization for Trajectory optimization

A discretization strategy on *joint-acceleration level* is used, because of the special structure of the underlying variational problem. Also, the set of optimal solutions is restricted to the linear space of cubic splines. This results in twice continuously differentiable cubic splines trajectories, which are known to be optimal in the sense

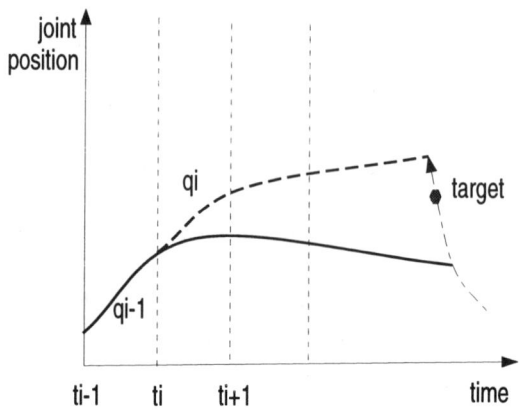

Figure 2: Trajectory-planning update strategy

of minimizing the average curvature yielding highly smooth joint position rate beha-viour [5], [3]. The resulting parameter optimization problem can be efficiently solved using the sequential quadratic programming method [10], [9]. The advantages of a discretization on joint-acceleration level are:

- Almost all relevant constraints (collision avoidance, goal configuration, box-constraints on joint positions and velocities) occur on joint position and joint velocity level, which can be transformed *linearly* to constraints on joint-acceleration level.

- Joint accelerations occur *linearly* in the dynamic equations of motion, in con-trary to joint positions and velocities. Hence for any time instant and state vector (q, q') a *linear* (bijective) transform exists and can be explicitly formu-lated. This linear transformation maps the rectangleoid of all admissible joint torques *exactly* to an equivalent polyhedra of all admissible joint accelerations. Thus torque box-constraints can be considered as general linear inequality con-straints on joint acceleration level. However, these linear inequalities depend on the actual state vector. The (unknown) state vector is approximated by the solution of the prior optimization problem. If the iterative trajectory planning process is fast enough then the actual values of joint position and velocity are close to the prior ones and the error of this approximation is negligible. Since the gradients of all linear constraints are constant, they have to be evaluated only once before starting an optimization run.

- The complete dynamical system is given as a purely algebraic equation (i.e all joint motor torques are functions of joint positions, velocities and accelerati-ons).

This concept of implicitly incorporating torque box-constraints on joint-acceleration level saves a lot of computation time and yields improved convergence properties, because the highly nonlinear robot system dynamics is completely eliminated in the optimization run.

The performance criteria to be minimized have to be formulated in Lagrange (i.e. time-integral) representation with integrands depending explicitly on joint positions, velocities, accelerations, and torques.

As this joint acceleration approach allows to use the robot's dynamic equations of motion as a purely algebraic equation, numerical quadrature [17] can be employed for evaluating these time-integrals, instead of using more computation-expensive numerical integration methods.

With a continuously piecewise linear approximation on joint acceleration level the time interval $[t_0, t_f]$ is replaced by grids:

- each joint acceleration $q_j'', j = 1, ndof$ ($ndof$ denotes the total number of joints) is allocated a (rather crude) grid

$$t_0 =: T_0^j < T_1^j < .. < T_{n(j)}^j := t_f,$$

- all constraints are discretized on a (rather fine) grid

$$t_0 =: t_1 < t_1 < .. < t_m^j := t_f.$$

The discretization results in a nonlinear, low-dimensional optimization problem subject to nonlinear constraints:

Minimize the Chebyshev norm of the slack-parameter vector $||p||_\infty$ subject to:

$$J_j(x) := \int_{t_0}^{t_f} L_j(q(x,t), q'(x,t), q''(x,t), u(x,t), t)dt \leq J_j^{max} + p_{m+1},$$

$$(j = 1, .., l),$$
$$q(x, t_0) = q_0, \quad q(x, t_f) = q_f,$$
$$q'(x, t_0) = q_0', \quad q'(x, t_f) = q_f',$$
$$d(x, t_i) \geq 0,$$
$$q_{min} \leq q(x, t_i) \leq q_{max},$$
$$-p_i + q_{min}' \leq q'(x, t_i) \leq q_{max}' + p_i,$$
$$-p_i + q_{min}'' \leq q''(x, t_i) \leq q_{max}'' + p_i,$$
$$-p_i + u_{min} \leq M(q_-(t_i)) \cdot q''(x, t_i) - \chi(q_-(t_i), q_-'(t_i)) - g(q_-(t_i)) \leq u_{max} + p_i,$$
$$(i = 1, .., m);$$

where

$x := [q_1''(T_1^1), q_1''(T_2^1), .., q_{ndof}''(T_{n(ndof)}^{ndof})), p]$ denote the optimization parameters and $p := [p_1, .., p_{m+1}]$ the slack-parameter vector;

M, χ, g denote dynamic quantities (i.e massmatrix, Coriolis and centrifugal terms, gravitational terms);

q, q', q'', u denote joint positions, velocities, accelerations, and motor torques;

q_{min}, q_{max} minimum, maximum value for each joint position component,

q_{min}', q_{max}' minimum, maximum value for each joint velocity component,

q_{min}'', q_{max}'' minimum, maximum value for each joint acceleration component,

u_{min}, u_{max} minimum, maximum value for each joint motor torque component,

q_-, q_-' are estimates for the actual state vector (solution of the previous problem),

l is the number of different performance indices, d is the distance function.

The final time t_f is not explicitly minimized. Instead, a heuristic is used to adapt the final time for each optimization run. The reason is, that linear constraints can only be obtained if t_f itself is not an optimization parameter. The heuristic is as follows:

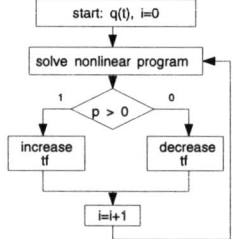

where $\|p\|$ accounts for dynamic constraint violations and guarantees a non-empty feasible set for the above optimization problems, (when all kinematic constraints are stated reasonably, i.e the feasible set with regard to box-joint constraints and collision avoidance is not empty)

The norm of the slack-parameter vector indicates a time-minimal solution, when changing from zero to a positive value.

In order to obtain fast convergence properties during an optimization procedure, two main aspects have to be taken into account: First, a good initial estimate must be provided. This is done by successively solving closely related optimization problems, where the actually calculated trajectory is used as an initial estimate for the next optimization run. Second, special scaling techniques [10] are realized, in order to achieve well-scaled optimization problem formulations.

4 Multi-criteria optimization

Considering a vector-valued criterion, the meaning of the 'best' solution is expressed by Pareto-optimality conditions [18], [13]. Due to the computing time constraint, it is in general not possible to compute a Pareto-optimal solution if the problem is posed in a multi-criteria setting. Therefore a generalization of multi-criteria design is defined, which results in sufficiently fast convergence because special functionalities of sequential quadratic programming methods are exploited: The user specifies an upper-bound demand level for each criterion $J_j, j = 1, 2, ..., l$. Then the goal of optimization is that all criteria become better (smaller) than the required demand levels. We call such a solution an *acceptable solution*.

A feasible point $\hat{x}$, satisfying all above constraints for $p = 0$, is called *acceptable*, if

$$J_j(\hat{x}) \leq J_j^{max}, \quad \forall j \in \{1, 2, .., l\},$$

with prescribed demand levels $J_j^{max}, \forall j \in \{1, 2, .., l\}$.

All sequential quadratic programming methods generate a sequence of points, converging to a so-called Kuhn-Tucker point, where necessary first-order conditions for a local mimimum are fulfilled. A point x is a (first-order) *Kuhn-Tucker* point [10], [9] if the following conditions hold:

1. x is feasible;

2. there exists a vector λ such that

$$\nabla_x J_0(x) = \sum_{j=1}^{m_e} \lambda_j \nabla_x h_j(x) + \sum_{j=m_e+1}^{m_e+r} \lambda_j \nabla_x h_j(x);$$

3.

$$\lambda_j \leq 0, \quad \forall j = 1, .. m_e,$$

$$\lambda_j h_j = 0, \quad \forall j = 1, .. m_e + r,$$

where $h_j, \quad j = 1, m_e$ denote the inequality constraints, $h_j, \quad j = m_e + 1, m_e + r$ denote the equality constraints, and J_0 denotes the cost. It is assumed that all these functions are to be at least twice-continuously differentiable.

In this vein the acceptable solution is a well defined optimality condition:

Theorem
Each acceptable solution is a Kuhn-Tucker point.

Proof:
To solve the above stated optimization problem an additional parameter $p_{bnd} := \|p\|_\infty$ and linear inequality constraints are to be introduced:

$$-p_{bnd} \leq p_i \leq p_{bnd}, \ i = 1, .., m + 1.$$

$$p_{bnd} \geq 0.$$

Further, denote by $\bar{x} := [x, p_{bnd}]$ the extended optimization-parameter vector and be x an acceptable solution. Per definition, the solution x is feasible with $p = 0$. This requires the inequality $h_1(\bar{x}) := -p_{bnd} \leq 0$ to be active. Furthermore, define

$$\lambda_1 := -1 \text{ and } \lambda_j := 0 \ \forall j \neq 1.$$

Thus λ satisfies the (Kuhn-Tucker-)sign condition and $\lambda_j h_j = 0, \forall j$.
So we get

$$\nabla_{\bar{x}} J_0(\bar{x}) = (0, .., 0, 1) = -\nabla_{\bar{x}} h_1(\bar{x})$$

$$= \sum_{j=1}^{m_e} \lambda_j \nabla_{\bar{x}} h_j(\bar{x}) + \sum_{j=m_e+1}^{m_e+r} \lambda_j \nabla_{\bar{x}} h_j(\bar{x}).$$

Hence all Kuhn-Tucker conditions are fulfilled.

5 On-line simulation

The *dynamic* behaviour of the proposed robot trajectory optimization scheme is
verified within ANDECS [11]. The required numerical and graphical computer power
is achieved by integrating distributed resources in a heterogeneous unix-computer
network using PVM (Parallel Virtual Machine) [7].

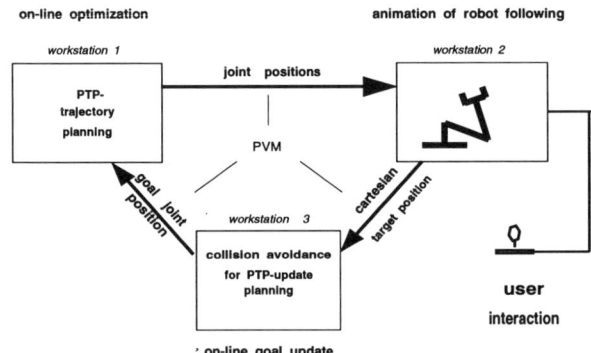

Figure 3: Computing environment for on-line simulation/animation

The dynamics simulation and the trajectory optimization is performed on a
IBM RS6000/250 workstation (20 MFlops). The sensor evaluation is performed
on another IBM RS6000/530 workstation. A Silicon Graphics high performance
graphics workstation is used for animation of the manipulator movement in the
work cell, figure 3. Position and orientation of the tool-centerpoint frame, which
is fixed in the manipulator hand, can be interactively moved via the DLR steering
ball (space mouse). Hereby, information on the actual teach-increment vector is
sent from the graphics- to the optimization workstation, the computed trajectory
is transmitted back to the visualization system for on-line animation by 3D-solid
models. With that configuration the experimentator gets a realistic look and feel of
the dynamic on-line behaviour of this trajectory planning system.

6 Application example

The above approach is validated in a simulation using a kinematically redundant ten-axes manipulator model with complete dynamics. The operator teaches interactively position and orientation of the tool-centerpoint frame in the manipulator hand in a three-dimensional workspace. There are two obstacles in the workspace, figure 6. Self-collision between non-adjacent manipulator links is avoided by modeling them as moving obstacles. The performance indices to be made as small as possible are the time, and the maximum norm of motor torques. The average computation time per optimization problem is less than 0.1 s (20 Mflops computing power), which yields a smooth, human controlled motion.

Workstation 1 displays the dynamic behaviour with respect to the given constraints on joint positions, -velocities, -accelerations, as well as motor torques, figure 5. Workstation 2 displays 3D-solid model animation of interactive steering and robot following, figure 4.

7 Conclusions

A kinematically redundant ten-dof manipulator model is used to demonstrate the feasibility of a method for on-line dynamic trajectory optimization where user-defined multi-performance criteria can be applied. Additional constraints like collision avoidance, box-constraints on joint positions, -velocities, -accelerations, and motor torques, are explicitly considered. Singularity avoidance is obtained implicitly by minimum time motions under dynamic constraints. With an $O(n)$ inverse dynamic modell description, the trajectory planning problem can be formally described as a point-to-point variational problem without differential equations. A discretization on joint acceleration level yields a numerical problem formulation, which can be successively solved by closely related sequential quadratic programming parameter optimizations. Hereby, a twice continuously differentiable cubic spline-trajectory representation on joint position level is realized, being especially efficient, as splines are known to be optimal in the sense of minimizing the average curvature and the maximum joint acceleration effort.

Furthermore, the trajectory planning approach is independent of special kinematic and dynamic robot model formulations. Therefore any kind of model, produced by some model building environment, can be used without adapting the algorithm. To achieve efficient solution processes, optimization is performed in minimal coordinate parameter spaces. Collision avoidance and detection is realized simultanously in Cartesian space using a forward kinematic approach for implicitly casting Cartesian space obstacles in joint space.

An on-line simulation is demonstrated where the operator interactively teaches position and orientation of the endeffector tool-centerpoint frame in moving through a work cell populated by obstacles.

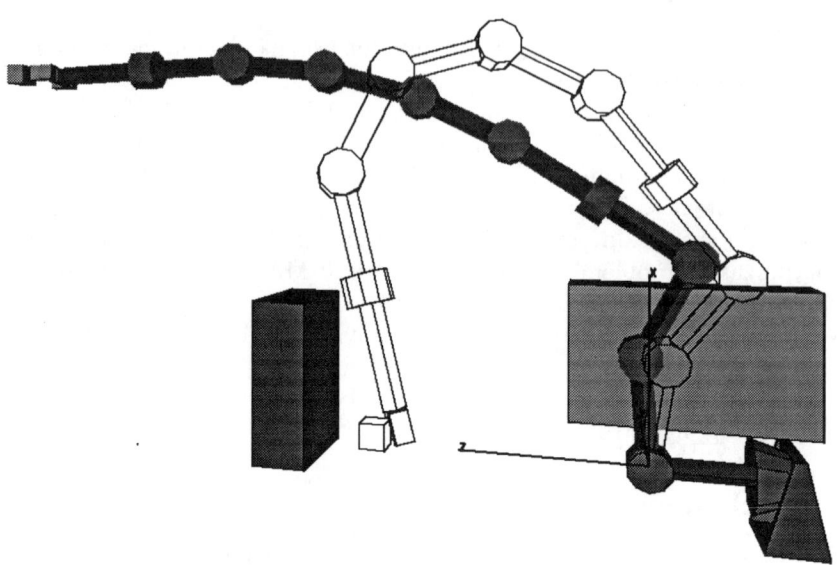

Figure 4: Real-time simulation/animation of a ten-dof robot with two obstacles: the wireframe model shows the starting configuration of the robot; the solid model shows the actual configuration while following the target cubicle.

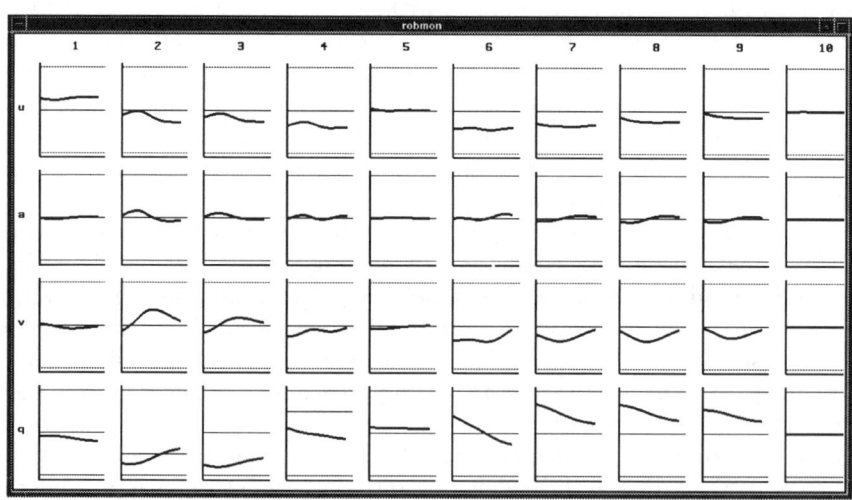

Figure 5: Time behaviour of joint positions (q), -velocities (v) , -accelerations (a), and motor torques (u) with respect to figure 4.

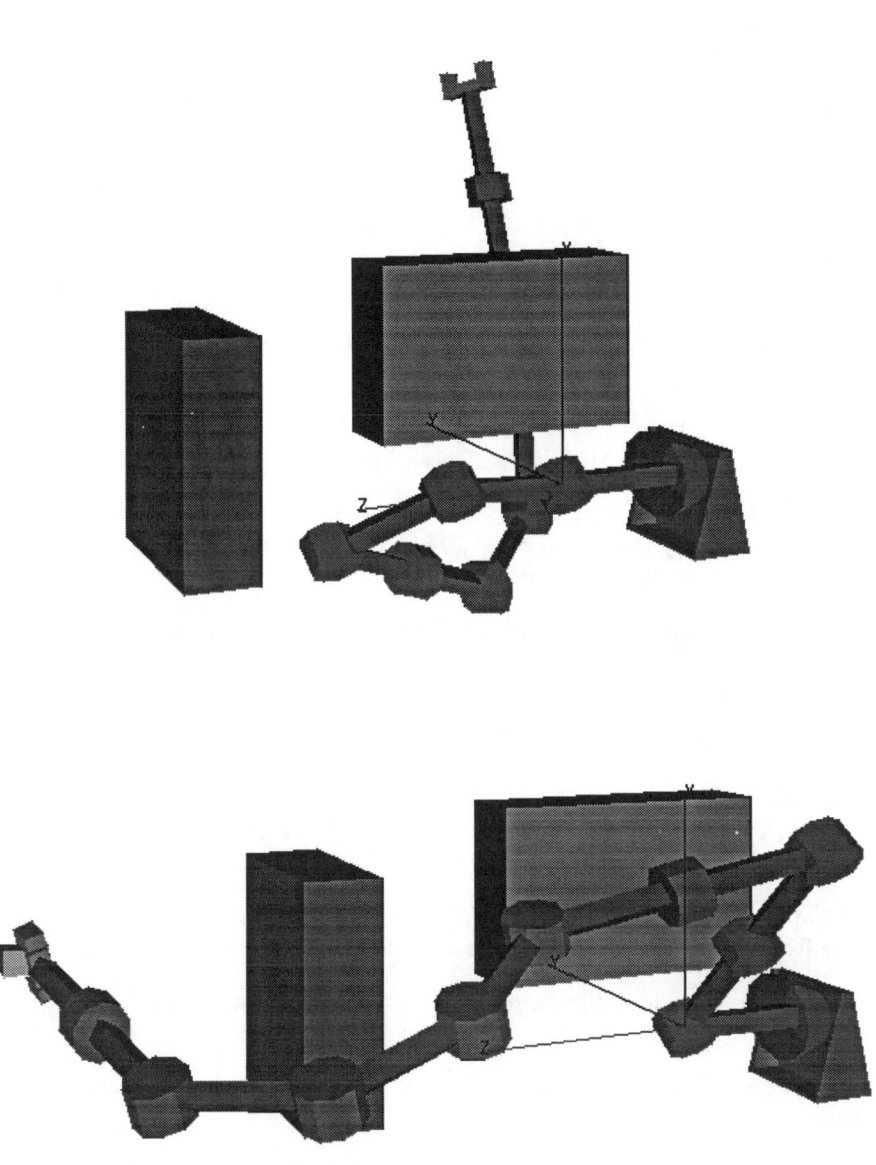

Figure 6: Two examples for collision avoidance

References

[1] Bellman, R.E. (1957): " Dynamic Programming ", Priceton University Press.

[2] Bryson, A.E. and Ho, Y.-C.: Applied Optimal Control. *Blaisdell Publishing Company*, 1969.

[3] Bulirsch, R., Stoer, J. (1978): " Introduction to Numerical Analysis ", Springer-Verlag, New York.

[4] Bulirsch, R. *Die Mehrzielmethode zur numerischen L"osung von nichtlinearen Randwertproblemen und Aufgaben der optimalen Steuerung.* Report der Carl-Cranz Gesellschaft.

[5] De Boor, C., Stoer, J. (1978): *A practical guide to splines*, Springer-Verlag, New York.

[6] Bobrow, J.E., Dubowsky, S., Gibson J.S. (1985): "Time-optimal control of robotic manipulators along specified paths", *The International Journal of Robotics Research*, Vol.4, No.3, Fall 1985.

[7] A. Beguelin, J. Dongarra, G. Geist, R. Manchek, K. Moore, and V.S. Sunderam: "PVM and HeNCE", MIT Press, December 1992.

[8] Donald, B.R, Xavier, P. (1989): "A provably good approximation algorithm for optimal-time trajectory planning", *Proc. of IEEE Int. Conference on Robotics and Automation '94*, Cincinnati, OH, 1536-1541.

[9] Fletcher, R.: Practical Methods of Optimization. *John Wiley & Sons*, 2nd edition, 1990.

[10] Gill, P., Murray, W., and Wright, M.H.: Practical Optimization. *Academic Press, Inc.*, 1988.

[11] Grübel, G., Finsterwalder, R., Joos, H.-D., Lewald, S., and Otter, M.: "AN-DECS: A Computation Environment for Robot-Dynamics Design Automation". *Proc. of IEEE Int. Conference on Robotics and Automation '94*, San Diego, 1994.

[12] Khatib, O. (1986): " Real-Time Obstacle Avoidance for Manipulators and Mobile Robots ", *International Journal of Robotik Research*, 5(1) 90-98.

[13] Leitmann, G.: The Calculus of Variations and Optimal Control. *Plenum Press*, 1981.

[14] Latombe, J-C. (1991): " Robot Motion Planning ", Kluwer Academic Publishers.

[15] Otter, M. and Grübel, G.: "Direct Physical Modeling and Automatic Code Generation for Mechatronics Simulation". *Proc. of Second Conference on Mechatronics and Robotics '93*, Duisburg/Moers, 1993.

[16] Pesch, H.J., M. Schlemmer, O. von Stryk: *Minimum-energy and minimum-time control of three-degrees-of-freedom robots. Part 1: Mathematical model and necessary conditions, Part 2: Numerical methods and results for the Manutec r3 robot.* In Vorbereitung.

[17] Piessens, R.,: " Quadpack, a subroutine package for automatic integration", Springer-Verlag, 1983.

[18] Sawaragi, Y., Nakayama, H., Tanino, T. (1985): "Theory of multiobjective optimization", Academic Press Inc.

[19] Stryk, O., Schlemmer, M.: Optimal Control of the Industrial Robot Manutec r3. In: Control Application of Optimization, R. Bulirsch, D. Kraft, eds. *International Series of Numerical Math.* (ISNM), Birkhäuser Publication, Basel, Boston, Stuttgart.

[20] Shin, K.G., McKay, N.D. (1986): " A Dynamic Programming Approach to Trajectory Planning of Robot Manipulators", *IEEE Trans. Auto. Control*, AC-31(6):491-500.

[21] Shin, S.K., Leu, M.C. (1991): "Manipulator Motion Planning in the Presence of Obstacles and Dynamic Constaints", *The International Journal of Robotics Research*, Vol.10, No.2, April 1991.

[22] Shiller, Z., Dubowsky, S. (1988): "Global time-optimal motions of robotic manipulators in the precense of obstacles", *The International Journal of Robotics Research*, Vol.4, No.3, Fall 1985.

[23] Vukobratovic, M., Kircanski, M. (1986): " Kinematics and Trajectory Synthesis of Manipulation Robots", Springer-Verlag, Tokyo.

[24] Wang, D., Hamam, Y. (1992): "Optimal Trajectory Planning of Manipulators with Collision Detection and Avoidance", *The International Journal of Robotics Research*, Vol.11, No.5, October 1992.

[25] Welzl, E. (1985): "Constructing the Visibility Graph for n Line Segments in $O(n^2)$ Time". *Information Processing Letters*, 20, 167-171

ECMI-PUBLICATIONS

**Vol. 1 – Proceedings of the First European Symposium
on Mathematics in Industry**
Amsterdam (1985). Ed.: Hazewinkel, Mattheij, van Groessen
1988. XIV, 238 pages. ISBN 3-519-02170-6
Bound DM 88,– / ÖS 687,– / SFr 88,–

Vol. 2 – Case Studies in Industrial Mathematics
Ed.: Engl, Wacker, Zulehner
1988. X, 218 pages. ISBN 3-519-02171-4
Bound DM 88,– / ÖS 687,– / SFr 88,–

**Vol. 3 – Proceedings of the Second European Symposium
on Mathematics in Industry**
Oberwolfach (1987). Ed.: Neunzert
1988. VII, 359 pages. ISBN 3-519-02172-2
Bound DM 98,– / ÖS 765,– / SFr 98,–

**Vol. 4 – Proceedings of the Second Workshop on Road-Vehicle-
Systems and Related Mathematics**
ISI Torino (1987). Ed.: Neunzert
1989. VI, 235 pages. ISBN 3-519-02173-0
Bound DM 82,– / ÖS 640,– / SFr 82,–

Vol. 5 – Third European Conference on Mathematics in Industry
Glasgow (1988). Ed.: Manley, McKee, Owens
1990. XII, 564 pages. ISBN 3-519-02174-9
Bound DM 180,– / ÖS 1404,– / SFr 180,–

**Vol. 6 – Proceedings of the Fourth European Conference
on Mathematics in Industry**
Strobel (1989). Ed.: Wacker, Zulehner
1991. X, 425 pages. ISBN 3-519-02175-7
Bound DM 190,– / ÖS 1482,– / SFr 190,–

**Vol. 7 – Proceedings of the Fifth European Conference
on Mathematics in Industry**
Lahti (Finland) (1990). Ed.: Heiliö
1991. 400 pages. ISBN 3-519-02176-5
Bound DM 198,– / ÖS 1545,– / SFr 198,–

**Vol. 8 – Proceedings of the Sixth European Conference
on Mathematics in Industry**
Limerick (1991). Ed.: Hodnett
1992. 340 pages. ISBN 3-519-02177-3
Bound DM 168,– / ÖS 1311,– / SFr 168,–

**Vol. 9 – Proceedings of the Seventh European Conference
on Mathematics in Industry**
Montecatinti Terme (1993). Ed.: Fasano, Primicerio
1994. X, 434 pages. ISBN 3-519-02178-1
Bound DM 156,– / ÖS 1217,– / SFr 156,–

Vol. 10 – Inverse Problems and Optimal Design in Industry
Philadelphia, USA (1993). Ed.: Engl. McLaughlin
1994. 269 pages. ISBN 3-519-02179-X
Bound DM 168,– / ÖS 1311,– / SFr 168,–

Vol. 11 – Numerical Methods in Multibody Dynamics
By: Eich, Führer
1995. approx. 300 pages. ISBN 3-519-02601-5
Bound approx. DM 44,– / ÖS 343,– / SFr 44,–

B. G. Teubner Stuttgart